Heat Transfer

Heat Transfer

Helmut Wolf
UNIVERSITY OF ARKANSAS, FAYETTEVILLE

HARPER & ROW, PUBLISHERS, New York
Cambridge, Philadelphia, San Francisco,
London, Mexico City, São Paulo, Sydney

1817

Sponsoring Editor: Cliff Robichaud
Project Editors: Bob Greiner and Pamela Landau
Production: Delia Tedoff
Photo Researcher: Mira Schachne
Compositor: Science Typographers, Inc.
Printer and Binder: The Maple Press Company
Art Studio: J & R Art Services, Inc.
Cover Design: Caliber Design Planning, Inc.

Heat Transfer

Copyright © 1983 by Helmut Wolf

Library of Congress Cataloging in Publication Data

Wolf, Helmut, 1924–
    Heat transfer.

    Includes bibliographical references and index.
    1. Heat—Transmission.   I. Title.
QC320.W64 1983   536′.2   82-11840
ISBN 0-06-047181-6

This book is dedicated to the three most important women in my life:

to Janey: my life's companion

to Cristine: who will write much more interesting things; and

to Martha: who will do much more interesting things

Contents

Preface

This introduction to the basic phenomena of heat transfer is intended for both students and practicing engineers. It is assumed that the readers are familiar with the concepts of thermodynamics and basic calculus. Therefore, it is assumed that readers know what an integral sign means and what the physical significance of the derivative of a function is. However, having a facility in manipulation is not necessary. A knowledge of differential equations is helpful but not essential.

It has been my feeling for some time now that heat transfer books have come to include more and more advanced material to the point where they are no longer intelligible to most undergraduate engineering students. They are generally not ready to appreciate, much less absorb, the subtleties and complexities of the conservation equations for solids and fluids expressed in partial differential equations. Those students who are ready will take advanced work to obtain a thorough background in first principles. Consequently, from the point of view of the beginning student, those aspects of a book dealing with the details of analytical solutions of partial differential equations are a source of bewilderment and confusion. The book has been written without mathematical embellishments and with the emphasis placed on understanding the physical principles involved. Thus, the book will be useful to anyone who desires a compact overview of heat transfer.

As stated above, the student or reader should be familiar with the basic concepts of calculus. Such a background is essential from the teaching-learning viewpoint in attacking the phenomena of conduction (which, it must be admitted, is largely applied mathematics). To understand the applicability of the formulas describing conduction situations, one really needs to know the boundary conditions and assumptions made in the formulation of the result. Generally the mathematical details of the solution per se do not restrict the applicability of the results at this introductory level. In more complicated situations, one has to cover the waterfront. It would be possible to disguise the assumptions and the boundary conditions in the form of descriptive "recipe" details given with the result, but it would be very hard for me to convince myself that doing so would be the best approach. Rather, I feel that the basic concepts of the à propos mathematics could be understood if taught with emphasis on the physical understanding of the situation at hand. How well this can be done depends on the student's curiosity and will to learn. These remarks apply largely to the subject of conduction and, to a lesser degree, to radiation, convection, and phase change. The subjects of convection and phase change can be thoroughly discussed in an introductory course without covering the intricacies of the boundary layer equations, Navier-Stokes equations, or Blasius solution.

The evolution and growth of computational equipment have matured to the point where most engineering organizations have an installation in-house or have ready access to one; consequently, having a facility in computer use is important to the engineer. Accordingly, several applications of FORTRAN programming of heat transfer problems have been included. The FORTRAN language has been selected because of its wide applicability and use in university and other computer installations. All of the programs have been written with simple standard FORTRAN library functions. A particular effort has been made to avoid specialized functions or subroutines that are not universally available; therefore, the programs in the text should run on any standard FORTRAN or WATFIV compiler when the particular JCL language required by the installation is used.

The material in the book is organized so that for classroom use it can be covered in a period of fifteen or sixteen weeks in a three-hour lecture course. It is also possible to split the material into three one-hour minicourses covering conduction; convection and heat exchangers; and radiation and phase exchange. The book uses both SI and British units; the results presented in the book can, of course, be evaluated with any consistent set of units.

In assembling the material for this book, I have been influenced by the very fine reference texts of Jacob, Schneider, Kreith, Parker, and Boggs, Holman, Kays, and the learning text of Myers. Stimulating discussions with interested students and colleagues in the department have sharpened my thinking on many aspects. A quotation attributed to Confucius, with appropriate paraphrasing, perhaps best sums up my philosophy of the teaching-learning process:

What I hear I forget,
    (Lecture format, no class notes taken)

What I see I remember,
  (Class notes taken by the student)
What I hear, see, and discuss I understand.
  (The ideal learning environment)

I would like to express my appreciation to Drs. S. deSoto, M. K. Jovanovic, and B. K. Hodge for their painstaking and detailed review of the manuscript and their helpful comments, and to Mrs. Virginia Swaim and Mrs. Jane Wolf for their dedicated help in transforming my notes into a polished manuscript. I would also like to acknowledge the fine cooperation of Cliff Robichaud and Bob Greiner.

HELMUT WOLF

Heat Transfer

Chapter 1
Basic Mechanisms

1.0 GENERAL REMARKS

The usual progression in the educational process of thermal systems engineering generally starts with the study of energy in a beginning physics course, progresses to the thermodynamics sequence, followed by a beginning heat transfer course, and then goes on to specialized courses in more advanced treatment of the basic modes of heat transfer and/or to applications oriented courses. In the thermodynamics approach, heat is identified as a quantity only at the system boundary and is either given or calculated from the first law. Therefore, a determination of the heat transfer rate for a real system by the methods of thermodynamics requires a complete application of the first law (conservation of energy) to a system or control volume. The determination of heat transfer as energy crossing a boundary by virtue of a finite temperature difference can also be accomplished by considering the mechanism of transport. In our examination, however, we will not look in such detail that we lose the comfort and generality of the macroscopic* point of view, a usual and legitimate engineering approach.

---

*As distinguished from the microscopic point of view where one looks at the behavior of atoms or molecules on an individual basis.

One generally identifies three basic mechanisms of transport: convection, radiation, and conduction. Of these mechanisms, conduction and radiation can be considered as pure in the sense that they can take place as the only propagating mechanisms. Convection, on the other hand, is a mixture of conduction and mass transport of energy, with radiation present in significant or insignificant amounts depending on the fluid present and the temperature levels. One might also want to include phase change as a basic mechanism; however, even more so than convection, phase change is a mixture of conduction and complicated mass transport processes in the fluid portion, in addition to the actual change of phase mechanism. It is, therefore, generally considered to be in a category by itself, as it is in this book. Since most heat transfer occurrences involve more than one mode, it will be necessary to critically examine the basic modes before considering more general happenings. In the discussion to follow, then, we focus only on the three basic modes and their concepts and definitions.

1.1 THE CONVECTION MODE

When a fluid at rest or in motion is in contact with a surface at a temperature different from the fluid, energy flows in the direction of the lower temperature as required by the principles of thermodynamics. Figure 1.1 shows some possibilities.

For both situations shown in Figure 1.1, we define the flux of energy leaving (or entering) the surface as

$$q_c = hA(T_s - T_\infty) \text{ W, Btu/hr} \tag{1.1}$$

The temperature T_s is that directly at the surface in contact with the fluid, and the temperature T_∞ is the fluid temperature far enough away from the surface so that no influence of the surface is evident. The area A is the surface area in contact with the fluid and we should note that A is perpendicular to the direction of the heat flux q. The proportionality factor h is called the heat transfer coefficient (also the unit area conductance or the convective conductance) and depends on the geometrical arrangement, orientation, and surface

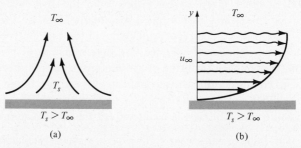

Figure 1.1 Convection concepts for a fluid in contact with a solid surface: (a) fluid far from surface at rest, and (b) fluid far from surface in motion with respect to the surface.

Table 1.1 REPRESENTATIVE VALUES FOR THE HEAT TRANSFER COEFFICIENT

CONDITION	h (Btu/hrft^{2}F)	h (kW/m^{2}K)
Free convection, air	1–6	0.006–0.035
Forced convection, air	5–150	0.028–0.851
Free convection, water	30–200	0.170–1.14
Forced convection, water	100–4000	0.570–22.7
Boiling water	1000–15,000	5.70–85
Condensing steam	10,000–30,000	57–170
Forced convection, sodium	20,000–40,000	113–227

condition (smooth or rough), as well as on the properties and velocity of the fluid. Table 1.1 lists some representative values for the heat transfer coefficient under a variety of engineering conditions. We will, in the later chapters on convection, learn how to calculate h for specific geometries and fluid conditions. With the value of h known, we can compute the heat flux involved for the cases like those illustrated in Figure 1.1. To satisfy our curiosity a little more, we observe that directly at the surface, the fluid is at rest when there are enough molecules to consider a continuum such that there is no slip; the energy flux q_c must therefore pass through this stagnant *layer* of fluid at rest by the mechanism of conduction. We further observe that there can be no chemical reaction between the fluid and the surface material for the concept expressed in Eq. 1.1 to hold. Consequently, the type of material comprising the surface does not enter into the matter and as long as the surface smoothness (or roughness) and the temperature are the same, the convective phenomena will be identical. We must, of course, recognize that the wall material is important in transporting the quantity q_c to or from the surface in contact with the fluid, but that has nothing to do with the convection phenomena. Let's conclude this description of the convection phenomena with the observation that in closed channel flow, the fluid temperature T_∞ is not a simple constant value. In such a case, one uses a temperature that is indicative of the energy (enthalpy) content of the fluid, the bulk temperature. The bulk temperature can best be visualized as the temperature one would obtain if the channel were cut and the fluid collected for a short period of time in a thermally insulated container, and well stirred until uniform in temperature.

1.2 THE CONDUCTION MODE

Energy propagation in a solid material, or in a fluid that is not in chaotic (turbulent) motion, is accomplished by the transmission of vibrational energy from molecule to molecule in the structure. Again we look only at the macroscopic effects and such propagation phenomena are described by Fourier's law:

$$q_k = -kA \text{ (temperature gradient) (W, Btu/hr)} \tag{1.2}$$

where the temperature gradient, in degrees per unit length, is the slope of the

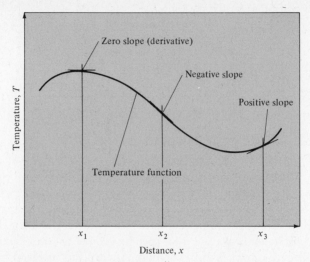

Figure 1.2 Arbitrary temperature profile in a homogeneous solid illustrating the physical meaning of the derivative.

temperature-distance relationship. The temperature gradient, or slope of the temperature profile, is the tangent to the curve at the point of interest. In calculus notation, this slope or tangent is called the derivative of the temperature function and when the temperature is a function of only one variable (distance x, for example), we can express the gradient or slope by means of the differential operator d as dT/dx. Figure 1.2 shows a region of a solid with an arbitrary temperature profile to illustrate the concept. The figure shows a zero slope at x_1, a negative slope at x_2, and a positive slope at x_3. The reader should note that Figure 1.2 shows an instantaneous picture of a transient situation, a more complicated situation that we will discuss in detail in a later chapter. At position x_3, for example, the energy flux is in the negative x direction as dictated by the second law of thermodynamics, which states that energy flows only from a higher temperature to a lower temperature when there is no other external agency (work) acting. We can then express Eq. 1.2 in the following form:

$$q_k = -kA\frac{dT}{dx}\,(\text{W, Btu/hr}) \tag{1.3}$$

The thermal conductivity k is the proportionality constant between the unit area flux q/A and the temperature gradient dT/dx, and is defined by Eq. 1.3 with consistent units for the system being used. In the Système International (SI) system, k is expressed in W/mK and in the British engineering (BE) system, k is expressed in Btu/hrftF.

The thermal conductivity k is generally a function of temperature. For solids, the variation with temperature is usually not a strong one, but for liquids and gases, k can be a strong function of temperature. Appendix C tabulates values of k for different substances. It is important for the reader to become familiar with the contents of the appendixes as sources for needed property information.

1.3 THE RADIATION MODE

Propagation of energy by radiation requires no carrier; thus radiant energy can be transmitted through a vacuum. The energy travels as discrete packets called quanta or photons whose energy content depends on their wavelengths or frequencies. We are interested in the sum total over all frequencies, so that we need only consider the Stefan-Boltzmann relationship at this introductory stage, which states that the maximum energy that can leave a surface of area A at absolute temperature T, is

$$q_{rad} = \sigma A T^4 \text{ (W, Btu/hr)} \tag{1.4}$$

where σ is a natural constant that depends on the units used:*

$$\sigma = 5.675 \times 10^{-8} \text{ W/m}^2\text{K}^4$$

$$\sigma = 0.1714 \times 10^{-8} \text{ Btu/hrft}^2\text{R}^4$$

For a diffuse surface the intensity of radiation is transmitted equally in all directions to the hemisphere above the surface; the unit area flux q_{rad}/A is called the emissive power E_b. A surface that emits the maximum amount of radiant energy like the one above we call a black surface or a perfect radiator. Real surfaces emit at a rate E which is less than the rate E_b at a given absolute temperature level, and we note that the ratio $\varepsilon = E/E_b$ is called the emittance of the surface.

In order to assess all the radiant energy that could leave a surface, we must first identify the concept of incident energy. We determine the irradiation G (W/m^2 or Btu/hrft^2) at a location just above the surface under consideration. The quantity G is the sum of all radiation striking the surface from all possible sources. A portion α of the incident radiation is absorbed; a portion ρ is reflected; and a portion τ could also be transmitted if the surface material were transparent. Conservation of energy requires that

$$G = \alpha G + \rho G + \tau G \tag{1.5}$$

or

$$1 = \alpha + \rho + \tau \tag{1.6}$$

The sum of absorptance α, reflectance ρ, and transmittance τ, must always be unity.

Figure 1.3 shows the phenomena that takes place for a transparent material, the most general situation. We note that there is irradiation G_1 striking surface ① and irradiation G_2 striking surface ②, and at the instant of observation the

---

*Some more recent measurements of σ according to Siegel (5) indicate values about 1 percent higher at 0.173 Btu/hrft^{2}R^4 or 5.729 W/m^{2}K. The values given above are satisfactory for most engineering applications.

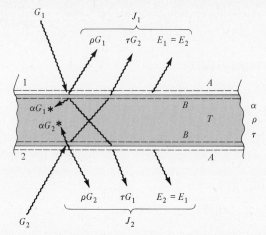

Figure 1.3 Illustration showing the disposition of radiant energy for a material that absorbs (α), reflects (ρ), and transmits (τ).

plate is at an absolute temperature level T such that it emits $E_1 = E_2$ (we assume here that internal resistance to energy transport is not an important consideration and that surfaces ① and ② are both at temperature T). We further observe that each surface has three components of energy leaving, a reflected component, a transmitted component, and an emitted component. Accordingly, we define the radiosity, J, to be the sum of these components, thus:

$$\left. \begin{aligned} J_1 &= \rho G_1 + \tau G_2 + E_1 \\ J_2 &= \rho G_2 + \tau G_1 + E_2 \end{aligned} \right\} \tag{1.7}$$

If we take as a system boundary the dotted lines A in Figure 1.3, we can observe that if $G_1 + G_2 > J_1 + J_2$, the temperature T of the plate must be increasing with respect to time; and conversely, if $G_1 + G_2 < J_1 + J_2$, or putting it another way, energy in is less than energy out, then the plate temperature T is decreasing with respect to time. We can also analyze the thermal state of the plate by considering system B, the boundary just below the surface of the plate. The energy into the system then becomes $\alpha(G_1 + G_2)$ and the energy out $E_1 + E_2$ with similar consequences. The concepts of radiosity, irradiation, and emissive power are especially useful tools and they will be utilized later in calculating exchange between real surfaces.

We shall conclude our introductory remarks concerning radiant transfer with the concept of the interchange factor, F_{12}. Consider two diffuse black surfaces A_1 and A_2 arbitrarily oriented with the provision that a surface is completely visible from each and every location on the other surface. We can then define F_{12} as the fraction of black diffuse radiation leaving surface A_1 and striking surface A_2. The radiant flux from A_1 to A_2 could be written as

$$q_{12} = (E_{b1}A_1)F_{12} \tag{1.8}$$

and the flux from A_2 to A_1 as

$$q_{21} = (E_{b2}A_2)F_{21} \tag{1.9}$$

The net flux then is $q_{12} - q_{21}$ or

$$q_{net} = E_{b1}A_1F_{12} - E_{b2}A_2F_{21} \tag{1.10}$$

We can intuitively understand that when two surfaces A_1 and A_2 are at the same temperature, there can be no net radiant transfer; also $E_{b1} = E_{b2}$ so that Eq. 1.10 reduces to $A_1F_{12} - A_2F_{21} = 0$ or

$$A_1F_{12} = A_2F_{21} \tag{1.11}$$

Equation 1.11 is called the reciprocity theorem for black diffuse interchange factors. The net flux can then be written as

$$q_{net} = \sigma A_1 F_{12}(T_1^4 - T_2^4) \tag{1.12}$$

We can observe that it is also permissible to apply the F_{12} concept to radiosity and note that

$$q_{net} = A_1 F_{12}(J_1 - J_2) \tag{1.13}$$

represents the net flux between surfaces A_1 and A_2 if the surfaces A_1 and A_2 are nonblack but still diffuse. The latter observation was made so that we can develop the analogy between heat transfer and simple dc electrical circuits for the phenomena of conduction, convection, and radiation.

1.4 THE ELECTRICAL ANALOGY

There are many situations in heat transfer analysis which can be analyzed more simply by the use of an electrical counterpart that can be easily solved by simple dc circuit principles. In this section, we want to formulate the concepts of comparison and look at a few simple examples.

The basic electrical principle that we will refer to states that a potential difference (volts) applied to a resistance (ohms) causes a current or flux (amperes) to flow through the resistance or conductor. Similarly in heat transfer, we can observe that a potential difference (temperature difference) applied to a material (thermal resistance) causes a heat flux *current* (watts, or Btu/hr) to flow through the material. In this instance, the SI unit watt is singularly appropriate. We can rewrite Eq. 1.1 in the manner described above for convection phenomena such that

$$q_c = \frac{T_s - T_\infty}{1/hA} = \frac{\text{temperature difference}}{R_c} \tag{1.14}$$

thereby defining the convective resistance $R_c = 1/hA$ in appropriate units (K/W, hrF/Btu). The convective conductance K_c is defined as hA, and one sometimes sees the heat transfer coefficient h referred to as the unit of convective conductance. The electrical analogy representing the situation pictured in Figure 1.1 would be as follows:

$$\begin{array}{c} q_c \\ T_s \quad \xrightarrow{\quad} \quad T_\infty \\ \text{o}\!-\!\!\!\bigwedge\!\bigwedge\!\bigwedge\!-\!\text{o} \\ R_c = 1/hA \end{array}$$

Heat transfer by conduction in a simple one-dimensional solid with thermal conductivity not a function of temperature can be found by simple integration of Eq. 1.3 (the technique of working with Eq. 1.3 will be explained in detail in Chapter 2) as follows:

$$q_k = \frac{T_1 - T_2}{\Delta x/kA} = \frac{T_1 - T_2}{R_k} \tag{1.15}$$

The physical situation is shown in Figure 1.4; the figure shows that the temperature difference $T_1 - T_2$ drives the heat energy q_k through the thermal resistance $R_k = \Delta x/kA$ offered by the solid material of thickness $\Delta x = x_2 - x_1$.

We run into a small discrepancy when we consider the radiation situation, in that radiation requires no substance or carrier to support photon transport. Therefore, the concept of resistance becomes nebulous. Let us, therefore, simply look at Eq. 1.13 and recognize that the driving potential is the radiosity difference $J_1 - J_2$, and that the geometrical orientation and separation of the areas A_1 and A_2 represent a *space* resistance, even though there need be no

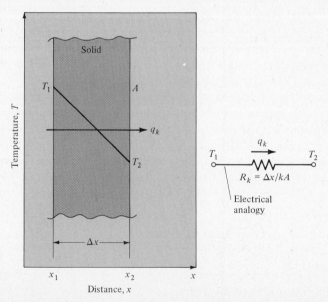

Figure 1.4   Sketch showing physical meaning of terms in the conduction Eq. 1.15 together with the corresponding electrical analogy.

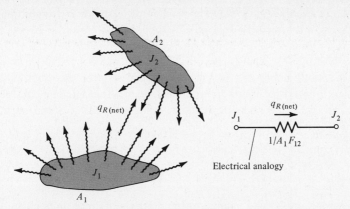

Figure 1.5   Sketch showing the physical meaning of terms in Eq. 1.16 in relation to the electrical analogy.

material present for radiation transfer. Therefore, we can write

$$q_{\text{rad}} = \frac{J_1 - J_2}{1/A_1 F_{12}} = \frac{J_1 - J_2}{R_R} \qquad (1.16)$$

as the defining equation for the radiation space resistance $R_R = 1/A_1 F_{12}$. Figure 1.5 shows the physical orientation and compares the analogous electrical circuit.

The discussion above can perhaps best be summarized by discussing an application. Consider a plate supported by a holder that is positioned in the open air during a bright sunny day as shown in Figure 1.6. When we consider an

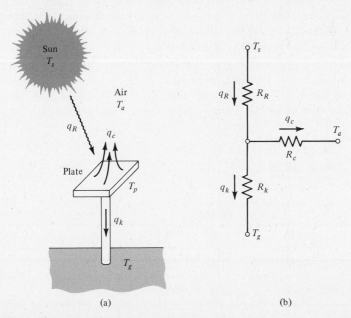

Figure 1.6   Physical arrangement of a flat plate irradiated by the sun (a) and associated electrical analogy (b).

energy balance on the plate, we note that for the steady-state condition, net radiant energy input from the sun, q_R, must be dissipated to the air, q_c, and to the ground (for ground temperature less than plate temperature) so that $q_R = q_k + q_c$. This illustration could also be applied to the analysis of the air conditioning load for an automobile or truck, or a building, with the exception that there are more heat transfer paths and internal heat sources that must be considered.

We shall conclude our introductory chapter with a short section on units.

1.5 UNITS

There is a strong pressure to change to the metric system of units in the United States. Several large corporations have already made the changeover for those items manufactured for export, and we are also seeing kilometer (km) distances given on turnpike signposts.

The unit of heat in the metric system (SI) is the joule (J), and the unit of power is the watt (W), corresponding to the British thermal unit (Btu/hr) and the horsepower (HP). Actually in this modern jet age, the term *horsepower* seems oddly out of date. The choice of another name would probably be more difficult to accomplish than changing to the metric system. Table 1.2 gives the base units and attendant symbols that we shall use in our work.

Table 1.3 presents some of the more commonly used derived units as we will encounter them in this book, together with symbols and conversion factors. For example, the heat transfer coefficient (see Eq. 1.1) has no particular name as such in either the SI or BE systems, and since we are in transition from the BE to the SI system, a multiplier M is given such that when the heat transfer coefficient h in Btu/hrft^{2}F is multiplied by $M = 5.67826$, one obtains h in W/m^{2}K. The basis of Table 1.3 is the 1956 definition by the International Steam Table Conference of the British thermal unit in terms of the joule as 1 Btu = 1055.056 J, as given in Reference 4. The definition was based on a statistical evaluation of the most reliable data available to that date on the

Table 1.2 BASE UNITS

QUANTITY	NAME OF SI BASE UNIT	SI SYMBOL	NAME OF BE UNIT	BE SYMBOL	M (BE $\times M$ = SI)
Length	meter	m	foot	ft	0.30480
Mass	kilogram	kg	pound mass	lb$_m$	0.453592
Time	second	s	second	s	1
Temperature	degree Kelvin	K	degree Rankine	R	0.55556
Electric current	ampere	A	ampere	A	1
Amount of substance	mole	mol	mole	mol	—

mechanical equivalent of heat. Some conversion tables in the literature are based on the earlier value of 1055.04 J per Btu. The table also gives pertinent helpful definitions and other information in the column under the heading of comments. The recommended abbreviations for multiples of units given by the International Standards Organization (3) are given in Appendix A.

We conclude our short discussion on units with a review of basic relationships of mechanics (Newton's law) and some useful relationships among derived units of both systems. It is unfortunate that the needs of commerce required a standardization of mass before the laws of motion and acceleration were clarified. Since it was logical to say that the force required to support a unit mass at rest in a standard gravitational field was also a unit force, we have the result that pounds force in the BE system is numerically the same as pounds mass under these conditions. When we consider Newton's second law, which states that force is proportional to mass times acceleration, we find that we do not have a unit constant of proportionality for unit quantities involved. It has been customary to write the constant of proportionality as $1/g_c$ so that Newton's law is expressed as

$$F = \frac{1}{g_c} ma \tag{1.17}$$

Experimentally it has been shown that one pound force (lb_f) when acting on one pound mass (lb_m) in the BE system will produce an acceleration of 32.174 ft/s^2. In Eq. 1.17 the constant of proportionality g_c must have the numerical value of 32.174 and the units of $lb_m ft/lb_f s^2$ in order for the equality sign to hold. This is a good spot to emphasize the fact that all equations, algebraic, differential, integral, or any other kind, must have equality in units as well as numerical equality, a fact that is unfortunately often overlooked. For the SI system, the magnitude of the newton is not numerically equal to the kilogram of mass when supported at rest in a standard gravitational field; it would take 9.80665 N of force to do so. When the force F in Eq. 1.17 is expressed in newtons (N), the mass m in kilograms (kg), and the acceleration in m/s^2, then $g_c = 1$ kgm/Ns^2.

The association of a scale of numbers to a certain quantity is simply a psychological matter and one can adjust easily to an abrupt change in a matter of a few months. For example, it is just as easy to accept 100 km/hr as a relatively comfortable driving speed as it is to accept 62.5 miles/hr when the sensation of scenery flying by is the same for both numbers. Also, a person who has a mass of 75 kg could just as well think of himself as weighing 735 N (a frightfully high number, but lower than 1000 N, one must admit!) as easily as he could 165 lb_f.

We have discussed only two systems of units, the SI and the BE; there are others and Van Wylen and Sonntag (1) give a concise summary of the different systems. Table 1.4 shows the relationships for the constant of proportionality in Newton's equation for the two systems of units used in this book.

Table 1.3 DERIVED UNITS. SI = SYSTÈME INTERNATIONAL (METRIC UNITS); BE = BRITISH ENGINEERING SYSTEM OF UNITS

QUANTITY	NAME OF SI UNIT	SI SYMBOL	SI UNITS	NAME OF BE UNIT	BE SYMBOL	BE UNITS	M (BE $\times$ M = SI)	COMMENTS
Frequency	hertz	Hz	s^{-1}	hertz	Hz	s^{-1}	1	Hz recently adopted in the BE system
Force	newton	N	kgm/s^2	pound force	F	$lb_m ft/s^2$	4.44822	1 lb_f acts on 1 lb_m to produce 32.174 ft/s^2 acceleration
Pressure	pascal	Pa	N/m^2	psi	p	$lb_f/in.^2$	6894.76	Must distinguish between psia and psig (absolute and gauge)
Energy	joule	J	Nm	—	—	$ftlb_f$	1.35582	(As work)
Energy	joule	J	Nm	British thermal unit	Btu	Btu	1055.056	British thermal unit as heat (International Steam Table Conference 1956)
Power	watt	W	J/s	—	HP	$ftlb_f/s$	745.701	550 $ftlb_f/s$ is 1 HP
Electrical charge	coulomb	C	As	coulomb	—	As	1	1 A is 1 C/s
Electrical potential	volt	V	J/C	volt	V, E	W/A	1	1 W is 1 V $\times$ 1 A
Electric capacitance	farad	F	C/V	farad	F	C/V	1	Usually expressed in microfarads
Electric resistance	ohm	Ω	V/A	ohm	R	V/A	1	Volts = amperes $\times$ resistance (Ohm's law)
Electric conductance	siemens	S	Ω^{-1}	mho	G	A/V	1	
Heat rate	watt	W	J/s	—	q	Btu/hr	0.293071	3.41214 Btu/hr is 1 W
Heat transfer coefficient	—	h	W/m^2K	—	h	$Btu/hrft^2R$	5.67826	Note ΔT in R same as ΔT in F

Quantity		Symbol	SI Units		Symbol	Engineering Units	Conversion	Notes
Thermal conductivity	—	k	W/mK	—	k	Btu/hrftR	1.72958	—
Specific heat capacity	—	c_p, c_o	J/kgK	—	c_p, c_o	Btu/lb$_m$R	4186.80	At constant pressure or constant volume
Specific energy	—	—	J/kg	—	h, u	Btu/lb$_m$	2326	Enthalpy or internal energy, flow, kinetic, or potential energy
Specific latent heat	—	—	J/kg	—	h_{fg}, h_{ig}	Btu/lb$_m$	2326	Liquid-vapor or solid-vapor change
Density	—	ρ	kg/m^3	—	ρ	lb$_m$/ft^3	16.01846	Mass density
Thermal diffusivity	—	α	m^2/s	—	α	ft^2/s	0.092903	$\alpha = k/\rho c_p$
Kinematic viscosity	—	ν	m^2/s	—	ν	ft^2/s	0.092903	$\nu = \mu/\rho$
Dynamic viscosity	—	μ	Ns/m^2	—	μ	lb$_f$s/ft^2	47.8802	(poise in cgs system)
Surface tension	—	σ	N/m	—	σ	lb$_f$/ft	14.5939	In equilibrium with vapor of the liquid
Wavelength	micrometer	λ	μm micrometer	—	λ	μm	1	$1\ \mu\text{m} = 10^{-6}$ m
Radiant intensity	radiance	—	W/srm^2 intensity	—	I	Btu/hrft^{2}sr	3.15459	Area is perpendicular to direction of propagation
Emissive power	—	E	W/m^2	—	E	Btu/hrft^2	3.15459	Can also be monochromatic, gray, or black
Irradiation	—	G	W/m^2	—	G	Btu/hrft^2	3.15459	Also monochromatic
Radiosity	—	J	W/m^2	—	J	Btu/hrft^2	3.15459	Also monochromatic

Table 1.4 COMPARISON OF SI AND BE SYSTEMS

SYSTEM	MASS	LENGTH	TIME	PROPORTIONALITY CONSTANT g_c
SI	kg	m	s	$g_c = 1 \text{ kgm}/\text{Ns}^2$
BE	lb_m	ft	s	$g_c = 32.174 \text{ lb}_m \text{ft}/\text{lb}_f \text{s}^2$

Let us conclude our discussion of units by noting some useful conversion factors:

$$778.168 \text{ ftlb}_f = 1 \text{ Btu}$$
$$745.701 \text{ W} = 1 \text{ HP}$$
$$2544.44 \text{ Btu}/\text{hr} = 1 \text{ HP}$$
$$3412.14 \text{ Btu}/\text{hr} = 1 \text{ kW}$$
$$2.20462 \text{ lb}_m = 1 \text{ kg}$$
$$444822 \text{ dyn} = 1 \text{ lb}_f$$
$$0.45359 \text{ kg} = 1 \text{ lb}_m$$
$$10^5 \text{ dyn} = 1 \text{ N}$$
$$10^7 \text{ dyncm} = 1 \text{ J}$$

Other conversion factors are given in Table 1.3. Zimmerman and Levine (2) give an exhaustive list of conversion factors in a compact format.

PROBLEMS

The section numbers in parentheses after the problem number indicate that the subject matter of the problem primarily relates to that section. It is included as an aid to the student as well as the instructor in selecting a balanced set of application experiences.

1. (1.1) The sun heats the top of a closed automobile to about 160F on a calm summer day when the air temperature is 85F. If the heat transfer coefficient at 30 miles/hr is 15 Btu/hrft^{2}F, what is the initial cooling rate per unit area on the roof? What type of heat transfer phenomena is involved?

2. (1.1) Sketch all the configurations you can think of for which the heat transfer phenomena are similar in free convection when $T_s > T_\infty$ and $T_s < T_\infty$. For example, Figure 1.1(a) shows a heated plate facing upward $(T_s > T_\infty)$ which would be similar to a cooled plate $(T_s < T_\infty)$ facing downward.

3. (1.1) Consider the flow of a fluid in a round pipe such that the heat transfer coefficient is h, the inside surface area is A, and the length L. Do a first-law analysis (first law of thermodynamics) for the open system or control volume you have chosen and relate the temperature rise of the fluid in passing through the pipe to the heat transfer coefficient, mass flow rate of the fluid, dimensions, and pertinent fluid properties.

4. (1.1) The surface of a small ceramic kiln is at 130F when the kiln is in operation. If the kiln is an 18-in. cube and is supported in air at 70F with negligibly small legs, with a heat transfer coefficient of 2 Btu/hrft^{2}F (assume to be an average for all surfaces), how many watts are required to keep the kiln in steady-state operation? What type of heat transfer phenomenon is involved?

5. (1.2) Examine Figure 1.2. What is the direction of heat flux at locations x_1, x_2, and x_3? Why?

6. (1.2) When the thermal conductivity of a material is not temperature dependent, then the temperature profile is a straight line (linear). Consider a 1-in.-thick slab of material (assume constant thermal conductivity) with one side at 50F and the other side at 100F. If the material were cork board, concrete, lead, cast iron, or copper, what would be the respective heat fluxes per unit area through the slabs?

7. (1.2) Sketch a temperature profile (temperature as a function of distance) for a slab that is W units thick and large compared to thickness, and has the following variation of thermal conductivity with temperature:
   (a) Increasing with increasing temperature.
   (b) Constant.
   (c) Decreasing with increasing temperature.
   Explain how you arrived at your result.

8. (1.2) Consider a plate of stainless steel that is 0.5 in. thick and intimately bonded to a plate of nickel 0.75 in. thick. Assuming constant conductivity, determine the temperature of the interface if the stainless-steel surface is at 32F and the free surface of the nickel is at 110F under steady-state conditions. What is the unit area heat flux through this sandwich?

9. (1.3) A flat plate is irradiated on top at the rate of 1892 W/m^2 and on the bottom at the rate of 1261 W/m^2. At the instant of observation the plate is at such a temperature that the top and bottom surfaces each emit 100 Btu/hrft^2. The plate has an absorptance of 0.3 and a reflectance of 0.4. What is the radiosity at the top and bottom surfaces and is the plate temperature increasing or decreasing?

10. (1.3) Consider two concentric spheres having radii of 1 ft and 2 ft for the inner and outer spheres. If the surfaces of the spheres were black and diffuse, what fraction of the radiation leaving the inner sphere strikes the outer sphere? What fraction of the radiation leaving the inner surface of the outer sphere strikes the inner sphere?

11. (1.3) Consider the sun to be a black spherical surface with a diameter of 1 million miles. The mean radius of earth's orbit is 93 million miles and the irradiation at earth's orbital distance due to the sun is about 440 Btu/hrft^2 before scattering and absorption by the earth's atmosphere takes place. Estimate the temperature of the sun under these conditions.

12. (1.3) Consider the human body to have about 15 ft^2 of surface area. Calculate the net radiation loss for a human clad only in brief swim trunks

(neglect) standing in a large room whose walls are at 70F. Consider the skin to be black and diffuse.

13. (1.4) Consider a refrigerated space to be a rectangular parallelepiped. One wall is covered by the evaporator coils and a thermocouple is suspended in the center of the enclosure, supposedly to measure the air temperature. Designating the evaporator temperature T_e, the walls T_w, the air temperature T_a, and the thermocouple temperature T_t, sketch an electrical analog circuit for the heat transfer situation, labeling convective resistances R_c and radiation resistances R_R.

14. (1.4) A composite wall is made up of sheetrock, 2×6 in. studs on 15-in. centers with glass wool insulation between, cellotex exterior sheathing, followed by cedar siding. Draw the equivalent electrical circuit for the heat flow situation and identify the resistances involved.

References

1. Van Wylen, G. and Sonntag, R. E., *Fundamentals of Classical Thermodynamics*, Wiley, New York, 1973, p. 27.
2. Zimmerman, O. T. and Levine, I., *Conversion Factors and Tables*, Industrial Research Service, Inc., Dover, N. H., 1961.
3. International Standard 1000, "SI Units and Recommendations for the Use of Multiples and of Certain Other Units," ISO 1000-1973 (E), American National Standards Inst., 1430 Broadway, New York, N.Y. 10018.
4. Mechtly, E. A., "The International System of Units," 2d Revision, NASA SP-7012, 1973.
5. Siegel, R. and Howell, J. R., *Thermal Radiation Heat Transfer*, 2d Edition, McGraw-Hill, 1981, p. 772.

Part I
CONDUCTION

Chapter 2
Steady Conduction

2.0 INTRODUCTION

The solution to the steady conduction problems which we will consider requires the application of Fourier's law to the geometry in question, and by application of the appropriate boundary conditions we will obtain, by mathematical means, the relationship between heat flux and the temperature potential for the given situation. Our purpose will be to see how the principles of calculus, in the simplest sense, can be put to work to obtain the results we seek. It is important that engineers be able to understand the basic mathematical concepts so that when they encounter more complicated situations, they can describe them intelligently to the mathematicians in their organization. With a good basic understanding of mathematics, engineers are also better able to put the solution they obtain from the mathematics section (or literature) into perspective with the goal they desire to accomplish.

In this chapter we will consider only situations where the temperature at any location in our model does not vary with time. Such an aspect comprises the usual definition of steady state, and it further reduces the mathematical aspects to simple ordinary differential equations for one-dimensional cases.

2.1 REVIEW OF MATHEMATICAL CONCEPTS

We will first consider some basic notation customs in mathematics and see how they translate or relate to our physical situation in heat transfer. Generally it is advantageous to retain as much similarity as possible between the notation in our heat transfer application and the notation customarily employed in mathematics courses. For example, in Cartesian coordinates x, y, and z are the three directions; in cylindrical coordinates they are r, θ, and z, and in polar coordinates r, ψ, and ϕ are used. These ideas are illustrated in Figure 2.1, which shows the location of a point P in the three coordinate systems. Accordingly, in steady state we will want to know the temperature T at point P (or any other location). In one-dimensional cases, such as those we will treat in this chapter, the temperature T will be a function of the distance x or radial position r. Rather than having the functional notation $y = f(x)$ as is the case in most math books, we will have $T = f(x)$ or $T = f(r)$ for the physical situations we will consider. We might note also that many mathematicians like to use u for the unknown, but engineers generally use u for specific internal energy and T for the temperature unknown, which makes more sense to us. Also in mathematical notation, the slope of the curve $y = f(x)$ at location x_1 was given by the value of the derivative of the function at x_1 and was expressed as $dy/dx = f'(x_1)$ where $f'(x)$ was evaluated with $x = x_1$. The reader will recall that the prime notation f' means differentiation with respect to the independent variable x. Since we will be concerned with $T = f(x)$, the slope of the temperature-distance relationship $T' = dT/dx$ is an important quantity that determines the heat flux at location x (see discussion of Fourier's law, Section 1.2, Eq. 1.3). To illustrate these concepts, suppose the temperature function could be expressed simply as

$$T = x^2 - 3x + 4 \quad (C, F)$$

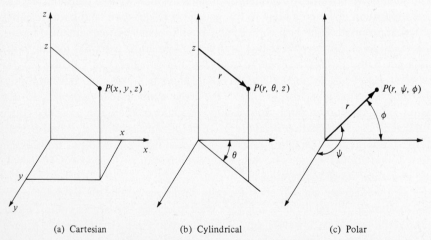

(a) Cartesian   (b) Cylindrical   (c) Polar

Figure 2.1 Relationship of spatial coordinates for the (a) cartesian, (b) cylindrical, and (c) polar coordinate systems.

or

$$T = ax^2 - bx + c \tag{2.1}$$

The coefficients of the x^2 and x terms ($a = 1$, $b = 3$, and $c = 4$) all must have units, specifically: $a = $ C/m^2, F/ft^2; $b = $ C/m, F/ft, and $c = $ C, F for the equation to be dimensionally correct. The temperature gradient (the slope) T' would then be expressed as

$$T' = 2x - 3 \quad \text{(C/m, F/ft)}$$

or

$$T' = 2ax - b \tag{2.1a}$$

by the differentiation formula $d(x^n)/dx = nx^{n-1}$ (see Appendix B).

The function given by Eq. 2.1 is shown in Figure 2.2; we should note, however, that such a temperature profile for a solid in one-dimensional heat flow would be physically real only in a transient situation of the type we will discuss

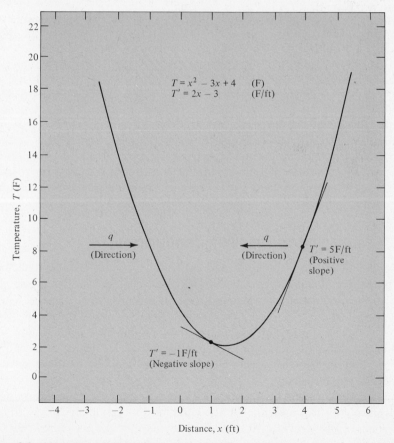

Figure 2.2   Variation of temperature with distance as given by Eq. 2.1, showing the direction of related heat flux.

in Chapter 3, but at this point we only want to establish a sense of orientation and understanding of how the mathematical language describes the physical phenomena of heat flow.

We can see from Figure 2.2 that to the left of the minimum point at $x = 1.5$ ft the energy flows to the right (in the positive x direction), because Fourier's law (Eq. 1.3) requires $q = -kA\,dT/dx$ or $q = -kAT'$. The derivative $T' = dT/dx$ is negative in the region to the left of $x = 1.5$; therefore, q is positive, which agrees with our physical reasoning; but in the region to the right of the minimum point (at $x = 1.5$), the energy flows to the left in the direction of decreasing temperature, and we note that the slope of the temperature profile is positive so that Fourier's law requires q to be negative, that is, in the negative x direction as it must be.

To complete this discussion, we can note that Eq. 2.1 could be made dimensionless by taking the distance variable as x/L where L is some important dimension like total length or maximum thickness, for example, and the temperature variable could be written as T/T_0, where T_0 might be the temperature at $x = 0$. We could then express Eq. 2.1 as follows:

$$\frac{T}{T_0} = a_0 \frac{x^2}{L^2} + b_0 \frac{x}{L} + 1 \tag{2.2}$$

Such a dimensionless expression would hold for any set of units, and the convenience of working with dimensionless equations should be apparent to the reader.

In the following sections, we will be applying Fourier's law $q = -kA(dT/dx)$ or $q = -kA(dT/dr)$ to plates, cylinders, and spheres in steady state in order to find out how much energy has passed into or out of a particular system or to determine the temperature at a particular location. Such calculations are an important aspect of the energy management problems that confront our society.

2.2 PHYSICAL PROPERTIES FOR CONDUCTION

The ability of a solid material to transport energy depends on a combination of the properties of thermal conducitivity (k), the density (ρ), the specific heat (c_p), and a combination of properties called the thermal diffusivity ($\alpha = k/\rho c_p$). Phenomenologically, the diffusion of energy through a liquid in laminar flow is similar to the diffusion of momentum into a liquid in laminar flow. The momentum diffusivity $\nu = \mu/\rho$ has the same units as α, that is, m^2/s or ft^2/s, and we will discuss this aspect in more detail later on in the chapter on convection mechanisms.

In general, we will briefly note that thermal energy is transported through solids mainly by two mechanisms: the vibrations of the atomic lattice structure and clouds of free electrons. At moderate temperatures and up, the free electrons account for practically all the energy transport in metals so that the same mechanism that transports charge also transports energy. Consequently, the ratio of thermal to electrical conductivity is nearly constant and can be used to predict thermal conductivity. For nonmetals the lattice vibrations are the chief means of energy propagation at moderate temperatures. The free electrons do not come into play until fairly high temperature levels are reached. Therefore, nonmetals, or nonconductors, are poor conductors of heat at low temperatures, or one might say they are good insulators.

The density ρ we take as the amount of mass per unit volume for homogeneous materials. In the case of porous material we must take a large enough volume to obtain a density that does not change with volume. We note that density for porous materials is an important factor because the air (or other gas) in the interstitial spaces influences the thermal conductivity of the aggregate material. For the smallest δv, then, where we get consistent values we can define density as

$$\rho = \lim_{\Delta v \to \delta v} \frac{\Delta m}{\Delta v} \quad (\text{kg/m}^3, \text{lb}_\text{m}/\text{ft}^3) \tag{2.3}$$

Representative values for density are tabulated in Appendix C.

For solids we need not distinguish between the constant-volume and constant-pressure specific heats because the effect of moderate pressure on the volume change of most solid material is negligible. Most tabulations for specific heat are, however, for constant pressure because that is the easiest way to carry out the experiments. Conceptually, the specific heat is the energy absorbing capability of a material per unit mass and per unit of temperature change. Therefore

$$c_p = \frac{\text{energy}}{\text{unit mass, temperature}}$$

$$c_p = \text{J/kg K, Btu/lb}_\text{m}\text{F} \tag{2.4}$$

Values for specific heat are tabulated in Appendix C.

The property thermal conductivity is derived from Fourier's law (Eq. 1.3) so that:

$$k = \frac{\text{unit area flux}}{\text{temperature gradient}}$$

$$k = \text{W/mK, (Btu/hrftF)} \tag{2.5}$$

Values for thermal conductivity for solids range, for example, from about 0.87 W/mK (0.5 Btu/hrftF) for glass to 380 W/mK (220 Btu/hrftF) for copper. Much higher values can be obtained if special care is taken in promoting crystal or fiber alignment in the material. Appendix C tabulates representative values for several metals and nonconductors for use in calculating numerical results in examples and problems.

Many times engineers are confronted with situations where they need values for physical properties that are not readily available. Sources such as References 2, 3, and 4 are helpful when no direct information is available from the manufacturer. The National Bureau of Standards also has published selected sources of physical properties in their National Standard Reference Data Series (5). We have noted above that the ratio of thermal and electrical conductivities can be a useful means for predicting values of k for metals. Specifically, the Wiedermann-Franz law states that the Lorenz number

$$\mathcal{L} = \frac{k}{\sigma T} \quad \text{(volts/degree)}^2 \tag{2.6}$$

where

k = thermal conductivity, W/mK, (Btu/hrftF)

σ = specific conductance, $(\Omega m)^{-1}$, $(\Omega ft)^{-1}$

T = absolute temperature, K (R)

is theoretically a constant at temperatures above the superconducting region. Bosworth (1) has compiled values for $\mathcal{L}$ for several different metals as shown in Table 2.1.

The metals listed in Table 2.1, with the exception of tin and bismuth, show a slight increase in $\mathcal{L}$ with increasing temperature. Since the units associated with the Lorenz number are rather unusual, it may be helpful to explore their composition:

$$\mathcal{L} \equiv (W/mK) \cdot (\Omega m/K)$$
$$\equiv (E^2/\Omega) \cdot (\Omega/K^2) \equiv (E/K)^2 \tag{2.7}$$

where Ohm's law has been employed to replace watts by E^2/Ω. As an example, Reference 3 gives the specific resistance (note that conductance is the reciprocal of resistance) of copper at 20C (293K) to be 1.7×10^{-6} Ωcm and k for copper at that temperature is about 385 W/mK so that we obtain

$$\mathcal{L} = 385(1.7)10^{-8}/293$$
$$= 2.23 \times 10^{-8}(E/K)^2 \tag{2.8}$$

Table 2.1 VALUES OF THE LORENZ NUMBER FOR VARIOUS METALS

METAL	$\mathcal{L} \times 10^8 (E/K)^2$		METAL	$\mathcal{L} \times 10^8 (E/K)^2$	
	0C	100C		0C	100C
Copper	— — 2.23	2.33	Iron	2.47	2.56
Silver	— — 2.31	2.37	Nickel	1.77	2.28
Gold	— — 2.35	2.40	Molybdenum	2.61	2.79
Zinc	— — 2.31	2.33	Palladium	2.59	2.74
Cadmium	— — 2.42	2.43	Tungsten	3.04	3.20
Aluminum	— — 2.12	2.23	Rhenium	2.57	2.57
Tin	— — 2.52	2.49	Iridium	2.49	2.49
Lead	— — 2.47	2.56	Platinum	2.51	2.60
Bismuth	— — 3.31	2.89			

SOURCE: Bosworth, R. C. L., *Heat Transfer Phenomena*, Wiley, New York, 1952. Reprinted with permission of Horwitz Group Books Pty. Ltd.

as given in Table 2.1. With a value for σ and k known at room temperature, a simple measurement of σ at other temperatures will permit a reasonable prediction of the variation of k with temperature.

When we encounter materials whose thermal conductivity is a function of temperature, it will be necessary to describe that variation mathematically in order to apply Fourier's law to determine heat flux or temperatures. In some cases a simple linear function such as

$$k = k_0(1 + aT) \tag{2.9}$$

can be used to fit the data over the range of applicable temperatures. The quantity k_0 is the conductivity at $T_0 = 0F$ and k_0a is the slope of the line. We must exercise caution with units and we also note that the quantity a can be either positive or negative. When the variation of k is not linear enough to use Eq. (2.9), sometimes an exponential expression such as

$$k = k_0 + a(T - T_0)^n \tag{2.10}$$

can be used over a certain range in temperature depending on the choice of the constants k_0, a, and n. Figure 2.3 illustrates an application of the above concepts

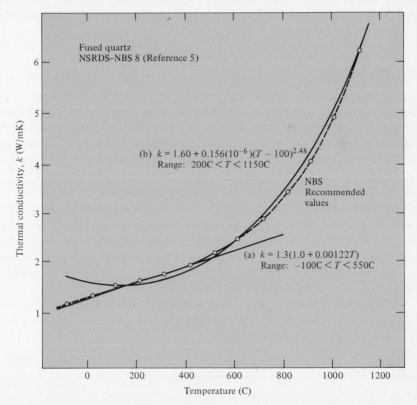

Figure 2.3   Variation of thermal conductivity of fused quartz from Reference (5) with temperature: (a) linear approximation, and (b) exponential approximation.

by comparing the recommended values of thermal conductivity for fused quartz (Reference 5) with a linear and an exponential approximation. The figure shows that in the range from about -100 to 550C a linear approximation as shown by curve (a) is reasonable, but over a wider range such as from 200 to 1150C, for example, an exponential expression [as shown by curve (b)] is required to give a reasonably accurate representation of the data. Curves (a) and (b) in Figure 2.3 are only two examples of what might be done; other more sophisticated curve fitting techniques can be employed if necessary. The example shown in Figure 2.3 also demonstrates the hazards involved in using empirical equations outside the range for which the constants in the equations were determined. Extrapolation is risky!

2.3 ONE-DIMENSIONAL CONDUCTION, NO SOURCES

In this section we will analyze the steady-state temperature distribution and corresponding heat flux for the cases where the heat flux, or the temperature, is a function of only one distance variable such as large plane walls, long cylinders, short cylinders with the ends insulated, or hollow spheres. We will see how to obtain results for constant thermal conductivity and for the thermal conductivity as a function of temperature.

2.3.1 Plane Walls

The requirement of one-dimensional heat flux limits our application to slabs of material with edges insulated or to plane walls that are large in extent compared to their thickness. The physical situation has to be such that whatever heat flux leaks out the ends of the slab or wall, it is a negligible fraction of the heat flux that passes through the material in the direction perpendicular to the face of the slab. Figure 2.4 shows the physical picture we are discussing in Cartesian coordinates; the wall is L units thick in the x direction ($L = x_2 - x_1$) and extends indefinitely in the y and z directions. We are assuming that the temperature is the same in all planes perpendicular to the x axis and that no heat flows in the y or z directions within the wall. Accordingly, the temperature T is a function of x only as shown in Figure 2.4(b), and we can apply Fourier's law (Eq. 1.3) to obtain the heat flux or the temperature profile for constant or variable thermal conductivity.

We will begin with complete detail in the mathematical manipulations and solutions so that later on we will not need to repeat similar steps for other geometries. Fourier's law states that

$$q = -kA \frac{dT}{dx} \tag{1.3}$$

or

$$\frac{q}{A} dx = -k \, dT \tag{2.11}$$

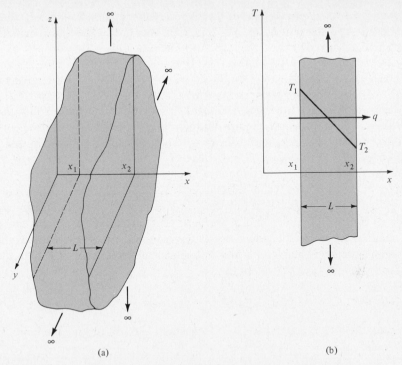

Figure 2.4   Physical arrangement of the homogeneous plane wall:
(a) orientation in cartesian coordinate system, and (b) orientation on the
one-dimensional T, x plane.

after separating the variables. The choice of arrangement is made as shown in
Eq. 2.11 because we can recognize that k may be a function of temperature so
that we must keep it on the side of the equation with dT. In this case the flux per
unit area is a constant so that q/A could be on either side of the equation and
we arbitrarily place it with dx. The next step is to integrate both sides of the
equation between the limits of x_1 and x_2 and the corresponding limits of T_1 and
T_2. Thus

$$\frac{q}{A} \int_{x_1}^{x_2} dx = - \int_{T_1}^{T_2} k \, dt \qquad (2.12)$$

We observe that since q/A is a constant and not a function of x, we have taken
it outside of the integral sign and, for the case where k can be considered
constant (that depends on the material and the temperature interval T_1 to T_2),
we can also take it outside the integral and then perform the actual integration
to obtain

$$\frac{q}{A}(x_2 - x_1) = -k(T_2 - T_1) \qquad (2.13)$$

according to calculus principles. The steps we show here may seem overly simple
to the reader who is very familiar with calculus ideas; we will encounter aspects

with more challenge in the paragraphs to follow. After we recognize that $x_2 - x_1$ is the thickness of the wall L, we can express the total heat flux q through the entire surface area A as

$$q = \frac{kA}{L}(T_1 - T_2) = \frac{T_1 - T_2}{L/kA} \tag{2.14}$$

We recall from Chapter 1 (Section 1.4) that kA/L is the conductance and L/kA is the thermal resistance for the temperature driving potential $T_1 - T_2$. Since we are discussing only conduction heat flux, we do not distinguish between q_k and q in this chapter.

When the thermal conductivity is a strong function of temperature, it is necessary to keep k under the integral sign in Eq. 2.12. The solution then requires that k be expressed as a function of temperature before the integration can be performed. In Section 2.2 we saw that sometimes a simple linear approximation for thermal conductivity as given by Eq. 2.9 is reasonable if the temperature range is not too wide. After we substitute Eq. 2.9 into the right-hand side of Eq. 2.12, we obtain

$$-\int_{T_1}^{T_2} k\, dT = -\int_{T_1}^{T_2} k_0(1 + aT)\, dT \tag{2.15}$$

Performing the integration, noting k_0 is a constant, and substituting the limits T_1 and T_2 yields

$$-\int_{T_1}^{T_2} k\, dT = -k_0\left[(T_2 - T_1) + \frac{a}{2}(T_2^2 - T_1^2)\right] \tag{2.16}$$

We can factor out $-(T_2 - T_1) = (T_1 - T_2)$ on the right side of Eq. 2.16 to obtain

$$-\int_{T_1}^{T_2} k\, dT = (T_1 - T_2)k_0\left[1 + \frac{a(T_1 + T_2)}{2}\right] \tag{2.17}$$

For the linear expression for k that we are discussing, k_0 times the term in brackets has a special significance; it represents the thermal conductivity at the mean or average temperature $(T_1 + T_2)/2$. Thus

$$-\int_{T_1}^{T_2} k\, dT = k_m(T_1 - T_2) \tag{2.18}$$

where

$$k_m = k_0[1 + a(T_1 + T_2)/2]$$

a result that we can use whenever we encounter the integral of $k\, dT$ for a linear variation in thermal conductivity.

With linearly varying thermal conductivity, we can then express the result for the solution of Eq. 2.12 as

$$q = \frac{k_m A}{L}(T_1 - T_2) \tag{2.19}$$

where k_m represents the mean conductivity at temperature $(T_1 + T_2)/2$ as defined previously. We can see that the result is very similar to Eq. 2.14 in form; however, had we used an expression other than a linear one for the variation of k with T, we would not have obtained a similar result, an aspect that is illustrated in Example 2.1.

Example 2.1

An experimental engine has a quartz window in the cylinder head that is 2.5 cm thick. During engine operation, the average inside surface temperature is estimated to be 1100C. The ambient temperature is 25C and the heat transfer coefficient at the outer face of the window is 20 Btu/hrft²F. We want to determine the unit area heat flux and the outside surface temperature of the quartz under these conditions.

SOLUTION

We assume a one-dimensional situation so that we can use the results developed above. Doing so will give us the maximum amount of heat transmitted through the quartz window since we do not have information on how the cylinder head is cooled. Since the thermal conductivity of quartz is a strong function of temperature (as shown in Figure 2.3) in the temperature range of interest, it will be informative to look at both a simplified and a more accurate solution. Not knowing the outer surface temperature requires an iterative (trial and error) solution when considering k variable with temperature but with the electronic hand calculators available, we need not throw in the towel and head for the shower.

We will first consider a simple approach using the electrical analog as pictured in Figure 2.5. Accordingly, we visually estimate an "average" value of

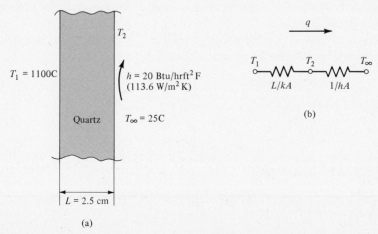

(a)

(b)

Figure 2.5 Thermal model (a) and corresponding electrical analog (b) for Example 2.1.

thermal conductivity to be about 3.5 W/mK from Figure 2.3. The electrical analogy pictures the flux q as flowing through the two series resistances L/kA and $1/hA$ from potential T_1 to potential T_∞; therefore, we can write the heat flux q as the potential difference divided by the equivalent resistance. Thus

$$q = \frac{T_1 - T_\infty}{L/kA + 1/hA}$$

or, since we are interested in the unit area heat flux,

$$\frac{q}{A} = \frac{T_1 - T_\infty}{L/k + 1/h}$$

From Table 1.2 we find $M = 5.67826$ in order to convert h units from BE to SI so that $h = 5.67826(20) = 113.6$ W/m^{2}K, and for unit area flux we calculate

$$\frac{q}{A} = \frac{1100 - 25}{0.025/3.5 + 1/113.6}$$

$$= \frac{1075}{0.01595}$$

$$= 67{,}380 \text{ W/m}^2 \ (21{,}380 \text{ Btu/hrft}^2)$$

We observe that the quartz resistance (0.00714 m^{2}K/W) is about the same as the convective resistance (0.00881 m^{2}K/W) so that we would expect the outer surface temperature to lie about halfway between T_1 and T_∞. Specifically, we find the temperature drop across the quartz

$$T_1 - T_2 = 67{,}380(0.00714) = 481\text{K}$$

so that $T_2 = 1100 - 481 = 619$C, since on a difference basis the size of the Kelvin degree is the same as the Celsius degree. It would be very important for the reader who is new to the topic to carefully check out all the units in the example so far, and below.

We now look at a more accurate solution by utilizing k as a function of temperature from Figure 2.3 where

$$k = k_0 + a(T - 100)^{2.48}$$

$$k_0 = 1.60 \text{ W/mK}$$

$$a = 0.156(10^{-6}) \text{ W/mK(C)}^n$$

The unit area flux can be calculated by using Eq. 2.12 and inserting the expression for k from above:

$$\left(\frac{q}{A}\right)_k (x_2 - x_1) = -\int_{T_1}^{T_2} \left[k_0 + a(T - T_0)^n\right] dT$$

We will integrate the general expression observing that k_0, a, and T_0 are constants so that we can more easily follow the manipulations

$$\int_{T_1}^{T_2} k \, dT = \int_{T_1}^{T_2} k_0 \, dT + a \int_{T_1}^{T_2} (T - T_0)^n \, dT$$

$$= k_0(T_2 - T_1) + \frac{a}{n+1} \left[(T_2 - T_0)^{n+1} - (T_1 - T_0)^{n+1} \right]$$

Since $x_2 - x_1 = L$, the thickness of the quartz, the expression for heat flux becomes, remembering the minus sign on the conductivity integral,

$$\left(\frac{q}{A} \right)_k = \frac{k_0(T_1 - T_2)}{L} + \frac{a}{L(n+1)} \left[(T_1 - T_0)^{n+1} - (T_2 - T_0)^{n+1} \right]$$

and

$$\left(\frac{q}{A} \right)_c = h(T_2 - T_\infty)$$

In principle we would equate $(q/A)_k$ with $(q/A)_c$ and solve for T_2, but that cannot conveniently be done so we have to resort to a trial-and-error solution. The most efficient way to solve iterative problems is to plot the value of the two functions against the unknown quantity (T_2 in this case); then the intersection of the two curves gives the value of the unknown which satisfies both equations. Let us calculate the two fluxes at $T = 650C$ and see where we stand:

$$(q/A)_k = 1.60(1100 - 650)/0.025$$

$$+ 0.156(10^{-6})\left[(1100 - 100)^{3.48} - (650 - 100)^{3.48} \right]/0.025(3.48)$$

$$= 72{,}019 \text{ W/m}^2$$

$$(q/A)_c = 113.6(650 - 25)$$

$$= 70{,}875 \text{ W/m}^2$$

The convection flux shows our temperature estimate to be low; at $T_2 = 660C$ we obtain

$$\left(\frac{q}{A} \right)_k = 70{,}979 \text{ W/m}^2$$

$$\left(\frac{q}{A} \right)_c = 72{,}009 \text{ W/m}^2$$

The heat fluxes are plotted in Figure 2.6 as a function of T_2; a linear interpolation shows the solution to be slightly over 655C. The reader should verify that the fluxes are the same at that temperature. We further note that the variable property solution gives a heat flux of 71,500 W/m^2 and a surface temperature of 655C as compared to 67,380 W/m^2 and a surface temperature of 619C, assuming an average constant thermal conductivity, a result that is significantly higher. This completes Example 2.1.

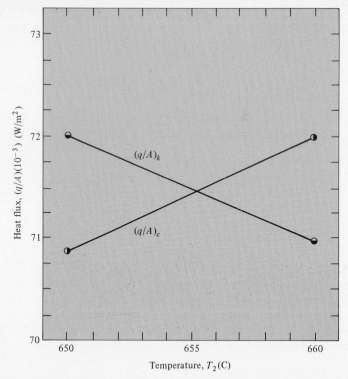

Figure 2.6   Variation of heat flux with surface temperature T_2 for the iterative solution discussed in Example 2.1.

2.3.2 Composite Walls

In the paragraphs above we discussed one-dimensional heat transfer through a single plane wall with constant and variable thermal conductivity. We will now examine situations where plane walls of different materials are placed in intimate contact so that the heat flows through them in either series or parallel paths. In our study of such situations, we make two limiting assumptions: first, that the contact resistance* between the different materials is negligible and second, that the heat energy always flows one-dimensionally. The former can be a serious limitation if air (or other gas) gaps of any appreciable size exist between the different materials. The latter limitation is not serious if the thermal conductivities of the various materials are not widely different, otherwise two-dimensional effects must be taken into account.

Figure 2.7(a) shows a composite plane wall made up of three different materials, and having a convective coefficient h_0 on the left face and h_4 on the right face. Figure 2.7(b) shows the equivalent electrical analog from which we can observe that the heat flux is equal to the overall potential difference divided

*We will discuss the aspect of contact resistance in more detail in Section 2.6.

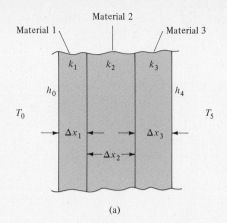

Material 2

Material 1

Material 3

k_1 k_2 k_3

h_0 h_4

T_0 T_5

Δx_1 Δx_3

Δx_2

(a)

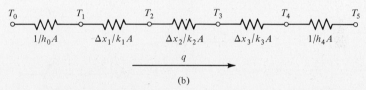

T_0 T_1 T_2 T_3 T_4 T_5

$1/h_0 A$ $\Delta x_1/k_1 A$ $\Delta x_2/k_2 A$ $\Delta x_3/k_3 A$ $1/h_4 A$

q

(b)

Figure 2.7 Composite plane wall with three different materials in series (a) and the corresponding electrical analogy (b).

by the equivalent resistance

$$q = \frac{T_0 - T_5}{R_{equiv}} \tag{2.20}$$

In this case the equivalent resistance is simply the sum of the individual resistances in series.

$$R_{equiv} = \frac{1}{h_0 A} + \frac{\Delta x_1}{k_1 A} + \frac{\Delta x_2}{k_2 A} + \frac{\Delta x_3}{k_3 A} + \frac{1}{h_4 A} \tag{2.21}$$

We should also observe that in the steady state, the same amount of heat flux in Btu/hr or watts which enters the left face passes through the different materials and leaves the right face at any given instant of time.

A more complex arrangement is shown in Figure 2.8 in that the heat flux paths are not in a simple series circuit; the figure shows an idealized sketch of a concrete block wall with a coating of stucco on the left and a coating of plaster on the right. The holes in the concrete block could just contain air or be filled with an insulating material such as vermiculite (expanded mica). We shall develop the expression for heat flux in the same general manner as we did for the simple composite wall in that the flux is the potential divided by the equivalent resistance. Thus

$$q = \frac{T_1 - T_5}{R_{equiv}} \tag{2.22}$$

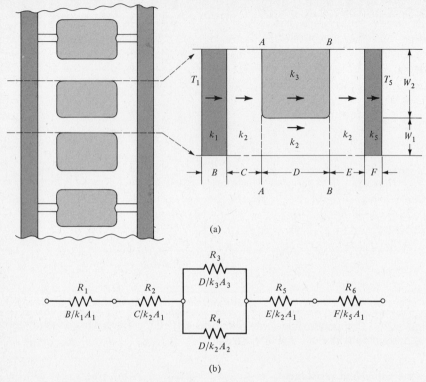

(a)

(b)

Figure 2.8 Composite plane wall (a) with series and parallel heat paths, and corresponding electrical analogy (b).

and the equivalent resistance becomes

$$R_{equiv} = R_1 + R_2 + \frac{R_3 R_4}{R_3 + R_4} + R_5 + R_6 \tag{2.23}$$

where $R_1 = B/k_1 A_1$, $R_3 = D/k_3 A_3$, and so forth. Care must be exercised in evaluating the areas involved; $A_1 = (W_1 + W_2)H$, $A_2 = W_1 H$, and $A_3 = W_2 H$, where H is height of the wall [perpendicular to the plane of Figure 2.8(a)] so that $A_1 \neq A_2 \neq A_3$.

We might observe the consequences of our one-dimensional assumption in concluding our discussion on this topic. The heat flux at plane A in Figure 2.8(a) is required to split into two parts, according to the relative magnitudes of R_3 and R_4, and then recombine at plane B to continue through the last two materials. The heat flux does not actually behave in that manner, but (as we emphasized earlier) if the conductivities are not widely different, the electrical analog method gives reasonable results.

2.3.3 Conduction in Long Cylinders

We now turn our attention to radially symmetric systems such as infinite cylinders and hollow spheres. When we separate variables in Fourier's law, as we did in Eq. 2.11, we place the distance variables and flux on the left and the temperature variables on the right side of the equal sign. Thus

$$\frac{q}{A(r)}\,dr = -k\,dT \tag{2.24}$$

We note that our distance variable is the radial distance r (instead of x) and the temperature variable T is the same; however, we must recognize that the area $A(r)$ through which the flux q passes is a function of r. We can see that $A(r)$ increases as r increases in the sketch illustrated in Figure 2.9. During steady-state conditions, all the heat flux q that enters the inside area $A(r_1)$ also passes through the outer surface area $A(r_2)$, but the unit area flux at r_2 will be less than that at r_1. If we substitute $A(r) = 2\pi r L$ into Eq. 2.24, we obtain

$$\frac{q}{2\pi L}\int_{r_1}^{r_2} r^{-1}\,dr = -\int_{T_1}^{T_2} k\,dT \tag{2.25}$$

for the condition that the inner surface at r_1 is at temperature T_1 and the outer surface at r_2 is at T_2. These are the mathematical boundary conditions for this situation. We can perform the indicated integration (see Appendix B for assistance if necessary) for constant properties to get the following result.

$$q = \frac{2\pi k L (T_1 - T_2)}{\ln(r_2/r_1)} \tag{2.26}$$

Comparing with Eq. 1.15, we can see that the thermal resistance R_k of a hollow

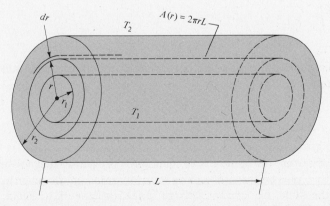

Figure 2.9   Nomenclature and boundary conditions for the infinite cylinder geometry.

cylinder can be expressed as

$$R_k = \frac{\ln(r_2/r_1)}{2\pi kL} \tag{2.27}$$

so that concentric hollow cylinders of different materials can be treated with the electrical analogy in a manner similar to composite plane walls. The equivalent resistance is the sum of the individual resistances in the series circuit, but, of course, the form is given by Eq. 2.27 instead of by R_k in Eq. 1.15.

If we need to consider thermal conductivity to vary with temperature and can use a linear approximation, the right side of Eq. 2.25 becomes

$$-\int_{T_1}^{T_2} k \, dT = -\int_{T_1}^{T_2} k_0(1 + aT) \, dT \tag{2.28}$$

which we have already evaluated in Eqs. 2.15 to 2.17 so that we can express the heat flux for linearly varying thermal conductivity as

$$q = \frac{2\pi k_m L(T_1 - T_2)}{\ln(r_2/r_1)} \tag{2.29}$$

where k_m was defined by Eq. 2.18 as the conductivity at the mean temperature $(T_1 + T_2)/2$. If the material under consideration has a variation of k with temperature (in the range T_1 to T_2) which cannot be considered linear, then that function required to express the k variation with temperature must be substituted into the right side of Eq. 2.25 before the integration procedure is carried out.

In order to obtain the temperature distribution in the solid, we change the upper limits of integration on the integrals in Eq. 2.25 to the indefinite values r and $T(r)$ so that

$$\frac{q}{2\pi L}\int_{r_1}^{r} r^{-1} \, dr = -\int_{T_1}^{T(r)} k \, dT \tag{2.30}$$

After the integration is performed for constant k as before, we can substitute for $q/2\pi Lk$ from Eq. 2.26 to obtain

$$T(r) = T_1 - \frac{(T_1 - T_2)\ln(r/r_1)}{\ln(r_2/r_1)} \tag{2.31}$$

for the temperature distribution. We mentioned previously that in steady state, the heat flux q passing through each cylindrical area from the inner to the outer surface is the same (like current through series resistances). It would, however, be helpful to actually see the variation of $q/A(r)$ and $T(r)$ with radial distance and we will make that calculation in the next section.

2.3.4 Computer Evaluation of *T(r)*

Facility in computer programming and use varies widely with different people. The readers and users of this text have probably had an introductory FORTRAN course, but little experience in programming. Accordingly, we will use the simple task of programming Eq. 2.31 to get our feet wet (so to speak), so that we can do some more sophisticated programming later on that would not be so easy to do on a hand-held calculator. Our purpose is then twofold: to review some basic programming ideas, and to learn something of heat flow characteristics in the cylindrical geometry.

PROGRAM OBJECTIVE
We desire to compute the heat flux q, the unit area heat flux $q/A(r)$, and the radial temperature profile $T(r)$ as a function of radius ratio r/r_1 for the cylindrical geometry shown in Figure 2.9.

PHYSICAL DESCRIPTION
Let the cylinder be 1 ft long with an inner radius of 1 ft with the end faces insulated. We choose the thermal conductivity $k = 2$ Btu/hrftF, the radius ratio $r_2/r_1 = 2$, the inner surface temperature $T_1 = 300F$, and the outer surface temperature $T_2 = 100F$ as convenient representative values for the parameters.

FORTRAN PROGRAM
The following represents the step-by-step procedure of what one would do at the key punch to get a deck ready to toss into the hopper at the computer, or type into an interactive CRT terminal. Capitals represent the individual symbols punched on FORTRAN cards and should be recognizable as FORTRAN language. Explanatory text is in lowercase. Most installations have interactive terminals that can be used instead of card readers. In the discussion above and to follow, the word "line" can be substituted for "card" if a terminal is being used.

• *$JOB CONTROL CARDS.* Since each installation has its own peculiarities, no specific information can be given here on control cards; find out what your friendly computer wants and do accordingly! We suggest requesting a control card that will result in only the FORTRAN list and output being printed in order to conserve paper.

Always include a program dictionary:

```
C   PROGRAM DICTIONARY:
C      Q = HEAT FLUX, BTU/HR
C   QAR = UNIT AREA FLUX, BTU/HRFT2
C      XL = LENGTH, FT
C      TR = TEMPERATURE AT RADIUS R, F
C    T12 = TEMP DIFFERENCE (T1 - T2), F
C    R21 = RADIUS RATIO, R2/R1
C      RR = RADIUS RATIO, R/R1
C      XK = THERMAL COND, BTU/HRFTF
```

Next in the program we have to supply the data that we need. Data could also be read in, but for this first attempt, we will stay as uncomplicated as we can.

```
10    T1=300.
20    R21=   2.0
30    T12=200.
40    XL=   1.0
50     PI=   3.1415926
60    XK=   2.0
```

We note that we are assigning statement numbers to each card; doing so serves as card identification as well as possible use in logic control. We could also number the cards on the right side in the identification columns 73 through 80, but using statement numbers is another way of accomplishing the same purpose. Using steps of 10 in the statement numbers is convenient in case we want to, perhaps, make some changes later on by adding cards in between or correcting mistakes while keeping the statement numbers consecutive. It's not necessary that statement numbers be consecutive; the computer would accept the program if they were all mixed up, but we're using them for the dual purpose of identification as we said above. Statements 10 through 60 supply the needed physical information, or data. They are incorporated directly into the program rather than read in by a READ statement from separate data cards because we do not anticipate a large number of calculations with different input.

```
70     RR=1.1
80   AR21=ALOG(R21)
```

The value set up for the radius ratio $RR = r/r_1$ is the initial value in a series of values that will increment in steps of 0.1 in the repetitive sequence below. The natural log of $R21 = r_2/r_1$ need only be calculated once.

```
 90   PRINT 500, R21
500   FORMAT (1H1,///10X,'TEMPERATURE AND FLUX VALUES FOR'/11X,
      'ONE-DIMENSIONAL  HEAT TRANSFER'/15X,'IN A HOLLOW CYLINDER'/18X,
      'WITH R21=',F4.0///11X,'RR',6X,'TR',7X,'Q',6X,'Q/(AR)'//)
```

The FORMAT statement number 500 prints the output table heading and the headings for the columns of numerical results RR, radius ratio, TR, temperature at RR; Q, the heat flux, and Q(AR), the unit area heat flux. We note that 1H1 is a carriage control that gives a new page and is not printed in the output, and that we have not shown a symbol in column 6 for continuation of the statement. The programmer can supply such symbols as needed.

```
100   IF (RR − R21) 110, 110, 170
110   TR=T1−(T12*ALOG(RR)/AR21)
120   Q=2.*PI*XK*XL*(T1−TR)/ALOG(RR)
130   QAR=Q/(2.*PI*RR*XL)
140   PRINT 510, RR, TR, Q, QAR
150   RR=RR+.1
160   GOTO100
170   STOP
510   FORMAT (10X, F4.1, F8.0, F9.0, F10.0)
      END
```

The statement sequence 100 through 160 is the main calculation loop for TR, Q,

Table 2.2 COMPUTER OUTPUT

TEMPERATURE AND FLUX VALUES FOR
ONE-DIMENSIONAL HEAT TRANSFER
IN A HOLLOW CYLINDER WITH R2/R1 = 2

RR	TR	Q	Q/A(R)
1.1	272.	3626.	525.
1.2	247.	3626.	481.
1.3	224.	3626.	444.
1.4	203.	3626.	412.
1.5	183.	3626.	385.
1.6	164.	3626.	361.
1.7	147.	3626.	339.
1.8	130.	3626.	321.
1.9	115.	3626.	304.
2.0	110.	3626.	289.

and QAR. The calculation sequence first checks the value of RR with the IF statement in 100. In this program, if RR is less than 2.0, the sequencing goes to 110, the calculations for TR, Q, and QAR are made and printed, the value of RR is updated by 0.1, and control is returned to statement 100 for repeating. When RR = 2.1, the IF statement in 100 is positive and the control sequence goes to 170, which stops the program. The reader will recall that FORMAT statements can appear anywhere in a program above the END statement and below any DIMENSION or EQUIVALENCE statements. The output from statement 500 is shown in Table 2.2.

The table format in Table 2.2, given in statement 500, can easily be altered and adapted to other programs the reader may write. The results shown in Table 2.2 are also plotted in Figure 2.10 which demonstrates that the temperature varies logarithmically with radius and that the unit area heat flux decreases toward the outer surface of the cylinder.

The above program presents a simple problem with easily understandable logic and a somewhat sophisticated output that can be traced back to the FORTRAN statement that produced it. The reader will recognize that a well-organized computer output table can be directly incorporated into a technical report and as such represents a considerable savings in time and effort in organizing results.

2.3.5 Conduction in Hollow Spheres

For the spherical geometry, we again apply Fourier's law (Eq. 2.24) except that in this case $A(r) = 4\pi r^2$ (a spherical shell of radius r) so that integration of Eq. 2.24 yields

$$q = \frac{r_1 r_2 4\pi k (T_1 - T_2)}{r_2 - r_1} \tag{2.32}$$

where T_1 is the temperature on the inner surface at r_1, and T_2 is the temperature

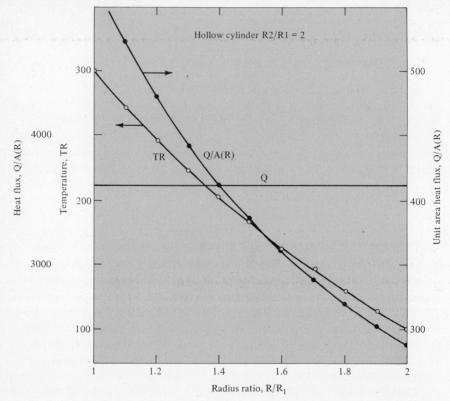

Figure 2.10 Variation of temperature TR, unit area heat flux $Q/A(R)$, and heat flux Q with radial position for the infinite hollow cylinder analyzed by computer solution.

on the outer surface of the sphere at r_2. We can see that a linear variation for thermal conductivity would again yield the mean conductivity k_m [evaluated at the mean temperature $T_m = (T_1 + T_2)/2$] in Eq. 2.32 instead of k; details of this analysis are explored in the home problems. The temperature variation with radial position can be obtained by integrating Eq. 2.24 between the limits of r_1 and r with the corresponding temperature T_1 and $T(r)$, then substituting for $q/4\pi k$ from Eq. 2.32 to obtain

$$T(r) = T_1 - \frac{[1 - (r_1/r)](T_1 - T_2)}{1 - (r_1/r_2)} \qquad (2.33)$$

One of the application problems asks the reader to include Eq. 2.33 in the computer program for the cylinder in order to make a physical comparison between the temperature profiles, heat flux, and unit area flux for the cylinder and sphere for the same radius ratio and inner surface temperature.

This section completes our discussion of conduction phenomena in simple geometries wherein we have applied Fourier's law in a straightforward integration process. We will now turn our attention to the situation where energy is generated uniformly inside the solid under consideration.

2.4 ONE-DIMENSIONAL CONDUCTION, UNIFORM SOURCES

In the steady-state situations we have discussed so far, the energy that has been conducted through the solid material has come from outside the material substance. We want to now consider those situations where the energy can also come from inside (or "disappear" into) the substance as well. Some familiar phenomena that can produce (by produce we, of course, must mean a change in energy form that results in heat) thermal energy within substances are Joulian heating (i^2R or resistance heating), atomic fission or fusion, high kinetic energy particle absorption such as neutrons or gamma rays (for example, high frequency radar radiation), and chemical reactions. Of these phenomena, only chemical reactions can be endothermic as well as exothermic, that is, absorbing energy or acting as an energy sink; the energy, of course, does not disappear but is utilized in rearranging molecular bonds and stored in the new arrangement.

We will consider steady-state situations for the simple geometries of plane plate, infinite (long) cylinder, and the sphere having source (or sink) strengths $\dot{q}$ in W/m^3 or $Btu/hrft^3$. In general, $\dot{q}$ may have a different value at each spatial location in the solid; we will here restrict our discussion and analysis only to those situations where $\dot{q}$ can be considered constant throughout the solid. It is very important to keep the idea or concept of $\dot{q}$, the generation rate, separate from the concept of heat flux rate q. Once the energy in the form of $\dot{q}$ has been deposited at a particular location, it joins (or leaves, if the "deposit" was negative) the stream of energy q that is flowing past the location under consideration, and the quantity q is governed by Fourier's law of conduction as we have seen before. It is important also to realize that $\dot{q}$ is deposited at each and every spatial location in the solid and not just at any point (we do not consider point or line sources in this chapter). We can then observe that in the steady state, no energy can be stored thermally (not the same as negative $\dot{q}$) in the solid being analyzed, so that all the energy generated ($\dot{q}$ times the volume) must be conducted to the surface and then given off to the surroundings by either convection or radiation or both mechanisms.

2.4.1 Plane Walls with Generation

We will begin our study of generation phenomena with the basic differential equation as obtained by Myers (6) and others from an energy balance applied to a differential element in the solid. For steady one-dimensional conduction with constant properties and constant generation, we can write

$$k\frac{d^2T}{dx^2} + \dot{q} = 0 \tag{2.34}$$

The first term represents the net conduction out of a small incremental slice of thickness Δx, and the second term is the energy deposited in that slice by the generation phenomena of strength $\dot{q}$. The equation is called Poisson's equation after the French mathematician. The formulation and origin of Poisson's equation from the general conduction equation is discussed extensively in a first graduate course in conduction or in texts such as Schneider (7), Myers (6), or Jakob (8). We should remember at this point to emphasize the fact that the equal sign in an equation demands numerical equality as well as equality in units. The first term in Eq. 2.34 has the units $(W/mK)(K/m^2) = W/m^3$, which agrees with the units for the generation term $\dot{q}$, also in W/m^3. The reader should verify the units for the British engineering system for Eq. 2.34.

Equation 2.34 can be solved by separating variables so that

$$d\,\frac{dT}{dx} = -\frac{\dot{q}}{k}\,dx \tag{2.35}$$

Integrating once indefinitely gives

$$\int d\,\frac{dT}{dx} = -\frac{\dot{q}}{k}\int dx$$
$$\frac{dT}{dx} = -\frac{\dot{q}x}{k} + C_1 \tag{2.36}$$

Separating variables and integrating indefinitely again yields the general solution with two arbitrary constants as required by a second-order equation. Thus

$$\int dT = -\frac{\dot{q}}{k}\int x\,dx + C_1\int dx$$
$$T = \frac{-\dot{q}x^2}{2k} + C_1 x + C_2 \tag{2.37}$$

We observe that the constant C_1 has units of K/m and C_2 has just degree units.

Figure 2.11 illustrates the geometry we're working with and the boundary conditions we'll apply to the general solution in order to get the desired particular solution. The figure shows the more general case where the surface temperatures are unequal for the slab of thickness $2L$ (choosing the thickness equal to $2L$ gives a little neater form for the result, a benefit of hindsight). We take the temperature of the left face at $x = 0$ to be fixed at T_1, and the right face at $x = 2L$ to be fixed at $T_2 > T_1$ for the temperature profile shown. The boundary conditions are stated mathematically as follows for convenient reference:

boundary condition [1]: $x = 0$, $T = T_1$
boundary condition [2]: $x = 2L$, $T = T_2$

We now apply these boundary conditions to Eq. 2.37 and evaluate the constants C_1 and C_2 to get the particular solution for the situation we are looking at. At this point it might be interesting to observe that often engineers mistakenly express disinterest in a topic once the language switches from familiar English jargon to mathematical terms because mathematical language is "too abstract." Hopefully,

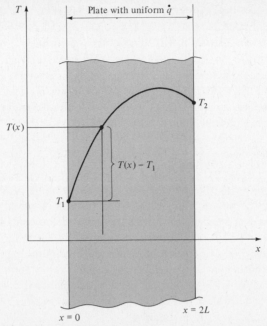

Figure 2.11 Geometry and boundary conditions for the infinite plane plate of thickness 2L with uniform internal generation.

we can show in the following paragraphs that a great deal of physical insight, or engineering understanding, can be wrung out of a mathematical analysis if one understands the concepts the language represents. We will deal only with straightforward mathematical ideas that can be related directly to the situation at hand.

As we mentioned above, after we get the particular solution, we can determine maximum or minimum temperatures, and with the aid of Fourier's law, the heat flux at any location. Boundary condition [1] applied to Eq. 2.37 (inserting $x = 0$ into the equation) gives us $C_2 = T_1$ and the application of boundary condition [2] gives, after some algebra,

$$C_1 = \frac{T_2 - T_1}{2L} + \frac{\dot{q}L}{k} \qquad (2.38)$$

Now we can put C_1 and C_2 into the general solution (Eq. 2.37) and after rearranging terms, we obtain the local temperature excess $T(x) - T_1$ as follows:

$$T(x) - T_1 = \frac{\dot{q}L^2}{2k}\left[2\frac{x}{L} - \left(\frac{x}{L}\right)^2\right] + \frac{(x/L)(T_2 - T_1)}{2} \qquad (2.39)$$

The physical interpretation of this temperature excess was pictured in Figure 2.11. We can observe by looking at Eq. 2.39 that when $T_2 = T_1$ the profile in temperature is symmetrical about $x/L = 1$. The quantity $\dot{q}L^2/2k$ has a special significance when $T_2 = T_1$ in that it represents the maximum temperature excess; it is called the generation temperature because the group has the dimensions of temperature.

Differentiation of Eq. 2.39 with respect to x/L gives

$$\frac{dT}{d(x/L)} = \frac{\dot{q}L^2}{2k}\left(2 - 2\frac{x}{L}\right) + \frac{T_2 - T_1}{2} \tag{2.40}$$

Setting Eq. 2.40 equal to zero will let us determine the location of the maximum (or minimum) temperature. Thus

$$\left(\frac{x}{L}\right)_M = 1 + \frac{T_2 - T_1}{4\left(\dot{q}L^2/2k\right)} \tag{2.41}$$

We can see immediately that when $T_2 = T_1$, the extremum in temperature is at the centerline of the slab at $x/L = 1$ as we intuitively observed above. Further, when $\dot{q}$ is positive, the extremum lies to the right of the centerline for $T_2 > T_1$ and to the left for $T_2 < T_1$; similar observations can be made for negative $\dot{q}$; and finally, as the value of $\dot{q}$ increases, the maximum or minimum approaches the centerline. All these gems of engineering insight came from looking at the mathematical analysis without making a lick of calculations!

To obtain the local heat flux $q(x)/A$ at any location we write Fourier's law, after multiplying dx by L/L, as follows

$$\frac{q(x)}{A} = -\frac{k}{L}\frac{dT}{d(x/L)} \tag{2.42}$$

and substituting for the derivative of T from Eq. 2.40, we obtain

$$\frac{q(x)}{A} = -\dot{q}L\left(1 - \frac{x}{L}\right) - \frac{k(T_2 - T_1)}{2L} \tag{2.43}$$

The heat flux will be zero at the location of the extremum

$$\frac{dT}{d(x/L)} = 0$$

and will be positive (flowing to the right) or negative (flowing to the left) depending on the location, the generation rate $\dot{q}$, and the value of $T_2 - T_1$.

To actually see what the profile and heat fluxes look like, we need to make some calculations and they are most easily accomplished with a computer as we did in Section 2.3.4.

2.4.2 Computer Evaluation of $T(x) - T_1$ and Flux

We want to calculate values for $T(x) - T_1$, the temperature gradient $dT/d(x/L)$, the location of the extremum $(x/L)_M$, and the local heat flux $q(x)/A$ for a given value of $T_2 - T_1$ with the generation of temperature $\dot{q}L^2/2k$ as a parameter for the geometry shown in Figure 2.11.

PHYSICAL DESCRIPTION

Let the plate be 2 ft thick so that x/L ranges from zero to 2 and let $T_2 - T_1 = 1000F$. We will let the generation temperature range from -500 to $+500F$ in steps of 250 as representative values of $\dot{q}L^2/2k$, and assume a value of 10 Btu/hrftF for thermal conductivity.

FORTRAN PROGRAM

• *$JOB CONTROL CARDS.* Supply the cards or terminal ID numbers required by your installation that will handle the program in FORTRAN language. Follow the control cards by a dictionary as shown below which defines the terms you are using; it helps to use acronyms.

```
C     XL=X/L DIMNLESS LENGTH
C      Q=QDOT LSQD/2K
C       L=HALF SLAB THICKNESS
C    XLM=MAX LOCATION (X/L)M
C    T21=(T2−T1)
C    TX1=T(X/L)−T1
C  DTDX=DERIV OF T(X/L)
```

Next comes relevant data

```
10   T21=1000.0
20    Q=  500.0
```

These are the initial values of Q and the parameter T21. Next we print the output table heading.

```
 30   PRINT 510
510   FORMAT (1H1, 19X, 'TEMPERATURE DISTRIBUTION'//)
 40   PRINT 520, T21, Q
520   FORMAT (15X, '(T2−T1)=', F7.0, 4X, 'Q=',
      F6.0,//15X,'X/L',8X,'(TX−T1)',8X,'DT/D(X/L)'//)
```

Statement 30 supplies a new page (1H1) and prints the heading for the table of calculated results; FORMAT 520 prints out the values of T21 and Q that were used in the calculations printed in the table and also the column headings. Now we are ready to write the actual program logic.

```
 50   DO 100 I=1, 21
 60   XL=FLOAT (I−1)/10.0
 70   TX1=Q*(2.*XL − XL*XL) + XL*T21/2.
 80   DTDX=Q*(2. − 2.*XL) +T21/2.
 90   PRINT 530, XL, TX1, DTDX
100   CONTINUE
530   FORMAT (14X, F4.1, 9X, F6.0, F18.3)
```

The main calculations for XL, TX1, and DTDX are done in this DO loop. Note that the value of XL is keyed to the loop index I and updates by a value of 0.1

starting from 0.0 at I = 1 and stopping at 2.0 when I = 21. We continue and calculate the location of the extremum.

```
110   IF(Q) 140, 120, 140
120   PRINT 500
500   FORMAT (10X//27X,'(X/L)M AT SURFACE OF SLAB')
130   GOTO 170
140   XLM = 1.+ T21/(4.*Q)
150   IF(XLM*XLM − 4.0) 160, 120, 120
160   PRINT 540, XLM
540   FORMAT (10X//27X,'(X/L)M = ', F6.2)
170   Q = Q + 250.
180   IF(Q − 600.) 30, 190, 190
190   STOP
      END
```

It is necessary to monitor the value of Q because the calculation of XLM contains Q in the denominator of the last term; so when Q is zero XLM becomes a monstrously large number (overflow) which gives the computer indigestion and the machine will rebel and refuse to continue. Physically, all this means from a heat transfer standpoint is that for Q = 0, the concept of XLM has no significance other than that the maximum temperature occurs on the face of the slab with the highest temperature, either T_2 or T_1. The IF(Q) check takes care of this possibility and the case where the value of Q could be such that the maxima of the curve described by XLM would fall outside the slab is taken care of by statement 150. After the calculations have been completed, statement 170

Table 2.3 COMPUTER RESULTS FOR THE SLAB WITH INTERNAL GENERATION

TEMPERATURE DISTRIBUTION FOR Q = 500. AND (T2 − T1) = 1000.		
X/L	(TX − TL)	DT/D(X/L)
0.0	0.	1500.000
0.1	145.	1400.000
0.2	280.	1300.000
0.3	405.	1200.000
0.4	520.	1100.000
0.5	625.	1000.000
0.6	720.	900.000
0.7	805.	800.000
0.8	880.	700.000
0.9	945.	600.000
1.0	1000.	500.000
1.1	1045.	400.000
1.2	1080.	300.000
1.3	1105.	200.001
1.4	1120.	100.000
1.5	1125.	0.000
1.6	1120.	− 99.999
1.7	1105.	− 200.000
1.8	1080.	− 299.999
1.9	1045.	− 400.000
2.0	1000.	− 500.000
	(X/L)MAX = 1.50	

updates Q by 250, the IF statement checks to see if Q is less than 600; if so, the program is repeated with the new Q; if not, the program stops. The procedure described above gives results for values of Q of -500, -250, 0, $+250$, and $+500F$. The results for Q = 500 are shown in Table 2.3 and the entire set of results are plotted in Figure 2.12 for temperature and Figure 2.13 for heat flux. Figure 2.12 shows that as the magnitude of Q increases, the location of maximum temperature approaches the slab centerline as we had observed above in our consideration of Eq. 2.41. We can also observe, for example, that for Q = $-250F$ energy enters the slab only on the right face and is absorbed in the slab material by some endothermic reaction. The figure shows the steady-state case, but in reality such a situation for $-Q$ could not occur indefinitely. Figure 2.13 shows the flux variation, Q/A, in the slab. We can see that most fluxes are negative, that is, flowing to the left, because the temperature profiles are sloped to the left due to the right face being 1000F higher than the left face. Only for the cases Q = ±500 does energy flow to the right in a small portion of the slab. For example, at Q = -500, energy *enters* the slab at the left face, but for other

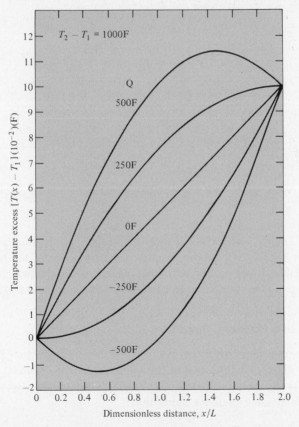

Figure 2.12   Variation of temperature excess with distance in the infinite plane plate of thickness 2L and for $T_2 - T_1 = 1000F$.

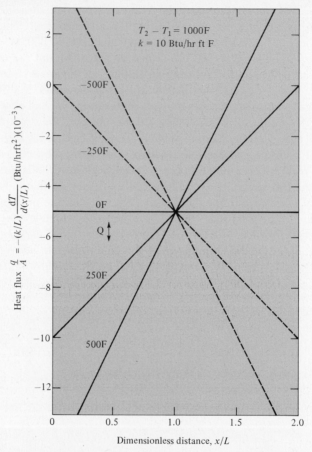

Figure 2.13   Variation of heat flux with distance for the infinite plane plate with uniform internal generation and thickness 2L.

values of Q, energy *leaves* the left face of the slab, an observation that can be verified from Figure 2.12 by our intuitive understanding of the second law of thermodynamics, which states that heat energy always flows from a higher temperature to a lower temperature when no other agency is acting.

2.4.3 Solid and Hollow Cylinders

We shall continue our treatment of one-dimensional conduction with constant generation with a brief discussion of the solid and hollow cylinder, and conclude with the solid sphere. The basic differential equations that must be solved for these cases (Reference 6) are for the cylinder

$$\dot{q}r + k\frac{d}{dr}\left(r\frac{dT}{dr}\right) = 0 \tag{2.44}$$

and for the sphere

$$\dot{q}r + \frac{kd^2(rT)}{dr^2} = 0 \tag{2.45}$$

which we will discuss in the next section.

The solid cylinder has a radius of r_1 and a surface temperature T_1 at $r = r_1$. After separating variables in Eq. 2.44, we can integrate twice to get the general solution as follows:

$$\frac{\dot{q}r^2}{2k} + \frac{r\,dT}{dr} - C_1 = 0 \tag{2.46a}$$

$$\frac{\dot{q}r^2}{4k} + T(r) - C_1\ln r - C_2 = 0 \tag{2.46}$$

Since there are two constants we need two boundary conditions to solve for C_1 and C_2. We know that at $r = r_1$ the temperature is T_1, and we recognize that along a diameter the heat flux decreases toward the center and at $r = 0$, the heat flux is zero (no energy can come from zero volume), then increases again with radial position along the same diametral line. Consequently, from Fourier's law, the only way for the heat flux to be zero is to have the temperature gradient $dT/dr = 0$ at $r = 0$, which is our second boundary condition. Setting $dT/dr = 0$ at $r = 0$ in Eq. 2.46a gives $C_1 = 0$ so that C_2 in Eq. 2.46 can be evaluated from the T_1 boundary condition at r_1. Thus

$$C_2 = \frac{\dot{q}r_1^2}{4k} + T_1 \tag{2.47}$$

The particular solution for this case can be obtained by putting C_1 and C_2 into Eq. 2.46 so that

$$T(r) - T_1 = \frac{\dot{q}r_1^2}{4k}\left(1 - \frac{r^2}{r_1^2}\right) \tag{2.48}$$

represents the temperature excess over T_1. When $r = 0$ then $T(r) = T_c$, the maximum centerline temperature; therefore, the maximum temperature excess is given by

$$T_c - T_1 = \frac{\dot{q}r_1^2}{4k} \tag{2.49}$$

so that we can express the temperature variation in the cylinder as a ratio of the temperature excess to the maximum excess as

$$\frac{T(r) - T_1}{T_c - T_1} = \frac{r_1^2 - r^2}{r_1^2} \tag{2.50}$$

We can differentiate Eq. 2.48 to get the slope of the temperature profile as

$$\frac{dT}{dr} = -2r\frac{\dot{q}}{4k} \tag{2.51}$$

so that the heat flux q at any radial location is given by Fourier's law as

$$q = -kA(r)\frac{dT}{dr}$$

$$= -k(2\pi rL)\frac{-2r\dot{q}}{4k} \tag{2.52}$$

where L is the length of the cylinder. We can check our analysis with our physical understanding of the situation because we know all the energy generated in the cylinder must come out through the surface. We can make that determination by setting $r = r_1$ in Eq. 2.52 to obtain the surface flux q_1 as

$$q_1 = k(2\pi r_1 L)\frac{2r_1\dot{q}}{4k}$$

$$q_1 = (\pi r_1^2 L)\dot{q} \tag{2.53}$$

We recognize $\pi r_1{}^2 L$ as the volume of the cylinder, so that the surface flux q_1, in Btu/hr or watts is equal to the generation rate $\dot{q}(vol)$ also in Btu/hr or watts.

Example 2.2

Consider a structural steel tie rod 1 in. in diameter in a reactor environment. Gamma heating causes a generation rate $\dot{q} = 1.894 \times 10^6$ Btu/hrft^3 or 1.959×10^7 W/m^3 in the rod. We would like to know the maximum temperature in the rod if the surface temperature is at 200F (367K). The thermal conductivity of steel is about 25 Btu/hrftF (43.2 W/mK) so that we can calculate T_c from Eq. 2.49 as follows.

$$T_c = T_1 + \frac{\dot{q}r_1^2}{4k}$$

$$T_c = 200 + \frac{1.894(10^6)(0.5)^2}{144(4)25}$$

$$T_c = 200 + 33 = 233F$$

This value for center temperature is not high enough to require any special provision for cooling the rod. This completes the example.

The hollow cylinder with an adiabatic inner surface has application in reactor fuel rod design. The reader may note that internal generation phenomena are very important in reactor operation and design. Fast oxide fuel pins are comprised of a thin walled, stainless-steel tube with the uranium oxide fuel pellets in the form of short circular cylinders. After a few hours of operation following reactor startup, densification of the fuel occurs and a hollow cylindrical void forms along the pellet centerline. Figure 2.14 illustrates the geometry that we will consider in our analysis to model the fuel pellet. We consider the cylinder to be adiabatic at the inner surface and on the ends so that heat flux flows only in the radial direction. The boundary condition at the inner surface

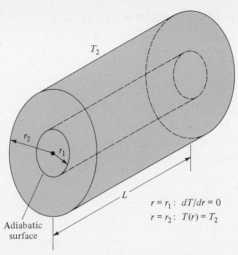

Figure 2.14   Hollow cylinder with adiabatic inner surface and adiabatic ends.

$r = r_1$ is, therefore, $dT/dr = 0$; and at the outer surface, the temperature $T(r) = T_2$. We now have to determine the constants C_1 and C_2 in the general solution (Eq. 2.46) for these boundary conditions. Equation 2.46a yields for the first constant

$$C_1 = \frac{\dot{q} r_1^2}{2k} \tag{2.54}$$

and Eq. 2.46 gives the second constant when $T(r) = T_2$ at the outer surface $r = r_2$. Thus

$$C_2 = \frac{\dot{q} r_2^2}{4k} + T_2 - \frac{\dot{q} r_1^2}{2k} \ln r_2 \tag{2.55}$$

and when C_1 and C_2 are inserted in the general solution, Eq. 2.46, we can then get the temperature excess at any point over the outer surface temperature as

$$T(r) - T_2 = \frac{\dot{q} r_2^2}{4k} \left[\left(1 - \frac{r^2}{r_2^2} \right) + 2 \left(\frac{r_1}{r_2} \right)^2 \ln \frac{r}{r_2} \right] \tag{2.56}$$

As we have noted above, all the energy generated in the solid must come out through the outer surface so that q_2 must equal the quantity $\dot{q}(\pi L)(r_2^2 - r_1^2)$. Energy conservation principles thus provide a good check on our solution. We will, therefore, calculate the flux from Fourier's law to make sure that we agree with our physical understanding. Fourier's law requires

$$q_2 = -kA(r_2) \left(\frac{dT}{dr} \right)_{r_2} \tag{2.57}$$

We get the derivative from Eq. 2.56. Thus

$$\frac{dT}{dr} = \frac{\dot{q}r_2^2}{4k}\left(\frac{-2r}{r_2^2} + \frac{2r_1^2}{rr_2^2}\right)$$

(2.58)

$$\left(\frac{dT}{dr}\right)_{r_2} = \frac{\dot{q}}{4k}\left(-2r_2 + \frac{2r_1^2}{r_2}\right)$$

(2.59)

The outer surface area $A(r_2) = 2\pi r_2 L$ so that putting this area and the derivative into Eq. 2.57, we get

$$q_2 = -k(2\pi r_2 L)\frac{\dot{q}}{4k}\left(-2r_2 + \frac{2r_1^2}{r_2}\right)$$

$$q_2 = \pi\dot{q}L(r_2^2 - r_1^2)$$

(2.60)

as we predicted previously. Example 2.3 shows the effect of source strength $\dot{q}$ on the magnitude of the maximum temperature excess.

Example 2.3

A fuel rod in a fast oxide reactor has annular pellets that are 6 mm OD with 2-mm-diameter cylindrical voids; a geometry very similar to that shown in Figure 2.14. The pellets are an 80 percent uranium oxide and 20 percent plutonium oxide mixture with a loading of fissionable material such that the rod power is 15 kW/ft. The thermal conductivity of the 80–20 mixture is approximately 1.25 Btu/hrftF (Reference 9). We want to calculate the generation rate $\dot{q}$, and the maximum temperature excess in the pellet.

SOLUTION
We can calculate the generation rate from Eq. 2.60 as follows.

$$\dot{q} = \frac{q_2/L}{\pi(r_2^2 - r_1^2)}$$

$$= 15,000 \ (\text{W/ft})(100 \ \text{mm}^2/\text{cm}^2)(2.54^2 \ \text{cm}^2/\text{in.}^2)\frac{(144 \ \text{in.}^2/\text{ft}^2)}{\pi(9-1)}$$

$$= 5.545 \times 10^7 \ \text{W/ft}^3$$

$$= 55.45 \ \text{MW/ft}^3$$

$$= 1.893(10^8) \ \text{Btu/hrft}^2$$

The maximum temperature difference is between the adiabatic inner face at

$r_1 = 1$ mm and the outer face at $r_2 = 3$ mm so that Eq. 2.56 gives

$$T_1 - T_2 = \frac{\dot{q}r_2^2}{4k}\left[1 - \left(\frac{r_1}{r_2}\right)^2 + 2\left(\frac{r_1}{r_2}\right)^2 \ln\frac{r_1}{r_2}\right]$$

$$= \frac{\dot{q}r_2^2}{4k}\left(1 - \frac{1}{9} + 2\frac{1}{9}\ln\frac{1}{3}\right)$$

$$= 0.6448\frac{\dot{q}r_2^2}{4k}$$

Using the value of $\dot{q}$ calculated above and converting r_2 into feet gives

$$r_2 = 3 \text{ mm}/10(2.54)12 = 0.009842 \text{ ft}$$

$$\frac{\dot{q}r_2^2}{4k} = 1.893(10^8)(0.009842)^2/4(1.25)$$

$$= 3668\text{F}$$

Therefore, $T_1 - T_2 = 0.6448(3668) = 2365\text{F}$. Generally the cladding runs at about 900 or 1000F so that the maximum temperature in the fuel would be about 3400F, which is below the estimated melting temperature of 5100F by a good margin. We can observe that the maximum temperature excess is directly proportional to the generation rate $\dot{q}$. In a later example, when we have looked at convection aspects, we will be able to predict the maximum temperature level with more certainty.

2.4.4 The Solid Sphere

We will conclude our section on conduction with uniform sources with a look at the solid sphere. The general solution for the sphere can be obtained by separating variables in Eq. 2.45 and integrating twice. Thus

$$\frac{\dot{q}r^2}{2k} + \frac{d(rT)}{dr} + C_1 = 0 \tag{2.61a}$$

$$\frac{\dot{q}r^3}{6k} + rT(r) + C_1 r + C_2 = 0 \tag{2.61}$$

We will look at the boundary conditions of $T(r) = T_0$ at the center where $r = 0$, and $T(r) = T_1$ at the surface $r = r_1$. In order to evaluate C_1 and C_2 we need to expand the derivative in Eq. 2.61a as follows

$$\frac{\dot{q}r^2}{2k} + \frac{r\,dT}{dr} + T(r) + C_1 = 0 \tag{2.62}$$

so that at $r = 0$, we get $C_1 = -T_0$ and at $r = r_1$, Eq. 2.61 gives

$$C_2 = \frac{-\dot{q}r_1^3}{6k} - r_1(T_1 - T_0) \tag{2.63}$$

and when C_1 and C_2 are then put back into Eq. 2.61, we get for the temperature distribution

$$T(r) - T_0 = \frac{\dot{q}r_1^2}{6k}\left[\frac{r_1}{r} - \left(\frac{r}{r_1}\right)^2\right] + \frac{r_1}{r}(T_1 - T_0) \tag{2.64}$$

Without going into the algebraic details we can show by equating energy generated to flux at the surface that the maximum temperature difference in the sphere is $T_0 - T_1 = \dot{q}r_1^2/6k$, thereby giving us a very simple result for the nondimensional temperature excess ratio

$$\frac{T_0 - T(r)}{T_0 - T_1} = \left(\frac{r}{r_1}\right)^2 \tag{2.65}$$

Such a simple result ought to make an engineer suspicious, especially after wrestling with reams of algebra. It is always wise to check with an alternate method or concept. We know that in the steady state all the energy generated in the solid comes out through the surface (as we have observed before), so that $q_1 = \dot{q}(\text{vol}) = \dot{q}(\frac{4}{3})\pi r_1^3$. When we calculate q_1 from Fourier's law and evaluate the derivative at $r = r_1$, we find that q_1 checks our observation; the actual details are outlined in a home problem.

Some of the other boundary conditions that can be worked out for the cylindrical and spherical geometries are explored in the home problems. We therefore, conclude our discussion of steady conduction with uniform generation with this section and turn our attention to extended surfaces.

2.5 EXTENDED SURFACES

We can increase the heat transfer from a surface at a given temperature in a particular surrounding fluid at temperature T_∞ and heat transfer coefficient h, by increasing the surface area. Important applications include fins on solid-state devices, fins on air-cooled cylinders of internal combustion engines, and fins on electrical transformers. The optimization of lengths or sizes for extended surfaces is a complicated procedure because it depends heavily on manufacturing and material costs. In this section we will treat only the heat transfer performance.

2.5.1 Fundamental Analysis

We will analyze the fin of constant cross-sectional area A with perimeter P, as shown in Figure 2.15, for three different physical situations (boundary conditions). Myers (6) has shown that the basic differential equation describing heat transfer in a constant area fin can be obtained from an energy balance on the differential slice dx units in thickness (shown in Figure 2.15). The energy balance requires that the energy conducted into the element, q_x, must leave the element by convection q_c, with the remainder conducted out the opposite face as q_{x+dx}. Thus

$$\frac{d^2T}{dx^2} - \frac{hP}{kA}(T - T_\infty) = 0 \tag{2.66}$$

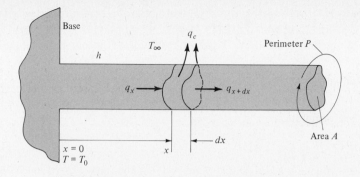

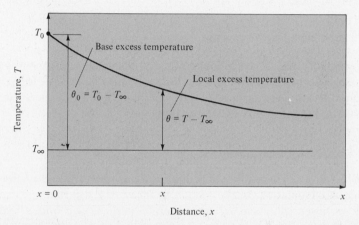

Figure 2.15   Geometry of the generalized cylindrical fin of constant cross-sectional area A and perimeter P with the corresponding temperature distance relationship.

The formulation of Eq. 2.66 required the following assumptions:

1. Steady-state conditions in the fin and the environment prevail and environment temperature T_∞ is not a function of distance x.
2. The thermal conductivity k of the fin material does not vary with temperature.
3. The temperature at any location x is the same over the cross section of the fin.
4. As we have stated, the perimeter P and cross-sectional area A are not functions of the distance x.
5. The heat transfer coefficient h does not vary with temperature or location. This is the most severe restriction on the analysis. As we will see later in the chapter on convection, the heat transfer coefficient is not constant along the fin.

Manipulation and solution of Eq. 2.66 can be simplified by defining the temperature excess $\theta = T - T_\infty$ so that $dT/dx = d\theta/dx$ since T_∞ is a constant;

consequently $d^2T/dx^2 = d^2\theta/dx^2$. We will outline the solution using operator notation $D \equiv d(\ )/dx$ and $D^2 \equiv d^2(\ )/dx^2$. The operator is linear and obeys the laws of algebra. If we let $m^2 = hP/kA$, we can write our basic equation as

$$\frac{d^2\theta}{dx^2} - m^2\theta = 0 \tag{2.67}$$

and in operator notation

$$(D^2 - m^2)\theta = 0 \tag{2.68}$$

for which the auxiliary equation $D^2 - m^2 = 0$ has the roots $D = \pm m$. When the roots of the auxiliary equation are real and distinct, the general solution can be written as

$$\theta = C_1 e^{mx} + C_2 e^{-mx} \tag{2.69}$$

The constants C_1 and C_2 are evaluated for the boundary conditions describing the physical phenomena of interest. We will look at three cases that are encountered in engineering practice; the very long fin, the finite fin with end insulated, and the finite fin with end uninsulated.

2.5.2 Case A: The Very Long Fin

The results for this case can be utilized for real fins when the end is close to the environment temperature. In the section to follow, we will find out that when $mL \equiv 3$, we essentially have the same thermal performance from a real fin of length L that we would have from a fin of infinite length. The boundary conditions for case A are $x = 0$, $\theta = \theta_0$. For $x = \infty$, $\theta = 0$ so that at $x = 0$, Eq. 2.69 gives $C_1 + C_2 = \theta_0$. At $x = \infty$ the only way we can have $\theta = 0$ is if $C_1 = 0$, so that leaves $C_2 = \theta_0$. The temperature function then becomes

$$\frac{\theta}{\theta_0} = e^{-mx} \tag{2.70}$$

and the heat transfer to the base of the fin (at $x = 0$) from the wall to which it is attached can be calculated from Fourier's law when the temperature derivative $dT/dx = d\theta/dx$ is evaluated at $x = 0$. Thus

$$\frac{d\theta}{dx} = \theta_0(-m)e^{-mx}\Big]_{x=0} = -m\theta_0 \tag{2.71}$$

The base heat flux then becomes, from Eq. 1.3,

$$q_0 = mkA\theta_0 \quad (\text{W, Btu/hr}) \tag{2.72}$$

We note that the flux that enters the base in steady state is dissipated to the environment all along the surface of the fin so that we could also calculate the base flux by integrating the surface convection flux over the length of the fin.

$$q_0 = \int_0^A h\theta \, dA \quad (A = \text{surface area})$$

$$= \int_0^\infty h\theta P \, dx$$

$$= hP\theta_0 \int_0^\infty e^{-mx} \, dx$$

$$= \frac{-hP\theta_0}{m} \left[e^{-mx} \right]_0^\infty$$

$$= \frac{hp\theta_0}{m}$$

But we recognize that $m^2 = hP/kA$ or $m = hp/kAm$ so we have

$$q_0 = mkA\theta_0$$

as a check on our analysis above.

2.5.3 Case B: Finite Length, End Insulated

The boundary conditions for this case are at $x = 0$, $\theta = \theta_0$ and at $x = L$, $d\theta/dx = 0$ because there must be zero flux at the end of the fin (at $x = L$). Since the area A is constant, Fourier's law requires the derivative to be zero because k cannot be zero. At the base of the fin, we again obtain $C_1 + C_2 = \theta_0$ from Eq. 2.69 but to get the other equation for the constants at $x = L$, we need the derivative of θ from Eq. 2.69 with respect to x. Thus

$$\frac{d\theta}{dx} = mC_1 e^{mx} - mC_2 e^{-mx} \tag{2.73}$$

We now evaluate $d\theta/dx$ at $x = L$ and set the result equal to zero as required by the second boundary condition which gives us two equations in C_1 and C_2. Thus

$$C_1 e^{mL} - C_2 e^{-mL} = 0$$

$$C_1 + C_2 = \theta_0$$

The solution for C_1 and C_2, and insertion into Eq. 2.69 involves a considerable amount of algebraic manipulation. The exponential terms can be judiciously grouped to conform to the definition of the hyperbolic functions so that the result can be expressed in compact form as

$$\frac{\theta}{\theta_0} = \frac{\cosh\left[m(L - x)\right]}{\cosh(mL)} \tag{2.74}$$

where the hyperbolic functions are defined by Burrington (10) or other math handbooks as

$$\left.\begin{aligned}
\cosh(z) &= \frac{e^z + e^{-z}}{2} \\
\sinh(z) &= \frac{e^z - e^{-z}}{2} \\
\tanh(z) &= \frac{\sinh(z)}{\cosh(z)}
\end{aligned}\right\} \tag{2.75}$$

Evaluation of Eq. 2.74 is most easily accomplished with an electronic slide rule that has exponential function keys or the hyperbolic functions keyed in directly. Handbooks such as Burrington (10) give tables of values.

The base heat flux can be calculated from Fourier's law if we observe that

$$\frac{d[\cosh(z)]}{dx} = \sinh(z)\frac{dz}{dx}$$

Evaluating the temperature derivative from Eq. 2.74 gives

$$\frac{d\theta}{dx} = \frac{\theta_0}{\cosh(mL)} \sinh[m(L - x)](-m) \tag{2.76}$$

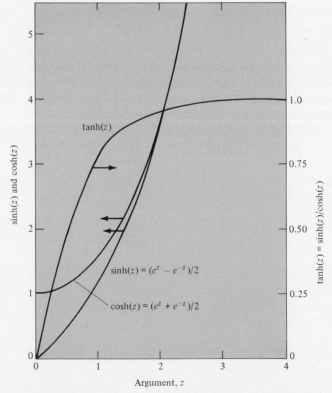

Figure 2.16 Variation of hyperbolic functions tanh(z), sinh(z), and cosh(z) with the argument z.

and when we evaluate $d\theta/dx$ at $x = 0$, the base heat flux becomes

$$q_0 = -kA\left(\frac{d\theta}{dx}\right)_{x=0}$$

$$q_0 = mkA\theta_0\tanh(mL) \quad (\text{W, Btu/hr}) \tag{2.77}$$

The result we just obtained gives us a means for estimating the length of a particular fin that will behave as though it were infinite in length. If we compare Eq. 2.77 with 2.72 we can see that when $\tanh(mL)$ is essentially one, the fin of length L [with mL so that $\tanh(mL) = 1$] dissipates the same flux as in case A; therefore, slopes and temperatures are essentially the same. Figure 2.16 shows how the hyperbolic functions vary with their argument up to $z = 4$ and we can see that when $mL = z = 3$ the $\tanh(3)$ is very near to 1. Since $m^2 = hP/kA$, we can note the physical effect of the variables in that high h, low k, and A give the shortest length L for case B that will behave like case A. For example, with $m = 3$ ft^{-1} and a length L of 1 ft for $mL = 3$, we would calculate at the midpoint of the fin $(L/2)$ a temperature ratio of $\theta/\theta_0 = 0.223$ for case A and a ratio of $\theta/\theta_0 = 0.234$ for case B which agrees within 5 percent. We will now look at the last case where the fin is uninsulated at the end.

2.5.4 Case C: Finite Length, End Uninsulated

When the fin is not insulated at the end, there will be a coefficient h_L on the end surface and a convection flux at $x = L$ that is equal to the conduction flux at that location so that the slope of the temperature profile at $x = L$ must be $d\theta/dx = -(h_L/k)\theta_L$. Evaluating the derivative from Eq. 2.73 and the temperature function from Eq. 2.69 at $x = L$ and noting at $x = 0$, we again obtain $C_1 + C_2 = \theta_0$ as before. We obtain the required two equations in C_1 and C_2 as

$$C_1 + C_2 = \theta_0$$

$$C_1e^{mL} - C_2e^{-mL} = -\frac{h_L}{mk}\left(C_1e^{mL} + C_2e^{-mL}\right) \tag{2.78}$$

The algebra required to evaluate C_1 and C_2 above is even worse than that for case B. If the reader has a weekend with bad weather ahead, the simultaneous solution for C_1 and C_2 could have two very significant benefits; for one, doing so would provide an excellent algebra review and two, the activity would certainly keep a person out of mischief. As in case B we can group the exponential terms from C_1 and C_2 into recognizable hyperbolic functions after the constants have been inserted into the general solution given by Eq. 2.69. The temperature ratio for case C then becomes

$$\frac{\theta}{\theta_0} = \frac{\cosh[m(L-x)] + (h_L/mk)\sinh[m(L-x)]}{\cosh(mL) + (h_L/mk)\sinh(mL)} \tag{2.79}$$

The heat flux at the base can be obtained in the same manner as we did for case

A and case B so that

$$q_0 = mkA\theta_0 \frac{\tanh(mL) + h_L/mk}{1 + (h_L/mk)\tanh(mL)} \tag{2.80}$$

Taking a look at Eqs. 2.80, 2.77, and 2.72, we can see that the expressions for base heat flux (the energy dissipated by the fin) all contain the factor $mkA\theta_0$ times a multiplier. The multipliers on $mkA\theta_0$ depend on the conditions at the end of the fin and decrease the flux from the maximum depending on the values of m and h_L. Before looking at fin design, we will consider an application of the above results.

Example 2.4

We would like to estimate the energy input required to solder together two very long pieces of bare number 14 copper wire (0.064 in. diameter) with a solder that melts at 385F. The wires are positioned horizontally in air at 75F and the heat transfer coefficient on the wire surface is 3 Btu/hrft²F. The thermal conductivity of this particular wire alloy is 195 Btu/hrftF.

SOLUTION

We approximate the real physical situation described above as two infinite fins with a base temperature of 385F in an environment at 75F with the given h. Accordingly, the base temperature excess $\theta_0 = 385F - 75F = 310F$. Summarizing the pertinent data to calculate the parameter m we have:

$$h = 3 \text{ Btu/hrft}^2\text{F} \qquad\qquad (17.02 \text{ W/m}^2\text{K})$$

$$k = 195 \text{ Btu/hrftF} \qquad\qquad (337 \text{ W/mK})$$

$$A = \pi D^2/4 = 2.234(10^{-5})\text{ft}^2 \qquad [2.075(10^{-6})\text{m}^2]$$

$$P = \pi D = 0.0168 \text{ ft} \qquad\qquad (0.00511 \text{ m})$$

Therefore, we obtain

$$m = \left(\frac{hP}{kA}\right)^{0.5} = \left[\frac{3(0.0168)}{195(2.234)10^{-5}}\right]^{0.5}$$

$$= 3.40 \text{ ft}^{-1}(11.15 \text{ m}^{-1})$$

The reader should not have any difficulty in distinguishing the m in $m^2 = hP/kA$ from the unit m for meter. Accordingly, the base flux for one wire is

$$q_0 = mkA\theta_0 = 3.40(195)(2.234)(10^{-5})310$$

$$= 4.59 \text{ Btu/hr } (1.34 \text{ W})$$

so that two wires would require 2.68 W to maintain the junction at the melt temperature of the solder. We should note that the effectiveness of the energy transfer before the junction is wetted by the solder is very poor. Consequently, most soldering pencils or guns are rated anywhere from 20–100 W in order to bring the junction up to melt temperature in a reasonably short period of time. Bringing the junction up in temperature quickly requires more energy input than is needed at steady-state conditions, as we have just calculated. We will discuss transient phenomena in more detail in the next chapter.

2.5.5 Fin Design

The developments outlined above yielded the equations for temperature and base heat flux for the constant area fin under the conditions outlined in cases A, B, and C. Similar results for variable area fins require much more complicated mathematical techniques such as Bessel functions to work out the results. Solutions for different configurations can be found in References 6, 7, 8, and 11, for example.

The actual determination of optimum performance of fins is a complicated procedure, the deciding factor in most cases being the manufacturing costs involved. Since we do not have fabrication cost information, we can discuss only thermal performance of the extended surface. Our purpose is to become familiar with the concept of fin efficiency as it is defined in the literature on the subject and to understand its physical meaning.

The fin efficiency η_F is defined as the ratio of the energy dissipated by the fin, q_0, to the energy that would be dissipated *by the fin* if the *entire fin* were at the base temperature θ_0. Accordingly,

$$\eta_F = q_0 / \left(hA_{\text{surface}} \theta_0 \right) \tag{2.81}$$

Since the surface area is involved, we recognize that the concept can be applied only to finite length fins. We could use the results for case B as a close approximation for case C if we add to the actual length of case B an amount such that the cylindrical area involved will give an amount of heat flux equal to what leaves the end of the fin in case C. For example, if the convective coefficients on surface and end are equal, using $L' = L + (D/4)$ for case B would give essentially the same performance for the end insulated fin as for the end uninsulated fin of length L.

For case B then, the fin efficiency as given by Eq. 2.81 becomes

$$\eta_F = \left[\tanh(mL) \right] / mL \tag{2.82}$$

which can be evaluated at $L = 0$ by using L'Hospital's rule to be $\eta_F = 1$. With increasing L, the fin efficiency decreases continuously as shown in Figure 2.17 so that the most effective fin is the fin of zero length, a confusing result at first glance. We must recognize that fin efficiency is useful only in comparing the performance of different fins. The aspect that makes comparison difficult is that there is no simple single criterion for comparison. Generally the comparison is based on the volume of material involved, since volume is related to cost. Such a comparison is made after the engineer has determined that the surface area without fins cannot be made to dissipate the required flux. When very short fins

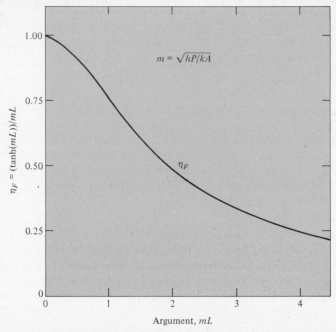

Figure 2.17 Relationship of fin efficiency with the argument *mL* for the finite fin of constant cross-sectional area insulated on the end.

are involved, the analysis requires knowledge of both two-dimensional convection effects (which are beyond our scope) and two-dimensional conduction effects.

2.6 CONTACT CONDUCTANCE

When two solids are placed into contact, the surfaces are never perfectly joined for a number of reasons. The most prevalent reason is that the surfaces are not perfectly flat or matched in profile so that only the high spots are in contact. Particles of dust or other foreign matter may be present on the surface to prevent contact from taking place, also formation of oxides may take place at different rates in different places, again causing surface irregularities. Accordingly, the transfer of energy across the contact joint then takes place by pure conduction at those spots where the two solids are in intimate contact and by conduction and convection through the interstitial medium that separates the solids where contact is not made. Usually the interstitial medium is air whose conducting abilities are poor, but on occasion it may be possible to substitute a fluid of higher conductivity which can tolerate the expected temperature levels and not cause corrosion problems.

The analysis of the exact thermal phenomena at a contact surface is virtually impossible to assess because of the difficulty in describing the surface topography. Compounding this difficulty is the fact that the topography is an unknown function of the pressure holding the surfaces together. Figure 2.18 illustrates the description of contact phenomena and the attendant temperature variations. The figure shows two bars of material 12 and material 34 held in position by the contact pressure P. Under steady-state conditions neglecting any losses when $T_1 > T_2$, the heat flux flows from left to right and we note that $q_{12} = q_j = q_{34}$ where q_j is the heat flux through the junction of cross-sectional area A. We can make measurements of temperature at locations x_1, x_2, x_3, and x_4 as shown in Figure 2.18, and when the conductivities k_{12} and k_{34} can be assumed independent of temperature, we can obtain the junction temperatures T_2' and T_3' by graphical extrapolation. The contact conductance for the joint, K_j, can then be determined from the energy balance as

$$K_j(T_2' - T_3') = \frac{k_{12}A}{x_2 - x_1}(T_1 - T_2) \tag{2.83}$$

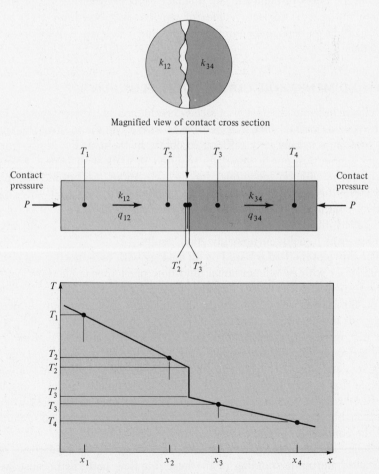

Figure 2.18   Physical arrangement and temperature variation for the contact conductance phenomena.

The contact conductance $K_i = h_i A$, where h_i is the unit area contact coefficient or unit area contact conductance with the units W/m^2K or $Btu/hrft^2F$. One may also define the contact resistance $R_i = 1/h_i A$ with the units K/W or hrF/Btu, depending on the application or the source in the literature. A simplified analysis of contact resistance phenomena is given by Holman (12) in terms of contact and void areas but it is not possible to evaluate or predict these areas in Holman's analysis.

Experiments (Reference 16) along the lines indicated by Figure 2.18 have shown that the contact conductance is at first a strong function of the pressure applied to the joint, but after a significant portion of the joint is in intimate contact, the effect of pressure diminishes. An important application of contact conductance occurs in the fuel rods of nuclear reactors. Fuel pellets are inserted in stainless-steel tubes and during operation the energy produced by fission must pass from the pellet through the gap (contact joint) between the pellet and tube and into the reactor coolant. Values of gap conductance from 1000 to 2000 $Btu/hrft^2F$ have been encountered in practice (Reference 13) for such an application.

2.7 TWO-DIMENSIONAL CONDUCTION, FLUX PLOTS

We will conclude our discussion of steady conduction phenomena with a look at a very versatile method that is applicable to irregularly shaped two-dimensional geometries. It is usually very difficult to obtain mathematical solutions for such irregular geometries; one is generally required to work with partial differential equations. The solution techniques for such equations are beyond our present scope. Also there are cases where the geometry is such that the required boundary conditions make a solution impossible. There exists a physical treatment, called flux plotting, of the heat flow phenomena in such irregular geometries which yields surprisingly good results.

The flux plot method is based on an analogy with two-dimensional electrical phenomena or with two-dimensional fluid flow phenomena. The flow or flux of heat corresponds to the flow of electrons in the electrical analogy, and to the flow of fluid particles in the hydraulic analogy; the temperature potential corresponds to either the voltage potential or the velocity potential, the potential being the agency that causes the flux to flow. Schneider (7) gives a very complete description of the continuous electrical analogy and how one can make voltage measurements in such a case and determine the shape of the isotherms, a technique that is very helpful in cases where the geometry unduly complicates matters. Our discussion in this section will consider only those cases where we can reasonably draw the flux lines and isotherms from our physical understanding of the conduction phenomena.

The hydraulic analogy is invaluable in aiding the "coagulation" of our visualization and understanding of the heat flux flow patterns because we can literally "see" what goes on in our mind's eye. Figure 2.19 shows a picture of water flow from the outside boundaries of a large square inward to the smaller

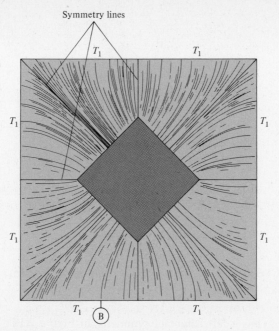

Figure 2.19   Hydraulic analogy of two-dimensional conduction due to Moore as given by Schneider (7).

black square. The water in the picture flows between a sheet of glass spaced about 2 or 3 mm above a light colored, flat opaque surface that models the geometry of interest. Tiny crystals of potassium permanganate were sprinkled on the bottom surface before the top cover glass was put in place. As the water flowed inward from the outer edge of the large square to the smaller black inner square, the paths of the fluid particles were made visible by the traces of permanganate behind the specs (crystals) in the picture. These traces represent the streamlines of the flow or the motion of fluid mass flux that corresponds to heat flux in a geometrically similar conduction situation. The visualization of the water particle flow paths thus helps us visualize heat flow paths in a similar geometry.

To complete the analogy we must recognize that the streamlines and velocity potentials are orthogonal families of curves that satisfy a partial differential equation that has the same form as the equation describing the heat transfer situation. That equation is called Laplace's equation (References 6, 7, 8, 11), but that knowledge is not essential to the application of the flux plot method unless the reader really wants to excavate all details of the entire background of what we are doing here. Accordingly, the heat flux lines and the isotherms are also families of orthogonal curves, and if we can construct the net of such curves for a given geometry, we can then calculate the heat flux for the entire geometry from information about the net. We proceed as follows.

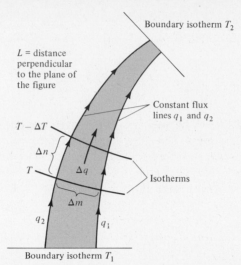

Figure 2.20 Arrangement of arbitrary isotherms and constant flux lines corresponding to the conduction phenomena in octant *B* shown in Figure 2.19.

Consider the lower left octant of symmetry in Figure 2.19 (just above the letter *B* in the figure) and pick two lines of constant flux q_1 and q_2 (as shown in Figure 2.20) which extend from the isothermal boundary T_1 on the outside of the square to the isothermal boundary T_2 on the smaller square inside. These constant flux lines will be spaced Δm units apart at a certain location; we can also pick two isotherms, T and $T - \Delta T$, which are spaced Δn units apart at the same location, as shown in the figure. We then use Fourier's law to express Δq, the flux flowing between q_1 and q_2, in terms of the pertinent variables.

$$\Delta q = kA \frac{\Delta T}{\Delta n} \tag{2.84}$$

We should observe that the minus sign has been included in ΔT and that we have made the approximation $\partial T / \partial n \simeq \Delta T / \Delta n$. The cross-sectional area through which Δq flows is $A = \Delta mL$, where L is the distance perpendicular to the plane of Figure 2.20, so that

$$\Delta q = k \frac{\Delta m}{\Delta n} L \Delta T \tag{2.85}$$

If there are N of the Δn increments, we can approximate ΔT by $(T_1 - T_2)/N$. Of course we recognize that such an approximation will improve as Δn gets smaller and when Δn is zero; ΔT will be exactly equal to $(T_1 - T_2)/N$ because then N will be infinitely large. The flux Δq in the single-flow tube can then be written as

$$\Delta q = k \frac{\Delta m}{\Delta n} \frac{L(T_1 - T_2)}{N} \tag{2.86}$$

All that remains now is to count the number of flow tubes or flow channels M in order to get the total flux $q = M \Delta q$ that flows from isotherm T_1 to isotherm T_2. Thus

$$q = k\frac{ML}{N}\frac{\Delta m}{\Delta n}(T_1 - T_2) \tag{2.87}$$

The only way Eq. 2.87 can be useful is if we make our construction of the orthogonal net such that $\Delta m = \Delta n$; doing so eliminates the necessity of evaluating the factor $\Delta m/\Delta n$ which would be cumbersome to arrive at for each flow tube. With $\Delta m = \Delta n$ the result is a network of curvilinear squares that can be sketched by hand with physical intuition as guide as we will discuss more in detail below.

The quantity ML/N depends only on the geometry or the shape of the solid between the isotherms T_1 and T_2 and is called the shape factor S. Thus

$$S = \frac{ML}{N} \tag{2.88}$$

where M = number of flow channels

N = number of temperature steps, or the number of spaces between the isoterms

L = distance into the plane of the two-dimensional region

The shape factor as we define it here has the dimensions of feet or meters, whereas some sources in the literature define a shape factor per unit depth as simply the ratio of M to N. After determining M and N by counting the flow channels and temperature steps on a sketch that has been drawn to scale, the heat flux can be calculated as

$$q = Sk(T_1 - T_2) \tag{2.89}$$

The procedure we have discussed above is a radical departure from the way we started the topic of conduction in this chapter. One might even say it's like going from the ridiculous to the sublime or vice-versa, depending on how one gets along with mathematical analysis. The approach we discuss here belongs more to the art of engineering which leans heavily on the understanding of the nature or behavior of the phenomena.

To make use of the shape factor method one starts with a mental picture like Figure 2.19 and begins mentally to construct the network of curvilinear squares. There are a few preliminary steps that are helpful to observe:

1. Note the isothermal boundaries and also any adiabatic boundaries.
2. Note lines of symmetry, break the geometry down to the maximum number of symmetrical regions, and select one to work out in detail.
3. Use a full-sized sheet of paper and make a *large drawing to scale* of the symmetry region determined in 2. It is absolutely essential to draw the region to scale and large enough to see what is going on.

4. Start the construction lightly in pencil with a few constant flux lines (fairly wide apart) in a part of the region where it's fairly obvious which way they go and what the shape should be.
5. Fill in the isotherms perpendicular to the flux lines to form the orthogonal net, making sure all intersections form curvilinear squares.
6. The result in 5 may not be very good on the first try so discard it and make another one, benefiting from the experience. Usually two or three sketches will produce a useful result.

There are a number of observations we can make about curvilinear squares which are also helpful in making a flux plot sketch and in deciding when a sketch is good or not. The curvilinear squares should have the following characteristics:

1. Tangents at the corners must be perpendicular to each other; isotherms and constant flux lines intersect perpendicularly.
2. Diagonals of the curvilinear squares bisect the angles at the corners.
3. Diagonals of the curvilinear squares intersect each other at right angles.

As a consequence of having an orthogonal net we note that the flux lines meet the isothermal boundaries perpendicularly and never cross, and isotherms meet adiabatic boundaries perpendicularly and also do not cross each other.

After counting the flux channels M, and the temperature steps N, we can calculate the shape factor S for the thickness L, and finally the heat flux q from Eq. 2.89. The flux plot corresponding to the region B in Figure 2.19 is sketched out in Figure 2.21. If we take a close look at some of the curvilinear squares in the figure, we can see that squares $(3,4)$ and $(6,2)$, for example, look pretty good; corners intersect at right angles and diagonals cross at right angles and bisect the corner angles. However, the only saving grace about square $(1,1)$ is that the diagonals intersect at right angles. But then only one sour note out of 56 doesn't ruin the melody. We count the number of flux channels to be 7 so that M is then 8 times 7 or 56, since there are 8 regions of symmetry, each having 8 temperature steps giving a shape factor of $S = 7$ ft for a depth of 1 ft into the plane of Figure 2.21. Then, for example, if $T_1 = 500F$ and $T_2 = 100F$ with a conductivity of $k = 1$ Btu/hrftF, we would calculate a heat flux of

$$q = Sk(T_1 - T_2) = 7(1)(400)$$
$$= 2800 \text{ Btu/hr (for 1-ft depth)}$$

The above discussion defined the shape factor in terms of M and N, but we could also consider Eq. 2.89 to define the shape factor and arrive at an expression for S analytically. Table 2.4 presents a summary of some analytical expressions for the shape factor taken from the literature by Kreith (14) and Holman (15). We can take a brief look at the origin of the first item in Table 2.4, which refers to the hollow cylinder of radius ratio r_2/r_1 by comparing Eq. 2.26,

$$q = \frac{2\pi L}{\ln(r_2/r_1)} k(T_1 - T_2)$$

with Eq. 2.89

$$q = Sk(T_1 - T_2)$$

Table 2.4 SHAPE FACTOR FOR DIFFERENT GEOMETRICAL ARRANGEMENTS*

DESCRIPTION	GEOMETRY	SHAPE FACTOR, S (ft)	EXPLANATION
(a) Hollow cylinder of length L, no heat loss at ends		$\dfrac{2\pi L}{\ln(r_2/r_1)}$	Ends insulated or $L \gg r_2$
(b) Hollow sphere		$\dfrac{4\pi r_2 r_1}{r_2 - r_1}$	
(c) Constant temperature sphere of radius r buried z units below isothermal surface		$\dfrac{4\pi r}{1 - r/2z}$	

Table 2.4 (CONTINUED)

DESCRIPTION	GEOMETRY	SHAPE FACTOR, S (ft)	EXPLANATION
(d) Constant temperature cylinder of radius r and length L buried z units below an isothermal surface.		$\dfrac{2\pi L}{\cosh^{-1}(z/r)}$ $\dfrac{2\pi L}{\ln(2z/r)}$ $\dfrac{2\pi L}{\ln(L/r) - \ln(L/2z)}$	$L \gg r$ (very long cylinder) $z > 3r$ (very long cylinder) $z \gg r$ $L \gg z$
(e) Constant temperature cylinder of radius r and length z perpendicular to isothermal surface as shown		$\dfrac{2\pi z}{\ln(2z/r)}$	$z \gg 2r$ (long cylinder)
(f) Two very long parallel cylinders D units apart in an infinite medium		$\dfrac{\pi 2L}{\cosh^{-1}\left[\dfrac{(D/r_1)^2 - 1 - R^2}{2R}\right]}$	$R = \dfrac{r_2}{r_1}$ $L \gg r$ $L \gg D$

(g) Plane wall

A

Δx

$\dfrac{A}{\Delta x}$

Edges insulated

(h) Edge intersection

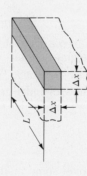

L

Δx

Δx

$0.54L$

Ends insulated, and
$L > 0.2\,\Delta x$

(i) Corner intersection of three walls of thickness Δx

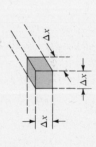

Δx

Δx

Δx

$0.15\,\Delta x$

Distance between corners $> 0.2\,\Delta x$

*As summarized by References 14 and 15.

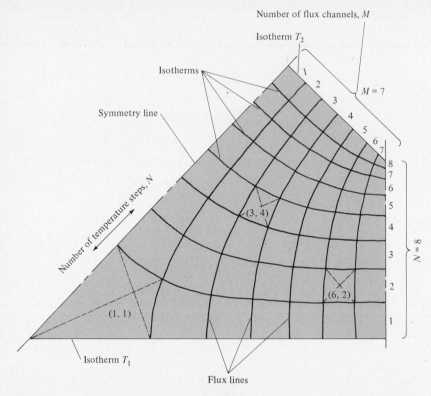

Figure 2.21 Flux plot sketch for the symmetry region B in Figure 2.19 drawn to a scale approximately three times larger.

and observing that for the hollow cylinder geometry

$$S = \frac{2\pi L}{\ln(r_2/r_1)}$$

as given in the table.

The following example gives some indication of the precision that can be achieved with the flux plot method.

Example 2.5

Consider the hollow cylinder with a radius ratio $r_2/r_1 = 2$ and determine the shape factor S by the flux plot method and from the analytical expression in Table 2.4.

SOLUTION

We construct a large sketch to scale of the geometry as shown in Figure 2.22. In assessing symmetry we note that all radial lines are lines of symmetry, so that we arbitrarily choose $1/32$ as a convenient division so that we can have reasonably sized curvilinear squares. The sketch shows that each of the 32 flux channels will

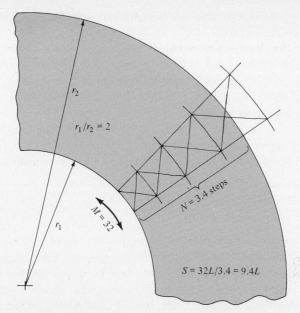

Figure 2.22 Flux plot for the hollow cylinder of radius ratio $r_2/r_1 = 2$ as described in Example 2.5.

have about 3.4 temperature steps for a shape factor $S = ML/N = 32L/3.4 = 9.4L$. When we use the expression from Table 2.4 part (a) we obtain

$$S = \frac{2\pi L}{\ln(r_2/r_1)} = \frac{2\pi L}{\ln 2} = 9.06L$$

In this example the flux plot gave a result that was about 4 percent high, which is acceptable for most engineering work. Generally speaking, we can say that a carefully drawn flux plot should give a shape factor within about 5 percent of the correct value. Using the above value for S we can calculate the heat flux through the 1-ft-long cylinder described in the computer example in Section 2.3.4

$$q = Sk(T_1 - T_2) = 9.4(2)(200)$$
$$= 3760 \text{ Btu/hr}$$

Compared to the calculated flux of 3626 Btu/hr given in Table 2.2 the result is 3.7 percent high, which is within the 5 percent figure discussed above.

2.8 NUMERICAL ANALYSIS

In most engineering applications the geometries encountered are not simple slabs, cylinders, or spheres, but rather complicated shapes depending on the purpose the design is to serve. Modeling such shapes with the solutions for the simple shapes can give only a very rough approximation of the actual thermal performance to be expected. A more precise prediction requires either a more complicated mathematical analysis (usually not solvable) or a numerical approximation to the problem.

In a numerical approach we depart from analyzing the region of interest as a continuum (and solving differential equations) and consider a series of discrete parts whose mass is all lumped at a single point in that part called a node. All the nodes are connected by fictitious thermal resistances to form a grid or network representing the region under study. By considering a finite number of discrete parts (nodes), we reduce the continuum problem (differential equation) to a series of finite algebraic equations that can be solved by hook or by crook. There are presently three general methods utilized in a numerical analysis:

1. The energy balance method.
2. The finite-difference method.
3. The finite-element method.

The energy balance method applies conservation of energy to each node such that the sum of all heat fluxes to a node represents the energy stored in the node. In steady-state, there can be no storage of energy so that the summation of heat fluxes to a node, then, is zero; in a transient situation (which we will consider in Chapter 3, Section 3.5) the storage term relates the variation of nodal temperature with time. The conservation equation at each node then yields an algebraic equation for each node involving the temperature of the node and the temperature of the nodes adjacent. We then solve the set of algebraic equations simultaneously (in principle).

The finite-difference method arrives at a set of algebraic equations by approximating the derivatives in the differential equation by finite differences depending on the size of the nodal network selected. For example, we would set $dT/dx \simeq (T_n - T_{n-1})/\Delta x$ and obtain a finite-difference equation at each node (also at the boundaries as dictated by the boundary conditions) which is essentially the same as was obtained in the energy balance method. We will see a little of how this method works in Section 3.6 when we look at graphical solutions to time dependent problems. As a matter of fact, these two methods are equivalent because the differential equations were obtained from an energy balance on an element of finite but small dimensions, and then taking the limit as the dimensions of the element are made to approach zero. In both the energy balance and finite-difference methods, it is extremely convenient to take the distances between nodes the same and also to work with a rectangular grid or network of nodes.

The finite-element method does not need the restriction of equal nodal distances or rectangular grids, but the method is more complicated because the function (the solution) itself is approximated rather than approximating the derivatives and the approximation is optimized by using the calculus of variations. The flexibility is well worth the additional mathematical complexity. In this introductory treatment, we will look only at the energy balance method and see how it is applied to a very simple situation. Finite-difference methods are very well described by Myers (6), the finite-element method is treated in depth by Zienkiewicz (17), and an excellent overview of the spectrum of numerical methods is given by Carnahan and coauthors (18).

2.8.1 Nodal Equations

The choice of node separation distance depends to a large extent on the computational facilities on hand as well as the amount of work one is willing to expend in the description of the node equations. In general, the larger the number of nodes, the closer one comes to the exact solution. There are basically only three ways energy can enter or leave a node: by conduction, by convection, or by radiation. There can, of course, be several mechanisms acting at a node. For example, at a three-dimensional interior node there would be six conducting paths connected to a node in a rectilinear network, or at a surface node in a two-dimensional situation, there would be three conducting paths (two if the node is on an outside corner and four if on an inside corner), one convection path, and possibly one radiation path if temperatures are such that radiation must be considered independently of convection. It's easy to see that node equations are custom made for each situation that is to be analyzed. It would be worthwhile to go back and quickly review the ideas presented in Section 1.4 of Chapter 1 before continuing. Figure 2.23 shows the three possibilities just discussed together with a general form of the applicable equation. Note that the node separation distance is $2(\Delta x/2)$ or Δx for the nodes inside the solid.

We will illustrate the numerical method by looking at a simple application of the energy balance method to the plane plate with uniform generation that we modeled and solved the differential equation for in Section 2.4.1.

2.8.2 Discretization of the Plane Plate with Internal Generation

We arbitrarily subdivide the plane plate into five nodes as shown in Figure 2.24 with one node on each surface and three interior nodes. Since we have five nodes in such an arrangement, the spacing between nodes $\Delta x = 2L/4$ or $L/2$, and in the general case for N nodes, the spacing $\Delta x = 2L/(N-1)$ where L is the half-thickness of the plate. The energy balance requires the sum of the fluxes at each node to be zero. Thus

$$
\left.
\begin{aligned}
N = 1: && q_1 + q_{k2 \to 1} + q_{\text{gen}} &= 0 \\
N = 2: && q_{k1 \to 2} + q_{k3 \to 2} + q_{\text{gen}} &= 0 \\
N = 3: && q_{k2 \to 3} + q_{k4 \to 3} + q_{\text{gen}} &= 0 \\
N = 4: && q_{k3 \to 4} + q_{k5 \to 4} + q_{\text{gen}} &= 0 \\
N = 5: && q_5 + q_{k4 \to 5} + q_{\text{gen}} &= 0
\end{aligned}
\right\} \qquad (2.90)
$$

The fluxes q_1 and q_5 are the heat fluxes at the surface of the slab and are determinable when the boundary conditions are specified. In this case we take the surface temperatures as fixed at $T_1 = 0\text{F}$ and $T_5 = 1000\text{F}$ to correspond

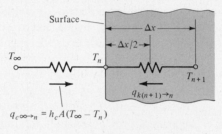

(a) Conduction

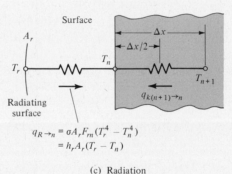

(b) Convection

Surface

$$q_{R \to n} = \sigma A_r F_{rn}(T_r^4 - T_n^4)$$
$$= h_r A_r (T_r - T_n)$$

(c) Radiation

Figure 2.23   Nodal equations for (a) conduction, (b) convection, and (c) radiation paths to node n.

with the temperature difference between the surfaces of 1000F as we did in Section 2.4.2. It would also be possible to specify fluid temperatures and convective coefficients on both sides of the plate, and in that case q_1 and q_5 would be convective fluxes and T_1 and T_5 would be unknowns. The generated flux is the unit volume generation rate $\dot{q}$ times the volume associated with the node. At node $N = 1$ and node $N = 5$,

$$q_{\text{gen}} = \frac{\dot{q} A \, \Delta x}{2} \tag{2.91}$$

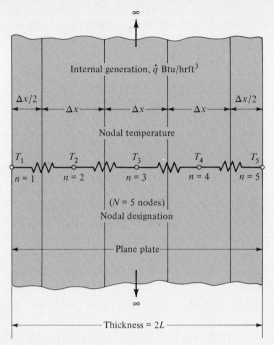

Figure 2.24   Discretization of plane plate into five nodes, $N = 5$.

and at the interior nodes $N = 2, 3, 4$,

$$q_{\text{gen}} = \dot{q} A \, \Delta x \tag{2.92}$$

The internal conductive fluxes all have the same general form, for example

$$
\left.
\begin{aligned}
q_{k1 \to 2} &= kA(T_1 - T_2)/\Delta x \\
q_{k4 \to 3} &= kA(T_4 - T_3)/\Delta x \\
q_{k5 \to 4} &= kA(T_5 - T_4)/\Delta x
\end{aligned}
\right\} \tag{2.93}
$$

and so forth. The energy balance Eq. 2.90 then can be written as follows:

$$
\left.
\begin{aligned}
N = 1: &\qquad q_1 + kA(T_2 - T_1)/\Delta x + \dot{q} A \, \Delta x/2 = 0 \\
N = 2: &\quad kA(T_1 - T_2)/\Delta x + kA(T_3 - T_2)/\Delta x + \dot{q} A \, \Delta x = 0 \\
N = 3: &\quad kA(T_2 - T_3)/\Delta x + kA(T_4 - T_3)/\Delta x + \dot{q} A \, \Delta x = 0 \\
N = 4: &\quad kA(T_3 - T_4)/\Delta x + kA(T_5 - T_4)/\Delta x + \dot{q} A \, \Delta x = 0 \\
N = 5: &\qquad q_5 + kA(T_4 - T_5)/\Delta x + \dot{q} A \, \Delta x/2 = 0
\end{aligned}
\right\} \tag{2.94}
$$

The nodal equations can be simplified by multiplying each equation by $\Delta x/kA$

so that we have expressions with just temperature terms and a constant. Thus

$$
\left.
\begin{aligned}
N = 1: &\quad \frac{q_1 \Delta x}{kA} + T_2 - T_1 + \frac{\dot{q} \Delta x^2}{2k} = 0 \\[6pt]
N = 2: &\quad T_1 + T_3 - 2T_2 + \frac{\dot{q} \Delta x^2}{k} = 0 \\[6pt]
N = 3: &\quad T_2 + T_4 - 2T_3 + \frac{\dot{q} \Delta x^2}{k} = 0 \\[6pt]
N = 4: &\quad T_3 + T_5 - 2T_4 + \frac{\dot{q} \Delta x^2}{k} = 0 \\[6pt]
N = 5: &\quad \frac{q_5 \Delta x}{kA} + T_4 - T_5 + \frac{\dot{q} \Delta x^2}{2K} = 0
\end{aligned}
\right\}
\qquad (2.95)
$$

With temperatures T_1 and T_5 known, we have three unknown temperatures T_2, T_3, and T_4 that require three equations for solution. Such a solution can, of course, be obtained algebraically, but when the number of nodes increases, other methods must be used.

The simultaneous solution of a number of algebraic equations can be accomplished by several different methods. We will look at a simple application of repeated iteration methods, the Jacobi and the Gauss-Seidel. The Jacobi method converges somewhat slower than the Gauss-Seidel method, but the rate of convergence can be improved considerably by applying over-relaxation techniques to the Gauss-Seidel method; the Jacobi method becomes unstable and diverges with over-relaxation.

2.8.3 Gauss-Seidel and Jacobi Iteration Methods

In order to compare with the analytical results of Section 2.4, we select the case where $Q = \dot{q} L^2/2k = 500F$, a slab half-thickness $L = 1$ ft, and a thermal conductivity of 10 Btu/hrftF; from Q we then determine the generation rate $\dot{q} = 10,000$ Btu/hrft^3. With a slab width of 2 ft for $N = 5$ nodes, the spacing $\Delta x = 2L/(n - 1) = 2/(5 - 1) = 0.5$ ft so that the generation "temperatures" in Eq. 2.95 become

$$
\left.
\begin{aligned}
\dot{q} \Delta x^2/k &= 10,000(0.5)^2/10 = 250F \\[4pt]
\dot{q} \Delta x^2/2k &= 10,000(0.5)^2/20 = 125F
\end{aligned}
\right\}
\qquad (2.96)
$$

Since we have three unknown temperatures ($T_1 = 0F$ and $T_5 = 1000F$), we work with the central three equations in the set 2.95 and after inserting the values for generation temperature from Eqs. 2.96 and T_1 and T_5, we obtain

$$
\left.
\begin{aligned}
T_3 - 2T_2 + 250 &= 0 \\[4pt]
T_2 + T_4 - 2T_3 + 250 &= 0 \\[4pt]
T_3 - 2T_4 + 1250 &= 0
\end{aligned}
\right\}
\qquad (2.97)
$$

These three equations can easily be solved to yield $T_2 = 625F$, $T_3 = 1000F$, and $T_4 = 1125F$. In this example with only three equations (three internal nodes), the numerical solution agrees exactly with the analytical solution, which is a parabolic equation (see Eq. 2.39).

We will now look at the application of the Jacobi and Gauss-Seidel iteration methods applied to the set of algebraic equations that we obtained from the energy balance on the nodes. First, however, we should observe that if we were to increase the number of nodes, the form of the node equations 2.95 would not change, only more of the central node type equations would appear. We will develop the iteration methods for N nodes in general so that we have some flexibility in calculating results other than for our example of $N = 5$ nodes.

The node equations for a general internal node $N = n$ can be expressed as

$$T_{n-1} + T_{n+1} - 2T_n + \frac{\dot{q}\,\Delta x^2}{k} = 0 \tag{2.98}$$

which can be solved for T_n. Thus

$$T_n = \frac{1}{2}\left(T_{n-1} + T_{n+1} + \frac{\dot{q}\,\Delta x^2}{k}\right) \tag{2.99}$$

We can observe that if we have an initial set of values T_n^0, we can calculate new improved values T_n^1 (values closer to the actual solution of the set of equations) by repeated application of Eq. 2.99:

$$T_n^1 = \frac{1}{2}\left(T_{n-1}^0 + T_{n+1}^0 + \frac{\dot{q}\,\Delta x^2}{k}\right) \tag{2.100}$$

A more general way to express the iteration scheme would be to iterate

$$T_n^j = \text{the new estimated temperature}$$

and

$$T_n^{j-1} = \text{the old estimated temperature}$$

at a particular node $N = n$. Then Eq. 2.100 would become

$$T_n^j = \frac{1}{2}\left(T_{n-1}^{j-1} + T_{n+1}^{j-1} + \frac{\dot{q}\,\Delta x^2}{k}\right) \tag{2.101}$$

which is the Jacobi method.

As we work through the set of equations, we observe that the new value of nodal temperature at the node just preceding the one we were working on is always available; if we use T_{n-1}^j instead of T_{n-1}^{j-1} in Eq. 2.101, we then have the Gauss-Seidel method which has the advantage of not requiring all terms to be stored.

$$T_n^j = \frac{1}{2}\left(T_{n-1}^j + T_{n+1}^{j-1} + \frac{\dot{q}\,\Delta x^2}{k}\right) \tag{2.102}$$

Let's look at the application of these ideas before we consider over-relaxation. We take as a first guess a simple linear increase in temperature from node $N = 1$

Table 2.5 DEVELOPMENT OF TEMPERATURE VALUES IN THE APPLICATION OF THE JACOBI METHOD (EQ. 2.101) TO THE SLAB WITH $N = 5$ NODES

NODE	T_n^0	T_n^1	T_n^2	T_n^3	$\cdots$	T_n^{43}
1	0	0	0	0	$\cdots$	0
2	250	375	437.5	500	$\cdots$	625
3	500	625	750	812.5	$\cdots$	1000
4	750	875	937.5	1000	$\cdots$	1125
5	1000	1000	1000	1000	$\cdots$	1000

to node $N = 5$ and, recalling that $T_1 = 0$F and $T_5 = 1000$F (fixed temperatures), then $T_2 = 250$F, $T_3 = 500$F, and $T_4 = 750$F would be our first guess. Table 2.5 shows the evolution of the numerical values as they were calculated by the computer program described below. The arrows in the table help to identify the procedure, for example:

$$T_2^1 = \left(T_1^0 + T_3^0 + \dot{q}\,\Delta x^2/k\right)/2$$

$$T_2^1 = (0 + 500 + 250)/2 = 375$$

and

$$T_3^2 = \left(T_2^1 + T_4^1 + \dot{q}\,\Delta x^2/k\right)/2$$

$$T_3^2 = (375 + 875 + 250)/2 = 750$$

and so forth. The last column in Table 2.5 does not represent exact values; those are the values that did not differ by 0.0001 degree from the previously calculated values. We accept that result as being close enough to the actual solution which in this case we know.

The development of the Gauss-Seidel method is traced in Table 2.6. In this case only T_2^1 is the same as in Table 2.5; all other values of T_n are larger and converge to the solution faster. The calculations go as follows:

$$T_2^1 = (0 + 500 + 250)/2 = 375$$

$$T_3^1 = (375 + 750 + 250)/2 = 687.5$$

$$T_4^1 = (687.5 + 1000 + 250)/2 = 968.7$$

and so forth; T_1 and T_5 remain unchanged since they are fixed as boundary

Table 2.6 DEVELOPMENT OF TEMPERATURE VALUES FOR THE GAUSS-SEIDEL METHOD (EQ. 2.102) AS APPLIED TO THE SLAB WITH $N = 5$ NODES

NODE	T_n^0	T_n^1	T_n^2	T_n^3	$\cdots$	T_n^{23}
1	0	0	0	0	$\cdots$	0
2	250	375	468.7	546.8	$\cdots$	625
3	500	687.5	843.7	921.8	$\cdots$	1000
4	750	968.7	1046.8	1085.9	$\cdots$	1125
5	1000	1000	1000	1000	$\cdots$	1000

conditions. It is significant to note that only 23 iterations are required to converge to 0.0001 degree between successive values. When the slab is divided into nine nodes, the Jacobi method requires 169 iterations and Gauss-Seidel 85 iterations to converge to 0.0001 degree between successive values. Of course, subdivision into smaller Δx's (more nodes) requires more iteration to converge. When the number of nodes becomes large, say, 150 or 200, the computational time becomes appreciable, even on a fast machine. Clearly it would be advantageous to speed things up a bit.

Over-relaxation accomplishes that purpose in the following manner. At any node the Gauss-Seidel procedure gives an increment $T_n^j - T_n^{j-1}$ in the temperature iteration. We could speed things up by multiplying this increment by a number ω greater than 1 and adding that value to the old estimate T_n^{j-1}. Thus

$$\overline{T}_n^j = \omega\left(T_n^j - T_n^{j-1}\right) + T_n^{j-1} \tag{2.103}$$

where $1 < \omega < 2$ is the over-relaxation factor and must be less than 2 for the procedure to be stable. The value $\overline{T}_n^j$ is the one used in the iteration procedure at node n instead of T_n^j after we move to the next node. We now look at how one might go about programming such a procedure.

2.8.4 Computer Program of Iteration Methods

The ideas described above and embodied in Eq. 2.101, 2.102, and 2.103 are programmed for the IBM 370/155 computer according to the flow diagram presented in Figure 2.25 and the FORTRAN listing given in Table 2.7. The program dictionary explains the symbols, but a few additional comments would probably be helpful. The number of nodes was designated as N so, therefore, the nodal iterations in stepping from one node to another is done with the index K that ranges from 2 to L = N $-$ 1 because node 1 and node N are fixed at 0F and 1000F, respectively. The over-relaxation factor ω is designated as W in the program and W = 1 gives no over-relaxation (see Eq. 2.103). The acronyms employed for the data should be easily recognizable; for example, DX for Δx, QDOT for $\dot{q}$, and so forth. Statement 4 sets the initial temperatures linearly from T_1 to T_5 and prints the values along with the heading in FORMAT 500. The main iteration is done in the DO10 loop, convergence is checked in the DO15 loop, and the DO25 loop replaces the old temperatures with the newly calculated temperatures if the convergence criteria is not met at each node and the GOTO8 change of control repeats the procedure after updating the iteration counter M by 1. When the iteration is complete, the heat fluxes QAl and QAN on both sides of the slab are calculated and printed together with the number of iterations required to converge. The DO30 loop repeats the entire procedure for values of the over-relaxation factor $1.0 \le W \le 1.9$ in steps of 0.1.

A typical output is shown in Tables 2.8 and 2.9 where Table 2.8 is for W = 1, or no over-relaxation, and Table 2.9 is for W = 1.2 which is near the optimum value for W which gives the least number of iterations; the convergence criteria for both tables is CONV = 0.0001 degrees F. Table 2.8 shows that

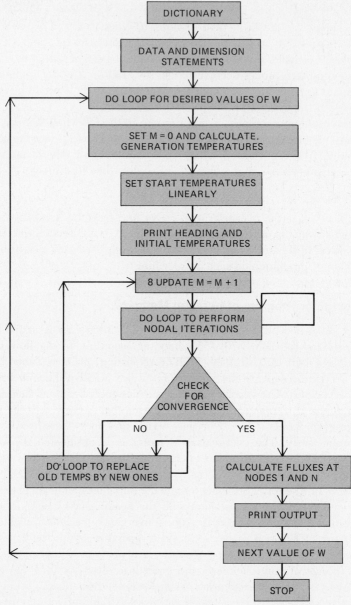

Figure 2.25   Flow diagram for numerical iteration computer program of section 2.8.4.

Table 2.7 FORTRAN LISTING FOR GAUSS-SEIDEL ITERATION PROGRAM

```
C     THIS PROGRAM CALCULATES A GAUSS-SEIDEL ITERATION OF
C     THE NODAL EQUATIONS REPRESENTING A NUMERICAL
C     APPROXIMATION TO THE STEADY STATE SOLUTION FOR THE
C     INFINITE SLAB WITH UNIFORM GENERATION.
C
C     PROGRAM DICTIONARY:
C
C                N=NUMBER OF NODES
C                M=NUMBER OF ITERATIONS
C                W=RELAXATION FACTOR
C                   (W=1 STRAIT GAUSS-SEIDEL METHOD)
C                   (W BET 1 AND 2 GIVES OVERRELAXATION)
C        TEMP(J,1)=THE OLD TEMPERATURES
C        TEMP(J,2)=THE NEW TEMPERATURES
C        TEMP(1,1)=TEMP NODE 1, F, FIXED
C        TEMP(N,1)=TEMP NODE N, F, FIXED
C               XK=THERMAL CONDUCTIVITY, BTU/HRFTF
C               DX=NODE SEPARATION, FT
C               XL=SLAB HALF THICKNESS, FT
C             CONV=CONVERGENCE CRITERIA
C           DIF(I)=DIF BETWEEN OLD AND NEW TEMPS
C               TQ=QDOT*DXSQD/XK THE GENERATION TEMP
C              QA1=UNIT AREA HEAT FLUX AT NODE 1
C              QAN=UNIT AREA HEAT FLUX AT NODE N
C
C     DATA AND DIMENSION STATEMENTS
C
      DIMENSION TEMP(50,2), DIF(50)
            N=5
           XN=N
         CONV=0.0001
           XK=10.
           XL=1.
           DX=2.*XL/(XN-1.)
      TEMP(1,1)=0.
      TEMP(1,2)=0.
      TEMP(N,1)=1000.
      TEMP(N,2)=1000.
         QDOT=10000.
      DO 30 KK=10,19
      W=KK/10.
C
C         CALCULATE GENERATION TEMPS
C             M=0
      TQ=QDOT*DX*DX/XK
      TQ2=TQ/2.
C
C             SET START TEMPS LINEARLY
C
        L=N-1
        DO 5 I=2,L
        XI=I
    4 TEMP(I,1)=TEMP(1,1)+((XI-1.)/(XN-1.))*
      A(TEMP(N,1)-TEMP(1,1))
    5 CONTINUE
      PRINT 500,N,W,M,(TEMP(I,1),I=1,N)
C
```

Table 2.7 CONTINUED

```
C                 PERFORM NODAL ITERATIONS
C
      8  M = M + 1
         DO 10 K = 2, L
         TEMP(K, 2) = (TEMP(K − 1, 2) + TEMP(K + 1,1) + TQ)/2.
         TEMP(K, 2) = W∗(TEMP(K, 2) − TEMP(K,1)) + TEMP(K,1)
     10  CONTINUE
         PRINT 505, M, (TEMP(I, 2), I = 1, N)
C
C                 CHECK FOR CONVERGENCE
C
         DO 15 I = 2, L
         DIF(I) = ABS(TEMP(I,1) − TEMP(I, 2))
         IF(DIF(I).GT.CONV) GOTO 20
     15  CONTINUE
C
C                 CALC FLUXES AT NODES 1 AND N.
C
         QA1 = XK∗(TEMP(1, 2) − TEMP(2, 2) − TQ2)/DX
         QAN = XK∗(TEMP(N, 2) − TEMP(N − 1, 2) − TQ2)/DX
         PRINT 510, M, QA1, QAN
         GOTO 30
C
C                 REPLACE OLD TEMPS BY THE NEW ONES
C
     20  DO 25 I = 2, L
         TEMP(I,1) = TEMP(I, 2)
     25  CONTINUE
            GOTO 8
     30  CONTINUE
         STOP
    500  FORMAT(1H1///34X, 'NODAL TEMPERATURE PREDICTIONS FOR A'/
         A35X, 'GAUSS-SEIDEL ITERATION OF THE SLAB'/
         A35X, '(USING SUCCESSIVE OVERRELAXATION)'/39X, 'WITH',
         B'UNIFORM GENERATION'/45X, 14, 'NODES'/45X, 'W = ', F4.2///
         B23X, 'ITERATION',
         C3X, 'N = 1', 8X, 'N = 2', 8X, 'N = 3', 8X, 'N = 4', 8X, 'N = 5'//
         D26X, 13, (5F11.3))
    505  FORMAT(26X, 13, (5F11.3))
    510  FORMAT(//46X, 'ITERATION HAS CONVERGED'/ 55X, 'M = ', 13/
         A49X, 'QA1 = ', F12.3/49X, 'QAN = ', F12.3)
         END
```

23 iterations were needed for the 0.0001-degree convergence criteria and the converged values agree with those presented in Table 2.3 of Section 2.4.2. Table 2.9 shows that only 13 iterations were required to reach convergence with an over-relaxation factor $W = 1.2$. With only five nodes the number of iterations is clearly not a significant concern, but with an increase in the number of nodes the situation changes rapidly. For example, Table 2.10 compares the number of iterations to converge (for same criteria of 0.0001 degree) for the subject example for $N = 5$ and $N = 9$ nodes. For $N = 5$ nodes the optimum W is about 1.2 and the iterations decrease from 23 to 13, whereas at $N = 9$ nodes, the optimum W is about 1.5 and the iterations required to converge decrease from 85 to 24. In general, the optimum value of W lies somewhere about 1.6 to 1.8 for

Table 2.8 NODAL TEMPERATURE PREDICTIONS FOR A GAUSS-SEIDEL
ITERATION OF THE SLAB (USING NO OVER-RELAXATION)
WITH UNIFORM GENERATION

		5 NODES W = 1.00			
ITERATION	N = 1	N = 2	N = 3	N = 4	N = 5
0	0.0	250.000	500.000	750.000	1000.000
1	0.0	375.000	687.500	968.750	1000.000
2	0.0	468.750	843.750	1046.875	1000.000
3	0.0	546.875	921.875	1085.938	1000.000
4	0.0	585.938	960.938	1105.469	1000.000
5	0.0	605.469	980.469	1115.234	1000.000
6	0.0	615.234	990.234	1120.117	1000.000
7	0.0	620.117	995.117	1122.559	1000.000
8	0.0	622.559	997.559	1123.779	1000.000
9	0.0	623.779	998.779	1124.390	1000.000
10	0.0	624.390	999.390	1124.695	1000.000
11	0.0	624.695	999.695	1124.847	1000.000
12	0.0	624.847	999.847	1124.924	1000.000
13	0.0	624.924	999.924	1124.962	1000.000
14	0.0	624.962	999.962	1124.981	1000.000
15	0.0	624.981	999.981	1124.990	1000.000
16	0.0	624.990	999.990	1124.995	1000.000
17	0.0	624.995	999.995	1124.998	1000.000
18	0.0	624.998	999.998	1124.999	1000.000
19	0.0	624.999	999.999	1124.999	1000.000
20	0.0	624.999	999.999	1125.000	1000.000
21	0.0	625.000	1000.000	1125.000	1000.000
22	0.0	625.000	1000.000	1125.000	1000.000
23	0.0	625.000	1000.000	1125.000	1000.000

ITERATION HAS CONVERGED
M = 23
QA1 = −14999.992
QAN = − 4999.992

Table 2.9 NODAL TEMPERATURE PREDICTIONS FOR A GAUSS-SEIDEL
ITERATION OF THE SLAB (USING SUCCESSIVE OVER-RELAXATION)
WITH UNIFORM GENERATION

		5 NODES W = 1.20			
ITERATION	N = 1	N = 2	N = 3	N = 4	N = 5
0	0.0	250.000	500.000	750.000	1000.000
1	0.0	400.000	740.000	1044.000	1000.000
2	0.0	514.000	936.799	1103.280	1000.000
3	0.0	609.280	990.176	1123.449	1000.000
4	0.0	622.249	999.384	1124.940	1000.000
5	0.0	625.180	1000.196	1125.129	1000.000
6	0.0	625.081	1000.087	1125.026	1000.000
7	0.0	625.036	1000.020	1125.006	1000.000
8	0.0	625.004	1000.002	1125.000	1000.000
9	0.0	625.000	1000.000	1125.000	1000.000
10	0.0	625.000	999.999	1125.000	1000.000
11	0.0	625.000	1000.000	1125.000	1000.000
12	0.0	625.000	1000.000	1125.000	1000.000
13	0.0	625.000	1000.000	1125.000	1000.000

ITERATION HAS CONVERGED
M = 13
QA1 = −14999.992
QAN = − 4999.992

Table 2.10 NUMBER OF ITERATIONS TO CONVERGE TO
WITHIN 0.0001 DEGREE FOR GAUSS-SEIDEL METHOD
WITH OVER-RELAXATION (SLAB WITH
INTERNAL GENERATION).

W	N = 5	N = 9
(1.0)*	(23)	(85)
1.1	18	73
1.2	13	60
1.3	15	49
1.4	19	37
1.5	25	24
1.6	32	32
1.7	44	50
1.8	64	64
1.9	132	126

*W = 1.0 indicates no over-relaxation.

a large number of nodes. We should note the significance of the minus signs on QA1 and QAN in Tables 2.8 and 2.9; a negative value simply indicates energy out of a node.

We conclude this discussion with the observation that convergence of the iteration scheme is a delicate matter and is generally settled by physical reasoning and understanding of the nature of the problem. Usually, if there are no physically unreal conditions (such as boundaries having different temperatures that meet), an exploration with successively smaller values of convergence criteria will give an indication of what can be expected when mathematical uniqueness cannot be shown.

2.9 CONCLUDING REMARKS

The phenomena of conduction requires a heavy dose of applied mathematics to understand what goes on. We might note that some of the early giants in mathematics, such as Biot and Fourier, worked in the area of conduction heat transfer. The operational mathematics of Laplace (which we did not discuss) had beginnings in the solution of differential equations associated with time dependent phenomena. Unfortunately, many engineers view such mathematical topics with awe and fright, and consequently topics associated with such mathematics are unpopular. Perhaps this would be a good spot to again emphasize (last chance might be a better way to put it) the fact that engineers need not be masters of the art of mathematics. They do, however, need to know the language and be familiar with the basic mathematical ideas pertaining to the work they are involved with so that when the mathematical row is hard to hoe (the weeding and seeding may also be tough) they can call for help and make their wants known intelligibly. Further, and equally important, when the help is delivered to them, they can assimilate and make use of it and be sure that it fits the situations they are wrestling with.

We have emphasized in the above sections the simple application of separation of variables and the application of boundary conditions to proceed from the general solution to the particular solution for the given situation. It is the boundary conditions that govern the application of the particular solution. Engineers cannot just grab a formula out of the book (really the literature) and use it without thinking. They must make sure that the boundary conditions reasonably approximate the real world they are working with. Engineers who catch on to this idea will run into little difficulty in using the vast store of information available on the scientific-engineering bookshelf.

On a more cheerful note we can state that the hardest part, mathematically speaking, has been covered. Transient conduction (the subject matter of the next chapter) requires a new dimension (time) of physical understanding, but the mathematical details are so complex that we will deal only with the results in the form of equations and graphs. The emphasis of our approach will lean more and more toward understanding the physical phenomena and we will not be forced to dwell as heavily on the mathematical details of solution as we have in this chapter.

PROBLEMS

1. (2.1) Consider a solid concrete (119 lb_m/ft^3 density) wall 6 ft thick to have a variation in temperature with distance (at a given instant) as follows:

$$T(x) = 70 - 27x + x^3 \quad (F)$$

The temperature at the left face where $x = 0$ is 70F, the right face is at $x = 6$ ft. Determine the location and magnitude of the maximum temperature in the wall, and the magnitude and direction of the unit area heat flux at each face.

2. (2.1) A hollow cylinder of a material that has $k = 2.0$ Btu/hrftF has an inner radius of 3 in. and an outer radius of 6 in. The temperature on the inner surface is 300F and the temperature profile is given by

$$T(r) = 300 - 288 \ln \frac{r}{3} \quad (F)$$

where r is in inches. Calculate the temperature at the outer face and the heat flux per unit length at the inner and outer faces.

3. (2.1) Draw the temperature profile indicated by the equation given in Problem 1 over the range $x = 0$ to $x = 6$ ft as precisely as possible on 10×10 to the $\frac{1}{2}$ in. graph paper. Construct the tangents to the curve at $x = 0$ and $x = 6$ ft, determine $T'(x)$, estimate unit area heat flux, and compare with your result in Problem 1 (including the sign).

4. (2.2) Predict a value of thermal conductivity for tungsten at a temperature of 4780F. Give a brief discussion of your approach.

5. (2.2) Calculate a value for the Lorenz number for molybdenum at 0C and compare it with the value given in Table 2.1.

6. (2.2) Examine the data for carbon steel (0.5 percent C) in Appendix C for the range in temperature from 32F to about 900F. Formulate an expression to represent the data as a function of temperature. Use either equation 2.9 or 2.10, whichever is appropriate.

7. (2.2) Plot the properties μ, k, ν, and the properties group $\beta g/\nu\alpha$ ($1/\text{Rft}^3$) for air and water over the temperature range from 32 to 600F. Put μ, k, ν on linear paper and $\beta g/\nu\alpha$ on log-log paper, choosing a scale so that the items can be read to three significant figures. These curves will be very convenient to have on hand for home problems and exams on the chapters on convection.

8. (2.3) Calculate the heat flux per unit area through a plane wall 6 in. thick that is made of: (a) low-density concrete, (b) sandstone, and (c) white pine, when one side of the wall is at 80F and the other side is at 20F.

9. (2.3) A 1-in.-thick plate of 0.5 percent C carbon steel has a temperature of 800F on one side and 150F on the other side. Calculate the most accurate value possible for the heat flux through the slab under these conditions.

10. (2.3) Foams make very good insulators as evidenced by their use as conveniently portable ice chests. A value of $k = 0.006$ Btu/hrftF is representative of such foams. A small chest $6 \times 8 \times 10$ in. (tall) inside dimensions with 1-in.-thick walls is filled with liquid nitrogen (at 1 atm pressure) to a depth of 9 in. The lid loosely covers the top. Estimate how long it would take for the level to drop 1 in. when the ambient temperature is 70F. Assume a convection coefficient on the outer surfaces to be 5 Btu/hrft^{2}F and that the chest is placed on supports such that 90 percent of the bottom is exposed to the ambient surroundings.

11. (2.3) Estimate the air conditioning load per square foot on the wall shown in the figure. Decide on symmetry planes and draw the electrical analog circuit using an inside $h = 4$ Btu/hrft^{2}F and an outside $h = 10$ Btu/hrft^{2}F to determine the heat flux.

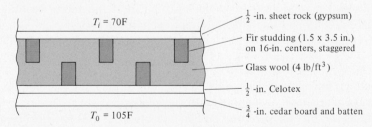

$T_i = 70F$

$\frac{1}{2}$-in. sheet rock (gypsum)

Fir studding (1.5 x 3.5 in.) on 16-in. centers, staggered

Glass wool (4 lb/ft^3)

$\frac{1}{2}$-in. Celotex

$T_0 = 105F$

$\frac{3}{4}$-in. cedar board and batten

12. (2.3) The experimental engine described in Example 2.1 is run at an increased load such that the inner surface temperature of the quartz window is increased to 1200C. Additional cooling on the outer surface of the quartz window increased the h to 250 W/m^{2}K with the ambient temperature remaining at 25C. Calculate the heat flux per unit area and the outer surface temperature of the quartz window for variable k under these conditions.

13. (2.3) A 4-in. schedule 80 carbon steel (1.5 percent C) steam line contains saturated steam at 125 psia and is covered with a 2-in. thick layer of 85 percent magnesia insulation. The line is 250 ft long and goes from the power plant to a manufacturing building and is suspended horizontally from supports 15 ft above the ground. During the winter time the ambient temperature is $-10F$ and the wind causes an $h = 30$ Btu/hrft^{2}F on the outside of the insulation. Calculate the mass of liquid condensate per hour which appears at the manufacturing plant end of the steam line. If the mass flow of the steam into the line is 200 lb$_m$/hr and 100 percent quality (going into the line), determine the quality at the manufacturing plant.

14. (2.3) A change in processing in the manufacturing plant in Problem 13 requires more steam for heating purposes. How much more steam could be obtained by doubling the thickness of the insulation on the steam line?

15. (2.3) Consider a hollow cylinder with the inside radius fixed and let the outer radius vary (variable wall thickness); let the heat transfer coefficient on the outer surface, h, and the ambient temperature, T_∞, also be fixed. The heat flux through the wall of this cylinder then depends on the sum of the conduction resistance R_k of the cylindrical wall and the convection resistance R_c on the outer surface.

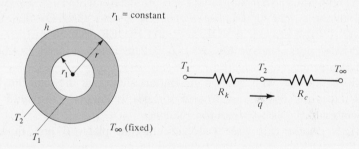

By examining R_k and R_c for a given length of the cylinder L, k, and h, we can observe that as r increases, R_k increases and R_c decreases. Accordingly, if r_1 is small enough, we could have a value of r that would give a maximum heat flux q (for a given L, k, and h). Show that the maximum q occurs at a value of $r = k/h$ by determining the derivative of q with respect to r and setting the result equal to zero.

16. (2.3) Calculate the maximum amount of heat per unit length that a 0.20-in.-diameter wire can dissipate if it is insulated with a layer of hard rubber with an $h = 2.0$ Btu/hrft^{2}F on the outside of the rubber. The surface temperature of the wire is 120F and ambient temperature is 40F. Plot the heat flux per unit length as a function of the insulation thickness on linear paper over the range in *thickness* from 0–0.7 in.

17. (2.3) Using the computer program described in Section 2.3.4., calculate and plot temperature profiles for the hollow cylinder with radius ratio $r_2/r_1 = 2$ as a function of radius for thermal conductivity XK = 1.6, 8, 40, and 200 as

parameter for a heat flux of 3626 Btu/hr in 1-ft length and an inside temperature of 300F. Use the program in Section 2.3.4 as much as practical, especially the output section, and include a GOTO loop to update XK by a factor of 5. Note that it will also be necessary to update T2 for each value of XK. What materials would be representative of the values of XK used here?

18. (2.3) An underwater submersible has an observation shell in the form of a lucite sphere ($k = 0.12$ Btu/hrftF) that has an inner diameter of 3 ft and a wall thickness of 76 mm. Assuming an $h = 100$ Btu/hrft^{2}F on the water side and an $h = 2$ Btu/hrft^{2}F on the inside, determine the heat load required to keep the inside at 70F in 34F seawater if 80 percent of the sphere is exposed to the sea.

19. (2.3) Beginning with Fourier's law as given by Eq. 2.24 and a linear variation in thermal conductivity like Eq. 2.9, formulate an expression for heat flux through a hollow sphere similar to Eq. 2.32.

20. (2.3) Include Eq. 2.33 in the computer program of Section 2.3.4 in order to compare profiles for the cylinder and sphere with equal radius ratios. Designate the temperature profile for the sphere TRS and insert the equation to calculate TRS just after statement 110 (call it 112) and add QS the flux for the sphere after statement 120. Note that you will need to add TRS and QS to the output heading and print statement in 140 and 500. The title for the output should also be changed to read 'ONE-DIMENSIONAL HEAT TRANSFER IN A HOLLOW CYLINDER AND SPHERE'.

21. (2.4) Do a units check on Eqs. 2.34, 2.36, 2.38, 2.41, and 2.43. Use both BE and SI units.

22. (2.4) For the slab with uniform generation, show that the total flux leaving both faces (as determined from Eq. 2.43) is equal to the total energy generated in the slab $\dot{q}$ times the volume.

23. (2.4) A reactor core uses enriched uranium plates 6 mm thick ($k = 20$ Btu/hrftF) and they are subjected to an internal generation rate of 8×10^7 Btu/hrft^3 due to the fission process. One of these plates is near the edge of the core and the coolant maintains one side of the plate at 600F and the other side at 700F. Calculate the maximum temperature in the plate under these conditions.

24. (2.4) Enriched uranium undergoes an undesirable phase transition at about 1200F which results in dimensional changes. Assuming a safe limit in temperature to be 1150F, what is the highest generation rate permissible for the plate in Problem 23 under the conditions described?

25. (2.4) Determine the current flow in amperes in a 3.175-mm-OD solid copper wire that would correspond to a 2F difference in temperature between the center and the surface of the wire. Remember that specific resistance ρ in Ωcm is related to wire dimensions and resistance by the relationship $R = \rho L/A$ where R is the resistance in ohms (we will need to qualify our answer here when we get to the chapter on convection).

26. (2.4) An over-power transient increases the rod power of the fuel rod described in Example 2.3 to 20 kW/ft. Determine the fuel temperature at

the inner surface of the annular pellet under these conditions if the outer surface is at 1000F.

27. (2.4) Determine the surface heat flux of the solid sphere described in Section 2.4.4 by evaluating the derivative of temperature with respect to radius at the surface $r = r_1$ and after inserting in Fourier's law, show that the surface flux is equal to the unit volume generation rate $\dot{q}$ times the volume of the sphere.

28. (2.4) Formulate an expression for the temperature profile in a hollow sphere of radius r_1 and r_2 (corresponding temperature T_1 and T_2) with uniform generation rate $\dot{q}$ and thermal conductivity k. Check your result by showing that the surface flux (adiabatic inner surface) is equal to the generation rate times the volume.

29. (2.5) Compare the heat flux dissipated by a 0.25-in.-diameter round iron fin ($k = 25$ Btu/hrftF) that is 1 in. long and not insulated on the end, to the heat flux that would be dissipated by the base area if the fin were not on the surface. The convection coefficient for all surfaces is 10 Btu/hrft^{2}F, and the base temperature is 300F at an ambient level of 75F.

30. (2.5) An empty nominal 3-in. schedule 40 steel pipe 4 ft long extends between two walls that are at 40F surface temperature. The ambient air is at -10F and the heat transfer coefficient due to the still air on the outside of the pipe is 2.0 Btu/hrft^{2}F. Determine the location on the pipe where the temperature is 32F.

31. (2.5) An induction coil is used to heat a spot on a long, $\frac{1}{2}$-in.-diameter, 1.5 percent C carbon steel reinforcing bar for an intricate bending operation. Assuming that 95 percent of the energy applied to the coil appears in the bar, estimate the required input watts to hold a spot on the bar at 1200F when $h = 15$ Btu/hrft^{2}F on the bar and the ambient temperature is at 60F.

32. (2.5) A round bronze bar, 2 ft long and $\frac{3}{4}$-in.-diameter extends between two walls, one at 100F and the other at 70F. Ambient air is at 30F with an $h = 2.0$ Btu/hrft^{2}F on the surface of the bar. Determine the location and magnitude of the minimum temperature in the bar.

33. (2.6) Two instrumented Duralumin bars are placed in an apparatus to measure contact conductance. The bars are 1 in.^{2} and are oriented as shown in Figure 2.18. The temperatures $T_1 = 100$F and $T_2 = 95$F were measured 2 in. apart as were $T_3 = 90$F and $T_4 = 85$F. The temperatures T_2 and T_3 were measured 0.2 in. either side of the contact plane of the two bars. Determine the unit area contact conductance h_i under these conditions.

34. (2.7) Determine the shape factor from a flux plot for a 12 in., square column with a 7 in., square duct inside and oriented such that the vertexes of the duct point to the middle of the column faces. The geometry is similar to that shown in Figure 2.19.

35. (2.7) A square duct has walls whose thickness is half the inside dimension, i.e., if the dimension of the inner square is L, then $L = 2T$, where T is the wall thickness. Determine the shape factor for a 1-ft length of the duct from a flux plot and compare with what you would calculate by using Table 2.4.

36. (2.7) Two long parallel cylindrical pipes 100 ft long and 1 ft in diameter are buried in an infinite medium (far from surface of medium) such that their centerlines are 2 ft apart. Determine the shape factor for heat transfer from one cylinder to the other by the flux plot method and compare with a result calculated from the formula given in Table 2.4.

37. (2.7) A very long and tall cinder concrete wall 3 ft thick has a 1 ft, square duct inside running parallel to the face of the wall and centered in the wall (with the side of the duct also parallel to the wall direction). The duct surface is at 100F and the wall faces are at 50F. Determine the heat loss from the duct per unit length.

38. (2.7) An octagonally shaped column has four opposite flats that are 1 ft wide; the column is 2 ft thick across these flats. The remaining four flats are somewhat less than 1 ft wide. A 1-ft-diameter round duct is centered inside the column. Determine the shape factor per unit length for the column and compare your result with the shape factor for an equivalent hollow cylinder of the same mass or volume.

39. (2.7) A solar collector utilizes a Silumin aluminum alloy plate shaped as shown in the figure. The plate is 1 ft wide by 4 ft long and has a solar flux of 1000 Btu/hr impinging on it. Estimate the temperature difference across the plate.

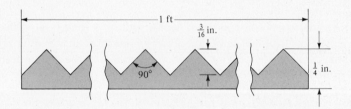

40. (2.7) A round column 2 ft in diameter has a triangular duct cast in the center. The duct is in the shape of an equilateral triangle with sides of 1 ft length. The triangular duct is centered about the axis of the column. Determine the shape factor per unit length for the column.

41. (2.7) Re-examine Problem 10 and estimate the heat gain, and consequently the time required for the level to drop 1 in., with the shape factor concept by including the corner edges and cubes as indicated in Table 2.4.

42. (2.8) Punch the cards as indicated in Table 2.7 or enter the lines on a terminal for the Gauss-Seidel iteration for the slab with generation and run the program with three different sets of initial conditions to determine the influence of initial condition on convergence (use $N = 5$ and CONV = 0.0001):

(a) All temperatures (except T_5) at 0F.

(b) All temperatures (except T_1) at 1000F.

(c) $T_2 = -5683F$, $T_3 = +9284F$, $T_4 = -6829F$

Summarize your observations in a paragraph or so.

43. (2.8) With the deck you punched in Problem 42 or your terminal program restored to linear initial conditions, make a run for N = 5, 9, 17, and 33 nodes and plot the resulting optimum value of W as a function of the number of nodes N. Note that you will need to change the PRINT 500 and PRINT 505 statements to index the print counter so that you will only have five temperatures on the output page. For example, for N = 9 the list would read (TEMP(I, 2) I = 1, N, 2) in order to print T_1, T_3, T_5, T_7, and T_9 and for N = 17 you would write I = 1, N, 4. You would also want to repunch card C of FORMAT 500 to reflect the correct value of N. It would be convenient to work in groups of four with your classmates and each take a value of N to minimize time spent in obtaining your "data." How about including N = 3 and 25? Plot your results on linear paper and discuss.

44. (2.8) The aspect of convergence is always of interest and since we know what values the Gauss-Seidel program for the slab (Table 2.7) converges to, we can take a look at how the convergence takes place at different nodes for different initial conditions. Convergence can be viewed as the difference between the analytical solution and the current value at the given iteration. If we let

ATEM(K)=Analytical solution at node K, then
ATEM(1)=OF
ATEM(2)=625F
ATEM(3)=1000F
ATEM(4)=1125F
ATEM(5)=1000F

The five values can be included in with the data. Then the difference

DTEM(K) = ATEM(K) − TEMP(K, 2)

represents the amount that we differ from the analytical value, and at convergence this value would be zero. The statement for DTEM(K) can be added to the program right before statement 10 in the modal iteration loop. Also, DTEM(N) and ATEM(N) must be added to the DIMENSION statement. Since the DTEM(K) values will be small, it would be best to print them out with an E format, say, E11.4 in FORMAT 505. Use CONV = 0.0000001 to get a good spread on the number of iterations. Plot DTEM(K) as a function of M with the node location as a parameter; semilog paper would probably be the most convenient to use with linear on the ordinate and log scale on the abscissa.

References

1. Bosworth, R. C. L., *Heat Transfer Phenomena*, Wiley, New York, 1952, p. 21.
2. Touloukian, Y. S., and Ho, C. Y., (Editors), *Thermophysical Properties of Selected Aerospace Materials*, TEPIAC/CINDAS, Purdue University, 1977.
3. Weast, R. C., and Selby, S. M., Eds., *Handbook of Chemistry and Physics*, 48th Edition, Chemical Rubber Publishing Co., Cleveland, Ohio, 1968.

4. Reid, R. C., and Sherwood, T. K., *The Properties of Gases and Liquids, Their Estimation and Correlation*, McGraw-Hill, New York, 1958.

5. National Standard Reference Data Series—National Bureau of Standards—8 (NSRDS-NBS8) Category 5—Thermodynamic and Transport Properties, Government Printing Office, Washington, D. C., 1966.

6. Myers, G. E., *Analytical Methods in Conduction Heat Transfer*, McGraw-Hill, New York, 1971.

7. Schneider, P. J., *Conduction Heat Transfer*, Addison-Wesley, Reading, Mass., 1957.

8. Jakob, M., *Heat Transfer*, Vols. I and II, Wiley, 1949.

9. Rubin, B. F., "Summary of $(U, Pu)O_2$ Properties and Fabrication Methods," General Electric Co., GEAP-13582, 1970.

10. Burrington, R. S., *Handbook of Mathematical Tables and Formulas*, 4th Edition, McGraw-Hill, New York, 1965.

11. Carslaw, H. S., and Jaeger, J. C., *Conduction of Heat in Solids*, Oxford University Press, London, 1959.

12. Holman, J. P., *Heat Transfer*, 3d Edition, McGraw-Hill, New York, 1972.

13. Rapier, A. C., Jones, T. M., and Mcintosh, J. E., "The Thermal Conductance of Uranium Dioxide/Stainless Steel Interfaces," *International Journal of Heat and Mass Transfer*, Vol. 6, 1963, pp. 397–416.

14. Kreith, F., *Principles of Heat Transfer*, 3rd Edition, Harper & Row, New York, 1973.

15. Holman, J. P., *Heat Transfer*, 3d Edition, McGraw-Hill, 1972, p. 39.

16. Miller, R. G., and Fletcher, L. S., "A Facility for the Measurement of Thermal Contact Conductance," *Proceedings of Southeastern Seminar on Thermal Science*, SESTS, Tulane University, New Oreleans, La., 1974, p. 263.

17. Zienkiewicz, O. C., *The Finite Element Method*, 3d Edition, McGraw-Hill, New York, 1977.

18. Carnahan, B., Luther, H. A., and Wilkes, J. O., *Applied Numerical Methods*, Wiley, New York, 1969.

Chapter 3
Transient Conduction

3.0 GENERAL REMARKS

In the work we have done up to this point, time as a variable has not been considered in the heat transfer situations we've described and discussed. All the situations we analyzed have been viewed as continuous happenings with no change in temperature at any spatial location inside or outside of the material under consideration. Such conditions satisfy the thermodynamic concept of steady-state analysis applied to a system or a control volume and are generally a good approximation of average operating conditions. Equipment must, however, be started up and shut down and also during operation load changes or other changes in operating conditions cause temperatures and heat fluxes to vary with time. Consequently, because there are many situations in the engineering application of heat transfer where steady-state conditions do not exist, we need to formulate an awareness and understanding of the response of a solid under transient conditions.

The term *temperature history* is often used to describe the results of a transient analysis, and fits the situation very well because we want to know what (temperature or heat flux) happens when (time) and where (spatial location) in the material we are examining. The what, when, where information is needed to determine possible thermal stress problems and to assess energy requirements on start-up or dissipation on shut-down of large equipment, for example.

When the conditions that cause the change in heat flux or temperature are simple and easy to describe, the transient response can usually be predicted mathematically (the mathematics may not be easy, but it can be done). For complicated situations engineers can use a graphical or a numerical approach to get the results they are after. For cases where the geometry is irregular or complicated, especially when there is more than one material to consider, the numerical or graphical method is the only practical way to approach a solution. The success of the numerical method was made possible by the high-speed, electronic digital computer of this age, which can process the large number of calculations required in numerical analysis in a reasonable period of time.

The central concept to understanding transient response is the relationship or interplay between the internal and external resistance to heat flow; those two aspects govern the resulting temperature response of the solid. We will examine and discuss these ideas in detail in the next section in order to provide a perspective on how to approach transient problems.

3.1 TRANSIENT BEHAVIOR OF SYSTEMS

We have introduced the concept of thermal resistance to heat flow in conduction and convection phenomena in Section 1.4, and we recall that a low resistance permits heat energy to flow easily and with small temperature differences; conversely, high thermal resistance makes it difficult for energy to flow and large temperature differences are required to get the energy to move. The way a solid behaves thermally then depends on how easy or difficult it is for the energy to move inside the solid as compared to the ease with which the energy can leave or enter the surface. We can observe that if it is difficult for the energy to leave the surface, for example, and relatively easy for the energy to move in the solid, then the temperature in the solid would change slowly with time and also be nearly the same at all locations. On the other hand, if it is easy for the energy to leave the surface and difficult to move in the solid, then there will be large changes in temperature with distance in the solid. How fast the temperature changes with time depends on the thermal diffusivity $\alpha = k/\rho c_p$ of the solid.

We have talked so far about these ideas in a qualitative sense, but now let's put them into a quantitative sense so that we can assign a number to what we have discussed. The surface resistance R_o we recognize as the convection resistance $R_o = 1/hA$ and the internal resistance of the solid R_i we recognize as the conduction resistance $R_i = L/kA$. We have to note here that the L in R_i requires a special interpretation in that it must be the volume of the solid divided by the surface area through which the heat energy passes, that is,

$L =$ volume of solid/surface area for heat transfer

We will consolidate the concept of $L = V/A_s$ later on in Section 3.2. The ratio of R_i/R_o then governs the type of thermal transient response that will result for a given situation.

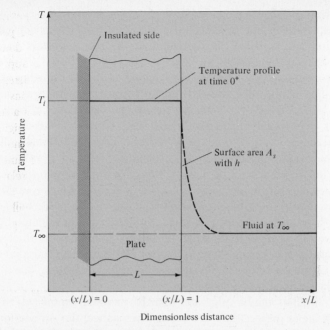

Figure 3.1 Infinite plane plate insulated on left face, h on right face, initially at uniform temperature T_i pictured at time 0^+

Let us consider the following physical situation and apply the resistance concepts described above to see how we can evaluate the ratio R_i/R_o. Figure 3.1 shows an infinite plate (very large so that heat transfer is only in the x/L direction) of thickness L exposed to a fluid at temperature T_∞ with the plate insulated on the left side. The temperature profile shown in Figure 3.1 is just at the time when the fluid has been brought into contact with the plate; we use the mathematical notation 0^+ to describe that time. At any time less than 0^+ the plate would not have been in contact with the fluid, and the plate temperature is the same, T_i, at all values of x/L. We are interested in what happens to the temperature in the plate at any location at times greater than 0^+ up to the time when there are no longer any changes taking place in the solid, that is, the steady-state situation.

The external resistance we noted above was $R_o = 1/hA_s$, but the internal resistance is not so simple to formulate; it depends on where or what location we are considering. We can, however, observe that the largest resistance would confront energy that was located at the insulated face. The energy there would have to go through the entire plate to get to the surface so that the resistance encountered would be $R_i = L/kA$. We recognize that for the one-dimensional case shown in Figure 3.1, the area A for conduction and the area A_s for convection are the same. The ratio R_i/R_o can then be expressed as follows:

$$\frac{R_i}{R_o} = \frac{L/kA}{1/hA_s} = \frac{hL}{k_{\text{solid}}} = N_{\text{Bi}} \qquad (3.1)$$

The subscript *solid* has been added to the k to emphasize the fact that the ratio is a conglomerate; h due to the fluid and L and k for the solid. The ratio is dimensionless and is called the Biot number, N_{Bi}, an important parameter in transient heat transfer analysis.

As we mentioned above, when the length L in the Biot number is evaluated as the volume of the solid divided by the surface area for heat transfer, the numerical level of the Biot number has the special significance of being the ratio of internal to external resistance. Specifically, when $N_{Bi} \leqslant 0.1$ or when the external resistance R_o is 10 times or more larger than the internal resistance R_i, then the temperature inside the solid is essentially uniform; it varies less than 5 percent. There must always be a slight slope for the energy to move out of or into the solid, but for the purposes of analysis and prediction we can assume the temperature *inside* the solid is essentially the same at all positions and changes only with time. Such a description is usually designated as a lumped parameter analysis; we discuss it in some detail in Section 3.2. When the Biot number is greater than 0.1 the temperature inside the solid cannot be considered uniform and we must use the results of the analyses described in Sections 3.3, 3.4, and 3.5 to determine temperature behavior.

When the Biot number is greater than 10, or when the *internal* resistance is 10 times or more greater than the external resistance, the surface temperature changes extremely rapidly. In such a case we can consider the surface temperature to change suddenly from its initial value to the ambient temperature of the medium surrounding the solid. Such an assumption gives a particular class of solutions, the results of which we will discuss and use in Sections 3.3 and 3.4. Let us summarize our observations and look at some examples for a simple system that is easy to visualize.

With regard to Biot number range, we can make the following observations:

$$N_{Bi} \leqslant 0.1 \qquad (R_o \geqslant 10R_i)$$

1. Temperature essentially uniform inside the solid, changes only with respect to time.
2. Surface temperature changes at the same rate as temperatures in the solid.

$$N_{Bi} \simeq 1 \qquad (R_o \simeq R_i)$$

1. Temperature varies with distance as well as time inside the solid.
2. Surface temperature changes more rapidly than temperatures in the solid.

$$N_{Bi} \geqslant 10 \qquad (R_o \leqslant 0.1R_i)$$

1. Temperature varies strongly with distance inside the solid, time variation depends on location.
2. Surface temperature changes very rapidly from initial value to very close to ambient temperature.

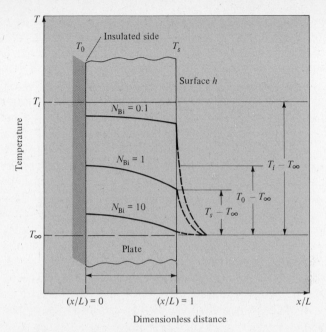

Figure 3.2 Influence of Biot number on the temperature response of a plate insulated on one side at a Fourier number of 1.0.

Table 3.1 TEMPERATURE RESPONSE FOR INFINITE PLATE OF FIGURE 3.2, T_o AT INSULATED FACE, T_s AT SURFACE FOR $N_{Fo} = 1$, ASSUMING A VALUE OF $T_i - T_\infty = 100$

$N_{Bi} = \dfrac{hL}{k}$	$\dfrac{T_o - T_\infty}{T_i - T_\infty}$	$\dfrac{T_s - T_\infty}{T_i - T_\infty}$	$T_o - T_\infty$	$T_s - T_\infty$
0.1	0.90	0.85	90	85
1	0.53	0.35	53	35
10	0.16	0.03	16	3

The effects of the above variation in Biot number on the slab of thickness L and insulated on the left face (same as the one in Figure 3.1) is shown in Figure 3.2 at a given time t such that the ratio* $\alpha t/L^2 = 1$. The figure shows the temperature profile for $N_{Bi} = 0.1$ to be nearly flat and those for $N_{Bi} = 1$ and 10 to vary with distance with the profile for $N_{Bi} = 10$, showing the largest change in surface temperature during the time t under consideration. The data for the construction of Figure 3.2 is given in Table 3.1 and was obtained from the charts discussed in Section 3.3.

It is difficult for the beginning student to visualize transient phenomena at the first encounter. As we progress into the topics of Sections 3.3, 3.4, and 3.5, it may be advantageous for the reader to peruse this section again.

*The ratio $\alpha t/L^2$ is the Fourier number and we will discuss its origin and significance in Section 3.2.

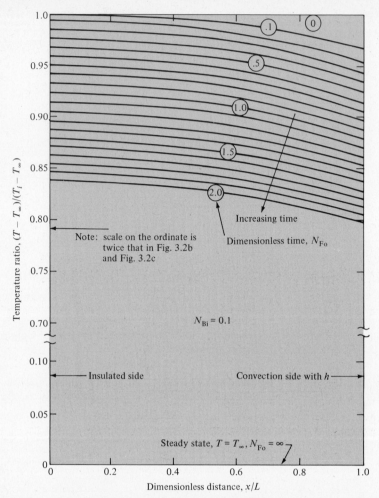

Figure 3.2(a)   Variation of dimensionless temperature ratio with distance for the semi-infinite plate, initially at uniform temperature T_i, insulated on the left face with convection on the right face such that $N_{Bi} = 0.1$, and with dimensionless time, N_{Fo}, as a parameter in steps of 0.1.

To consolidate the importance of the Biot number, the temperature response of the plate described in Figure 3.2 has been illustrated extensively in Figures 3.2(a), 3.2(b), and 3.2(c) for values of N_{Bi} of 0.1, 1, and 10, respectively; the three figures all have the Fourier number as a parameter. Each curve on Figures 3.2(a), 3.2(b), and 3.2(c) represents an equal increment in time, increasing in the downward direction on the figure; the bottom abscissa represents the steady-state profile in all three cases.

Figure 3.2(a) shows that the variation in temperature inside the plate is less than 5 percent at any given time for an $N_{Bi} = 0.1$ and that the surface temperature changes slowly with respect to time; whereas Figure 3.2(b) at an

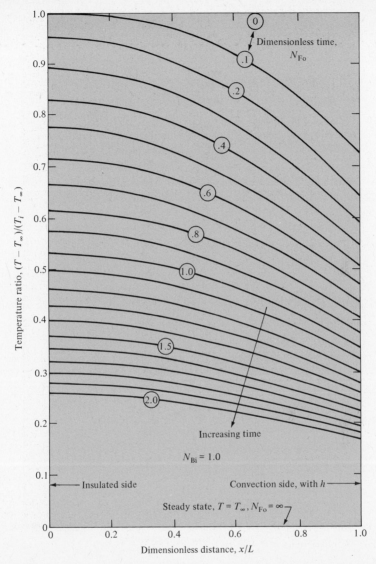

Figure 3.2(b)   Variation of dimensionless temperature ratio with distance for the semi-infinite plate, initially at uniform temperature T_i, insulated on the left with convection on the right such that $N_{Bi} = 1.0$, and with dimensionless time, N_{Fo}, as a parameter in steps of 0.1.

$N_{Bi} = 1$ shows about a 20 percent variation in temperature throughout the solid at early times and the surface temperature varies more rapidly than for the $N_{Bi} = 0.1$ case. Figure 3.2(c) illustrates the extreme variation of temperature inside the plate at early and moderate times for the case where the $N_{Bi} = 10$; the surface temperature change is the most rapid and approaches the ambient temperature very quickly. We should note that the analysis illustrated is for a

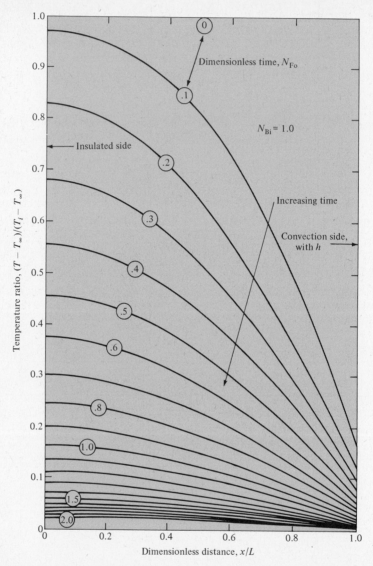

Figure 3.2(c) Variation of dimensionless temperature ratio with distance for the semi-infinite plate, initially at uniform temperature T_i, insulated on the left with convection on the right such that $N_{Bi} = 10.$, and with dimensionless time, N_{Fo}, as a parameter in steps of 0.1.

given slab of material. Therefore, changing the Biot number means that we are only changing the *external* resistance; the internal resistance remains the same. Consequently, the heat transfer coefficient for $N_{Bi} = 10$ is 100 times as large as it is when the $N_{Bi} = 0.1$. The internal variation in temperature for $N_{Bi} = 0.1$ is much less because the energy is being removed from the surface at a much slower rate than it is for N_{Bi} of 1 or 10.

The program used to calculate the numerical results for Figures 3.2(a), 3.2(b), and 3.2(c) is given in Appendix D. The reader can probably get the most benefit from the three figures by reviewing the descriptions of the Biot numbers' influence, and then visualizing those happenings by scanning the corresponding figure (i.e., Figure 3.2(b) for $N_{Bi} = 1$) from top to bottom to get the sensation of time dependency.

The sharp change in surface temperature at $N_{Bi} = 10$ with respect to time is further illustrated and compared to $N_{Bi} = 1$ and 0.1 in Figure 3.3. The figure shows only the surface temperature variation as a function of the N_{Fo} number (time) from the data that produced Figures 3.2(a), 3.2(b), and 3.2(c). Figure 3.3 shows how rapidly the surface temperature approaches the ambient temperature

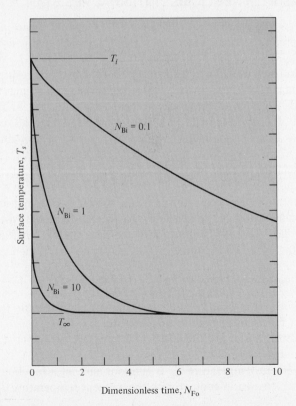

Figure 3.3 Variation of surface temperature with Fourier number (time) at Biot numbers of 0.1, 1, and 10 for the slab insulated on one face.

T_∞ when the Biot number $N_{Bi} = 10$. Whenever a solid is exposed to a different (but constant) ambient temperature, most of the action takes place in the real time required to give a $N_{Bi}N_{Fo}$ product value of about 3 to 5. Again, we will see more of the physical significance of this product in the next section.

We can conclude this section by noting that whenever we want to analyze a transient situation, we immediately calculate the Biot number with the characteristic length as the volume divided by the area for heat transfer. If that value of Biot number is 0.1 or less, we can use the simplified approach described in Section 3.2; if that Biot number is significantly greater than 0.1, we must use the results of analyses described in Sections 3.3 and 3.4 or go to a numerical or graphical analysis. Further insight into more complicated transient situations, like nonconstant initial conditions, for example, can be obtained from a study of the normalization of the governing differential equations. A great deal of insight can be obtained from such a process; Myers (1) gives a detailed discussion with examples on the application of the normalization process.

3.2 SYSTEMS WITH NEGLIGIBLE INTERNAL RESISTANCE ($N_{Bi} < 0.1$)

When we say negligible internal resistance we, of course, mean in comparison with the external resistance due to the heat transfer coefficient h. Such a situation prevails when the Biot number N_{Bi} is less than 0.1. We may also observe that for such a situation we can say that the external resistance governs the behavior of the physical system we are considering and since there is very little (negligible) change in temperature inside the solid, the solid behaves as though it were all lumped together in a blob. Some references call such a method of analysis the lumped parameter method as a result of the aforementioned observation.

We will analyze the arbitrary solid shown in Figure 3.4, and since the Biot number is < 0.1, the temperature in the solid is essentially the same at all locations at any time. The geometry or shape, therefore, does not enter into our analysis. The only aspect of the geometry that we must carefully ascertain is the amount of surface A_s that is subjected to the heat transfer coefficient h. It is possible that the entire surface area of the solid is involved; then A_s represents the entire outer surface. The physical situation shown in Figure 3.4 is subject to the following conditions:

1. The solid has mass m, density ρ, volume V, and specific heat c_p which are not functions of time or temperature.
2. The heat transfer coefficient h is a constant; it is not a function of temperature or time.
3. The ambient temperature T_∞ is constant and not a function of time.
4. The surface area in contact with the ambient temperature T_∞ is A_s.

We can apply the first law of thermodynamics to the mass m and we take the system boundary as the confining surface of the mass as shown in Figure 3.4.

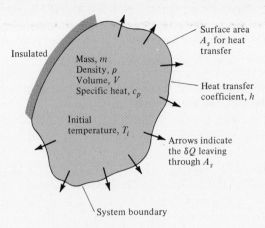

Ambient pressure P, constant

Insulated

Mass, m
Density, ρ
Volume, V
Specific heat, c_p

Initial
temperature, T_i

Surface area
A_s for heat
transfer

Heat transfer
coefficient, h

Arrows indicate
the δQ leaving
through A_s

System boundary

Ambient temperature, $T_\infty < T_i$

Figure 3.4   First-law application to an arbitrary solid to formulate temperature response.

The expression for the first law is given by

$$\delta Q = dU + \delta W \tag{3.2}$$

which states that the heat δQ added to a system must equal the change in internal energy dU plus the work δW done by the system. For the situation we are considering, the only work being done is expansion work $P\,dV$ as the solid either expands or contracts due to temperature changes. We should also recall that δQ and δW are inexact differentials or path functions, whereas dU and dV are exact differentials or state point functions. We can combine $dU + P\,dV$ by noting that the ambient pressure P is a constant so that we can write

$$dU + P\,dV = d(U + PV)$$
$$= dH \tag{3.3}$$

due to the linearity property of the differential operator such that $d(A + B) = dA + dB$. We recognize the property enthalpy $H = U + PV$ and since we are dealing with a constant-pressure process, we can relate enthalpy to temperature through the definition of constant-pressure specific heat c_p as follows:

$$dH = mc_p\,dT \tag{3.4}$$

The reader who has just completed a thermodynamics course will recognize that we are avoiding use of the symbol for specific enthalpy $h = H/m$ so that we have no conflict with the symbol h for the heat transfer coefficient.

The heat transferred δQ in the time interval dt can be expressed by Newton's equation (Eq. 1.1) as

$$\delta Q = -hA_s(T - T_\infty)\,dt \tag{3.5}$$

where the negative sign is required by the sign convention on δQ from thermodynamics. Now, when we put Eqs. 3.5, 3.4, and 3.3 into the first law (Eq. 3.2), we get an expression that will let us determine temperature as a function of time. Thus

$$-hA_s(T - T_\infty)\,dt = mc_p\,dT \tag{3.6}$$

It is helpful to reemphasize at this point that T is temperature and t is time. Separating the variables in Eq. 3.6 gives

$$\frac{dT}{T - T_\infty} = -\frac{hA_s}{\rho V c_p}\,dt \tag{3.7}$$

Since T_∞ is a constant, we can write $dT = d(T - T_\infty)$ and proceed with the integration of Eq. 3.7 for the following boundary conditions.

time zero: $\quad t = 0 \quad\quad T - T_\infty = T_i - T_\infty$

time t: $\quad\quad t = t \quad\quad T - T_\infty = T - T_\infty$

The above conditions may seem to be written peculiarly, unless we note that the temperature variable is $T - T_\infty$; of course, we recognize that at time zero the temperature is T_i and at time t we have temperature T.

According to physical conditions (1) through (4) above, we can write

$$\int_{T_i-T_\infty}^{T-T_\infty} \frac{d(T - T_\infty)}{T - T_\infty} = -\frac{hA_s}{\rho c_p V}\int_0^t dt \tag{3.8}$$

Since $\int dz/z = \ln z$, Eq. 3.8 becomes

$$\ln\frac{T - T_\infty}{T_i - T_\infty} = \frac{-hA_s t}{\rho c_p V} \tag{3.9}$$

or

$$\frac{T - T_\infty}{T_i - T_\infty} = \exp\left(\frac{-hA_s t}{\rho c_p V}\right) \tag{3.10}$$

The exponent in Eq. 3.10 has a particularly important significance. The ratio V/A_s has the dimensions of a length and gives the characteristic length L that we discussed earlier in Section 3.1. We are not going to give the characteristic length L a subscript so it will be necessary to keep the concept in mind since we will use L to designate other lengths. Repeating for emphasis, we use $L = V/A_s$ in the Biot number to determine the ratio of internal to external resistance R_i/R_o and, of course, we are discussing the domain $N_{Bi} < 0.1$ in this section.

If we multiply the exponent in Eq. 3.10 by L/L and k/k, we can regroup the terms as follows:

$$\frac{htLk}{\rho c_p L^2 k} = \frac{hL}{k}\,\frac{kt}{\rho c_p L^2}$$

$$= \frac{hL}{k}\,\frac{\alpha t}{L^2}$$

$$= N_{Bi} N_{Fo} \tag{3.11}$$

where we recognize the Biot number group $N_{Bi} = hL/k$ and the Fourier number group $N_{Fo} = \alpha t/L^2$. The Fourier number represents the dimensionless time for the physical situation since the thermal diffusivity $\alpha = k/\rho c_p$ and L change very little during the time interval of interest. We could then express Eq. 3.10 as

$$\frac{T - T_\infty}{T_i - T_\infty} = e^{-N_{Bi} N_{Fo}} \tag{3.12}$$

Figure 3.5 shows the temperature as a function of time for Eqs. 3.10 and 3.12. The upper part of the figure shows the physical meaning of the temperature excesses $T - T_\infty$ and $T_i - T_\infty$ and hence the interpretation of $(T - T_\infty)/$

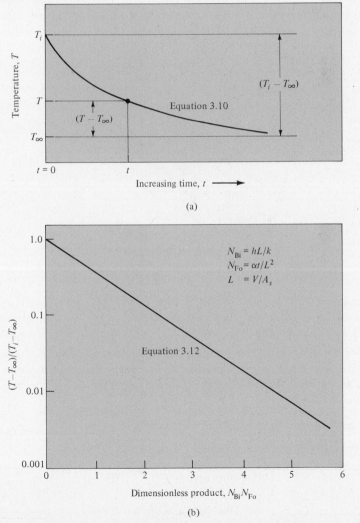

Figure 3.5   Variation of temperature and time for the lumped parameter case:
(a) dimensional representation, and (b) nondimensional representation.

$(T_i - T_\infty)$, while the lower part of the figure shows that Eq. 3.12 is a straight line on semilog paper. Figure 3.5(b) further shows that the system is within about 5 percent of equilibrium with ambient conditions at a value of $N_{Bi}N_{Fo} = 3$ which reinforces our observation in Section 3.1 that most of the thermal action takes place in the range of $N_{Bi}N_{Fo}$ product from about zero to 3 or 5.

Example 3.1

A bronze (75 percent Cu, 25 percent tin) bearing blank consists of half of a right circular cylinder as shown in the figure. The blank is removed from a heat treating furnace at 900F and is cooled in 75F air to 400F (in a vertical position) for a forging operation prior to finish machining. How long does it take to cool the sleeve if the heat transfer coefficient in the vertical position is 12.5 Btu/hrft^{2}F?

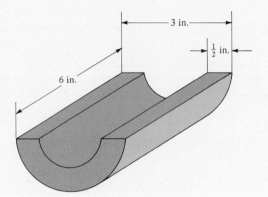

Figure for Example 3.1.

SOLUTION
In the vertical position the surface area for heat transfer comprises the inner- and outer-half cylindrical surfaces, the edge areas, plus the top half-washerlike surface. We assume no heat transfer to the air from the bottom surface. As in most transient situations, it is prudent to first calculate the characteristic length and then to check the Biot number in order to get an idea of the nature of the transient response.

$$\text{area (inside cylindrical surface)} = 0.5(2\pi6) = 6\pi \text{ in.}^2$$

$$\text{area (outside cylindrical surface)} = 0.5(3\pi6) = 9\pi \text{ in.}^2$$

$$\text{half-washer area} = 0.5\left(\frac{\pi}{4}\right)(9 - 4) = \frac{5}{8}\pi \text{ in.}^2$$

$$\text{edge area} = 2(0.5)6 = 6 \text{ in.}^2$$

$$\text{total area} = 21.6\pi \text{ in.}^2$$

$$\text{volume} = \text{base area} \times \text{height}$$

$$= \frac{5}{8}\pi 6 = 3.75\pi \text{ in.}^3$$

$$\text{characteristic length} = L = \frac{V}{A_s} = \frac{3.75\pi}{21.6\pi} = 0.174 \text{ in.}$$

To calculate the Biot number, we need the thermal conductivity representative of the temperature range under consideration. We estimate $k = 18$ Btu/hrftF from Appendix C since the other copper alloys have k increasing with temperature.

$$N_{\text{Bi}} = \frac{hL}{k} = \frac{12.5(0.174)}{12(18)}$$

$$= 0.010$$

The Biot number is less than 0.1; therefore, we know that the temperature changes in the solid we are considering will be very small and we can assume the temperature to be uniform (negligible internal resistance) in the solid at any instance of time. Accordingly, we recognize that we can obtain the time from the Fourier number in Eq. 3.12 since we know the temperature ratio. We estimate the thermal diffusivity associated with a k of 18 Btu/hrftF as

$$\alpha = \frac{18}{15}0.333 = 0.40 \text{ ft}^2/\text{hr}$$

and

$$T - T_\infty = 400 - 75 = 325$$
$$T_i - T_\infty = 900 - 75 = 825$$

Therefore, $(T - T_\infty)/(T_i - T_\infty) = 325/825 = 0.394$.
    From Eq. 3.12:

$$\ln\frac{T - T_\infty}{T_i - T_\infty} = -N_{\text{Bi}}N_{\text{Fo}}$$

$$N_{\text{Fo}} = \frac{-\ln 0.394}{0.010} = 92.7$$

$$t = \frac{L^2 N_{\text{Fo}}}{\alpha}$$

$$= \left(\frac{0.174}{12}\right)^2\frac{92.7}{0.40} = 0.0485 \text{ hr}$$

$$t = 2.9 \text{ min}$$

When we look at convection aspects in Chapter 6, we will be able to see what kind of conditions will give us the specified heat transfer coefficient.

3.2.1 Time Constant as an Indicator of Response

It is convenient to have a measure or an indicator of the speed with which a system will respond thermally to a change in ambient conditions. Such an indicator is particularly useful in assessing temperature measuring transducers or temperature measuring instruments. The time constant is always expressed in seconds (sometimes in milliseconds) rather than minutes or hours because the numbers are of a convenient magnitude since speed is the desired result.

We define the time constant from Eq. 3.10 as that time required to make the exponent on the base of natural logarithms equal to unity. Thus

$$\frac{hA_s t_{tc}}{\rho c_p V} = 1.0 \tag{3.13}$$

or

$$t_{tc} = \frac{\rho c_p V}{hA_s} \tag{3.14}$$

Making the mass or volume as small as possible and providing as high an h as possible minimizes the time constant. Choosing the characteristic length $L = V/A_s$ small also minimizes the time constant. During one time constant interval the temperature ratio would change to $1/e$ of its initial value; or as in the case of a measuring instrument, the sensing element would register 63.2 percent of the difference between the environment and its initial temperature. Figure 3.6 shows the response of a lumped system that is undergoing either a decay process [Figure 3.6(a)] like the temperature response for an object removed from a furnace as was discussed in Example 3.1; or a growth process such as the response of a thermocouple or thermometer after insertion into a hot fluid stream. In either case we can see that the process of achieving equilibrium is 63.2 percent completed in one time constant; 86.5 percent in two time constants; 95 percent in three time constants; and so forth. One can generally assume that equilibrium is essentially reached in five time constants (99.3 percent).

A note of caution is in order here. Reaching equilibrium in a measurement situation does not guarantee that the indicated quantity is the desired objective, since conduction or radiation aspects may produce an equilibrium measurement that is considerably in error from the actual temperature of the quantity being measured. We will cover this aspect in the chapters on radiation and convection.

We conclude our discussion of the time constant idea by noting the relationship between the growth and decay phenomena shown in Figure 3.6. The ordinate in Figure 3.6 involves the temperature T at the time ratio t/t_c, the initial temperature T_i at time zero, and the ambient temperature T_∞ (a constant) as we have learned in previous discussion. In order to have the nondimensional growth curve start at zero and the nondimensional decay curve start at one, we must use different temperature ratios to describe the two phenomena. The ratios

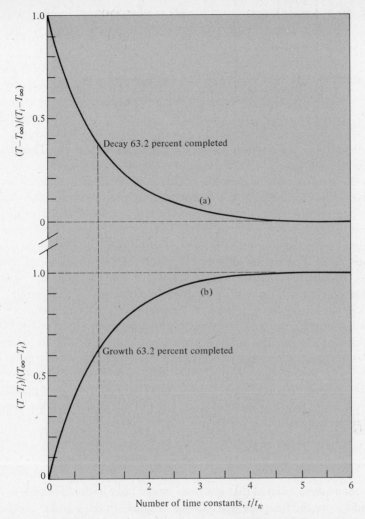

Figure 3.6   Exponential temperature decay (a) and exponential temperature growth (b) as a function of the number of time constant periods t/t_{tc}.

involve the same quantities and the relationship between them is given by

$$\frac{T - T_i}{T_\infty - T_i} = 1 - \frac{T - T_\infty}{T_i - T_\infty} \tag{3.15}$$

From the definition of the time constant Eq. 3.14 and Eq. 3.12, we can write

$$\frac{T - T_i}{T_\infty - T_i} = 1 - e^{-t/t_{tc}}$$

$$= 1 - e^{-N_{Bi}N_{Fo}} \tag{3.16}$$

to describe the temperature ratio for the growth process, as compared to Eq. 3.12 for the decay process.

A very extensive discussion of the application of the time constant concept to step-excited first-order mechanical systems is presented by Beckwith and Buck (2) in addition to measurement systems.

3.3 FINITE INTERNAL RESISTANCE, ONE-DIMENSIONAL SYSTEMS

When the internal resistance is *not negligible* compared to the external resistance for heat transfer ($N_{Bi} > 0.1$), we cannot make a simple application of the first law of thermodynamics to describe energy considerations for the entire object under consideration because the temperature is not uniform inside the object. Rather, we are forced to apply conservation principles to a very small region (differential in size) that is essentially uniform in temperature to obtain a partial differential equation to describe the temperature and heat flux happenings. The solution of the conduction differential equations (see Appendixes E and F) requires sophisticated mathematical treatment that is generally covered in graduate courses dealing with the topic (Reference 3). The results of such analytical work have been presented in graphical form (see References 4, 5, 6) and can be utilized without recourse to the equations as long as the boundary conditions for the solution are well understood and match the boundary conditions for the application (geometry) being considered. This latter approach is the one we will use in this chapter. We will work from graphs drawn from the transient solutions that have been worked out and given in Appendix F for the plate, cylinder, and sphere; and we will use combinations thereof for different geometries by the product solutions method.

3.3.1 The Infinite Plate

The mathematical solution that has been worked out in the literature is given in Appendix F for the flat plate that extends infinitely in the y and z directions (see insert sketch in Figure 3.7), and is $2L$ units thick in the x direction. The plate has convective heat transfer of identical magnitude on both surfaces (at $x = L$ and at $x = -L$) and the temperature inside the plate is uniform at all locations at T_i before the plate is brought into convective contact with the surroundings at time zero whose temperature is T_∞, a constant value. We can also express these ideas in mathematical language as follows:

a. Initial condition:

$$t \le 0 \qquad T = T_i \qquad -L \le x \le L \tag{3.17}$$

b. Boundary conditions:

$$t > 0^+ \qquad x = -L \qquad x = L$$
$$q_{convection} = hA(T_L - T_\infty) \tag{3.18}$$

We note there are two boundaries (at $x = L$ and $x = -L$), but the solution is for identical h on each face. The reader should observe that such conditions result in symmetry about the centerline of the plate with the temperature profile having a horizontal slope at the centerline (i.e., $dT/dx = 0$). Therefore, since $dT/dx = 0$, the heat flux is zero at $x = 0$ and we would have the identical result temperaturewise for a plate that is only L units thick but *insulated* on one face with convection heat transfer (h) on the other face.

The important parameters for the flat plate geometry are the Fourier number N_{Fo} (dimensionless time) and the Biot number N_{Bi} (ratio of internal to external resistance). The Fourier number, $\alpha t/L^2$, appears when the governing differential equation is normalized (Reference 7) and the Biot number hL/k appears when Eq. 3.18 is nondimensionalized. We have enough background to do the latter here so that we can dispel at least half of the mystery. We recognize that at the surface the conducted flux just inside the solid must be equal to the flux convected away from the surface so that we can write Eq. 3.18 in the following form since $q_k = q_c$ at the surface.

$$-kA\frac{dT}{dx} = hA(T_L - T_\infty) \tag{3.19}$$

To proceed with nondimensionalization, we recognize that since T_∞ is a constant then $dT = d(T - T_\infty)$, and for distance the variable is x/L so that $dx = Ld(x/L)$ where L also is a constant. Since the area A for conduction is the same area for convection, Eq. 3.19 becomes

$$\frac{d(T - T_\infty)}{d(x/L)} = -\frac{hL}{k}(T_L - T_\infty) \tag{3.20}$$

and we recognize the parameter hL/k as the Biot number. Equation 3.20 would be truly dimensionless were we also to divide both sides by the constant initial temperature excess $T_i - T_\infty$, but that in no way affects the Biot number. The above remarks could just as well apply to the cylinder or the sphere wherein the nondimensionalization process would produce the Biot number hR_o/k. Solution of the partial differential equation applicable to the plate gives the nondimensional temperature ratio $(T - T_\infty)/(T_i - T_\infty)$ as a function of x/L, N_{Fo}, N_{Bi}. The results cover a very large range and are most conveniently represented as the variation of temperature along the centerline $(T_o - T_\infty)/(T_i - T_\infty)$ as shown in Figure 3.7, and a position correction chart, Figure 3.8, which gives the ratio of the temperature at the desired location to the temperature at the center as a function of the Biot number. The desired temperature excess ratio is then a product of the quantities read from the two curves. Thus

$$\frac{T_{x/L} - T_\infty}{(T_i - T_\infty)_{plate}} = \left(\frac{T_o - T_\infty}{T_i - T_\infty}\right)_{Fig.\,3.7} \left(\frac{T_{x/L} - T_\infty}{T_o - T_\infty}\right)_{Fig.\,3.8} \tag{3.21}$$

The information contained in Figures 3.7 and 3.8 was first published by Heisler (4) in 1947 in a different format. Heisler presented the excess ratio as a semilog plot with the temperature ratio on the log scale and Fourier number on a linear

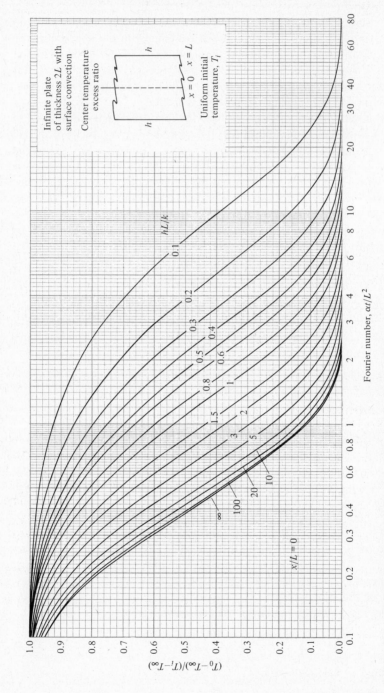

Figure 3.7 Center temperature excess ratio for the infinite flat plate.

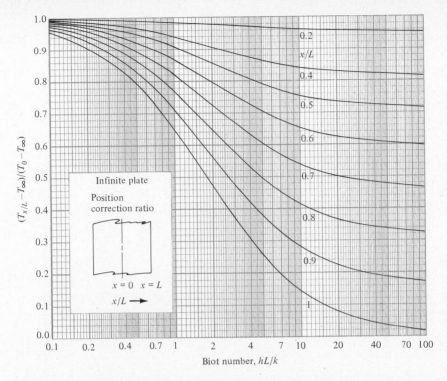

Figure 3.8 Position correction ratio for the infinite flat plate for use with Figure 3.7.

scale because there were no large-scale electronic calculating machines available at that time, and because at Fourier numbers larger than 0.5 only one term in the series solution was needed for three-figure accuracy. Hence the temperature relationship could be presented as a straight line on Heisler's charts. Unfortunately, the charts are very inconvenient to read in the range of Fourier number below 1 where a lot of the action takes place. It is also confusing to the beginner to use the reciprocal of the Biot number as a parameter, as was done in the Heisler charts. Schneider has recognized the above problems in his very extensive collection of temperature response charts (Reference 6), but he presents an excess ratio based on the initial temperature rather than on the ambient temperature. It is my personal conviction that it is physically more satisfying to think of the solid as being filled with an amount of energy (temperature being roughly proportional to energy level) and then emptying into the environment, that is, the excess ratio goes from 1 at time zero to 0 a long time after contact with the environment; however, one could argue either way.

3.3.2 The Infinite Cylinder

Information similar to that for the plate is presented for the cylinder of infinite length or finite length with ends insulated in Figures 3.9 and 3.10. As with the

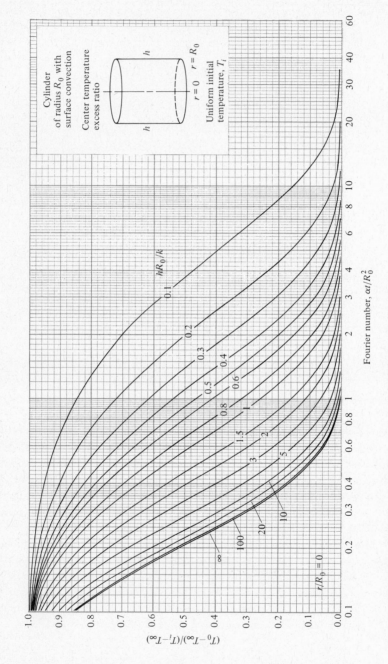

Figure 3.9 Center temperature excess ratio for the infinite cylinder.

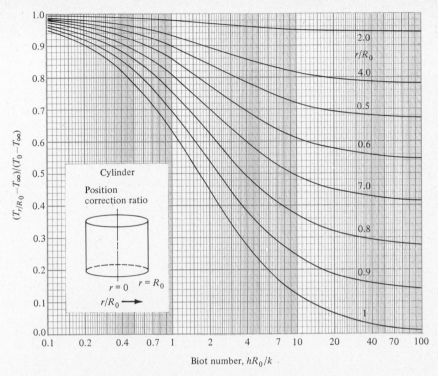

Figure 3.10   Position correction ratio for the infinite cylinder for use with Figure 3.9.

plate, the cylinder is assumed to be at a constant uniform initial temperature T_i at time zero and is then exposed to an environment at fixed temperature T_∞ under such conditions that the heat transfer coefficient h on the cylindrical surface is constant. The temperature excess ratio along the axis of the cylinder then depends on the Biot number hR_o/k and the Fourier number $\alpha t/R_o^2$. The reader should note the length criterion in N_{Bi} is $L = R_o \neq V/A_s$. To obtain the temperature excess at any radial location r/R_o, one reads the position correction ratio from Figure 3.10 and multiplies that fraction times the center excess temperature obtained from Figure 3.9. Thus

$$\frac{T_{r/R_o} - T_\infty}{(T_i - T_\infty)_{\text{cylinder}}} = \left(\frac{T_o - T_\infty}{T_i - T_\infty} \right)_{\text{Fig. 3.9}} \left(\frac{T_{r/R_o} - T_\infty}{T_o - T_\infty} \right)_{\text{Fig. 3.10}} \qquad (3.22)$$

3.3.3 The Sphere

Figure 3.11 presents the center temperature excess ratio for the sphere with uniform initial temperature T_i at time zero which is exposed to the ambient temperature T_∞ with constant h at the surface. As with the plate and cylinder, the solution shown in Figure 3.11 is for the case where h and T_∞ are constants.

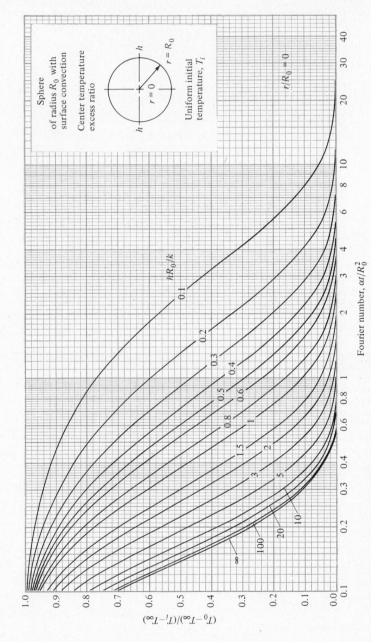

Figure 3.11 Center temperature excess ratio for the sphere.

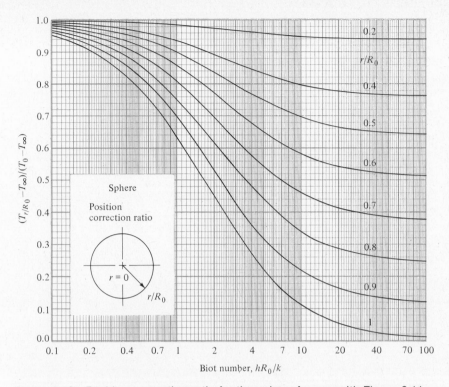

Figure 3.12 Position correction ratio for the sphere for use with Figure 3.11.

The center temperature then is a function of the Biot number hR_o/k and the Fourier number $\alpha t/R_o^2$.

The temperature at any radial position r/R_o for the sphere can be obtained from Figure 3.12 as a product of the position correction ratio and the center excess ratio. Thus we obtain

$$\frac{T_{r/R_o} - T_\infty}{(T_i - T_\infty)_{\text{sphere}}} = \left(\frac{T_o - T_\infty}{T_i - T_\infty}\right)_{\text{Fig. 3.11}} \left(\frac{T_{r/R_o} - T_\infty}{T_o - T_\infty}\right)_{\text{Fig. 3.12}} \tag{3.23}$$

the same way that we have done for the plate and the cylinder.

We are also interested in the amount of energy which enters or leaves a solid during the time of the transient. Gröber (8) has calculated the energy change ratio Q/Q_i for the plate, cylinder, and sphere and that information is presented in Figure 3.13. The quantity $Q_i = \rho c_p(\text{volume})(T_i - T_\infty)$ represents the energy level of the solid above (or below) the ambient temperature datum, and the quantity Q represents the energy that has left (or entered) the solid in the time interval of the transient from 0 to t. The quantity Q/Q_i is dimensionless and always less than 1.

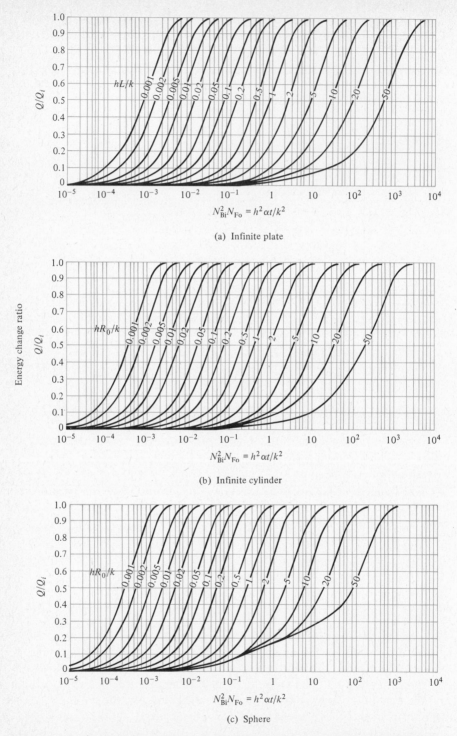

Figure 3.13 Energy change ratio, Q/Q_i, for (a) the infinite plate, (b) the infinite cylinder, and (c) the sphere with uniform initial temperature and convection at the surface according to Gröber (8). Reproduced with permission of McGraw-Hill Book Co.

3.3.4 Chart Utilization

It is instructive to work through an example by using the charts we have just discussed. In doing so we can also get some insight on the influence geometry plays in thermal response under similar external conditions.

Example 3.2

Let's look at a plate, cylinder, and sphere of such a material that $k = 10$ Btu/hrftF with an $h = 10$ Btu/hrft^{2}F on the outer surface, and let the plate be 2 ft thick ($L = 1$ ft), the cylinder 2 ft in diameter ($R_o = 1$ ft), and the sphere also 2 ft in diameter ($R_o = 1$ ft). Most materials have a ρc_p product in the range of 20 to 50 so that for a k of 10 Btu/hrftF the value would be near that for stainless steel, about 50 Btu/ft^{3}F; therefore, the thermal diffusivity $\alpha = k/\rho c_p = 10/50 = 0.2$ ft^2/hr. Summarizing:

$$h = 10 \text{ Btu/hrft}^2\text{F} \quad (56.7 \text{ W/m}^2\text{k})$$
$$k = 10 \text{ Btu/hrftF} \quad (17.3 \text{ W/mK})$$
$$\alpha = 0.2 \text{ ft}^2/\text{hr} \quad (0.0186 \text{ m}^2/\text{hr})$$

We want to determine the time required to reach a temperature excess ratio of 0.4 at a position halfway to the surface ($x/L = 0.5$ and $r/R_o = 0.5$) for the case where the three geometries are initially at a uniform temperature T_i, and where at time zero, they are put into contact with the environment at ambient temperature T_∞ with the h given above.

SOLUTION
In order to get the desired time, we need to determine the Fourier number from the appropriate chart at the stated temperature ratio and the existing Biot number.

The Biot numbers are calculated as follows,

$$\text{plate:} \quad N_{\text{Bi}} = \frac{hL}{k} = \frac{10(1)}{10} = 1$$

$$\text{cylinder and sphere:} \quad N_{\text{Bi}} = \frac{hR_o}{k} = \frac{10(1)}{10} = 1$$

and are dimensionless. In order to arrive at the Fourier number we must determine the center temperature excess from Eqs. 3.21, 3.22, and 3.23. We know that $(T_{0.5} - T_\infty)/(T_i - T_\infty) = 0.4$ so that we can write

$$\frac{T_o - T_\infty}{T_i - T_\infty} = \frac{0.4}{\text{position correction ratio}}$$

For example, for the plate (Eq. 3.21), we would read the position correction ratio from Figure 3.8 (at $N_{\text{Bi}} = 1$) to be 0.910 halfway to the surface ($x/L = 0.5$), so

Table 3.2 RESULTS FOR PLATE, CYLINDER, AND SPHERE OF EXAMPLE 3.2

GEOMETRY	$\dfrac{T_{0.5} - T_\infty}{T_i - T_\infty}$	$\dfrac{T_{0.5} - T_\infty}{T_o - T_\infty}$	$\dfrac{T_o - T_\infty}{T_i - T_\infty}$	N_{Fo} ($N_{Bi} = 1$)	t (hr)	$\dfrac{Q}{Q_i}$	$\dfrac{\text{volume}}{A_s}$
Plate	0.4 (given)	0.910 Fig. 3.8	0.440	1.25 Fig. 3.7	6.25	0.60 Fig. 3.13(a)	1
Cylinder	0.4 (given)	0.902 Fig. 3.10	0.443	0.638 Fig. 3.9	3.19	0.62 Fig. 3.13(b)	0.50
Sphere	0.4 (given)	0.899 Fig. 3.12	0.445	0.425 Fig. 3.11	2.12	0.67 Fig. 3.13(c)	0.333

$$N_{Bi} = 1 \qquad\qquad\qquad \alpha = 0.2 \text{ ft}^2/\text{hr}$$

then the center temperature ratio can be calculated as

$$\frac{T_o - T_\infty}{(T_i - T_\infty)_{\text{plate}}} = \frac{0.4}{0.910} = 0.440$$

Then from Figure 3.7 at $N_{Bi} = 1$ we can read off the associated Fourier number as $N_{Fo} = 1.25$. The time is calculated from the Fourier number $N_{Fo} = \alpha t/L^2$ as $t = 1.25(1)^2/0.2 = 6.25$ hr.

The parameter $N_{Bi}^2 N_{Fo} = h^2 \alpha t/k^2 = (1)^2(1.25) = 1.25$ so that Figure 3.13 gives the ratio $Q/Q_i = 0.60$ which means that 60 percent of the energy change required to bring the plate into equilibrium with the surroundings has been accomplished.

The procedure to obtain the desired time for the cylinder and sphere follows in similar manner; the results are tabulated in Table 3.2. The table shows that the sphere reaches the desired temperature level almost three times as fast as the plate does because it is much easier for the energy to get into or out of a sphere since the surface area (for heat transfer) to volume ratio is three times as great for the sphere as compared to the plate.

Example 3.2 was instructive in the understanding of time aspects of a transient happening, but it would also be helpful to look at a situation that involves actual temperature levels to see how the excess ratios are manipulated.

Example 3.3

A spherical bearing 1.5 in. in diameter is at a uniform initial temperature of 600F in an annealing or soaking furnace. The bearing is taken out of the furnace and allowed to cool in an air stream under conditions such that the heat transfer coefficient has a value of 140 Btu/hrft^{2}F and the air temperature is 85F; the bearing is made of a 5 percent chrome steel. How long will it be before the bearing can be safely handled without gloves if 120F is the limiting surface temperature from a physiological standpoint?

SOLUTION

We can obtain property values from Appendix C so that $k = 22$ Btu/hrftF for the temperature range we are concerned with. Accordingly we must adjust the thermal diffusivity $\alpha = k/\rho c_p = (22/23)0.430 = 0.41$ ft^2/hr for the change in conductivity. We assume that changes in density and specific heat are small enough to be neglected. We can now calculate the Biot number, get the position correction ratio, and the center temperature excess, and the corresponding Fourier number will then determine the time interval that we seek. Thus

$$N_{\text{Bi}} = \frac{140(0.75)}{12(22)} = 0.398$$

From Figure 3.12 we read the position correction ratio to be 0.824 at $r/R_o = 1$ (the surface) and a Biot number of 0.398. At the surface the excess temperature is $T_1 - T_\infty = 35$F so that we calculate the center temperature excess as $T_o - T_\infty = 35/0.824 = 42.5$ which when divided by the initial temperature excess $T_i - T_\infty = 600 - 85 = 515$F gives the center excess ratio

$$\frac{T_o - T_\infty}{T_i - T_\infty} = \frac{42.5}{515} = 0.0825$$

At a Biot number of 0.398 and the above center excess ratio, Figure 3.11 yields a Fourier number of 2.37 for those conditions. The time in hours then becomes $t = R_o^2 N_{\text{Fo}}/\alpha = (0.75/12)^2(2.37/0.41) = 0.023$ hr, or about 80 s, a minute and one-third.

3.4 SEMI-INFINITE GEOMETRY

The description of a solid as semi-infinite means that a surface exists and the directions parallel to and perpendicular to the surface extend indefinitely. There are no such actual geometries but the earth, for example, is a good approximation of such a mathematical model and also relatively thick plates at very short time intervals respond thermally as though they were infinitely thick. As long as the temperature wave has not penetrated a finite plate to the other boundary, the temperature response is essentially that predicted for the semi-infinite geometry at distances less than the plate thickness and times less than the time required for the temperature wave to penetrate the semi-infinite solid to a distance equal to the plate thickness. The temperature wave is an imaginary concept that can be visualized as the effect an observer would feel were he or she standing at a given location in the solid. As the temperature changes at the surface, the change makes its way into the solid with a speed proportional to the thermal diffusivity $k/\rho c_p$. The observer senses the initial temperature for a time but when a just perceptible increase or decrease in temperature (the change) has been felt, we say the temperature wave has arrived at the observer's location. Figure 3.14 shows a qualitative (not calculated) sketch of the temperature response we are talking about. The letters a, b, c, and so on, show the progression of the temperature wave into the solid.

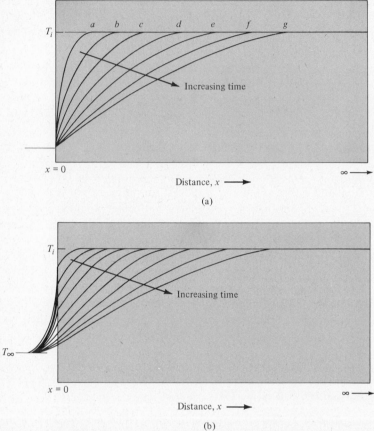

Figure 3.14   Transient response for the semi-infinite geometry. (a) Sudden jump in surface temperature, $h = \infty$. (b) Finite convection at the surface, $h < \infty$. Not to scale.

As was the case in Section 3.3, the mathematical description of the heat transfer phenomena gives a partial differential equation of the type shown in Appendix E which, together with appropriate boundary and initial conditions, gives a solution that we can work with. We will discuss only the uniform initial temperature situation and the two cases pictured in Figure 3.14: (a) the sudden jump in surface temperature [infinite (extremely large) convection coefficient], and (b) surface temperature, a function of time (finite convection coefficient at the surface).

3.4.1 Infinite Convection Coefficient at Surface

For the sudden change in surface temperature, the boundary and initial conditions are

1. At $x = 0$ (the surface), $T = T_\infty$ for all time greater than zero.
2. At $t \leqslant 0$ the temperature is T_i at all locations x into the solid.

Schneider (9) presents the solution to the conduction equation under these conditions as

$$\frac{T - T_\infty}{T_i - T_\infty} = \text{erf}\left(\frac{x}{2\sqrt{\alpha t}}\right) \tag{3.24}$$

where the function erf( ) is the Gauss error function defined as

$$\text{erf}(z) = \frac{2}{\sqrt{\pi}} \int_0^z e^{-\lambda^2} d\lambda \tag{3.25}$$

In the error function the variable of integration λ is really a dummy variable because the definite integral depends only on the value of the limits. Values for Eq. 3.25 are given in Table 3.3 from References 9 and 10.

Equation 3.24 can be differentiated with respect to x to form $(dT/dx)^*$ which when inserted into Fourier's law yields the heat flux at any given location x in the semi-infinite solid at time t so that

$$q_x = -kA\left(e^{-x^2/4\alpha t}\right)(T_i - T_\infty)/\sqrt{\pi \alpha t} \tag{3.26}$$

at the surface $x = 0$ so that the above equation reduces to

$$q_0 = -kA(T_i - T_\infty)/\sqrt{\pi \alpha t} \tag{3.27}$$

for the surface heat flux in Btu/hr or watts, depending on the choice of units. We may also want to know the total amount of energy Q_0 that has entered the surface of the semi-infinite solid in the time interval from 0 to t. The quantity Q_0 can be obtained by integrating q_0 over the required time interval as follows:

$$Q_0 = \int_0^t q_0 \, dt \tag{3.28}$$

The integration is easy to do, because when we look at q_0 closely, we find that the variable of integration t is a very simple function since in the integration we must consider all the other quantities π, α, T, A, and k to be constants. Thus

$$Q_0 = C_1 \int_0^t t^{-1/2} \, dt \tag{3.29}$$

---

*To form dT/dx requires operation on the definition of the error function (Eq. 3.25) according to theorem of Taylor (11): "The derivative of a definite integral with respect to the upper limit of integration is equal to the value of the integrand function at this upper limit." Accordingly,

$$d[\text{erf}(z)] = d\left(\frac{2}{\sqrt{\pi}} \int_0^z e^{-\lambda^2} d\lambda\right) = \frac{2}{\sqrt{\pi}} e^{-z^2} dz$$

But we have $z = ax$ where $a = 1/(2\sqrt{\alpha t})$ and we note that T is really a function of x and t so that what we are obtaining here is really the partial derivative of T with respect to x at some time t that we consider constant. So that

$$\frac{d[\text{erf}(z)]}{dx} = \frac{d[\text{erf}(z)]}{dz}\frac{dz}{dx} \quad \text{and} \quad \frac{dz}{dx} = \frac{1}{2\sqrt{\alpha t}}$$

therefore,

$$\frac{d[\text{erf}(z)]}{dx} = \frac{e^{-x^2/4\alpha t}}{\sqrt{\pi \alpha t}}$$

Table 3.3 VALUES OF THE ERROR FUNCTION ERF(z)

z	erf(z)	z	erf(z)	z	erf(z)	z	erf(z)
0.00	0.00000	0.35	0.37938	0.70	0.67780	1.05	0.86244
0.01	0.01128	0.36	0.38933	0.71	0.68467	1.06	0.86614
0.02	0.02256	0.37	0.39921	0.72	0.69143	1.07	0.86977
0.03	0.03384	0.38	0.40901	0.73	0.69810	1.08	0.87333
0.04	0.04511	0.39	0.41874	0.74	0.70468	1.09	0.87680
0.05	0.05637	0.40	0.42839	0.75	0.71116	1.10	0.88020
0.06	0.06762	0.41	0.43797	0.76	0.71754	1.11	0.88353
0.07	0.07886	0.42	0.44747	0.77	0.72382	1.12	0.88679
0.08	0.09008	0.43	0.45689	0.78	0.73001	1.13	0.88997
0.09	0.10128	0.44	0.46622	0.79	0.73610	1.14	0.89308
0.10	0.11246	0.45	0.47548	0.80	0.74210	1.15	0.89612
0.11	0.12362	0.46	0.48466	0.81	0.74800	1.16	0.89910
0.12	0.13476	0.47	0.49374	0.82	0.75381	1.17	0.90200
0.13	0.14587	0.48	0.50275	0.83	0.75952	1.18	0.90484
0.14	0.15695	0.49	0.51167	0.84	0.76514	1.19	0.90761
0.15	0.16800	0.50	0.52050	0.85	0.77067	1.20	0.91031
0.16	0.17901	0.51	0.52924	0.86	0.77610	1.21	0.91296
0.17	0.18999	0.52	0.53790	0.87	0.78144	1.22	0.91553
0.18	0.20094	0.53	0.54646	0.88	0.78669	1.23	0.91805
0.19	0.21184	0.54	0.55494	0.89	0.79184	1.24	0.92050
0.20	0.22270	0.55	0.56332	0.90	0.79691	1.25	0.92290
0.21	0.23352	0.56	0.57162	0.91	0.80188	1.26	0.92524
0.22	0.24430	0.57	0.57982	0.92	0.80677	1.27	0.92751
0.23	0.25502	0.58	0.58792	0.93	0.81156	1.28	0.92973
0.24	0.26570	0.59	0.59594	0.94	0.81627	1.29	0.93190
0.25	0.27633	0.60	0.60386	0.95	0.82089	1.30	0.93401
0.26	0.28690	0.61	0.61168	0.96	0.82542	1.31	0.93606
0.27	0.29742	0.62	0.61941	0.97	0.82987	1.32	0.93806
0.28	0.30788	0.63	0.62705	0.98	0.83423	1.33	0.94002
0.29	0.31828	0.64	0.63459	0.99	0.83851	1.34	0.94191
0.30	0.32863	0.65	0.64203	1.00	0.84270	1.35	0.94376
0.31	0.33891	0.66	0.64938	1.01	0.84681	1.36	0.94556
0.32	0.34913	0.67	0.65663	1.02	0.85084	1.37	0.94731
0.33	0.35928	0.68	0.66378	1.03	0.85478	1.38	0.94902
0.34	0.36936	0.69	0.67084	1.04	0.85865	1.39	0.95067

Table 3.3   (CONCLUDED)

z	erf(z)	z	erf(z)	z	erf(z)
1.40	0.95228	1.75	0.98667	2.20	0.998137
1.41	0.95385	1.76	0.98719	2.22	0.998308
1.42	0.95538	1.77	0.98769	2.24	0.998464
1.43	0.95686	1.78	0.98817	2.26	0.998607
1.44	0.95830	1.79	0.98864	2.28	0.998738
1.45	0.95970	1.80	0.98909	2.30	0.998857
1.46	0.96105	1.81	0.98952	2.32	0.998966
1.47	0.96237	1.82	0.98994	2.34	0.999065
1.48	0.96365	1.83	0.99035	2.36	0.999155
1.49	0.96490	1.84	0.99074	2.38	0.999237
1.50	0.96610	1.85	0.99111	2.40	0.999311
1.51	0.96728	1.86	0.99147	2.42	0.999379
1.52	0.96841	1.87	0.99182	2.44	0.999441
1.53	0.96952	1.88	0.99216	2.46	0.999497
1.54	0.97059	1.89	0.99248	2.48	0.999547
1.55	0.97162	1.90	0.99279	2.50	0.999593
1.56	0.97263	1.91	0.99309	2.55	0.999689
1.57	0.97360	1.92	0.99338	2.60	0.999764
1.58	0.97455	1.93	0.99366	2.65	0.999822
1.59	0.97546	1.94	0.99392	2.70	0.999866
1.60	0.97635	1.95	0.99418	2.75	0.999899
1.61	0.97721	1.96	0.99443	2.80	0.999925
1.62	0.97804	1.97	0.99466	2.85	0.999944
1.63	0.97884	1.98	0.99489	2.90	0.999959
1.64	0.97962	1.99	0.99511	2.95	0.999970
1.65	0.98038	2.00	0.995322	3.00	0.999978
1.66	0.98110	2.02	0.995720	3.20	0.999994
1.67	0.98181	2.04	0.996086	3.40	0.999998
1.68	0.98249	2.06	0.996424	3.60	1.000000
1.69	0.98315	2.08	0.996734		
1.70	0.98379	2.10	0.997020		
1.71	0.98441	2.12	0.997284		
1.72	0.98500	2.14	0.997525		
1.73	0.98558	2.16	0.997747		
1.74	0.98613	2.18	0.997951		

where $C_1 = -kA(T_i - T_\infty)/\sqrt{\pi\alpha}$ is a constant. With the aid of Appendix B we can proceed directly with the integration to obtain

$$Q_0 = C_1\left(\frac{t^{1/2}}{1/2}\right)_0^t = 2C_1 t^{1/2} \tag{3.30}$$

or

$$Q_0 = \frac{-2kA(T_i - T_\infty)}{\sqrt{\pi\alpha/t}} \tag{3.31}$$

the total flux in Btu or Ws. We observe in Eqs. 3.26, 3.27, and 3.31 that the sign on the flux gives the direction; negative, toward the surface and positive, away from the surface. For example, when the initial temperature T_i is less than the imposed environmental temperature T_∞ then $T_i - T_\infty$ is negative and the heat fluxes are positive in the direction of increasing x into the solid. Also Q_0 would be the amount that has entered the surface.

Example 3.4

A manufacturing process requires a thin veneer 1 mm in thickness to be glued onto a thick piece of particle board. The adhesive is thermal setting and requires a temperature of 120C to fuse and adhere to the base. Consider the veneer and base to have essentially the same conductivity $k = 0.14$ W/mK and $\alpha = 1.26 \times 10^{-7}$ m^2/s. If the particle board and veneer are initially at 25C, and the surface temperature is suddenly changed to 200C by a heated copper pressure plate, how long would it take for the adhesive layer to fuse and set?

SOLUTION

We assume that the heated pressure plate makes intimate contact with the veneer layer and instantaneously raises the surface temperature to 200C. The veneer and particle board then react as a semi-infinite solid as the temperature wave begins to penetrate. If the board is thick enough, we can predict the time required for the adhesive layer to reach reaction temperature by using Eq. 3.24. With low conductivity material such as wood composites, such an assumption is reasonable for the geometry assumed. The geometry and the temperature happenings are illustrated in Figure 3.15. For the data involved, the temperature ratio is calculated to be

$$\frac{T - T_\infty}{T_i - T_\infty} = \frac{120 - 200}{25 - 200} = 0.457$$

Therefore, the erf$(x/2\sqrt{\alpha t})$ must have the value 0.457 and Table 3.3 gives a value of about 0.43 for the argument $x/2\sqrt{\alpha t}$ for the nearest three decimals of the erf function. Accordingly,

$$\frac{x}{2\sqrt{\alpha t}} = 0.43$$

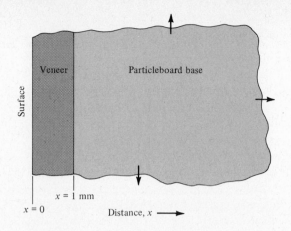

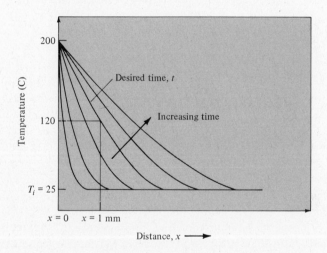

Figure 3.15   Temperature happenings for Example 3.4.

and solving for the time t yields

$$t = \frac{(x/0.86)^2}{\alpha}$$

To calculate a numerical value we must recognize that the distance x must be in meters to be consistent with the units for thermal diffusivity m^2/s, therefore

$$t = \frac{(0.001/0.86)^2}{1.26(10^{-7})} = 10.7 \text{ s}$$

We also recognize that in the actual situation the surface temperature is not changed instantaneously; therefore the 10.7 s represents the minimum time possible in which the bond at 1-mm depth could be brought up to 120C. In the next few paragraphs we will consider this situation in more detail.

3.4.2 Finite Convection Coefficient at Surface

In the paragraph above, we discussed the situation where the surface temperature was changed instantaneously due to an extremely high (infinite for the mathematical solution) convection coefficient. We now look at the situation where a finite convection coefficient exists and see what the temperature response in the semi-infinite solid is like. Schneider (9) presents the solution to the conduction equation for the semi-infinite solid with finite surface convection in terms of a temperature excess based on the initial temperature. It is more convenient to recast Schneider's result to be consistent with our other charts and use the excess based on the environment temperature. Thus

$$\frac{T - T_\infty}{T_i - T_\infty} = \text{erf}(X) + e^A \left[1 - \text{erf}\left(X + \frac{h\sqrt{\alpha t}}{k} \right) \right] \tag{3.32}$$

where

$$X = \frac{x}{2\sqrt{\alpha t}}$$

$$A = \left(\frac{hx}{k} + \frac{h^2 \alpha t}{k^2} \right)$$

$\text{erf}(z) = $ error function with argument (z) defined by Eq. 3.25

Values for the temperature excess ratio calculated from Eq. 3.32 are shown in Figure 3.16 as a function of the dimensionless distance $x/2\sqrt{\alpha t}$ with the local Biot number, hx/k, as a parameter. The sketch in Figure 3.16 of the physical arrangement shows that the distance x starts at the surface of the semi-infinite solid and increases into the solid in a direction perpendicular to the surface. Figure 3.17 presents the same temperature information as a function of dimensionless distance but with $h^2 \alpha t / k^2$ as a parameter. When temperature excess and time are known, location can be obtained directly from Figure 3.17 and when location and temperature ratio are known, time can be obtained directly from Figure 3.16. Example 3.5 illustrates the above concepts.

Example 3.5

Consider a copper water line buried in the earth a distance of x units below the surface and parallel to the surface. The earth is at a uniform initial temperature of 50F (for a considerable depth) and the ambient temperature suddenly drops to 5F as a weather front moves through. We could be concerned with two types of questions regarding such a situation:

1. How deep must the water line be buried so that the earth at that location does not reach the freezing temperature (32F) in 8 hr?
2. If the water line is 24 in. below the surface, how long would it take for the freezing temperature to penetrate the earth to that location?

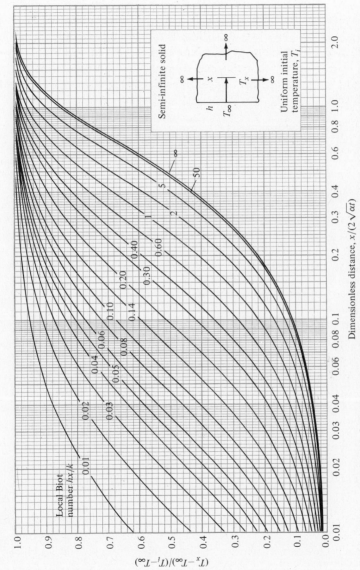

Figure 3.16 Temperature excess ratio for the semi-infinite solid with surface convection and local Biot number as parameter.

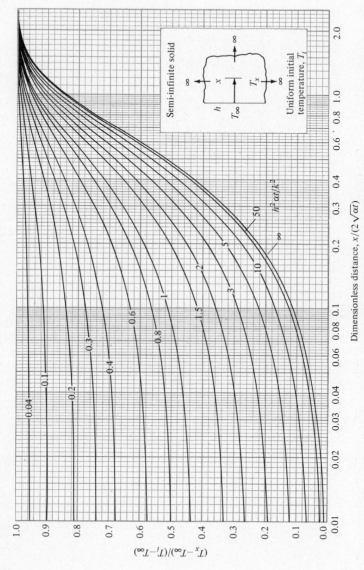

Figure 3.17   Temperature excess ratio for the semi-infinite solid with surface convection and $h\,\alpha t/k^2$ as parameter.

We assume wind conditions are such than an $h = 10$ Btu/hrft²F exists at the surface and that the earth is at an initially uniform temperature of 50F to a depth considerably below the pipe so that we can model the situation as a semi-infinite solid. The soil properties are as indicated in the figure showing the physical situation.

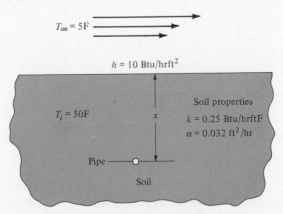

$T_\infty = 5F$

$h = 10$ Btu/hrft²

$T_i = 50F$

x

Soil properties
$k = 0.25$ Btu/hrftF
$\alpha = 0.032$ ft²/hr

Pipe

Soil

Figure for Example 3.5.

SOLUTION
In order to determine the depth x for part (1), we recognize that we must use Figure 3.17 because we can obtain the depth from the dimensionless distance $x/2\sqrt{\alpha t}$ directly for the known value of $h^2\alpha t/k^2$; using Figure 3.16 would require a trial-and-error procedure. Accordingly, the temperature excess and $h^2\alpha t/k^2$ are

$$\frac{T_x - T_\infty}{T_i - T_\infty} = \frac{32 - 5}{50 - 5} = 0.60$$

$$\frac{h^2\alpha t}{k^2} = \frac{(10)^2(0.032)8}{(0.25)^2} = 410$$

units check: $\dfrac{\text{Btu}^2}{\text{hr}^2\text{ft}^4\text{F}^2} \cdot \dfrac{\text{ft}^2}{\text{hr}} \cdot \dfrac{\text{hr}}{1} \cdot \dfrac{\text{hr}^2\text{ft}^2\text{F}^2}{\text{Btu}^2}$, dimensionless

From Figure 3.17 we read $x/2\sqrt{\alpha t} = 0.59$ so that

$$x = 2\sqrt{\alpha t}\,(0.59)$$

$$= 2\sqrt{0.032(8)}\,(0.59)$$

$$= 0.597 \text{ ft or about } \underline{7.2 \text{ in.}}$$

which means that at this depth it would take 8 hr for the temperature at the pipe to drop from 50 to 32F.

For part (2) we want to know how long it would take for the temperature to drop from 50 to 32F if the pipe were 24 in. from the surface. Since both $x/2\sqrt{\alpha t}$ and $h^2 \alpha t/k^2$ contain the unknown time t, use of Figure 3.17 would require a trial-and-error procedure; however, Figure 3.16 gives the dimensionless distance (that contains the time) directly at a given value of the local Biot modulus. Therefore

$$\frac{T_x - T_\infty}{T_i - T_\infty} = \frac{32 - 5}{50 - 5} = 0.60$$

$$\frac{hx}{k} = \frac{10(24/12)}{0.25} = 80$$

units check: $\dfrac{\text{Btu}}{\text{hrft}^2\text{F}} \dfrac{\text{ft}}{1} \dfrac{\text{hrftF}}{\text{Btu}}$, dimensionless

From Figure 3.16 we read $x/2\sqrt{\alpha t} = 0.60$ so that

$$t = x^2/\alpha[2(0.60)]^2 = 4/0.032(1.2)^2 = 86.8 \text{ hr or about 3.6 days}$$

This amount of time is not long enough to cover a week-long cold snap so, therefore, water lines are usually placed below a 3-ft depth in temperate climates.

We note in Example 3.5 that in parts (1) and (2) the Biot modulus and $h\sqrt{\alpha t}/k$ parameters were very close to the value for a sudden change in surface temperature; hence the dimensionless distance was about the same in both parts. Values of the local Biot number and $h^2\alpha t/k^2$ must be significantly below 50 to have appreciable variation in surface temperature with time. At a position close to the surface, such would be the case as Example 3.6 shows.

Example 3.6

Let's reexamine Example 3.4 and estimate a more realistic time to bring the adhesive to the required temperature. We could conceive of the hot plate acting like an ambient environment at 200C and the thin layer of air between the veneer and the hot plate to have a conductance of 100 $\text{W}/\text{m}^2\text{K}$, so that we could use Eq. 3.32 or Figure 3.16 to determine the time (note that Eq. 3.32 would require trial and error to solve for time). The local Biot number, then, would be

$$\frac{hx}{k} = \frac{100(0.001)}{0.14} = 0.71$$

$$\frac{T_x - T_\infty}{T_i - T_\infty} = 0.457 \text{ (same as before)}$$

and from Figure 3.16 we estimate the dimensionless distance $x/2\sqrt{\alpha t} = 0.19$ under these conditions. We calculate the time, then, as

$$t = x^2/4(0.19)^2\alpha$$
$$= (0.001)^2/4(0.19)^2 1.26(10^{-7})$$
$$= 55 \text{ s}$$

This is almost 1 min compared to the 10.7 s that we estimated in Example 3.3 and represents a more exact estimate of the actual time required.

We conclude the topic of semi-infinite solid with finite conductance at the surface by noting that erf(0) = 0 so that the surface temperature excess $T_s - T_\infty$ can be obtained from Eq. 3.32 as

$$\frac{T_s - T_\infty}{T_i - T_\infty} = e^{h^2 \alpha t / k^2}\left[1 - \text{erf}\left(\frac{h\sqrt{\alpha t}}{k}\right)\right]$$ (3.33)

Since the dimensionless distance scales on Figures 3.16 and 3.17 are logarithmic, the location $x = 0$ for the surface cannot be read from the figures to determine the surface temperature variation with time and Eq. 3.33 must be utilized.

3.5 PRODUCT SOLUTIONS

The geometries that have been studied so far, when $N_{\text{Bi}} > 0.1$, are the sphere, the infinite cylinder, the infinite plate, and the semi-infinite solid. Often times engineers are concerned with objects that can reasonably be approximated by (if not actually so) finite cylinders, cubes, or rectangular blocks. For such geometries it can be shown that the temperature response is a product of the solutions for the basic geometries given in Sections 3.3.1, 3.3.2, and 3.3.5. The requirements for a product solution to hold are that the basic partial differential equation is linear and homogeneous and that the initial and boundary conditions be the same. Such is the case for the basic solutions we have studied; all are based on a uniform initial temperature T_i and constant environment temperature T_∞. The results presented in Figures 3.7, 3.8, 3.9, 3.10, 3.16, and 3.17 can be used in the construction of product solutions for certain finite and semi-infinite rod and bar geometries.

In order to simplify the nomenclature somewhat in the explanations to follow, we define TR as the temperature excess ratio with suitable subscripts to identify the geometry as follows:

$$\text{TR}_{\text{PL}} = \frac{T_{x/L} - T_\infty}{T_i - T_\infty} \qquad \text{infinite plate, Figures 3.7, 3.8}$$

$$\text{TR}_{\text{CY}} = \frac{T_{r/R_o} - T_\infty}{T_i - T_\infty} \qquad \text{infinite cylinder, Figures 3.9, 3.10}$$

$$\text{TR}_{\text{SI}} = \frac{T_x - T_\infty}{T_i - T_\infty} \qquad \text{semi-infinite solid, Figures 3.16, 3.17}$$

Example 3.7 illustrates the application of the method.

Example 3.7

Determine the temperature at the center of a block of carbon steel (0.5 percent C) 6 in. on a side which is initially at a uniform temperature of 1000F with an $h = 50$ Btu/hrft^{2}F on five sides and insulated on the bottom side 5 min after exposure to ambient conditions at 100F.

SOLUTION

We consider the cube to be formed by the intersection of three infinite plates as shown in Figure 3.18. The plates $2L_1 = 6$ in. thick and $2L_2 = 6$ in. thick are identical, but the third plate $L_3 = 6$ in. thick is half of a $2L_3 = 12$ in. thick plate since the bottom face is insulated. We recognize that the central plane of a plate $2L$ units thick has no heat flux crossing it at any time due to symmetry of the temperature profile so that a plate L units thick and insulated on one side behaves thermally the same as half of the $2L$ plate with identical convection on both faces. We, therefore, formulate the solution as

$$\mathrm{TR}_{\mathrm{CUBE}} = \left(\mathrm{TR}_{\mathrm{PL}_1}\right)\left(\mathrm{TR}_{\mathrm{PL}_2}\right)\left(\mathrm{TR}_{\mathrm{PL}_3}\right)$$

For the first plate of thickness $2L_1 = 6$ in. and assuming an average $k = 25$ Btu/hrftF, the Biot number $hL/k = 50(\frac{3}{12})/25 = 0.5$, which is also the Biot number for the second plate. The corresponding thermal diffusivity is $\alpha = 0.46$ ft^2/hr for the k value used above and permits calculation of the Fourier number

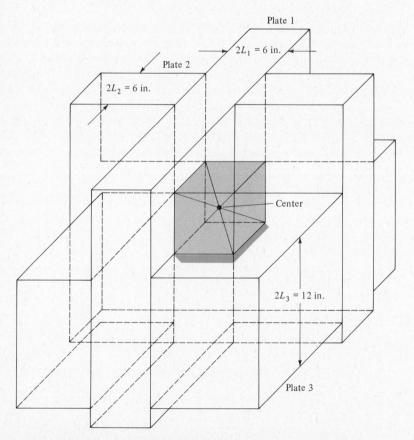

Figure 3.18 Sketch showing the intersection of three infinite plates $2L_1$, $2L_2$, and $2L_3$ to form the cube described in Example 3.7.

$\alpha t/L^2 = 0.46(\frac{5}{60})/(\frac{3}{12})^2 = 0.61$. Therefore, we read the temperature ratio from Figure 3.7 at $hL/k = 0.5$ and $\alpha t/L^2 = 0.61$ as

$$\mathrm{TR}_{PL_1} = \mathrm{TR}_{PL_2} = 0.821$$

For third plate $hL/k = 50(\frac{6}{12})/25 = 1$ and $\alpha t/L^2 = 0.46(\frac{5}{60})/(\frac{6}{12})^2 = 0.153$. However, the center of the cube lies at a position of $x = 3$ in. in the third plate so that we must use the position correction chart in Figure 3.8 in addition to Figure 3.7 in order to determine TR_{PL_3}. Thus we obtain

$$\mathrm{TR}_{PL_3} = \underbrace{\left(\frac{T_o - T_\infty}{T_i - T_\infty}\right)_{L_3}}_{\text{Figure 3.7}} \underbrace{\left(\frac{T_{0.5} - T_\infty}{T_o - T_\infty}\right)_{L_3}}_{\text{Figure 3.8}}$$

$$= (0.98)(0.91) = 0.89$$

and for the center position of the cube

$$\mathrm{TR}_{CUBE} = (0.821)(0.821)(0.89) = 0.601 = \frac{T_c - T_\infty}{T_i - T_\infty}$$

The center temperature after 5-min exposure to ambient conditions with $h = 50$ Btu/hrft^{2}F on five surfaces then can be determined as

$$T_c = 0.601(T_i - T_\infty) + T_\infty$$
$$= 0.601(1000 - 100) + 100 = 641\mathrm{F}$$

Other geometries that can be analyzed by the product solution method are shown in Figure 3.19 together with the appropriate product solutions for the cases shown. The following abbreviated nomenclature has been used in Figure 3.19:

$$\mathrm{TR} = \text{temperature excess ratio at the indicated point } P$$
$$= \frac{T_P - T_\infty}{T_i - T_\infty}$$

$\mathrm{TR}_{PL}(0) =$ center temperature excess ratio for the
infinite plate, Figure 3.7

$\mathrm{TR}_{PL}(x) =$ position correction ratio for the infinite
plate, Figure 3.8

$\mathrm{TR}_{CY}(0) =$ center temperature excess ratio for the
infinite cylinder, Figure 3.9

$\mathrm{TR}_{CY}(r) =$ position correction ratio for the infinite
cylinder, Figure 3.10

$\mathrm{TR}_{SI}(x) =$ temperature excess ratio for the semi-infinite
solid, Figure 3.16 or Figure 3.17

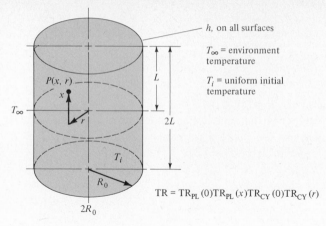

$$TR = TR_{PL}(0)TR_{PL}(x)TR_{CY}(0)TR_{CY}(r)$$

(a) Finite short rod.

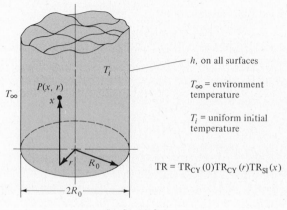

$$TR = TR_{CY}(0)TR_{CY}(r)TR_{SI}(x)$$

(b) Semi-infinite rod.

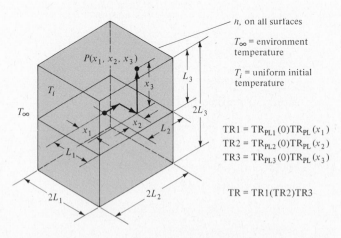

$$TR1 = TR_{PL1}(0)TR_{PL}(x_1)$$
$$TR2 = TR_{PL2}(0)TR_{PL}(x_2)$$
$$TR3 = TR_{PL3}(0)TR_{PL}(x_3)$$

$$TR = TR1(TR2)TR3$$

(c) Finite bar.

Figure 3.19 Geometric arrangements for which the product solution is applicable. (a) Finite short rod. (b) Semi-infinite rod. (c) Finite bar.

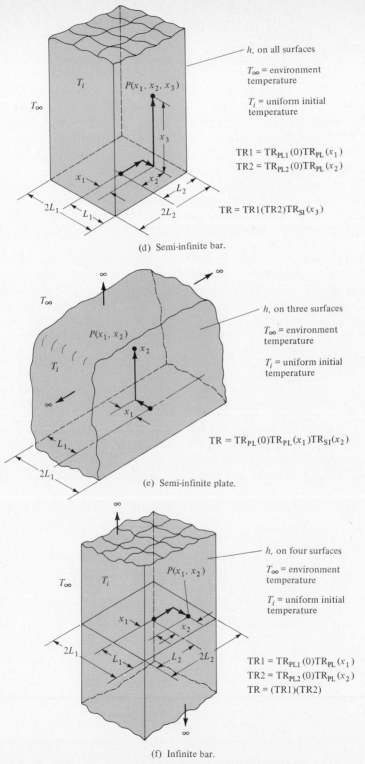

(d) Semi-infinite bar.

- h, on all surfaces
- T_∞ = environment temperature
- T_i = uniform initial temperature

$$TR1 = TR_{PL1}(0)TR_{PL}(x_1)$$
$$TR2 = TR_{PL2}(0)TR_{PL}(x_2)$$
$$TR = TR1(TR2)TR_{SI}(x_3)$$

(e) Semi-infinite plate.

- h, on three surfaces
- T_∞ = environment temperature
- T_i = uniform initial temperature

$$TR = TR_{PL}(0)TR_{PL}(x_1)TR_{SI}(x_2)$$

(f) Infinite bar.

- h, on four surfaces
- T_∞ = environment temperature
- T_i = uniform initial temperature

$$TR1 = TR_{PL1}(0)TR_{PL}(x_1)$$
$$TR2 = TR_{PL2}(0)TR_{PL}(x_2)$$
$$TR = (TR1)(TR2)$$

Figure 3.19 (continued) (d) Semi-infinite bar. (e) Semi-infinite plate. (f) Infinite bar.

3.6 NUMERICAL ANALYSIS

In our discussion of steady conduction phenomena, we examined a means for analyzing difficult situations by considering a discrete grid or network of nodes to represent the geometry under consideration. In Section 2.8 we did not consider energy storage in the node because in steady state there can be no storage of energy. In transient situations, storage (or depletion) of energy takes place at the node so that we must add a term to the nodal energy balance (Eq. 2.90) to account for that happening.

For example, Figure 3.20 shows a general interior node n from Figure 2.23(a) for which we write the energy balance as

$$\Sigma \text{ energy rate in} = \text{energy rate stored} \tag{3.34}$$

The energy stored in the node is that stored in the mass $m = \rho A \, \Delta x$ associated with the node where $A \, \Delta x$ is the volume of the node and ρ the density. The energy is stored during the time increment Δt from k to $k + 1$ so that the resultant node temperature after storage is T_n^{k+1}. In that time interval Δt the energy storage rate can be expressed as:

$$\begin{aligned} q_{\text{stored}} &= mc_p\big(T_n^{k+1} - T_n^k\big)/\Delta t \\ &= \rho A \, \Delta x c_p\big(T_n^{k+1} - T_n^k\big)/\Delta t \end{aligned} \tag{3.35}$$

Before we can evaluate Eq. 3.34 with Eq. 3.35, we must know the physical geometry under consideration. We will look at two examples, (1) the plane plate with uniform generation, and (2) the plane plate with no generation and a sudden change of surface temperature. These examples will illustrate the method and we can compare the steady-state results with results from Chapter 2 that we are familiar with.

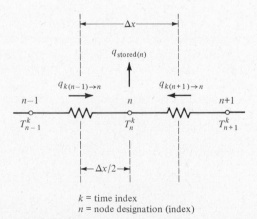

k = time index
n = node designation (index)

Figure 3.20   Generalized interior node n with connecting resistances R_k from Figure 2.23(a).

3.6.1 Plane Plate with Uniform Generation

Uniform generation means that there is a constant supply of energy to the nodes, due to $\dot{q}$ the generation rate, in the amount $q_{gen} = \dot{q}A\,\Delta x$ so that when we substitute Eq. 3.35 into Eq. 3.34, we first obtain symbolically

$$q_{k(n-1)\to n} + q_{k(n+1)\to n} + q_{gen} = q_{stored} \tag{3.36}$$

and in detail after the fluxes have been evaluated (see Eq. 2.94 for the left-hand side), the expression becomes

$$kA\left(T_{n-1}^k - T_n^k\right)/\Delta x + kA\left(T_{n+1}^k - T_n^k\right)/\Delta x + \dot{q}A\,\Delta x$$
$$= \rho A\,\Delta x c_p\left(T_n^{k+1} - T_n^k\right)/\Delta t \tag{3.37}$$

Let's keep sight of the fact that all (T_n^k)'s are the temperatures we know at the beginning, or at the beginning of each time interval Δt and that the quantity T_n^{k+1} is the new temperature after the end of the time interval, and can be expressed in terms of known quantities. Looking ahead, we then simply step through all the nodes and determine all the new temperatures at $k + 1$ from a known initial condition, then repeat the process again for $k + 2$, $k + 3$, and so on, as far as we want to go, or until we reach steady state.

To proceed, we simplify Eq. 3.37 by multiplying through by $\Delta x/kA$ and regrouping to obtain

$$T_{n-1}^k + T_{n+1}^k - 2T_n^k + \frac{\dot{q}\,\Delta x^2}{k} = \frac{\Delta x^2}{\alpha\,\Delta t}\left(T_n^{k+1} - T_n^k\right) \tag{3.38}$$

where we have let $\alpha = k\rho c_p$, the thermal diffusivity. The reader with a sharp eye will recognize $\alpha\,\Delta t/\Delta x^2$ as a type of Fourier number like we have seen before and it represents the dimensionless time increment for the procedure and the nodal geometry. Letting $M = \Delta x^2/\alpha\,\Delta t$ and solving Eq. 3.38 for the new temperature yields

$$T_n^{k+1} = \frac{1}{M}\left[T_{n-1}^k + T_{n+1}^k + \frac{\dot{q}\,\Delta x^2}{k} + (M - 2)T_n^k\right] \tag{3.39}$$

so that the new temperature (at the end of the time interval Δt) is given directly in terms of known quantities. The value of M depends on the magnitude of Δx and Δt and values are limited to $M \geqslant 2$; values less than 2 make the procedure unstable.

3.6.2 Plane Plate with No Generation

Equation 3.39 will represent the case for the plane plate with no generation when $\dot{q}$ is set equal to zero. The expression for the new temperature then becomes

$$T_n^{k+1} = \frac{1}{M}\left[T_{n-1}^k + T_{n+1}^k + (M - 2)T_n^k\right] \tag{3.40}$$

Of course, the same comments with regard to the value of M made above apply here also. The reader will note that $M = 2$ drops the last term and results in a simplification that we will make use of in the next section on graphical analysis. Taking $M = 2$ restricts the value of Δt once the node separation has been decided upon. Generally the approach is to select the node separation, then choose an $M \geqslant 2$ to get a convenient value of Δt. Since $\Delta t = \Delta x^2/\alpha M$, the largest Δt results when $M = 2$, which may or may not be convenient.

3.6.3 Computer Program of Transient Numerical Analysis

We will program Eq. 3.39 in as general a manner as possible so that we can use the result to calculate a numerical example for the two cases we mentioned above: (1) the plane plate with uniform generation, and (2) the plane plate with no generation and a sudden change in both surface temperatures, both cases (1) and (2) having uniform initial temperature.

The approach and the results will look very similar to the Gauss-Seidel iteration but there is no relationship between the Gauss-Seidel and the transient analysis other than that they will predict the same steady-state condition for the same physical phenomena and boundary conditions. We might note that the transient analysis under discussion must have physically real initial conditions whereas the Gauss-Seidel iteration technique simply requires starting temperatures. The closer those starting temperatures are to the steady state, the faster that procedure converges. In the transient analysis we might also be tempted to speak of convergence but that would not be correct. We are interested to know when the calculations have reached steady state from succeedingly correct values of temperature as functions of time. Arrival at steady state is a difficult question to ascertain when steady-state conditions are not known. Generally one uses very strict criteria to start out with, then less strict until a discrepancy is observed in the resulting steady-state temperatures. Once we have the program running we can take a look at this aspect in the home problems.

The flow diagram for the computer program is shown in Figure 3.21 and the FORTRAN list is given in Table 3.4. We have again selected the width of the plate as $2L = 2$ ft to compare with the analytical calculation made in Section 2.4.2 and corresponding to a thermal conductivity of $XK = 10$ Btu/hrftF, we select the thermal diffusivity $ALFA = 0.172$ ft^2/hr; the acronyms are self-explanatory. Other symbols are explained in the DICTIONARY.

The initial conditions are a uniform temperature of 0F throughout, and those temperatures are set in the DO 10 LOOP. The numerical method has a very large error in the first time step when the surface temperature takes a sudden jump to a new value. To ease this error, at the first time step we take only half of the initial jump as shown in lines 30 and 31; then revert back to the actual value for the rest of the time steps. Note lines 44 and 45 reestablish the values of surface temperature to the boundary conditions under consideration.

The output algorithm described in lines 32, 33, and 34 is a convenience that permits printing out the surface temperature and three equally spaced intermediate temperatures for values of $N = 5, 9, 13, ---$. That way we have some

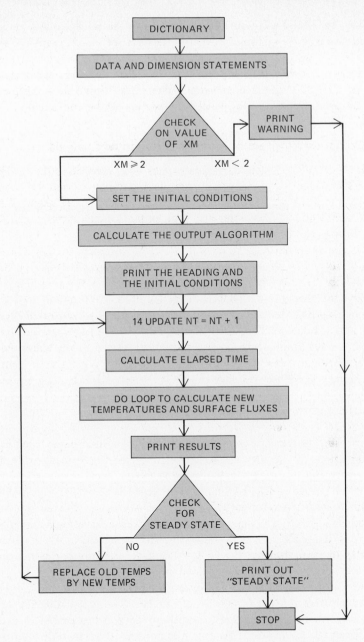

Figure 3.21   Flow diagram for computer program of transient numerical analysis.

Table 3.4 FORTRAN LIST FOR TRANSIENT ANALYSIS PROGRAM

```
        C          THIS PROGRAM CALCULATES THE TRANSIENT RESPONSE OF AN
        C          INFINITE PLANE PLATE WITH OR WITHOUT UNIFORM INTERNAL
        C          GENERATION BY A NUMERICAL ANALYSIS OF A FINITE NODAL
        C          APPROXIMATION TO THE GEOMETRY USING N NODES.
        C
        C          PROGRAM DICTIONARY:
        C
        C                 N=NUMBER OF NODES (USE 5, 9, 13, ---)
        C                NT=NUMBER OF TIME INTERVALS CALCULATED
        C                DT=TIME INTERVAL, HOURS
        C              QDOT=GENERATION RATE, BTU/HRFT3
        C                TQ=GENERATION TEMPERATURE
        C         TEMP(K,1)=THE OLD TEMPERATURES
        C         TEMP(K,2)=THE NEW TEMPERATURES AT THE END OF DT
        C                XK=THERMAL CONDUCTIVITY, BTU/HRFTF
        C                DX=NODE SEPARATION
        C                XL=SLAB HALF-THICKNESS, FT
        C               QA1=UNIT AREA FLUX AT NODE 1
        C               QA2=UNIT AREA FLUX AT NODE N
        C              ALFA=THERMAL DIFFUSIVITY, FT2/HR
        C                XM=DXSQD/(ALFA*DT)
        C               ELT=ELAPSED TIME, HR
        C               SSC=STEADY-STATE CRITERIA
        C             NX(K)=PRINTOUT ALGORITHM
        C
        C          DIMENSION AND DATA STATEMENTS
        C
0001               DIMENSION TEMP(50, 2), NX(5), DIF(50)
0002                    N=5
0003                    XN=N
0004                    XK=10.
0005                    DT=0.4
0006                    XL=1.
0007                    DX=2.*XL/(XN−1.)
0008                    QDOT=10000.00
0009                    ALFA=0.172
0010                    SSC=0.001
0011                    N4=(N−1)/4
0012           TQ=QDOT*DX*DX/XK
0013           TQ2=TQ/2.
0014           XM=DX*DX/(ALFA*DT)
0015           IF(2.−XM) 5, 5, 2
0016         2 PRINT 500, XM
0017           GOTO 35
0018         5 L=N−1
0019           NT=0
        C
        C          SET THE INITIAL CONDITIONS: UNIFORM TEMPERATURE
        C
0020           TEMP(1,1)=0.0
0021           TEMP(1, 2)=0.0
0022           TEMP(N,1)=1000.0
0023           TEMP(N, 2)=1000.0
0024           DO 10 I=2,L
0025           TEMP(I,1)=0.0
0026        10 CONTINUE
0027           ELT=0.0
0028           QA1=0.0
```

continued

Table 3.4   continued

```
0029        QAN = 0.0
0030        IF(NT.EQ.0) TEMP(1,1) = TEMP(1,1)/2.
0031        IF(NT.EQ.0) TEMP(N,1) = TEMP(N,1)/2.
      C
      C               CALCULATE OUTPUT ALGORITHM
      C
0032        DO 12 K = 1, 5
0033        NX(K) = 1 + (K−1)*(N−1)/4
0034     12 CONTINUE
0035        PRINT 505,QDOT,N,DT,DX,XM,NX(1),NX(2),NX(3),NX(4),NX(5),
      A         ELT, QA1, (TEMP(I,1), I = 1,N,N4), QAN
      C
      C               CALCULATE NEW TEMPS AT THE END OF DT
      C               AND THE HEAT FLUXES AT NODES 1 AND N.
      C
0036     14 NT = NT + 1
0037        ELT = NT*DT
0038        DO 20 J = 2, L
0039        TEMP(J, 2) = (1./XM)*(TEMP(J−1,1) + TEMP(J+1,1)
      A              + TQ + (XM − 2.)*TEMP(J,1))
0040     20 CONTINUE
0041        QA1 = XK*(TEMP(1, 2) − TEMP(2, 2) − TQ2)/DX
0042        QAN = XK*(TEMP(N, 2) − TEMP(N−1, 2) − TQ2)/DX
0043        PRINT 510, ELT, QA1, (TEMP(I, 2), I = 1,N,N4), QAN
0044     21 TEMP(1,1) = 0.0
0045        TEMP(N,1) = 1000.00
      C
      C               CHECK FOR STEADY STATE
      C
0046        DO 22 I = 2,L
0047        DIF(I) = ABS(TEMP(I, 2) − TEMP(I,1))
0048        IF(DIF(I).GT.SSC) GOTO 25
0049     22 CONTINUE
0050        PRINT 520
0051        GOTO 35
      C
      C               REPLACE THE OLD TEMPS BY THE NEW ONES
      C
0052     25 DO 30 I = 2,L
0053        TEMP(I,1) = TEMP(I, 2)
0054     30 CONTINUE
0055        GOTO 14
0056     35 STOP
0057    500 FORMAT(1H1,32X,'THE VALUE OF M = ', F6.2,
           A'WHICH IS LESS THAN 2.0')
0058    505 FORMAT(1H1////28X,'TRANSIENT TEMPERATURE RESPONSE FOR THE'
           A/26X, 'PLATE WITH UNIFORM QDOT = ', F8.0, 'BTU/HRFT3'
           B/21X, 'APPROXIMATED BY', I3, 'NODES USING A', F5.2,
           C'HOUR TIME STEP'//31X, 'DELTA X = ', F6.3, 'FT AND M = ',
           DF6.3///15X, 'TIME',4X,'QA1',7X,'T',I1,6X,'T',I2,6X,'T',I2,
           E 6X,'T',I2,6X,'T',I2,5X,'QAN'//14X,
           F F5.2, F8.0, (5F9.1), F9.0)
0059    510 FORMAT(14X, F5.2, F8.0, (5F9.1), F9.0)
0060    520 FORMAT(36X, 'STEADY STATE HAS BEEN REACHED')
0061        END
```

---

independence of node selection without getting a whole sheet full of numbers that would not fit on the output page. The evaluation of Eq. 3.39 is actually done in line 39 where TEMP(J, 2) represents T_n^{k+1}; the second subscript 2 represents the *new* temperatures; second subscript 1 represents the *old* temperatures.

We check for steady state in the DO 22 LOOP by comparing the new and old temperatures at a node; when the difference at all nodes is less than 0.001 degree we assume that steady state has been reached. That criterion is satisfactory for this example but may not be so for another more complicated situation. When the difference between subsequent temperature calculations at a node for

Table 3.5

TRANSIENT TEMPERATURE RESPONSE FOR THE PLATE WITH UNIFORM QDOT = 10000 BTU/HRFT3 APPROXIMATED BY 5 NODES USING A 0.40-HR TIME STEP							
DELTA X = 0.500 FT AND M = 3.634							
TIME	QA1	T1	T2	T3	T4	T5	QAN
0.0	0.	0.0	0.0	0.0	0.0	500.0	0.
0.40	− 3876.	0.0	68.8	68.8	206.4	1000.0	13372.
0.80	− 4873.	0.0	118.7	175.5	455.7	1000.0	8385.
1.20	− 5909.	0.0	170.4	305.8	597.2	1000.0	5556.
1.60	− 7092.	0.0	229.6	417.5	696.6	1000.0	3567.
2.00	− 8238.	0.0	286.9	511.4	772.1	1000.0	2058.
2.40	− 9271.	0.0	338.5	590.2	831.9	1000.0	862.
2.80	−10168.	0.0	383.4	656.2	880.4	1000.0	−109.
3.20	−10936.	0.0	421.8	711.7	920.4	1000.0	− 909.
3.60	−11586.	0.0	454.3	758.1	953.7	1000.0	−1574.
4.00	−12134.	0.0	481.7	797.1	981.4	1000.0	− 2128.
4.40	−12595.	0.0	504.7	829.8	1004.6	1000.0	− 2592.
4.80	−12982.	0.0	524.1	857.3	1024.0	1000.0	−2981.
5.20	−13307.	0.0	540.4	880.3	1040.3	1000.0	− 3307.
5.60	−13580.	0.0	554.0	899.6	1054.0	1000.0	− 3580.
6.00	−13809.	0.0	565.4	915.8	1065.4	1000.0	− 3809.
6.40	−14001.	0.0	575.0	929.3	1075.0	1000.0	− 4001.
6.80	−14162.	0.0	583.1	940.7	1083.1	1000.0	− 4162.
7.20	−14297.	0.0	589.8	950.3	1089.8	1000.0	− 4297.
7.60	−14410.	0.0	595.5	958.3	1095.5	1000.0	− 4410.
8.00	−14505.	0.0	600.3	965.0	1100.3	1000.0	− 4505.
8.40	−14585.	0.0	604.3	970.7	1104.3	1000.0	− 4585.
8.80	−14652.	0.0	607.6	975.4	1107.6	1000.0	− 4652.
9.20	−14708.	0.0	610.4	979.4	1110.4	1000.0	− 4708.
9.60	−14755.	0.0	612.8	982.7	1112.8	1000.0	− 4755.
10.00	−14795.	0.0	614.7	985.5	1114.7	1000.0	− 4795.
⋮	⋮	⋮	⋮	⋮	⋮	⋮	⋮
24.80	− 15000.	0.0	625.0	1000.0	1125.0	1000.0	− 5000.
25.20	− 15000.	0.0	625.0	1000.0	1125.0	1000.0	− 5000.
25.60	− 15000.	0.0	625.0	1000.0	1125.0	1000.0	− 5000.
26.00	− 15000.	0.0	625.0	1000.0	1125.0	1000.0	− 5000.
26.40	− 15000.	0.0	625.0	1000.0	1125.0	1000.0	− 5000.
26.80	− 15000.	0.0	625.0	1000.0	1125.0	1000.0	− 5000.
27.20	− 15000.	0.0	625.0	1000.0	1125.0	1000.0	− 5000.
27.60	− 15000.	0.0	625.0	1000.0	1125.0	1000.0	− 5000.
28.00	− 15000.	0.0	625.0	1000.0	1125.0	1000.0	− 5000.
STEADY STATE HAS BEEN REACHED							

a given time interval is greater than 0.001 degree (note that such need be the case at only one node), control in the program changes to statement 25 (line 52) wherein the "new" temperatures TEMP(I, 2) are placed into the location for the old temperatures TEMP(I, 1) in the DO 30 LOOP, the counter NT is updated by 1, and the procedure is repeated. When all nodes show less than 0.001 degree between subsequent calculations, PRINT520 in line 49 is executed and the program stops. The number of nodes, time interval, generation rate, and so forth, can easily be changed by replacing lines 1 through 9.

The output obtained from the program is listed in Table 3.5 and shows how the transient temperature history develops. Steady state is reached in 28 hr according to our definition in the program and agrees with Section 2.4.2, but for all practical purposes we are nearly there (within 1 degree) in about 15 hr. We can make that statement only because we know what the steady-state value is. The reader should be very cautious about declarations of arrival at steady state in other cases.

As the node spacing is decreased, the accuracy of the transient results calculated is improved, however, only to a point. Table 3.6 shows values for the center temperature (T at $x/L = 1$) for the program described in Table 3.4 for the slab with uniform generation using a criterion of $T^{k+1} - T^k \leqslant 0.001$ to determine whether or not steady state has been reached. The table shows a significant difference in the results for $N = 5$ and $N = 13$ nodes, especially at short values of time. Going to 17 nodes does not result in any improvement in accuracy (see Problem 34).

We will close our discussion of transient numerical analysis with a look at the plane plate with no internal generation, a uniform initial temperature and a sudden jump in temperature on both surfaces. We can compare our results with those shown in Figure 3.7 for the case where $h = \infty$ ($hL/k = \infty$). According to

Table 3.6 VALUES OF CENTER TEMPERATURE $T(x/L = 1)$ FOR THE PLANE PLATE WITH UNIFORM GENERATION (PROGRAM IN TABLE 3.4)

t (hr)	$N = 5$ NODES	$N = 9$ NODES	$N = 13$ NODES
2	511.4	509.3	507.5
4	797.1	793.5	790.1
6	915.8	913.1	910.5
8	965	963.4	961.9
10	985.5	984.6	983.7
12	994	993.5	993.1
14	997.5	997.3	997
16	999	998.9	998.7
18	999.6	999.5	999.4
⋮	⋮	⋮	⋮
(SS)*	1000 (28)	1000 (26.1)	1000 (23.7)

*SS = Steady state ($T^{k+1} - T^k \leqslant 0.001$)

Table 3.7

TRANSIENT TEMPERATURE RESPONSE FOR THE
PLATE WITH UNIFORM QDOT = 0 BTU/HRFT3
APPROXIMATED BY 13 NODES USING A 0.05-HR TIME STEP

DELTA X = 0.167 FT AND M = 3.230

TIME	QA1	T1	T4	T7	T10	T13	QAN
0.0	0.	500.0	0.0	0.0	0.0	500.0	0.
0.05	50712.	1000.0	0.0	0.0	0.0	1000.0	50712.
0.10	37887.	1000.0	0.0	0.0	0.0	1000.0	37887.
0.15	32113.	1000.0	14.8	0.0	14.8	1000.0	32113.
0.20	28346.	1000.0	46.6	0.0	46.6	1000.0	28346.
0.25	25676.	1000.0	80.8	0.0	80.8	1000.0	25676.
0.30	23644.	1000.0	114.2	0.9	114.2	1000.0	23644.
0.35	22032.	1000.0	145.8	3.8	145.8	1000.0	22032.
0.40	20711.	1000.0	175.0	9.0	175.0	1000.0	20711.
0.45	19604.	1000.0	201.9	16.4	201.9	1000.0	19604.
0.50	18657.	1000.0	226.7	25.9	226.7	1000.0	18657.
0.55	17836.	1000.0	249.7	37.1	249.7	1000.0	17836.
0.60	17114.	1000.0	271.0	49.6	271.0	1000.0	17114.
0.65	16473.	1000.0	290.9	63.2	290.9	1000.0	16473.
0.70	15897.	1000.0	309.6	77.7	309.6	1000.0	15897.
0.75	15376.	1000.0	327.2	92.9	327.2	1000.0	15376.
0.80	14899.	1000.0	343.9	108.5	343.9	1000.0	14899.
0.85	14461.	1000.0	359.8	124.4	359.8	1000.0	14461.
0.90	14054.	1000.0	375.0	140.5	375.0	1000.0	14054.
0.95	13675.	1000.0	389.6	156.7	389.6	1000.0	13675.
1.00	13318.	1000.0	403.6	172.8	403.6	1000.0	13318.
1.05	12982.	1000.0	417.1	189.0	417.1	1000.0	12982.
1.10	12663.	1000.0	430.1	205.0	430.1	1000.0	12663.
1.15	12358.	1000.0	442.8	220.9	442.8	1000.0	12358.
1.20	12067.	1000.0	455.0	236.6	455.0	1000.0	12067.
1.25	11787.	1000.0	467.0	252.2	467.0	1000.0	11787.
1.30	11518.	1000.0	478.5	267.5	478.5	1000.0	11518.
1.35	11259.	1000.0	489.8	282.5	489.8	1000.0	11259.
1.40	11007.	1000.0	500.8	297.3	500.8	1000.0	11007.
1.45	10764.	1000.0	511.5	311.9	511.5	1000.0	10764.
1.50	10528.	1000.0	522.0	326.2	522.0	1000.0	10528.
1.55	10298.	1000.0	532.2	340.2	532.2	1000.0	10298.
1.60	10075.	1000.0	542.2	354.0	542.2	1000.0	10075.
1.65	9857.	1000.0	551.9	367.5	551.9	1000.0	9857.
1.70	9645.	1000.0	561.4	380.8	561.4	1000.0	9645.
1.75	9438.	1000.0	570.7	393.8	570.7	1000.0	9438.
1.80	9236.	1000.0	579.8	406.5	579.8	1000.0	9236.
1.85	9039.	1000.0	588.8	419.0	588.8	1000.0	9039.
1.90	8846.	1000.0	597.5	431.2	597.5	1000.0	8846.
1.95	8658.	1000.0	606.0	443.1	606.0	1000.0	8658.
2.00	8474.	1000.0	614.3	454.9	614.3	1000.0	8474.
⋮	⋮	⋮	⋮	⋮	⋮	⋮	⋮

what we've learned above, we should take N = 13 and DT = 0.05 (lines 2 and 5) so that we keep M ⩾ 2 and get reasonable accuracy. We change lines 20, 21, and 44 to read

```
20 TEMP(1,1)=1000.0
21 TEMP(1,2)=1000.0
44 TEMP(1,2)=1000.0
```

After replacing line 8 with QDOT = 0.0, the program is ready to run; the results are shown in Table 3.7 for the first 40 time increments. At the end of 2 hr the temperature at the center of the plate $T_7 = 454.9$ corresponds to the T_o of our results in Figure 3.7. Since we have a sudden jump at the surface, the $N_{Bi} = \infty$ and the Fourier number $N_{Fo} = \alpha t/L^2 = 0.172(2)/(1)^2 = 0.344$. The dimensionless temperature excess ratio is calculated as

$$\frac{T_o - T_\infty}{T_i - T_\infty} = \frac{454.9 - 1000}{0 - 1000} = 0.545$$

and from Figure 3.7 we read at $N_{Bi} = \infty$, $N_{Fo} = 0.344$ the value 0.54 which is in good agreement as it must be. We now turn our attention to a graphical method of transient analysis.

3.7 GRAPHICAL ANALYSIS

In principle any situation that can be described graphically can be described numerically, but the ease of the description depends on the programming facility of the practitioner. In some moderately complex situations with time dependent boundary conditions and variable properties, it may be simpler for an engineer to make a graphical analysis of the problem, at least to see if it is necessary to go to more extensive efforts. Equation 3.39 can be solved on a graphical basis for any value of $M \geqslant 2$ as in numerical analysis, but the amount of labor involved is tremendous. By sacrificing some flexibility in the choice of Δx and Δt, one can choose $M = \Delta x^2/\alpha \Delta t = 2$ and considerably simplify the procedure. With a choice of $M = 2$ the last term in Eq. 3.39 drops out and the *new* temperature at the end of the time interval Δt is given by

$$T_n^{k+1} = \left(T_{n-1}^k + T_{n+1}^k\right)/2 + \dot{q}\,\Delta x^2/2k \tag{3.41}$$

The construction indicated by Eq. 3.41 is easy to do for equal Δx's because the term on the left of the equation is simply determined by connecting the temperature at node $n - 1$ with the temperature at node $n + 1$ with a straight line; the intersection of that straight line with the Δx line at node n gives the value of $(T_{n-1}^k + T_{n+1}^k)/2$. Adding the generation temperature $\dot{q}\,\Delta x^2/2k$ to the average of the adjacent *old* temperatures then gives the *new* temperature T_n^{k+1} at node n; Figure 3.22 illustrates the happenings just described. One proceeds through the network of nodal temperatures until all new temperatures T_n^{k+1} are determined and repeats for as many times increments as desired.

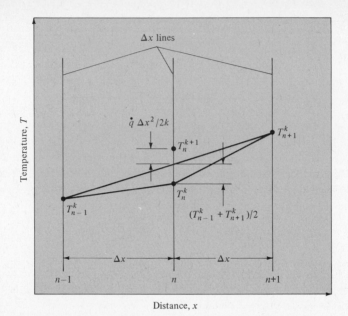

Figure 3.22 Graphical determination of the *new* temperature T_n^{k+1} at node *n* for a plate with uniform generation $\dot{q}$.

3.7.1 Plane Plate with Uniform Generation

To illustrate the procedure for the plane plate with internal generation, we model the same familiar situation that we looked at numerically and analytically. We take a node on each face and three equally spaced intervening nodes as shown in Figure 3.23 so that $\Delta x = \frac{2}{4} = 0.5$ ft for the plate 2 ft thick. Initial temperature is uniform $T_i = 0F$ and the right face changes to $T_2 = 1000F$ at time zero while the left face is kept at $T_1 = 0F$. Uniform internal generation $\dot{q} = 10,000$ Btu/hrft^3, conductivity $k = 10$ Btu/hrftF, and thermal diffusivity $\alpha = 0.172$ ft^2/hr as in Section 3.6.3. For the graphical analysis we use $M = 2$ to minimize the labor as mentioned above. Therefore, the resulting time increment $\Delta t = \Delta x^2/2\alpha = (0.5)^2/2(0.172) = 0.727$ hr and the generation temperature is $\dot{q}\,\Delta x^2/2k = 10,000(0.5)^2/2(10) = 125F$. The letters placed on the diagram in Figure 3.23 should help to trace the procedure. As in the numerical procedure, a sudden change in surface temperature produces a large error near the surface at the first time step; therefore, we take only half the surface jump at the first time step, as suggested in McAdams (12). Line $ABCOP$ represents the temperature at time zero (the old temperatures). To determine the "new" temperatures at the end of the first time step we connect point E at 500F and C with a straight line to determine $(T_3^k + T_5^k)/2$ at point D. We then add (with small dividers from a set of drawing instruments that every engineer has) the generation temperature of 125F to determine T_4^{k+1} at point F. Next we connect point B and O to locate point C and add 125F to determine T_3^{k+1} at point G. Similarly, for A and C to

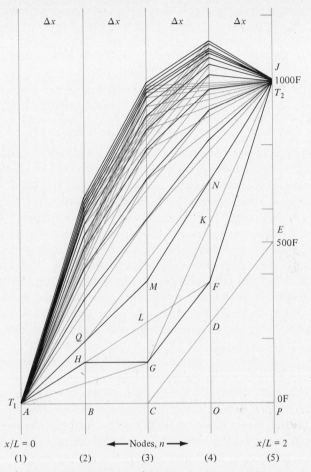

Figure 3.23   Graphical evaluation of Eq. 3.41 for the plane plate with uniform generation $\dot{q}$, uniform initial temperature, and a sudden jump of surface temperature at $x/L = 2$.

get B, add 125F to get T_2^{k+1} at H. The line $AHGFJ$ then represents the temperature profile at the end of one time interval (0.727 hr). All subsequent construction lines between nodes 3 and 5 go to J at 1000F; for example, connecting G and J to get K and adding 125F determines T_4^{k+2} at point N. Proceeding to the left we connect H and F to get L, adding 125 to get T_3^{k+2} at M, and so on. Line $AQMNJ$ represents the temperature profile at the end of two time intervals (1.45 hr). After about 12 or 14 time intervals, we approach steady state and the change in temperature at a node becomes about the size of the line width and precision is lost. The results on Figure 3.23 indicate steady-state temperatures of 625, 990 and 1120F at nodes 2, 3, and 4, which is in reasonable agreement with Section 2.4.2 and the figures shown in Table 3.4. The reader will readily observe that the numerical procedure using a computer has a large advantage in convenience and precision when such computational facilities are available.

3.7.2 Plane Plate with No Generation

When there is no generation, $\dot{q} = 0$, the second term in Eq. 3.41 drops out and the expression for the new temperature T_n^{k+1} becomes simply

$$T_n^{k+1} = \left(T_{n-1}^k + T_{n+1}^k\right)/2 \tag{3.42}$$

Therefore, the graphical technique becomes very simple and consists of drawing one construction line to locate the new temperature. We recall that selecting $M = 2$ restricts the relationship between time increment and nodal separation to $\Delta t = \Delta x^2/2\alpha$, but the simplicity in the method is well worth the sacrifice in convenience. We also noted in Figure 3.23 that there was considerable error near the surface at the first time increment which was investigated by taking only half of the initial surface temperature jump for the first time increment, Δt.

To illustrate the construction method we again consider the plane plate 2 ft thick ($x/L = 2$) but this time with $\dot{q} = 0$. The initial temperature T_i is uniform at 0F throughout, as before; but we will consider slightly different boundary conditions in that we consider both surfaces to suddenly change from 0 to 1000F. By considering such a situation, we will be able to compare our results with those presented in Figure 3.7 for $N_{\mathrm{Bi}} = \infty$ since the sudden jump in surface temperature is caused by an extremely large heat transfer coefficient h. We select five nodes for our construction so that for a plate 2 ft thick, we have $\Delta x = 0.5$ ft and for $\alpha = 0.172$ ft^2/hr, as before, we have a corresponding time increment $\Delta t = (0.5)^2/2(0.172) = 0.727$ hr. Figure 3.24 shows the construction for eight time increments; points A and G in the figure are located at half the initial jump in surface temperature for use at the first time increment only. Working from left to right we connect points A and D to locate L, C, and E to locate D, and D and G to locate J so that line $MLDJH$ represents our estimate of the temperature profile one time increment after the boundary conditions have been imposed. To continue the construction we connect M and D to locate N, L and J to locate K, and D and H to locate P, thus determining line $MNKPH$ as the temperature profile at the end of two time increments. Continuing the construction in the same manner determines line $MSOQH$ at the end of $3\,\Delta t$, line $MTRUH$ at the end of 4 Δt, and so forth until line MH is reached at steady state. The temperature T_o at point R is about 630F at a time 4 $\Delta t = 4(0.727) = 2.9$ hr. The Fourier number $N_{\mathrm{Fo}} = \alpha t/L^2 = 0.172(2.9)/(1)^2 = 0.5$ and $N_{\mathrm{Bi}} = \infty$, so that we can read 0.37 for the temperature excess ratio from Figure 3.7. The excess ratio at point R in Figure 3.24 is $(T_o - T_\infty)/(T_i - T_\infty) = (630 - 1000)/(0 - 1000) = 0.37$ which compares exactly in this case with the ratio predicted by Figure 3.7.

When one selects an odd number of nodes, the temperature profiles have the V shape at the center of the plate as shown in Figure 3.24, but for an even number of nodes, the profile is flat at the center of the plate as shown in Figure

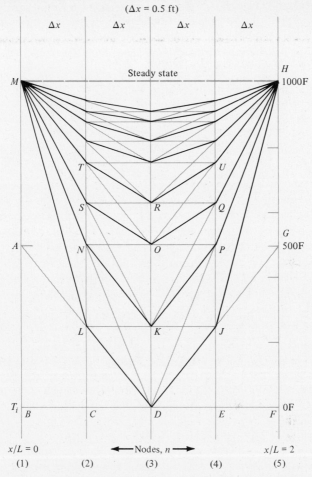

Figure 3.24 Graphical evaluation of Eq. 3.42 for the plane plate with no generation, uniform initial temperature, and sudden jump in temperature at both surfaces for five nodes.

3.25 for all six nodes. As the number of nodes is increased, Δx made smaller, the instantaneous estimates of the temperature profiles approach continuous curves and the construction labor increases.

3.7.3 Finite Convection at Surface

We will wind up our discussion of the graphical method with a look at how finite convection at the surface is handled and how one treats an insulated boundary. At the surface, at any instant of time, the conducted flux q_k must equal the convected flux q_c. Accordingly, we can approximate this condition by writing

$$ kA\left(\frac{\Delta T}{\Delta x}\right)_s \simeq hA(T_s - T_\infty) $$

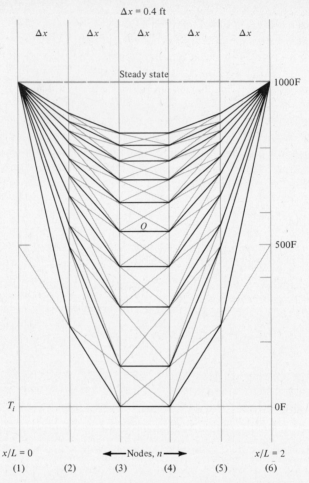

Figure 3.25 Graphical construction of temperature response for the plane plate with no generation, uniform initial temperature, and sudden temperature jump at both surfaces for six nodes.

or

$$\left(\frac{\Delta T}{\Delta x} \right)_s \simeq \frac{T_s - T_\infty}{k/h} \tag{3.43}$$

We write the approximately equal sign because we are letting $(\Delta t/\Delta x)_s$ approximate the derivative dT/dx in the solid at the surface. In order to obtain the condition specified by Eq. 3.43 in our graphical construction, we must shift the network $\Delta x/2$ to the left or right so that the nodal volume $A \Delta x$ is centered about the surface. This technique was first suggested by Schmidt and described by McAdams (12); therefore, the construction is often referred to as a Schmidt plot. The construction line $\Delta x/2$ outside the solid represents a fictitious Δx line that we will need to use as a reference line to determine either T_{n-1} or T_{n+1} for the old temperature depending on which surface we are near; this requirement will clarify itself as we proceed through an example.

In keeping with our illustrations above, we will continue to analyze the plane plate, however, this time 1 ft thick. The plate is insulated on the left face with the other face exposed to air at 1000F with an $h = 12.5$ Btu/hrft^{2}F and the initial temperature uniform at 0F as before. We select $\Delta x = 0.5$ ft and with $\alpha = 0.172$ ft^2/hr each time increment Δt represents 0.727 hr. With a $k = 10$ Btu/hrftF, we then lay out a construction line $k/h = (10/12.5) = 0.8$ ft from the right face of the slab. We next lay out the Δx grid lines by putting the first one $\Delta x/2$ distance outside the plate on the right face (in order to center the node volume at the surface as indicated above), then at distances Δx units apart and labeling the leftmost line (0) then (1), (2), and (3) as shown in Figure 3.26.

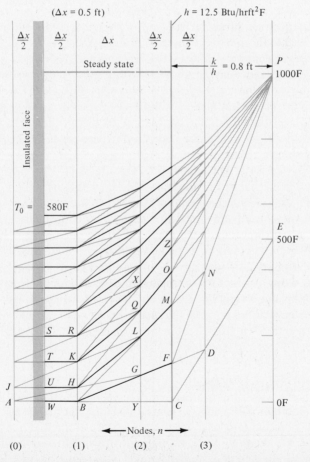

Figure 3.26   Graphical construction of temperature response for the plane plate with no generation, uniform initial temperature, insulated on one side, and finite convection on other side suddenly exposed to fluid at 1000F.

At the insulated face, Fourier's law, $q = -kA(dT/dx)$, requires that $dT/dx = 0$, which means in our construction $\Delta T/\Delta x = 0$ or the temperature line must be horizontal (parallel to the x axis) from line (0) to line (1) at each time interval.

We begin the construction by locating point E at $(T_i + T_\infty)/2 = 500F$ for the first time increment only. Connecting points E (on the k/h line) with point C (on the surface of the plate) locates reference point D on Δx line (3); connecting D and B determines T_2^1 at G. Connecting W and Y determines B so that $T_1^1 = T_1^0$ and no change in temperature has taken place in one time interval at Δx line (1). Line $WBGF$ then represents our graphical estimate of the temperature profile 1 Δt or 0.727 hr after exposure of the plate to the 1000F air. Continuing the construction, we connect P (at 1000F) to F, the surface temperature at 1 Δt, to locate reference point N. To determine T_2^2 and T_1^2 we connect B and N to determine L (which is T_2^2) and G and A to locate H (which is T_1^2). We then draw HU horizontally (and extend to locate J) because we require $\Delta t/\Delta x$ to be zero at the left (insulated) boundary. Line $UHLM$ represents the temperature profile at the end of two time intervals. Similarly, line $TKQO$ represents the profile at the end of 3 Δt, and line $SRXZ$ at the end of 4 Δt; Figure 3.26 shows the construction for 10 time intervals. We observe that the plate pictured and analyzed graphically in Figure 3.26 is half of a 2-ft plate with the same conditions on the left face as on the right face. At the end of 10 Δt the temperature of the left face $T_o = 580F$ corresponds to the center temperature of a $2L$ plate with finite convection on both sides. The excess ratio $(T_o - T_\infty)/(T_i - T_\infty)$ can then be compared to that predicted from Figure 3.7. For the graphical construction $(T_o - T_\infty)/(T_i - T_\infty) = (580 - 1000)/(0 - 1000) = 0.42$; and for a Biot number $hL/k = (12.5)(1)/10 = 1.25$ at a Fourier number (recalling 10 $\Delta t = 7.27$ hr) $N_{Fo} = \alpha t/L^2 = 0.172(7.27)/(1)^2 = 1.25$, we read from Figure 3.7 an excess ratio of 0.41 which agrees very well considering the fact that we took only three nodal points to work with in Figure 3.26.

We close our discussion of the graphical method by noting that it would be very easy to account for any describable variation of fluid temperature with time simply by shifting point P along the k/h line (see Figure 3.26) as required. One may just as easily account for a dependence of h with time by varying the distance k/h as the h variation prescribes. Doing so just slows the construction procedure somewhat, whereas in the analytical method variable surface conditions are very difficult to treat. This section completes our discussion of graphical procedures.

3.8 CONCLUDING REMARKS

The most important observation to make in looking back over our discussion of transient phenomena is to recognize the importance of the Biot number based on the characteristic length. For N_{Bi} less than 0.1 the temperature inside the solid is essentially uniform and the simple lumped parameter method describes the temperature response with time; while for N_{Bi} greater than 0.1 the results of analytical, numerical, or graphical solutions must be utilized to determine temperature response with time at a given position. In calculating the N_{Bi} number to make the above distinction, the characteristic length is the volume of the solid divided by the surface area through which the heat transfer takes place.

The usefulness of the graphical method has been seriously debated since the availability of high-speed digital computers has brought extensive computational power within reach of practically every engineer. The graphical method has one immediate, powerful benefit in that the temperature profiles are actually displayed and give an instantaneous picture of the temperature happenings. One could also watch a graphics terminal on a computer display such information, but the means and methods for commanding the graphics display are not as simple as the construction methods that we have discussed in Section 3.7. From the visualization standpoint, the graphical analysis is hard to equal.

PROBLEMS

1. (3.1) A large plate, 2 in. thick, which has a thermal conductivity of 10 Btu/hrftF and is at a uniform initial temperature of 100F is suddenly exposed to an air temperature of 200F on one side and an air temperature of 0F on the other side with a convection coefficient of 12 Btu/hrft^{2}F on both sides. (a) Sketch a few temperature profiles inside the plate between time zero and steady-state conditions. (b) What are the steady-state surface temperatures?

2. (3.1) The plate described in Problem 1 with the uniform initial temperature of 100F is exposed to the same air temperatures (200F on one side and 0F on the other side), but due to fans moving the air on each side, the convection coefficient is increased to 180 Btu/hrft^{2}F. (a) Sketch some temperature profiles in this case between time zero and steady state, and (b) determine the steady-state temperatures.

3. (3.1) The plate described in Problem 1 (uniform initial temperature 100F) is subjected to the same temperatures (200F on one side and 0F on the other side), but in this case the gas is helium instead of air and is at a much higher velocity so that the convection coefficient is raised to 2400 Btu/hrft^{2}F. (a) Sketch a few representative temperature profiles between time zero and steady-state conditions, and (b) determine the steady-state temperatures.

4. (3.1) A copper sphere 2.75 in. in diameter is at a uniform initial temperature of 350F and is gently submerged in a large amount of still water at 50F. Sketch the temperature profiles as a function of radius with time as a parameter. Discuss the physical happenings you envisioned to make your sketch in a paragraph or so.

5. (3.1) A long steel rod 1 in. in diameter is initially at 500F throughout. The rod is positioned in a moving stream of air at 50F so that the convective coefficient is 250 Btu/hrft^2. Assume that the thermal conductivity of the steel is about 24 Btu/hrftF. (a) Determine the characteristic length L for use in the Biot number; then calculate the Biot number, and (b) sketch some temperature profiles as a function of radial position with time as parameter.

6. (3.1) Study the plate that is insulated on one side and pictured in Figure 3.2. Note the plate is initially at temperature T_i throughout and at steady-state conditions the plate will be at $T_\infty < T_i$ throughout. Sketch the temperature T_o at the insulated face and the temperature T_s at the surface as a function of time (both temperatures on same plot) from time zero to steady-state conditions for the case where the Biot number is 10.

7. (3.1) Repeat Problem 6 for the case where the Biot number is 0.001. Compare your results with those of Problem 6 and comment.

8. (3.2) A carburetor body casting has a mass of 2.63 lb_m and comes out of the injection molder at a temperature of 650F. The mold operator picks the castings up with tongs and hangs them from a conveyor for transport to an inspection station. The surface area of the casting can be estimated to be about 0.8 ft^2 and the heat transfer coefficient about 10 Btu/hrft^2. If the air temperature around the conveyor is 65F, estimate, using the lumped parameter approach, how long before the casting can be safely handled without gloves (assuming most persons can tolerate 120F for a short period of time.

9. (3.2) A thin metal container 6 in. in diameter and 8 in. tall is filled with water to a depth of 6 in. The container and water are placed on the heating element of an electric stove whose temperature is 1000F. The water is initially at 50F and is well stirred during the heating process so that the internal resistance can be considered negligible. If it takes 6 min to increase the water temperature to 200F, estimate the conductance between the bottom of the container and the heating element.

10. (3.2) A mild steel slab, 1.5 in. thick, is heated to 1900F for a hot forming operation. If the minimum temperature for the hot forming operation is 1500F, how long can the plate be held in an environment at 80F with a heat transfer coefficient of 5 Btu/hrft^{2}F on the surfaces?

11. (3.2) A solar energy storage bin contains 1778 rocks that can be approximated as 3-in.-diameter spheres. Rock properties are $k = 2.2$ Btu/hrftF, $\alpha = 0.07$ ft^2/hr, $c_p = 0.2$ Btu/lb$_m$F, and density is 160 lb$_m$/ft^3. How long would it take to energy saturate the bin using 200F air from a solar collector? How much energy is stored in the bin under these conditions if the bin is initially at 79F and h for the bin is 5 Btu/hrft^{2}F?

12. (3.2) A thermocouple has a spherical tip that is 0.05 in. in diameter and initially at a temperature of 70F. The thermocouple is inserted into a stream of air whose temperature is 175F and which causes a heat transfer coefficient of 50 Btu/hrft^{2}F to exist at the tip of the thermocouple. Properties of the tip material are $k = 32$ Btu/hrftF and $\alpha = 0.6$ ft^2/hr. How long will it take the thermocouple to indicate within 1 degree of the air temperature; and what is the time constant?

13. (3.2) Determine the time constant for the bearing sleeve in Example 3.1 for the conditions described.

14. (3.2) The lumped parameter method is one way of easily estimating the heat transfer coefficient for a given physical geometry (when the Biot number is

less than 0.1) since only an accurate temperature-time measurement is required, which is simple to make with a good strip chart self-balancing potentiometer. Such a measurement was made with a pure copper sphere $\frac{1}{2}$ in. in diameter initially at 83F, which was gently tucked into still water at 63F and it was found to take 12.2 s for the temperature of the sphere to drop to 70.36F. Estimate the heat transfer coefficient between the sphere and the water during this time.

15. (3.3) A large steel plate (10 percent chrome) 2 in. thick has been hot rolled and is at 1200F. The plate is laid flat on an insulating layer of transite and subjected to a blast of air at 60F such that the heat transfer coefficient is 100 Btu/hrft^{2}F. How long will it take for the insulated side of the plate to drop to 200F?

16. (3.3) A board of yellow pine measuring $\frac{3}{4} \times 11.5$ in. is placed in a drying kiln where the gas temperature is 350F. If the board is initially at 40F, how long does it take to bring the center of the board up to 200F with the circulating hot gases causing an $h = 115$ W/m^{2}K on both sides?

17. (3.3) Two pieces of $\frac{1}{4}$-in. plate glass were joined with epoxy to form a plate $\frac{1}{2}$ in. thick. A very fine (No. 36 wire) thermocouple was laid in the epoxy before the two pieces of glass were sandwiched together so that a record of the centerline temperature history could be made. The plate, initially at 70F, was suspended in an oven where the air temperature was 300F and the thermocouple record indicated that the center plane temperature reached 249F in 6.31 min. Estimate the average heat transfer coefficient existing on the surfaces of the plate during this time.

18. (3.3) Determine the surface temperature of the $\frac{1}{2}$-in. plate glass sandwich of Problem 17 at the end of the 6.31-min time interval.

19. (3.3) A concrete (144 lb$_m$/ft^3) pillar 1 ft in diameter has a small copper water line running along the axis. The concrete is at a uniform temperature of 60F when suddenly the pillar is subjected to an ambient air temperature of -10F such that the heat transfer coefficient on the surface is 5 Btu/hrft^{2}F. Estimate how long it will take for the small copper line to reach the freezing temperature of water.

20. (3.3) A crossrib roast rolled can be approximated as a short cylinder 5 in. in diameter and 10 in. long. We can assume meat properties are near those of water so that $k = 0.335$ Btu/hrftF and $\alpha = 0.0054$ ft^2/hr. The roast is placed in an oven with an air temperature of 325F and the heat transfer coefficient on all sides is 10 Btu/hrft^{2}F. If the roast came out of a refrigerator at a uniform temperature of 40F, how long does it take to bring the center of the roast up to 200F?

21. (3.5) Square bars, 4 in. on a side, of 0.5 percent carbon steel are formed in a continuous molding process, then cut into convenient long lengths and allowed to cool before further processing. The bars are at a uniform temperature of 1700F as they are placed on racks to cool where the ambient air temperature is 100F and the resulting heat transfer coefficient on the

surfaces of the bar is 24 Btu/hrft^{2}F. Determine the temperature at the corner and at the centerline of a side 10 min after the bar has been set on the rack to cool.

22. (3.2) A ball bearing made of 2 percent manganese steel is 2 cm in diameter and receives a surface treatment process by quenching in oil from a uniform temperature of 800F. The oil is at 200F and the bearing should be removed from the bath when the temperature at a depth of 2 mm has reached 500F. If the heat transfer coefficient can be estimated to be 190 Btu/hrft^{2}F, how long should the bearing remain in the oil?

23. (3.4) A very strong blizzard suddenly reduces the surface temperature of the earth in an open area to −15F and stays that way for a 36-hr period of time. If the ground was initially at 60F, estimate how far the freezing temperature would penetrate, neglecting any latent heat effects due to the soil moisture, using a value of thermal diffusivity of 0.032 ft^2/hr.

24. (3.4) A styrofoam box has walls 2 in. thick and is filled with liquid nitrogen (LN2) at atmospheric pressure and a saturation temperature of 139R. The LN2 boils under these conditions so that the resulting heat transfer coefficient is very high and we may assume that the inside surface temperature of the styrofoam is suddenly changed from the initial temperature of 70F to 139R. The styrofoam has a thermal diffusivity of about 0.4 ft^2/hr. Estimate how long it takes for the temperature wave to reach the centerline of the box wall, assuming that a 1 degree drop indicates the wave has arrived.

25. (3.4) A thick steel slab is to be surface hardened by raising the temperature near the surface, then quenching. The slab is exposed to a gridwork of torches whose combustion temperature is 4000F and the result is a heat transfer coefficient at the surface of 200 Btu/hrft^{2}F. The slab has $k = 14$ Btu/hrftF and $\alpha = 0.26$ ft^2/hr, and is initially at 60F throughout. What is the surface temperature and the temperature at a $\frac{1}{4}$-in. depth 1 min after exposure to the torches?

26. (3.4) Reconsider Problem 23 for the situation where the weather front moves in, drops the temperature quickly to −15F, then becomes calm instead of a blizzard so that a heat transfer coefficient of 5 Btu/hrft^{2}F exists at the surface. The thermal conductivity of the earth can be taken to be 0.25 Btu/hrftF. Under these conditions, how far will the freezing temperature penetrate in 6 hr?

27. (3.4) Concrete starts to crumble and spall when the temperature rises much above 500F. How long would it take for the temperature, at a depth of 1 in., of a thick portland cement wall to rise to 500F from an initial temperature of 65F when the wall is suddenly exposed to hot gases from a fire at 1700F and $h = 10$ Btu/hrft^{2}F at the surface? Assume thermal diffusivity to be 0.007 ft^2/hr.

28. (3.5) A long Invar (36 percent nickel steel) rod, 10-cm-diameter has one end machined flat perpendicular to the axis. The rod is initially at 600F and is set to cool in 10C air such that the heat transfer coefficient on the surface is 43

W/m^2K. What is the temperature at a point 2 cm below the cylindrical surface and 4 cm in from the flat end after 20 min?

29. (3.5) A solid cube of material is 12 in. on a side, is at 1000F uniformly, and is brought into contact with air at 0F with an $h = 24$ Btu/hrft^{2}F on all sides. Thermal conductivity of the cube material is 12 Btu/hrftF and thermal diffusivity is 0.3 ft^2/hr. How long does it take for a point on the major diagonal halfway from the center to the corner to cool to 500F?

30. (3.5) A large flat plate has square edges so that a location near the middle of an edge could be considered to behave thermally like a semi-infinite plate. The plate is initially at 50F and is placed in an oven with an air temperature of 800F and an $h = 5$ Btu/hrft^{2}F on the surfaces. The plate material has $k = 2$ Btu/hrftF, $\alpha = 0.004$ ft^2/hr, and is 4 in. thick. What is the temperature of a point 1 in. in from the flat faces and 3 in. in from the edge 130 min after being placed in the oven?

31. (3.6) The flat plate analyzed in Section 3.6.3 had surface temperatures fixed. Consider the plate to be divided into five nodes with Δx, ρ, c_p, k, and α as the variables to be placed into contact with a fluid at fixed T_∞ and heat transfer coefficient h at the surface. Write the nodal energy balance, starting with Eq. 3.34, for node 5 on the right face of the plate and obtain an expression for T_5^{k+1} in terms of the pertinent variables and T_∞, T_4^k, and T_5^k.

32. (3.6) Consider a rod of diameter D which is attached to a wall at a fixed temperature T_w and is at a uniform temperature T_w along its length. The rod is then cooled by blowing air at a temperature $T_\infty < T_w$ over the length of the rod. To analyze this situation numerically, we would subdivide the rod into a number of equally spaced nodes Δx units apart and write energy balance equations for each node. Write the nodal energy balance for a node (say, $n = 8$) near the middle of the rod and obtain an expression for the new temperature T_8^{k+1} at that node in terms of the pertinent rod properties ρ, c_p, k, α, and the heat transfer coefficient h.

33. (3.6) In the program described in Table 3.4 for the plane plate, check to make sure you have set N = 5 nodes and make a few runs with different values for the criteria for steady state that is used in internal statement number (ISN) 48. Use 1.0, 0.1, 0.01, 0.001, 0.0001, and 0.00001 but first add the statement ERR = (TEMP(3,2) − 1000.) right after 22 CONTINUE then insert PRINT 525, ERR just ahead of GOTO 35. Put the statement 525 FORMAT (36X, 'ERR(3,2) = ', F9.6) in front of the end statement. Plot ERR as a function of SSC on suitable graph paper and choose a scale to best portray the results.

34. (3.6) Make a run with your program listed in Table 3.4 set up for DT = 0.04 and N = 17 nodes. Check and make sure your steady-state criterion is 0.001, then compare your results with Tables 3.5 and 3.6 at the same times and for steady state.

35. (3.7) A ladle for transporting molten alloy steel ($T = 2400$F) has refractory brick walls ($\alpha = 0.035$ ft^2/hr) that are 8 in. thick. The outer shell of the

ladle is water cooled so that the outer surface of the brick is held at 150F. Consider the brick to be at a linear initial temperature between 350F on the inner face and 150F on the outer face, and that the inner surface changes suddenly to 2400F when the metal is poured. Determine the temperature on the centerline of the brick 30 min after pouring. Use a graphical construction with $\Delta x = 1$ in.

36. (3.7) Consider the plane wall with uniform generation which is illustrated in Figure 3.23. Using the graphical method with $\Delta x = 0.5$ ft as shown in the figure, estimate the steady-state temperature at the centerline of the wall and the length of time required to reach the steady-state temperature for the case where both surfaces are held at 0F for the given generation rate of $\dot{q} = 10,000$ Btu/hrft^3, thermal conductivity $k = 10$ Btu/hrftF, thermal diffusivity $\alpha = 0.172$ ft^2/hr, and uniform initial temperature of 0F throughout the wall. Compare your result with what you would predict by using Eq. 2.39 of Section 2.4.1.

37. (3.7) The temperature at point O in Figure 3.25 can be estimated to be about 540F. For the time indicated by location O, compare the temperature at the center of the 2-ft plate with that you would predict from Figure 3.7 for the appropriate Fourier number and Biot number.

38. (3.7) A very thick concrete slab is initially at 50F throughout. A weather front comes through and drops the temperature at the rate of 28.8F/hr until 0F is reached; then the air stays at 0F from then on. The conductivity of the slab is 0.8 Btu/hrftF and an $h = 8$ Btu/hrft^{2}F exists at the surface of the slab. How long does it take for the temperature at a depth of 1.5 in. to drop to 32F for a thermal diffusivity $\alpha = 0.02$ ft^2/hr for the concrete?

References

1. Myers, G. E., *Analytical Methods in Conduction Heat Transfer*, McGraw-Hill, New York, 1971, p. 206.
2. Beckwith, T. G. and Buck, N. L., *Mechanical Measurements*, 2d Edition, Addison-Wesley, Reading, Mass., 1973, p. 21.
3. Myers, G. E., *Analytical Methods in Conduction Heat Transfer*, Chapter 3.
4. Heissler, M. P., "Temperature Charts for Induction and Constant Temperature Heating," *Transactions ASME*, Vol. 69, 1947, pp. 227–236.
5. Boelter, L. M. K., Cherry, V. H., Johnson, H. A., and Martinelli, R. C., *Heat Transfer Notes*, McGraw-Hill, New York, 1963 Anniversary Edition of 1941 Notes.
6. Schneider, P. J., *Temperature Response Charts*, Wiley, New York, 1963.
7. Myers, G. E., *Analytical Methods in Conduction Heat Transfer*, p. 217.
8. Gröber, H., Erk, S., and Grigull, U., *Fundamentals of Heat Transfer*, McGraw-Hill, New York, 1961, p. 50.
9. Schneider, P. J., *Conduction Heat Transfer*, Addison-Wesley, Reading, Mass., 1955, p. 240.
10. Abramowitz, M., and Stegun, I. A., Eds., *Handbook of Mathematical Functions*, NBS AMS-55, June 1964.
11. Taylor, A. E., *Advanced Calculus*, Ginn and Co., New York, 1955, p. 55.
12. McAdams, W. H., *Heat Transmission*, 3d Edition, McGraw-Hill, 1954, pp. 45–48.

Part II
CONVECTION

Chapter 4
Basic Convection Concepts

4.0 INTRODUCTION

The processes involving the transfer of heat between a solid surface and a fluid, liquid, or gas are called convection. Since a fluid is able to move, the process of heat transfer involves a transport of energy by mass motion in addition to the mechanism of Fourier conduction. The nature of the fluid motion present has therefore a profound effect on the magnitude of the heat energy transferred to or from the surface.

A fluid will move either due to buoyancy effects (density differences) or due to an enforced pressure drop. Buoyancy induced velocities are very small compared to velocities that can be obtained by an induced pressure drop; consequently, the heat fluxes due to the forced velocities are much greater than those due to buoyancy effects. Convection heat transfer from buoyancy caused velocities is called *free convection* and that due to pressure drop induced velocities is called *forced convection*. We shall describe these two phenomena in more detail below and discuss the means for calculating heat transfer in separate chapters for free and forced convection.

It will be necessary to review some aspects of thermodynamics as they relate to energy conservation for a flowing fluid as well as some fluid mechanics aspects relating to friction factor definitions and pressure drop calculations. After discussing similarity and introducing the dimensionless numbers we will be working with, we will look at a simple relationship between fluid friction and turbulent convective heat transfer which can be used to make estimations when no specific heat transfer correlations are available for the fluid and geometry being considered.

In Chapter 1 we set forth the basic defining equation for the convective heat flux in Eq. 1.1 as

$$q_c = hA(T_s - T_\infty) \tag{4.1}$$

which we repeat here for convenience. We defined h as the heat transfer coefficient, also called the convective coefficient, or the unit area conductance in either Btu/hrft^{2}F or W/m^{2}K, depending on the choice of BE or SI units, and we noted A was the area in contact with the fluid and perpendicular to the direction of q_c at the surface. The temperature T_s is the surface temperature and T_∞ is the temperature of the fluid far from the surface for external phenomena such as flow over a flat plate or across a single cylinder. For flow inside ducts T_∞ must be interpreted as that temperature which is indicative of the energy content of the fluid at the cross section under consideration. Such a temperature is usually called the bulk temperature or the mixed mean temperature and can be defined as the ratio of the total energy carried past a cross section by the fluid per unit time $\dot{E}$ to the energy capacity of the fluid per unit time $\dot{C}$. Mathematically the concept of bulk temperature is expressed as follows for flow in a duct of radius r_0:

$$T_b = \frac{\dot{E}}{\dot{C}} = \frac{\int_0^{r_0} \rho u c_p T 2\pi r \, dr}{\int_0^{r_0} \rho u c_p 2\pi r \, dr} \tag{4.2}$$

where ρ, u, c_p, and T are the density, velocity, specific heat at constant pressure, and temperature at radial distance r from the centerline of the round duct of radius r_0. We shall refer to Eq. 4.2 for understanding and visualization of the concept; we will not be making calculations with Eq. 4.2. An allied concept is expressed by the bulk velocity u_b which is representative of the mass motion past a given cross-sectional area A_c. The bulk velocity is sometimes referred to as the average velocity, especially for the case of isothermal flow. For a circular duct of radius r_0 the bulk velocity is defined mathematically as

$$u_b = \frac{\int_0^{r_0} 2\pi \rho u r \, dr}{\int_0^{r_0} \rho 2\pi r \, dr} \tag{4.3}$$

or as expressed by the continuity equation

$$u_b = \frac{\dot{m}}{\rho_b A_c} \tag{4.4}$$

Rather than compute u_b from the integral in Eq. 4.3 one generally evaluates ρ_b

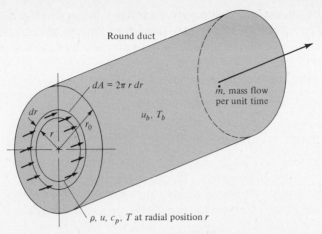

Figure 4.1 Representation of flow through a differential ring element dA for visualization of the integrals in Eqs. 4.2 and 4.3 for u_b and T_b.

from the equation of state for the substance using the temperature T_b; the bulk velocity can then be calculated from Eq. 4.4 and the known mass flow rate $\dot{m}$. For isothermal flow the density ρ_b is simply the density at the temperature of the fluid; for nonisothermal flow the density ρ_b is evaluated at the bulk temperature since that density is associated with the mass flow. A visualization of how we could determine u_b and T_b experimentally in steady state flow will help to put these ideas in perspective. The two items, u_b and T_b, can be visualized by a mental experiment wherein we cut the tube at the cross section of interest; collect the fluid in an insulated container for a given time; stir gently until the temperature is uniform; and then weigh the amount collected. The temperature measured is the bulk temperature T_b and the amount collected per unit time is $\dot{m}$, thus permitting calculation of u_b from Eq. 4.4 with the density at the temperature T_b. Figure 4.1 presents a visualization of density and flow through the differential ring area $dA = 2\pi r\, dr$ which can be integrated over the cross section in Eqs. 4.2 and 4.3 to obtain bulk temperature and bulk velocity when the variation of temperature and velocity with radius is known. Equation 4.4 is the one we will use to relate mass flow to bulk velocity.

Before we proceed to discuss the types of flow conditions that have an influence on the heat transfer coefficient h we should briefly describe the properties that have an important bearing on the subject.

4.1 PHYSICAL PROPERTIES

The properties of the fluid that are required in the calculation of h are the density ρ, the specific heat c_p, the thermal conductivity k, the volume expansivity β, and the viscosity μ. There are some combinations of properties that have special importance such as the momentum and thermal diffusivities, ν and α. The density and specific heat we have described in Chapter 2, Eqs. 2.3 and 2.4.

Thermal conductivity also follows from the defining equation for Fourier's law (Eq. 1.2 and 1.3); and we further note that k for a liquid represents the transport ability under absolutely still conditions, no aggregate fluid motion of any kind.

4.1.1 Volume Expansivity

The coefficient of volume expansivity β is defined as

$$\beta = \frac{1}{v}\left(\frac{dv}{dT}\right)_P \tag{4.5}$$

which is the rate of change of specific volume with temperature at constant pressure per unit specific volume. The derivative must be obtained from data or an equation of state. Equations of state are more difficult to specify for liquids than for gases. For perfect or ideal gases whose equation of state is $Pv = RT$ the derivative in Eq. 4.5 is easy to evaluate; since $v = (R/P)T$ we obtain $(dv/dT)_P = R/P$ and $1/v = P/RT$ so that substituting in Eq. 4.5 yields

$$\beta = \frac{1}{v}\left(\frac{dv}{dT}\right)_P = \frac{P}{RT}\frac{R}{P} = \frac{1}{T} \tag{4.6}$$

The units of β are either R^{-1} or K^{-1} in absolute temperature Rankine or Kelvin degrees. The reader should note that R in the gas law ($PV = RT$) is the specific gas constant $ftlb_f/lb_m R$ or Nm/kgK and should not be confused with Rankine degrees. The volume expansivity is an essential component of the Grashoff number that we will use in Chapter 5.

4.1.2 Viscosity

The viscosity μ is the proportionality constant between shearing stress and shearing strain for a fluid in laminar motion. Thus

$$\tau = \mu\frac{du}{dy} \tag{4.7}$$

where τ is the fluid shear stress and the velocity gradient du/dy is the fluid strain. When μ is a constant, the fluid is called a Newtonian fluid. The units for viscosity are defined by Eq. 4.7; the shear stress is lb_f/ft^2 and the shear strain is $1/s$ so that units for viscosity are $lb_f s/ft^2$ in BE units and Ns/m^2 in SI units. We will be using the viscosity of fluids in the calculation of Reynolds number (described in Section 4.4) which contains mass units in the density property. Accordingly, it is convenient to have viscosity also in lb_m units. The conversion is accomplished by using the constant of proportionality g_c from Newton's Eq. 1.17 such that

$$\begin{aligned}
\mu_m &\equiv g_c\mu_f \equiv \left(kgm/Ns^2\right)\left(Ns/m^2\right) \\
&\equiv kg/ms \text{ (SI units)} \\
&\equiv \left(lb_m ft/lb_f s^2\right)\left(lb_f s/ft^2\right) \\
&\equiv lb_m/fts \text{ (BE units)}
\end{aligned} \tag{4.8}$$

The kinematic viscosity ν is obtained when the dynamic viscosity μ is divided by density. Using consistent units one obtains

$$\nu \equiv \frac{g_c \mu_f}{\rho} \equiv \frac{\mu_m}{\rho}$$

(4.9)

$$\equiv \text{ft}^2/\text{s or m}^2/\text{s}$$

The kinematic viscosity has no force or mass units; it has the units of a diffusion coefficient in square measure/unit time, and can be interpreted as the diffusion coefficient for momentum away from a surface to a fluid in laminar flow.

4.1.3 Thermal Diffusivity

We have already encountered this property (as a combination of the properties k, ρ, c_p) in Section 2.2 and in Section 3.1 where we observed that the thermal diffusivity $\alpha = k/\rho c_p$ determined how fast a solid was able to respond to a temperature transient impressed on the surface. We note a similar interpretation in the understanding of α for a fluid in that α represents the rate at which thermal energy is diffused into (or out of) a fluid that is either not in motion or in laminar flow. The units of α are also square measure/unit time, either ft^2/s or m^2/s, and do not contain any force, mass, or thermal units as do the constituents. We might note that the product of ρc_p is mostly influenced by the density ρ since the specific heat ranges over a relatively narrow range (about 0.2 to about 1.4 $\text{Btu}/\text{lb}_m\text{F}$ in BE units for most gases and liquids) so that the range of values to be expected for α for fluids is quite large. For example, at 70F for air $\alpha = 0.83$ ft^2/hr and for water at the same temperature $\alpha = 0.0056$ ft^2/hr (0.0771 compared to 0.00052 m^2/s in SI units). In order to form an understanding of the influence thermal diffusivity exerts, we will look at an example of pure thermal diffusion in fluids (Example 4.1).

Example 4.1

In order to consider pure thermal diffusion in a fluid, we must arrange the physical situation such that there can be no mass motion due to buoyancy effects. To eliminate buoyancy effects, we orient the negative temperature gradient (decreasing temperature) in the same direction as the gravity vector. Accordingly, we will consider a vertical cylinder of air and a vertical cylinder of water, say, both 6 in. in diameter, contained in a long, thick-walled styrofoam pipe as shown in the figure. The columns of air and water are long enough so that no temperature effects are evident at the bottom. The air and water are initially at 40F throughout and at time zero we place a thermostatically controlled hot plate on top of each column and maintain the upper surface at a steady value of 100F. We then ask: How long does it take for the temperature in a plane 6 in. below the 100F surface to reach 70F, assuming that the styrofoam is a perfect insulator, that is, one-dimensional heat flow in the negative z direction?

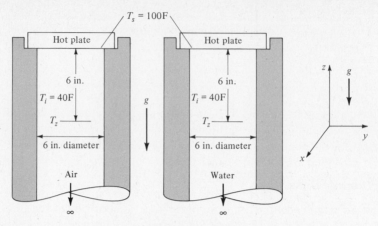

Figure for Example 4.1.

SOLUTION

Since the buoyancy forces are blocked and there is no mass motion, the heat transfer phenomena is one of pure conduction (thermal diffusion) in the direction normal to the surface and corresponds to the semi-infinite solid as described by Eq. 3.24 of Section 3.4.1. We can therefore determine that the temperature excess ratio $(T - T_\infty)/(T_i - T_\infty)$ corresponds to $(T_{6\ in.} - T_s)/(T_i - T_s)$ in the present situation (since we have suddenly changed the surface temperature from 40 to 100F), and is given by $(70 - 100)/(40 - 100) = 0.5$. Table 3.3 shows that the argument z for the error function $\text{erf}(z) = 0.50$ is $z = 0.475$, and since $z = x/2\sqrt{\alpha t}$ we can solve for the time $t = x^2/(4\alpha z^2)$. With $x = \frac{6}{12} = 0.5$ ft and at an average temperature of $(70 + 100)/2 = 85$F for air $\alpha = 0.873$ ft^2/hr, and for water $\alpha = 0.00575$ ft^2/hr so that we can calculate the times as follows

air: $\quad t = (0.5)^2/4(0.873)(0.475)^2 = 0.317$ hr

water: $\quad t = (0.5)^2/4(0.00575)(0.475)^2 = 48.2$ hr

These times represent minimum times in that we are not considering any energy that may be absorbed by the styrofoam walls. The reader will note that air takes about 20 min and water takes about two days to reach the 70F level 6 in. below the 100F surface under the conditions given. We could also calculate the variation of temperature with time at the location $x = 6$ in. and plot it as shown in Figure 4.2. The figure also emphasizes the fact that for air the temperature rises much more rapidly than it does for water because air has a much lower thermal capacity. It takes much less energy to raise a unit volume of air 1 degree than it takes for water.

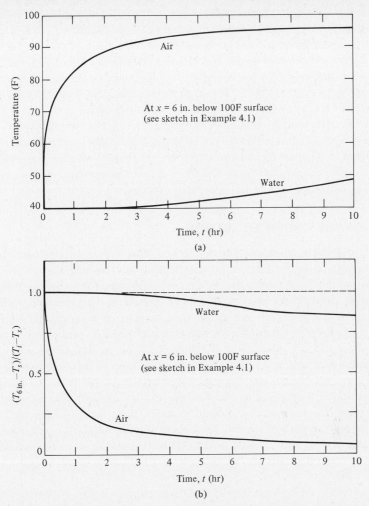

Figure 4.2   Temperature variation with time due to thermal diffusion in air and water according to Example 4.1 at a level 6 in. below the 100F surface.

4.1.4 Prandtl Number

The combination of properties $c_p \mu / k$ is also a property (but dimensionless) and is called the Prandtl number after Ludwig Prandtl, who was a pioneer in boundary layer studies. We can express the Prandtl number as $N_{Pr} = c_p \mu / k = \nu / \alpha$, the ratio of momentum diffusion to thermal diffusion in the fluid while at rest or during laminar flow. Under turbulent flow conditions other aspects also transport momentum and thermal energy in addition to the ν and α mechanisms. We should have some idea of the range in Prandtl number for different substances. For example, liquid metals have a very low Prandtl number; the value for liquid sodium ranges from 0.01 to 0.003 and decreases as the temperature increases. Gases have values that range from about 0.7 to 1 and are relatively constant over a wide range in temperature. Liquids exhibit the largest spread in Prandtl number values and range in value from about 1 upwards into the thousands. Glycerine, for example, has $N_{Pr} = 1.5$ at about 130F and increases to $N_{Pr} = 85,000$ at 32F. The large variation in Prandtl number for liquids is due mainly to the sensitivity of viscosity to temperature.

The Prandtl number is the key factor in relating the thermal and velocity boundary layers in forced flow along a surface, an aspect we will examine in more detail in Chapter 6, and consequently, the Prandtl number represents the property influence in the dimensionless heat transfer correlations that we will encounter and work with in the chapters to follow.

4.2 BUOYANCY INDUCED MOTION

For those physical situations where the heat flux vector (more precisely, the direction of maximum temperature gradient) is not parallel and in the same direction as the force vector due to gravitational attraction, the buoyant force vector can cause fluid motion. We have looked at the one case where a temperature gradient does not cause fluid motion in Example 4.1 and pure conduction exists. In Chapter 5 we will examine the heat transfer consequences of buoyancy induced motion; at this time we want to show only some major configurations in which it occurs and see what some of the flow patterns look like. Figure 4.3 shows the buoyancy induced flow patterns on horizontal cylinders and on horizontal and vertical plates that are heated and cooled. The arrows show the motion of the fluid particles near the surface which carry energy to or from the surface. The reader will note the similarity in flow patterns for (a) and (b), (c) and (d), and (e) and (f) in Figure 4.3; as a consequence the heat transfer correlations we will encounter in Chapter 5 will apply to those physical situations whose geometry and flow patterns are similar. In enclosed spaces between heated and cooled surfaces, a circulation pattern is set up involving flow patterns like those shown in Figure 4.3. For example, if one vertical side of an enclosed box is heated and the opposite side cooled, the fluid inside will move up the heated side and down the cooled side carrying energy from the hot wall to the cooler wall. Correlations for such configurations are not only geometry sensitive but also depend on the inclination of the surfaces with the vertical to a greater degree than do surfaces that do not enclose a fluid. Fluid motion is not always regular, but may split up into so-called convection cells. The phenomena that take place in the bottom of a shallow pan of water that is being brought up to the boiling point illustrate the concepts involved in convection cells, an experiment that can easily be done by the reader. The above brief description introduces a physical understanding of the happenings that take place in a fluid when the only mass motion present is due to density differences. We now turn our attention to visualization of the flow phenomena when pumps or blowers force a fluid over a surface or through a duct.

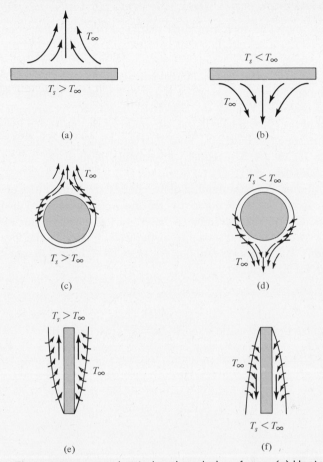

Figure 4.3   Flow patterns near heated and cooled surfaces. (a) Heated plate facing up. (b) Cooled plate facing down. (c) Heated horizontal cylinder. (d) Cooled horizontal cylinder. (e) Heated vertical plate. (f) Cooled vertical plate.

4.3 FORCED FLOW INSIDE DUCTS

A fluid will move through a duct when some agency imparts energy to the fluid; generally such an agency is a pump, a fan, a blower, or similar equipment that does work on the fluid. In real fluids that undergo viscous shear, there will be a pressure drop in the flow direction and the velocity will be zero at the wall when the fluid is at a pressure high enough to be considered a continuum. We will not deal with individual molecular interaction with the confining surface to or from which we transfer thermal energy.

We will review some concepts from thermodynamics, namely, the continuity equation and the first law and also some concepts from fluid mechanics which deal with the nature of the fluid motion. Aspects of the relationship between fluid friction and pressure drop will be covered in Section 4.5.

4.3.1 Laminar Flow

When a fluid moves in such a manner that a particle with the same density as the fluid describes a smooth line that is essentially parallel to the boundary, the flow phenomenon is called laminar. The name is indicative of a layerlike action since we speak of laminations as an agglomeration of layers. When plywood was first introduced, reference was often made to laminated plywood, but because the product is so well known, the adjective "laminated" is rarely, if ever, used. The motion of the fluid in layerlike sheets could be demonstrated by introducing a dye stream into the flow of a fluid like water and observing the motion as suggested in Figure 4.4. The reader will probably recall seeing such a sketch in a fluid mechanics text.

The criterion for ascertaining flow condition is the ratio of viscous to momentum forces as represented by the Reynolds number. For flow inside round ducts, the Reynolds number is based on the diameter (for external flow the length criterion is the distance from the leading edge to the location under consideration). The Reynolds number, N_{Re}, can be expressed in several different ways; the defining equation is

$$N_{Re} = \frac{D u_b \rho}{\mu} \tag{4.10}$$

If we recognize the kinematic viscosity $\nu = \mu/\rho$ we can write

$$N_{Re} = \frac{D u_b}{\nu} \tag{4.10a}$$

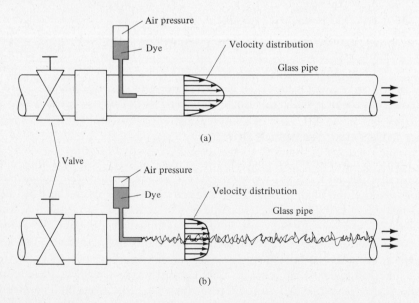

(a)

(b)

Figure 4.4   Behavior of dye streams in (a) laminar flow and (b) turbulent flow in a transparent duct.

a particularly convenient form because we can see that Du_b has dimensions of m^2/s or ft^2/s, which are the same as those for ν, an immediate visual dimension check. We can also use the continuity equation $\dot{m} = \rho u_b A$ (see Eq. 4.4) to substitute for ρu_b in Eq. 4.10 to obtain

$$N_{Re} = \frac{D\dot{m}}{\mu A} \tag{4.10b}$$

which for a round duct where $A = \pi D^2/4$ becomes

$$N_{Re} = \frac{4\dot{m}}{\pi D \mu} \tag{4.10c}$$

a form that visually demands that the units of μ must be the same as those of $\dot{m}/D$, lb_m/fts or kg/ms, for example. The flow rate per unit cross-sectional area $G = \dot{m}/A$ can also be used to delineate Reynolds number in Eq. 4.10b so that

$$N_{Re} = \frac{DG}{\mu} \tag{4.10d}$$

The reader will note that when the bulk velocity is known, Eqs. 4.10 or 4.10a are most convenient; but when the mass flow rate is known, then 4.10b, 4.10c, or 4.10d are more convenient to use.

Under normal circumstances $N_{Re} = 2300$ represents the upper limit for laminar pipe flow; at 2300 the flow generally starts to become turbulent and depending on the amount of vibration present, becomes fully turbulent at about $N_{Re} = 5000$ to 10,000. The range $2300 < N_{Re} < 5000$ is called the transition region and it is undesirable to operate in this region since the heat transfer aspects cannot be reliably predicted. For some flow situations having smooth walls, where there are no pumps and there is a lack of vibration, laminar flow and hence, the transition region, may be shifted to a higher Reynolds number range than given above.

In laminar flow, energy can be transmitted to the fluid only by conduction in the direction perpendicular to the wall. Some conduction can occur in the flow direction, but compared to the energy carried by the fluid, that is a negligible amount for fluids other than liquid metals, except for very low flow rates. The Peclet number, $N_{Pe} = DGc_p/k$, represents the ratio of the energy carrying capacity of the fluid to the conduction ability of the fluid. At Peclet numbers greater than 100, axial fluid conduction has a negligible influence on the bulk and local temperatures and on the heat transfer characteristics.

4.3.2 Turbulent Flow

At values of the Reynolds number above 5000, the nature of the flow is such that the momentum forces become predominant and the fluid motion departs from a smooth layerlike flow to a situation where there is significant motion in the direction transverse to the direction of the main mass flux. Figure 4.4(b) shows how the dye filament is broken up and spread throughout the channel. Such action is very effective in transferring heat from the wall into the mainstream of the fluid and we would expect heat transfer coefficients for turbulent flow to be much higher than those for laminar flow for a given fluid.

The relationship between shearing stress and shearing strain in turbulent flow is complex due to the chaotic nature of the flow. A simple model originally proposed by Reynolds (1) postulates a momentum diffusivity coefficient ε_M for turbulent motion with units m^2/s or ft^2/s like ν for laminar flow so that turbulent shear stress can then be expressed as a linear combination of the laminar and turbulent coefficients

$$g_c\tau = (\mu + \rho\varepsilon_M)\frac{du}{dy} \tag{4.11}$$

or by substituting from Eq. 4.9 $\mu = \rho\nu$ we can write

$$\frac{g_c\tau}{\rho} = (\nu + \varepsilon_M)\frac{du}{dy} \tag{4.12}$$

In a similar manner Reynolds proposed a turbulent exchange coefficient ε_H that enhanced the conduction mechanism in a linearly additive way so that turbulent heat transfer can be written as

$$\frac{q}{A} = -(k + \rho c_p\varepsilon_H)\frac{dT}{dy} \tag{4.13}$$

The turbulent exchange coefficient ε_H in ft^2/s or m^2/s is a measure of the diffusion of thermal energy due to transverse motion of the fluid to the main flow direction, like the thermal diffusivity α (also in ft^2/s or m^2/s) is a measure of thermal diffusion of a fluid in laminar flow. We can express these latter two ideas more clearly by factoring out ρc_p in Eq. 4.13 and remembering that $\alpha = k/\rho c_p$ so that

$$\frac{q}{A\rho c_p} = -(\alpha + \varepsilon_H)\frac{dT}{dy} \tag{4.14}$$

The combination $\alpha + \varepsilon_H$ then accounts for the diffusion of energy in the y direction transverse to the x direction of fluid motion (see coordinate axes in Figure 4.5). Equations 4.12 and 4.14 indicate that proportionality can be expected between turbulent shear (fluid friction) and heat transfer. We will explore that aspect in more detail in Section 4.7 after we discuss some ideas concerning external flow and fluid friction.

4.4 FORCED FLOW EXTERNAL TO SURFACES

The flow conditions discussed in the previous two subsections dealt with the case where the fluid was flowing in a duct such that the entire circumference was wetted. In this section we take a brief look at the physical happenings in the case where the fluid is not in a confining passage, but flows over a surface and exchanges momentum and energy with the surface. Some commonly recognized applications would be the flow of air over the surface of an automobile or an airplane. There the skin friction or surface shear stress gives rise to a drag force that in addition to the form drag force must be overcome by the power plant in the vehicle to maintain motion. In heat exchangers, fluids flow normally over the outside of tubes and exchange energy, as in air conditioning systems. We desire, therefore, to obtain a visualization of the external flow situation for flat plates and flow over cylinders.

Figure 4.5 shows a sketch of fluid flow phenomena near the surface of a flat plate. There is a uniform velocity field u_∞ and at temperature T_∞ far from the plate. The fluid is dragged to zero velocity at the surface of the plate ($y = 0$) through the action of viscous shear in the layers of fluid very near the surface. The force F_D is required to balance the wall shear on both sides of the plate and to aid in keeping the plate either in position if the fluid is in motion or in motion if the fluid is at rest. Near the front edge of the plate the interaction layer is thinner than at the back edge. The velocity variation with distance y (perpendicular to the plate surface) is shown in Figure 4.6. At location ⓪ there is no interaction, but at location ① shear at the surface slows down the fluid and the location $y = \delta_1$ where the velocity u_∞ starts to decrease is called the momentum boundary layer thickness. The boundary layer is a concept that describes the layer of fluid close to the plate where virtually all of the viscous interaction takes place; skin friction takes place at the surface where $y = 0$. A similar concept can be described for the energy interaction process as shown in Figure 4.7(b). The temperature T_∞ is greater than the surface temperature T_s (the fluid puts energy into the surface, or is cooled by the surface) and that location $y = \delta_t$ where the temperature just begins to drop is called the thermal boundary layer thickness. Eckert (11) has shown that the momentum boundary layer thickness and the thermal boundary layer thickness are related through a function of the Prandtl number ν/α such that the ratio $\delta/\delta_t = \sqrt[3]{N_{Pr}}$ for the laminar boundary layer.

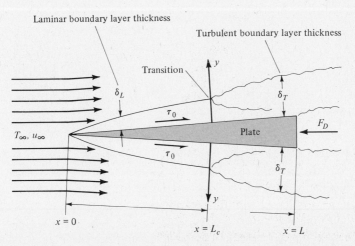

Figure 4.5 External flow phenomena on a flat plate showing boundary layer growth and transition from laminar to turbulent condition.

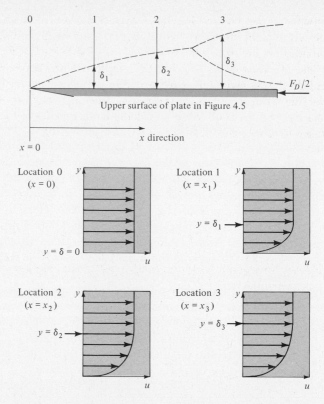

Figure 4.6 Formulation of boundary layer concept on a flat plate.

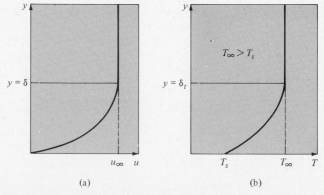

Figure 4.7 Velocity and temperature profiles at a particular location on a surface showing (a) momentum boundary layer thickness δ and (b) thermal boundary layer thickness δ_t.

The growth of the boundary layers depends on the distance from the leading edge of the plate, the fluid properties, and the free stream velocity u_∞. The flow conditions are described by the local Reynolds number $N_{\text{Re}x} = x u_\infty \rho_\infty / \mu_f$, and at values of $N_{\text{Re}x} < 5 \times 10^5$ the flow in the boundary layer is laminar. Schlichting (8) gives the solution of the governing partial differential equations for the laminar boundary layer, and from the variation of u with distance from the surface, Schlichting relates the laminar boundary layer thickness to distance ratio to the local Reynolds number as follows:

$$\frac{\delta_L}{x} = \frac{5}{\sqrt{N_{\text{Re}x}}} \tag{4.15}$$

where

$$N_{\text{Re}x} = \frac{u_\infty x}{\nu}$$

For the turbulent boundary layer Schlichting calculates the boundary layer thickness to distance ratio to be

$$\frac{\delta_T}{x} = \frac{0.370}{\left(N_{\text{Re}x}\right)^{0.2}} \tag{4.16}$$

based on the assumption of a $\frac{1}{7}$ power law velocity profile. Experiments have shown that the laminar boundary layer persists to a local Reynolds number of approximately 5×10^5, which for a given velocity u_∞ determines the critical length $x = L_c$ (shown in Figure 4.5), where the transition from a laminar to turbulent flow begins.

Example 4.2

Calculate the thickness of the boundary layer on a flat plate at 160 km/hr, 6 in. from the leading edge for an air temperature of 60F and 1 atm.

SOLUTION
To determine whether the boundary layer is laminar or turbulent, we calculate the local Reynolds number. The free stream velocity u_∞ is determined as

$$u_\infty = \frac{160}{1.6} \frac{5280}{3600} = 146.7 \text{ ft/s}$$

$$\left.\begin{array}{l} 5280 \text{ ft/mi} \\ 3600 \text{ s/hr} \\ 1.6 \text{ km/mi} \end{array}\right\} \text{conversion factors}$$

at 60F ν for air $= 0.5685 \text{ ft}^2/\text{hr}$

Therefore

$$N_{\text{Re}x} = \frac{u_\infty x}{\nu}$$

$$= \frac{146.7(0.5)3600}{0.5685}$$

$$= 464{,}500$$

a value just slightly less than 5×10^5 so that we are close to the transition point but still laminar. Accordingly,

$$\delta_L = \frac{5(6)}{\sqrt{464{,}500}} = .044 \text{ in.}$$

$$= 1.12 \text{ mm}$$

The reader will note upon close observation that the thicknesses calculated from Eqs. 4.15 and 4.16 do not coincide at $N_{\text{Re}x} = 5 \times 10^5$ due to the assumption of the $\frac{1}{7}$ power law velocity profile and the assumption that the turbulent boundary layer starts from the leading edge. The values of boundary layer thickness are very small, and are subject to inaccuracies as we saw above, so that we cannot expect prediction of the transition location from values of the thicknesses. The transition Reynolds number of 5×10^5 was estimated from measured values of drag coefficient as shown in Figure 4.8; the value of $N_{\text{Re}L} = 5 \times 10^5$ for transition represents an upper limit on the persistence of the laminar boundary layer.

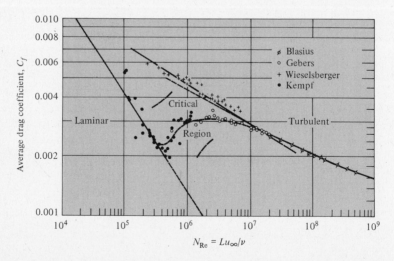

Figure 4.8 The average drag coefficient on a flat plate as a function of the Reynolds number at the end of the plate according to Rouse (12). Reproduced with permission of Dover Publications, Inc.

4.4.1 Skin Friction Coefficient

Determination of the drag force on the plate depends on the magnitude of the average wall shear stress over the entire surface. The wall shear is proportional to the kinetic energy of the fluid and is expressed as an equation by introducing the local skin friction coefficient. Thus

$$\tau_0 = C_f \frac{\rho u_\infty^2}{2g_c} \tag{4.17}$$

where the local coefficient C_f depends on the flow conditions. For laminar boundary layers

$$C_f = \frac{0.664}{(N_{\text{Re}x})^{0.5}} \tag{4.18}$$

and for turbulent boundary layers

$$C_f = \frac{0.0592}{(N_{\text{Re}x})^{0.2}} \tag{4.19}$$

The average wall shear over the plate surface $\bar{\tau}_0$ is obtained by an integration over the length of the plate.

$$\bar{\tau}_0 = \frac{1}{L} \int_0^L \tau_0 \, dx \tag{4.20}$$

If the plate has both laminar and turbulent boundary layers as is shown in Figure 4.5, the integration must be split into two parts:

$$\bar{\tau}_0 = \frac{1}{L} \left(\int_0^{Lc} \tau_0 \, dx + \int_{Lc}^L \tau_0 \, dx \right) \tag{4.21}$$

In the case where only a laminar boundary layer exists, we can perform the integration by substituting Eq. 4.17 into Eq. 4.20 and evaluating the integral as follows:

$$\bar{\tau}_0 = \frac{1}{L} \int_0^L 0.664 N_{\text{Re}x}^{-0.5} \frac{\rho u_\infty^2}{2g_c} \, dx$$

$$= \frac{1}{L} 0.664 \left(\frac{u_\infty}{\nu} \right)^{-0.5} \frac{\rho u_\infty^2}{2g_c} \int_0^L x^{-0.5} \, dx$$

$$= 0.664 \left(\frac{u_\infty}{\nu} \right)^{-0.5} \frac{\rho u_\infty^2}{2g_c} \frac{L^{0.5}}{0.5L}$$

$$= 1.328 \left(\frac{Lu_\infty}{\nu} \right)^{-0.5} \frac{\rho u_\infty^2}{2g_c} \tag{4.22}$$

Recognizing the Reynolds number based on length, $N_{\text{Re}L} = Lu_\infty/\nu$, we can write the average shear stress as

$$\bar{\tau}_0 = 1.328(N_{\mathrm{Re}L})^{-0.5}\frac{\rho u_\infty^2}{2g_c} \tag{4.23}$$

and by comparison with Eq. 4.17 we can define an average skin friction coefficient for the plate as

$$\bar{C}_f = \frac{1.328}{(N_{\mathrm{Re}L})^{0.5}}$$

for $L < L_c$ or for $N_{\mathrm{Re}L} < 5 \times 10^5$.

Example 4.3

Calculate the drag force due to skin friction if the air is in contact with both sides of the plate in Example 4.2 and the plate is 2 ft wide.

SOLUTION
The Reynolds number at the end of the plate at $x = L = 0.5$ ft was calculated to be 464,500. We therefore calculate the average skin friction coefficient as

$$\bar{C}_f = \frac{1.328}{(464,500)^{0.5}}$$

$$= 0.001948 \text{ dimensionless}$$

At 60F the air density $\rho = 0.07633 \text{ lb}_m/\text{ft}^3$ so that the kinetic energy per unit volume becomes

$$\frac{\rho u_\infty^2}{2g_c} = \frac{0.07633(146.7)^2}{2(32.174)}$$

$$= 25.53 \text{ ftlb}_f/\text{ft}^3$$

Thus, the average surface shear stress for the plate can be calculated as

$$\bar{\tau}_0 = \bar{C}_f \frac{\rho u_\infty^2}{2g_c}$$

$$= 0.001948(25.53)$$

$$= 0.0497 \text{ lb}_f/\text{ft}^2$$

The drag force per unit area is equal to the average shear stress so that $F_D = \bar{\tau}_0 A$ and for an area $A = 2[2(0.5)] = 2 \text{ ft}^2$ for both sides of the plate,

$$F_D = 2(0.0497) = 0.099 \text{ lb}_f$$

or about one-tenth pound force. This force is due only to shear action of the air flowing parallel to the surface. If the plate were oriented with the surface area perpendicular to the flow, the drag would be several orders of magnitude larger and consist of form drag instead of skin friction.

We are interested in the skin friction phenomena because it is analogous to heat transfer. We will consider such an analogy after we discuss fluid friction inside tubes and dimensionless numbers.

4.5 FLUID FRICTION AND PRESSURE DROP IN DUCTS

Since we are dealing with continuum fluids that experience shear, there will be a dissipation of energy due to that fluid shear similar to the dissipation of energy that takes place when two solid surfaces rub together. The frictional forces act through a distance, the product of which is really work, and that work is always transformed into thermal energy in the plane (or surface) where the friction takes place. Under conditions of moderate flow rate, the degradation of energy results in a pressure drop with virtually no measurable rise in fluid temperature. When the flow velocities are high enough in gases, it is possible that viscous dissipation can deposit enough energy to expand the gas and further increase the velocity so that sonic conditions are reached at the end of the duct, thus effectively choking the flow. We will be dealing with the flow of liquids where such phenomena do not occur and with gases at speeds low enough that we need not consider compressibility effects.

The pressure drop due to viscous shear is generally written as a proportionality between the kinetic energy of the fluid and the dimensionless length. For flow in a round tube as shown in Figure 4.9, the pressure drop is given by

$$\Delta P = P_1 - P_2 = f(KE)\frac{L}{D}$$

$$= f\frac{\rho u_b^2}{2g_c}\frac{L}{D}$$

(4.24)

where the constant of proportionality is the dimensionless friction factor f.

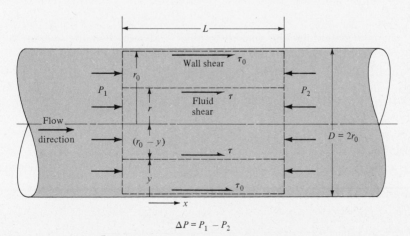

$$\Delta P = P_1 - P_2$$

Figure 4.9   Force balance on a fluid element.

A force balance on the fluid element shown in Figure 4.9 for the case where the fluid is in steady nonacceleration flow requires that the pressure forces be balanced by the shear force at the wall, τ_0. Thus

$$(P_1 - P_2)\frac{\pi D^2}{4} = \pi DL\tau_0 \tag{4.25}$$

so that

$$\tau_0 = (P_1 - P_2)\frac{D}{4L} \tag{4.25a}$$

The wall shear stress can be expressed in terms of the friction factor and kinetic energy by substitution from Eq. 4.24 so that

$$\tau_0 = \frac{f}{4}\frac{\rho u_b^2}{2g_c} \tag{4.26}$$

With this result we can determine the relationship between f and flow conditions for laminar flow. For turbulent flow we must look at experimental results because the nature of the flow is too complicated to describe analytically.

4.5.1 Laminar Flow Friction Factor

An equilibrium force balance for steady flow can be applied to any arbitrary fluid element of diameter $2r$ shown in Figure 4.9 so that

$$(P_1 - P_2)(\pi r^2) = 2\pi rL\tau \tag{4.27}$$

and dividing Eq. 4.27 by 4.25 yields

$$\frac{\tau}{\tau_0} = \frac{r}{r_0} \tag{4.28}$$

We can now determine the velocity profile (and also then the average bulk velocity) for a Newtonian fluid and then the friction factor; we proceed as follows. Equation 4.7 relates the shear stress in the fluid to the velocity gradient referred to the distance from the wall. Figure 4.9 shows that $r = r_0 - y$ so that $dr = -dy$; we then write Eq. 4.7 in terms of the velocity gradient with respect to radial distance as

$$\tau = \mu\frac{du}{dy} = -\mu\frac{du}{dr} \tag{4.7}$$

(Repeat)

From Eq. 4.28 we can relate the velocity gradient to the wall shear. Thus

$$\frac{du}{dr} = \frac{-\tau_0 r}{\mu r_0} \tag{4.29}$$

Separating variables and integrating gives

$$\int_0^u du = \frac{-\tau_0}{\mu r_0} \int_{r_0}^r r\, dr$$

$$u = \frac{-\tau_0}{\mu r_0} \left[\frac{r^2}{2} \right]_{r_0}^r \qquad (4.30)$$

$$u = \frac{\tau_0}{2\mu r_0} \left(r_0^2 - r^2 \right)$$

where we have applied the boundary conditions that at the wall $r = r_0$, the velocity $u = 0$, and at any radial location r the velocity is u in the x direction. The result given by Eq. 4.30 shows the velocity profile to be parabolic in shape for laminar flow. We have defined the average bulk velocity in Eq. 4.3 so that for isothermal flow of constant density we can write

$$u_b = \frac{\displaystyle\int_0^{r_0} 2\pi u r\, dr}{\displaystyle\int_0^{r_0} 2\pi r\, dr}$$

$$= \int_0^{r_0} \frac{2 u r\, dr}{r_0^2} \qquad (4.31)$$

Inserting the expression for velocity u from our result above and performing the integration steps yields

$$u_b = \int_0^{r_0} 2 \frac{\tau_0}{2\mu r_0} \left(r_0^2 - r^2 \right) \frac{r\, dr}{r_0^2}$$

$$= \frac{2\tau_0}{2\mu r_0^3} \int_0^{r_0} \left(r_0^2 r - r^3 \right) dr$$

$$= \frac{2\tau_0}{2\mu r_0^3} \left[\frac{r_0^2 r^2}{2} - \frac{r^4}{4} \right]_0^{r_0}$$

$$= \frac{2\tau_0}{2\mu r_0^3} \frac{r_0^4}{4}$$

$$u_b = \frac{2\tau_0 r_0}{8\mu} = \frac{\tau_0}{8\mu} D \qquad (4.32)$$

As a matter of interest we can observe that the center velocity, at $r = 0$, is given by $u_c = \tau_0 r_0 / 2\mu$ from Eq. 4.30 so that the average bulk velocity is one-half of the center velocity, $u_b = u_c/2$.

When we substitute for the wall shear in Eq. 4.26 from our result in Eq. 4.32, we obtain the friction factor as a function of the average bulk velocity in a very special form. Thus, for laminar flow

$$\frac{f}{4}\frac{\rho u_b^2}{2g_c} = \frac{8\mu u_b}{D}$$

$$f = \frac{64}{Du_b\rho/g_c\mu} \tag{4.33}$$

$$f = \frac{64}{N_{Re}}$$

The reader will note that units for the viscosity in the shear stress Eq. 4.7 must be in force units so that the product $g_c\mu$ is the viscosity in mass units as was shown in Eq. 4.8. The Reynolds number is generally given with the viscosity in mass units; therefore, $N_{Re} = Du_b\rho/\mu$ is consistent with our definition in Eq. 4.10. Some confusion may arise since we do not distinguish between symbols for viscosity in mass units and in force units; when in doubt, check the units out.

4.5.2 Turbulent Flow Friction Factor

We cannot analyze the fluid motion for turbulent flow as simply as we did for laminar flow since the variation of ε_M with radial position and flow conditions is difficult to determine precisely. Experimental measurements analyzed by Moody (2) are presented in Figure 4.10 where the friction factor, defined by Eq. 4.24, is shown as a function of the flow conditions (Reynolds number) and the relative roughness ε/D. Some typical roughness figures for various surfaces are given in the lower left corner of the figure. Moody (2) gives an approximation for the friction factor as a function of N_{Re} and ε/D as follows:

$$f \cong 0.0055\left[0.1 + \left(20,000\frac{\varepsilon}{D} + \frac{10^6}{N_{Re}}\right)^{1/3}\right] \tag{4.34}$$

For smooth tubes the friction factor can be approximated as a simple function of the Reynolds number as

$$f \cong 0.190 N_{Re}^{-0.2} \tag{4.35}$$

$5000 \leqslant N_{Re} \leqslant 5 \times 10^6$ to within ± 5 percent over the Reynolds number range given. Deissler (3) gives a variation of Von Karman's equation with constants evaluated from his data to relate friction factor to Reynolds number as

$$\frac{1}{\sqrt{f}} = -1.84 + 0.982\ln\left(N_{Re}\sqrt{f}\right) \tag{4.36}$$

or

$$N_{Re} = \exp\left(\frac{1/\sqrt{f} + 1.84}{0.982}\right)/\sqrt{f} \tag{4.36a}$$

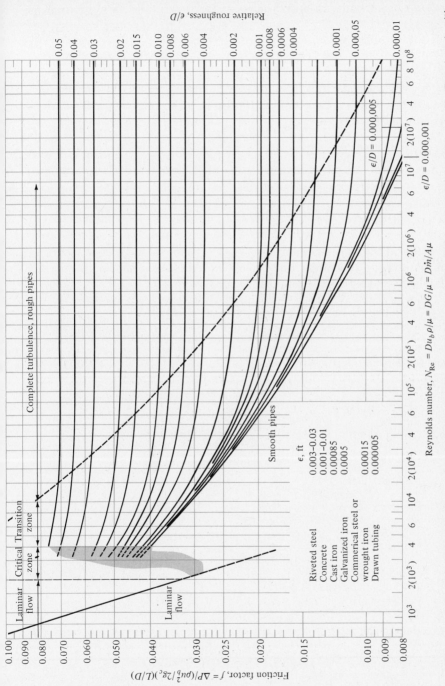

Figure 4.10 Variation of friction factor with flow conditions and roughness according to Moody (2). Reprinted with permission of Macmillan Publishing Co., Inc. from *Heat Transfer* by Alan J. Chapman, © 1960, Alan J. Chapman.

The latter two equations are not, however, very convenient to use for manual calculations.

Example 4.4

We want to calculate the pressure drop due to water at 70F flowing in a galvanized iron pipe with a 1.38-in. ID that is 25 ft long with a flow rate of 50 gpm (gallons per minute).

SOLUTION
For a galvanized iron pipe the roughness is 0.0005 ft and the relative roughness is

$$\frac{\varepsilon}{D} = \frac{0.0005}{1.38/12} = 0.00435$$

Calculation of the Reynolds number will let us determine the friction factor and then the pressure drop from Eq. 4.24. For the Reynolds number we need to determine the bulk velocity from the volume flow rate,

$$\dot{V} = 50 \text{ gpm } (0.1337 \text{ ft}^3/\text{gal})(\text{min}/60 \text{ s}) = 0.1114 \text{ ft}^3/\text{s}$$

The volume flow rate $\dot{V} = Au_b$; therefore, we calculate the bulk velocity as

$$u_b = \frac{\dot{V}}{A} = \frac{0.1114}{\pi(1.38)^2/4(144)}$$

$$= 10.73 \text{ ft/s}$$

The physical properties for water at 70F from Appendix C are

$$\rho = 62.27 \text{ lb}_m/\text{ft}^3$$
$$\mu = 2.37 \text{ lb}_m/\text{hrft}$$

so that the Reynolds number becomes

$$N_{Re} = \frac{Du_b\rho}{\mu}$$

$$= \frac{1.38}{12} 10.73 \frac{62.27}{2.37/3600}$$

$$= 116,700$$

The reader should carefully check the units for these quantities and verify that the Reynolds number is dimensionless. We can read the friction factor from Figure 4.10 at $N_{Re} = 116,700$ and $\varepsilon/D = 0.00435$ to be

$$f = 0.030$$

Calculating the kinetic energy per unit volume of the fluid and the length to diameter ratio

$$\frac{L}{D} = \frac{50}{1.38/12} = 434.8$$

$$KE = \frac{\rho u_b^2}{2g_c}$$

$$= \frac{62.27(10.73)^2}{2(32.174)}$$

$$= 111.4 \text{ lb}_f/\text{ft}^2 \text{ or ftlb}_f/\text{ft}^3$$

so that substitution into Eq. 4.24 for the pressure drop yields

$$\Delta P = f \frac{\rho u_b^2}{2g_c} \frac{L}{D}$$

$$= 0.030(111.4)(434.8)$$

$$= 1453 \text{ lb}_f/\text{ft}^2$$

$$= 10.1 \text{ lb}_f/\text{in.}^2$$

The power dissipated in moving the fluid through the pipe at 50 gpm is $\dot{V} \Delta P/550$ in horsepower so that

$$\text{power} = \frac{\dot{m} \Delta P}{\rho} = \frac{\dot{V} \Delta P}{550}$$

$$= \frac{(0.1114)1453}{550} = 0.294 \text{ HP}$$

This amount of power is dissipated by the fluid in viscous shear at the walls of the pipe. A motor would have to supply this power plus whatever inefficiencies were associated with the pump used to move the fluid. This completes Example 4.4.

We close our brief review of fluid friction and pressure drop with a comment on the influence of heat addition or heat extraction on the friction factor. Few experimental data on friction factors in diabatic flow have been reported in the literature; Wolf (4) measured friction coefficients for air and carbon dioxide in smooth tubes under turbulent flow conditions where the gas was heated and cooled over a range in wall to bulk temperature ratio from about 0.5 to 2.8 and found that the ratio $f/f_{\text{isothermal}}$ could be expressed as

$$\frac{f}{f_{\text{isothermal}}} = \left(\frac{T_s}{T_b} \right)^{-0.2}$$

$$0.5 < \frac{T_s}{T_b} < 2.8$$

(4.37)

over the Reynolds number range from about 40,000 to 200,000. The factor

$f_{\text{isothermal}}$ is the friction coefficient under constant temperature conditions (no heat transfer or $T_s/T_b = 1$) and can be obtained from the Moody diagram in Figure 4.10; the factor f is the friction coefficient with heat transfer or $T_s/T_b \neq 1$. For liquids Kreith (5) recommends the temperature ratio term be replaced by a viscosity ratio term so that

$$\frac{f}{f_{\text{isothermal}}} = \left(\frac{\mu_s}{\mu_b}\right)^{0.14} \tag{4.38}$$

where μ_s is evaluated at the surface temperature T_s and μ_b at T_b; the recommendation is most likely based on analogy with heat transfer for liquids, a topic we will discuss later on.

The friction coefficients discussed in this section all have to do only with wall shear phenomena. In some instances where the fluid (especially gases) is very strongly heated or cooled, flow acceleration or deceleration occurs with attendant forces caused by the acceleration changes. Those forces give rise to a momentum pressure drop that also contributes to the pressure drop along the duct as discussed in Hall (6), Zucrow (7), or a fluid dynamics book that covers flow with heat transfer.

4.5.3 Hydraulic Diameter

In all of our discussions and deliberations so far concerning flow inside ducts, we have considered only round or circular ducts. Round tubes or pipes are probably most often encountered in practice, but noncircular ducts are also of importance. The hydraulic diameter is defined as

$$D_h = \frac{4A_c}{W_P} \tag{4.39}$$

where A_c is the cross-sectional area of the fluid flowing in the duct and is usually the same as the inside cross-sectional area of the duct when the duct flows full, and W_P is the wetted perimeter of the duct, that is, the distance around the perimeter where the fluid is in contact with the wall and shearing action due to the flow takes place. The numerical factor 4 is included in the definition so that the hydraulic diameter for a circular tube coincides with the geometrical diameter.

To illustrate the concept we can calculate the hydraulic diameter for a square tube of side s flowing full. The cross-sectional area is s^2 and the wetted perimeter is $4s$ so that $D_h = 4s^2/4s = s$, the length of the side.

In pressure drop calculations for noncircular tubes, we will use the equations we have defined in the above sections with the hydraulic diameter, including the Reynolds number, and take the friction factor f from Figure 4.10. For situations where heat transfer is to be calculated all pertinent dimensionless parameters in the correlation equations will be evaluated with the hydraulic diameter.

4.6 SIMILARITY AND DIMENSIONLESS NUMBERS

We have already obtained some intuitive understanding of what we mean by similarity in the description of laminar flow in smooth tubes. We know that as long as the Reynolds number is, say, 2000, for example, the flow will be laminar regardless of the size of the duct or the magnitude of the velocity. Similar considerations apply to external flow situations with regard to the flow patterns around the object being considered. The Reynolds number, with the pertinent length criteria, determines the flow conditions because it represents the ratio of the momentum forces to the viscous forces.

Figure 4.11 shows a few of the physical situations we will be encountering

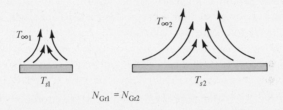

(a) Buoyancy induced flow, heated horizontal plate facing upward, $N_{Gr} = L^3 (\beta g/\nu^2) \Delta T$.

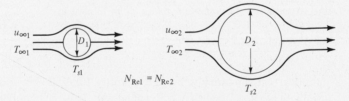

(b) Forced flow normal to a cylinder, $N_{Re} = Du_\infty/\nu$.

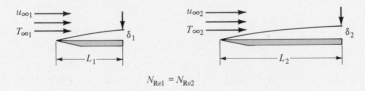

(c) Forced flow aver a flat plate, $N_{Re\,x} = \mu_\infty x/\nu$.

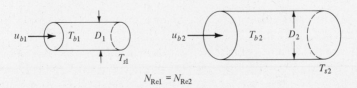

(d) Forced flow inside a round tube, $N_{Re} = Du_b\, \rho/\mu$.

Figure 4.11 Dynamically similar phenomena for flow with heat transfer.

in the next two chapters and shows the dimensionless quantity that determines similarity. We can note that for forced flow situations the Reynolds number determines similarity and for buoyancy induced flow, the Grashof number (ratio of buoyant forces to viscous forces) determines similarity because the momentum forces are negligibly small.

4.6.1 Considerations Near the Surface

In any convection situation involving a fluid that can be considered a continuum, the fluid at the surface is at rest even though a short distance away the fluid is moving vigorously. Energy is propagated through this stagnant layer by conduction and then by convection into the fluid. We can, therefore, observe that the surface $q_k = q_c$ so that we can equate Fourier's law (Eq. 1.3) to the convection Eq. 1.1 to obtain

$$-kA\frac{dT}{dy} = hA(T_s - T_\infty) \tag{4.40}$$

We note that dT/dy is the slope of the temperature profile at the wall ($y = 0$) and keep that in mind (a special designation would be cumbersome in the manipulation that we want to do). The area A in this case is the same and if we note that $dT = d(T - T_s)$, then we can multiply both sides by a characteristic length L and express the derivative in terms of y/L so that

$$\frac{d\big[(T_s - T)/(T_s - T_\infty)\big]}{d(y/L)} = \frac{hL}{k} \tag{4.41}$$

We then have thermal similarity when the quantity hL/k is the same for two given situations. The grouping hL/k is called the Nusselt number Nu and

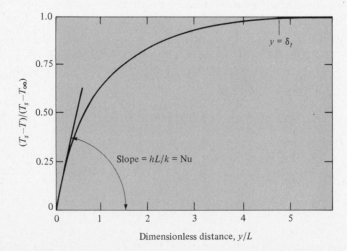

Figure 4.12   The Nusselt number as the slope of the dimensionless temperature profile at the surface, $y = 0$.

represents the ratio of convective energy to conductive energy in the fluid; Figure 4.12 shows the concept graphically. The Nusselt number describes thermally similar heat transfer situations and permits prediction of performance for different fluids.

4.6.2 Flow Similarity Criteria

We have already had ample discussion of the Reynolds number $N_{Re} = (length)u_{\infty}/\nu$ as a criterion for forced flow conditions.

When the movement of the fluid is due to buoyancy forces, the dimensionless criterion is the Grashof number $N_{Gr} = L^3\rho^2\beta g\,\Delta T/\mu^2$. The $\rho^2\beta g\,\Delta T$ portion represents the strength of the buoyant forces and the viscosity is indicative of the viscous forces. An analysis of the partial differential equations that govern fluid motion under the influence of buoyant forces shows that the ratio

$$\frac{N_{Gr}}{N_{Re}^2} > 1 \tag{4.42}$$

in order for the buoyancy effects to predominate.

The physical properties ν and α of the fluid determine the relative magnitude of the substance to propagate shear (ν) and to propagate thermal energy (α), and the ratio of the properties ν and α is called the Prandtl number, N_{Pr} as we saw in Section 4.1.4. In turbulent flow other factors due to the chaotic motion of the fluid particles enhance the interchange mechanism in a complicated manner. The ratio $\varepsilon_M/\varepsilon_H$ has been defined as the turbulent Prandtl number, but measurements are extremely difficult to make.

In the flow situations where heat conduction in the direction of the fluid motion can be important, such as in liquid metals, the Peclet number describes the similarity aspects. The Peclet number $N_{Pe} = Du_b\rho c_p/k$ is the ratio of the mass transported energy to the conducted energy and can be recognized as the product of the Reynolds and Prandtl numbers. Pearson (9) has shown that when the Peclet number is greater than 100, axial conduction effects in the flow direction are negligible.

The grouping $N_{Gr}N_{Pr}$ appears often and is an important parameter in free convection heat transfer; the combination is called the Rayleigh number $N_{Ra} = L^3\beta g\,\Delta T/\nu\alpha$ and represents the ratio of buoyancy effects to viscous and thermal diffusion effects. The part $\beta g/\nu\alpha$ of the Rayleigh number is all properties and can be plotted as a function of temperature for a given fluid to simplify calculations.

In laminar flow of a fluid the conduction mechanism is weak compared to the energy carrying capacity of the fluid so that the length of the channel is important. The Graetz number is defined as $N_{Gz} = N_{Re}N_{Pr}(D/L)$ and appears in laminar flow heat transfer correlations for flow inside ducts.

Table 4.1 summarizes the dimensionless similarity parameters we have discussed in a compact, convenient reference location. The table gives the name, the defining equation, together with alternate useful forms, the length criteria where applicable, and a brief statement of the physical meaning or

Table 4.1 DIMENSIONLESS PARAMETERS

NAME OF PARAMETER	SYMBOL	PHYSICAL MEANING	DEFINING EQUATION	ALTERNATE EQUATION	LENGTH DIMENSION	AREA OF APPLICATION
Nusselt	Nu	$\dfrac{\text{convected energy}}{\text{conducted energy}}$	$\dfrac{hL}{k}$	—	L, D, x	Free, forced convection
Reynolds*	N_{Re}	$\dfrac{\text{momentum forces}}{\text{viscous forces}}$	$\dfrac{Lu\rho}{\mu}$	$\dfrac{Lu}{\nu}, \dfrac{D\dot{m}}{A\mu}$	L, D, x	Forced flow situations
Prandtl	N_{Pr}	$\dfrac{\text{momentum diffusion}}{\text{thermal diffusion}}$	$\dfrac{c_p\mu}{k}$	$\dfrac{\nu}{\alpha}$	—	Property effects
Grashof	N_{Gr}	$\dfrac{\text{buoyant forces}}{\text{viscous forces}}$	$\dfrac{L^3\rho^2\beta g\,\Delta T}{\mu^2}$	$\dfrac{L^3\beta g\,\Delta T}{\nu^2}$	L, D, X	Free convection
Rayleigh	N_{Ra}	$\dfrac{\text{buoyant forces}}{\text{viscous and thermal forces}}$	$\dfrac{L^3\beta g\,\Delta T}{\nu\alpha}$	$N_{\text{Gr}}N_{\text{Pr}}$	L, D, X	Free convection
Mach	N_{Ma}	$\dfrac{\text{local velocity}}{\text{acoustic velocity}}$	$\dfrac{u}{a}$	$\dfrac{u}{\sqrt{kgRT}}$	—	Gas dynamics, compressible flow
Stanton*	N_{St}	$\dfrac{\text{convective ability}}{\text{enthalpy transport}}$	$\dfrac{h}{c_p\rho u}$	$\dfrac{h}{c_p G}$	—	Turbulent flow
Peclet*	N_{Pe}	$\dfrac{\text{momentum forces}}{\text{thermal diffusion}}$	$\dfrac{Du}{\alpha}$	$\dfrac{Du\rho c_p}{k}$	—	Liquid metal convection
Graetz*	N_{Gz}	$\dfrac{\text{enthalpy transport}}{\text{conducted energy}}$	$\dfrac{\pi N_{\text{Pe}} D}{4L}$	$\dfrac{\dot{m}c_p}{kL}$	L, D	Laminar flow convection
Biot†	N_{Bi}	$\dfrac{\text{internal resistance}}{\text{external resistance}}$	$\dfrac{hL_c}{k}$	—	$L_c = \dfrac{V}{A}$	Conduction in solids
Fourier†	N_{Fo}	Dimensionless time	$\dfrac{\alpha t}{L^2}$	$\dfrac{\alpha t}{r_0^2}$	L, r_0, D	Transient conduction

*The velocity u will be either free stream velocity u_∞ or bulk velocity u_b.
†Review from the chapters on conduction.

interpretation of the group. The Biot and Fourier numbers from Part I on conduction are also included as a review and a reminder that the phenomena they describe may at times be important in the phenomena we treat and discuss.

4.6.3 Compressibility Effects

In most commonly encountered industrial applications of the heat convection phenomena we have discussed, the velocities are low enough that compressibility effects are not significant. Temperature measurement, especially in gases, requires careful consideration of two aspects that can cause an indication by the measuring instrument to be different from the temperature sought. The first aspect is energy gain or loss due to thermal radiation; we will study radiation phenomena in detail in Part III. The second effect is due to a phenomena called compressibility. The terminology actually refers to the fact that the fluid specific volume is not constant during a particular process. In the temperature measurement process, the fluid is partly brought to rest at some point on the measurement instrument. If the fluid has a high enough velocity, the kinetic energy associated with that velocity then becomes significant as compared to the specific internal energy and when the fluid is slowed and brought to rest by the measuring instrument, that kinetic energy is transferred into internal energy and causes a higher temperature to be indicated than if the transducer were moving along with the fluid at the same velocity. The latter temperature (no relative motion between the measuring instrument and the fluid) is the temperature that must be used for thermodynamic and transport properties. The temperature measured by an instrument placed in the flow (that is stationary compared to the flow) measures a temperature that is somewhere between the static temperature and the theoretical stagnation temperature due to complete transformation of kinetic to internal energy.

The Mach number, $N_{Ma} = u/u_c$, is the ratio of the fluid velocity to the local acoustical velocity (speed of sound in the fluid) and is the criterion that indicates when stagnation temperature effects are significant. In general, when the Mach number is less than 0.2 (velocities less than 20 percent of the speed of sound) the stagnation (measured) temperature is very nearly identical to the static (desired) temperature. It can be shown from gas dynamics analysis (Reference 10) that the ratio of stagnation temperature T° to static temperature T is given by

$$\frac{T^\circ}{T} = 1 + \left(\frac{c_p}{c_v} - 1 \right) \frac{N_{Ma}^2}{2} \tag{4.43}$$

For air, for example, the ratio of specific heats c_p/c_v is about 1.4 so that for $N_{Ma} = 0.2$, the ratio $T^\circ/T = 1 + 0.4(0.2)^2/2 = 1.008$ which demonstrates that T° is essentially the same as T.

We conclude our brief introduction of dimensionless numbers with a comment on their origins. Dimensionless groups appear when the partial differential equations that describe the phenomena under consideration are nondimensionalized to make them easier to treat mathematically. In that context they become the important parameters for the solution. Dimensionless groups can also be constructed by a treatment of all the pertinent variables in a situation by an algebraic method called *dimensional analysis* and depending on the number of variables and the number and type of fundamentally independent units employed, a certain number of dimensionless groups will be produced. The method of dimensional analysis has a very fundamental disadvantage in that the practitioner must know all the variables involved and often that is not possible for complicated situations. The formulation of partial differential equations, however, generally begins with an analysis of a differential subsystem and the application of known conservation principles (energy, mass, and momentum) so that the result is on a much firmer footing than just trying to guess the variables involved. Such topics are beyond the scope of our introductory treatment in this book but are covered in advanced courses on conduction, convection, or radiation.

4.7 REYNOLDS ANALOGY

We have previously discussed in Section 4.3.2 the turbulent exchange models ε_M and ε_H that transport momentum and energy in the direction perpendicular to the direction of the bulk velocity. For fluids whose Prandtl number, ν/α, is around 1 the value of ε_M is very close to ε_H. Therefore, the turbulent mixing process transports momentum with about the same facility as energy; we can assume, therefore, that the turbulent momentum diffusion $\nu + \varepsilon_M$ in Eq. 4.12 is equal to the turbulent energy diffusion $\alpha + \varepsilon_H$ in Eq. 4.14. Accordingly, with the assumptions

(a) $N_{Pr} = 1$ or $\nu = \alpha$
(b) $\varepsilon_M = \varepsilon_H$

we can divide Eq. 4.12 by 4.14, thereby relating the velocity field to the temperature field from which we will obtain the useful result that the heat transfer coefficient is related to the friction factor (flow conditions). Thus

$$\frac{g_c \tau A \rho c_p}{\rho q} = \frac{-du}{dT} \tag{4.44}$$

In order to proceed with separation of variables, we assume that we are dealing with moderate size tubes so that we can neglect the variation of τ and q with radial position as suggested by Deissler (3); therefore

$$\frac{g_c \tau_0 A c_p}{q} \int_{T_s}^{T_b} dT = - \int_0^{u_b} du \tag{4.45}$$

The boundary condition at the wall is straightforward; the velocity is zero and the wall temperature is T_s. At the location where $u = u_b$, the bulk velocity, we must assume the temperature will be T_b. For fluids whose Prandtl number is near 1 such an assumption is very nearly correct because the dimensionless temperature and velocity profiles are similar. Performing the simple integration indicated yields

$$g_c\tau_0 c_p A(T_s - T_b)/q = u_b \tag{4.46}$$

Recognizing that $A(T_s - T_b)/q = 1/h$ (from Eq. 4.1), we obtain the simple result

$$g_c\tau_0 = \frac{hu_b}{c_p} \tag{4.47}$$

Somewhere in the course of a development, it is usually wise, and informative, to check the units before proceeding. Doing so helps spot manipulation mistakes and also points out conversion factors that may be needed. The left side of Eq. 4.47 has units

$$g_c\tau_0 \equiv \left(lb_m ft/lb_f s^2\right)\left(lb_f/ft^2\right) \equiv lb_m/fts^2$$

and with the customary units the right side of Eq. 4.47 has units of

$$\frac{hu_b}{c_p} \equiv \left(Btu/hrft^2 F\right)(ft/s)(lb_m F/Btu) \equiv lb_m/fthrs$$

so that we see a conversion factor of 3600 s/hr is required, or that we convert g_c to hour units $[g_c = 4.1698(10^8)\ lb_m ft/lb_f hr^2]$ when we make calculations. It is certainly advantageous to convert to dimensionless expressions, but we must still be careful of units to assure that the expressions we formulate as dimensionless are actually so.

The wall shear stress from Eq. 4.26 is given by $g_c\tau_0 = (f/8)\rho u_b^2$ so that Eq. 4.47 can be expressed as

$$\frac{f}{8} = \frac{h}{\rho u_b c_p} = N_{St} \tag{4.48}$$

The quantity $h/\rho u_b c_p$ is dimensionless and is called the Stanton number. The result of Reynolds analogy (that $\varepsilon_H = \varepsilon_M$) for a fluid with $N_{Pr} = 1$ is that the Stanton number is the same as $f/8$. This result has an important implication in that it predicts that roughening the surface increases the heat transfer coefficient for a given fluid and mass flow per unit cross-sectional area. The benefits of increased heat transfer coefficient must, however, be weighed against the increased pumping power costs due to the increased friction coefficient.

We might note that the Stanton number can also be expressed as $N_{St} = h/c_p G$ since the quantity ρu_b can be replaced by the mass flow per unit cross-sectional area from the continuity equation ($\dot{m} = \rho u_b A$) as $\rho u_b = \dot{m}/A = G$. Example 4.5 illustrates the above ideas.

Example 4.5

We want to heat the water flowing in the galvanized pipe of Example 4.4 and we would like to know what magnitude of heat transfer coefficient h to expect under the circumstances given in that example.

SOLUTION

The Reynolds number for the flow conditions was determined to be 116,700 which is well into the turbulent region. We can, therefore, use the results of Reynolds analogy to predict h. Using the result from Example 4.4, we calculate G as

$$G = \rho u_b$$

$$= 62.27(10.73)(3600)$$

$$= 2.405 \times 10^6 \text{ lb}_m/\text{hrft}^2$$

The friction factor f was read from Figure 4.10 to be $f = 0.030$ so that the Stanton number can be calculated from Eq. 4.48 as

$$N_{St} = \frac{f}{8} = \frac{0.030}{8} = 0.00375$$

The heat transfer coefficient then becomes

$$h = \frac{f}{8}c_p G = 0.00375(1)2.405(10^6)$$

$$h = 9020 \text{ Btu}/\text{hrft}^2\text{F}$$

Referring back to Table 1.1 in Chapter 1, the values given for water in forced convection range from 100 to 4000 in BE units. Our result in this example is for a rough pipe at a relatively high flow rate. The friction factor is almost twice that for a smooth tube so we would expect the value of h to be about twice the highest one given in Table 1.1.

This section concludes our introductory examination of the physical happenings basic to free and forced convection. In Chapter 5 we will consider correlations describing free convection heat transfer for different geometrical arrangements and in Chapter 6 we consider correlations for forced flow heat transfer situations.

PROBLEMS

1. (4.0) Occasionally the engineer deals with volumetric flow rates in cubic measure per unit time such as gallons per minute (gpm) for liquids or cubic feet per minute (cfm) for gases, in the BE system, for example. In the SI system we could consider cubic meters per second or minute (m^3/s or m^3/min). For a mass flow rate of 10 lb_m/s of water at 70F flowing in a 1.5-in. nominal schedule 80 pipe, determine the volumetric flow rate in gpm and in cfm.
2. (4.0) What is the bulk velocity for air at 16 psi and 80F flowing at 40 cfs in a rectangular 8 × 12 in. duct? What is the mass flow in lb_m/min?
3. (4.0) Light oil is transported in a refinery through a nominal 4-in. schedule 40 pipeline in isothermal flow at a temperature of 100F and a flow rate of 200 gpm. What is the bulk velocity in ft/s and the unit area mass flow rate G in lb_m/hrft^2 under these conditions?

4. (4.0) Freon-12 at 0C flows in a 4.0-cm-ID tube with a bulk velocity of 50 cm/s. Determine the mass flow rate in kg/s and the volumetric flow rate in m^3/s.

5. (4.1) The properties grouping $\rho^2 \beta g/\mu^2 = \beta g/\nu^2$ are needed to calculate the Grashof and Rayleigh numbers for free convection analysis. Plot the group for air and saturated water on 2×2 cycle log-log paper with overlapping coordinates so that the range of temperature for 1 to 3000F will fit onto the curve.

6. (4.1) To get acquainted with the property tables in the Appendix, calculate the volume expansivity for air at 500 and 2000F from Eq. 4.6 and compare with the value given in the table.

7. (4.1) Calculate the thermal and momentum diffusivity for water at 200F from the defining equations and compare with the values given in the property table; calculate the Prandtl number as $c_p \mu/k$ and also as the ratio of diffusivities calculated above. Check out units carefully on each calculation.

8. (4.2) Examine the results shown in Figure 4.2 and note that in the range from 70 to 100F, the thermal conductivity for air is about 0.015, and for water the value is around 0.350 in BE units, almost 24 times as large, yet it takes water about 150 times as long to reach 70F at the 6-in. depth as air does. Why is this so?

9. (4.2) Study the free convection phenomena shown in Figure 4.3 carefully and in detail. In a similar manner, sketch the physical happenings for the case of a cooled horizontal plate facing upward and a heated horizontal plate facing downward. The plates are flat and finite with no restrictions to flow at the edges; they are similar in physical appearance to the plates in Figures 4.3(a) and 4.3(b).

10. (4.2) Sketch the buoyancy induced flow pattern existing *inside* a horizontal tube whose wall temperature is suddenly increased in temperature over the fluid inside the tube. Sketch the flow pattern that exists *inside* the tube when the tube wall is suddenly cooled below the temperature of the fluid inside. Note that the sketches are at a given instant of time (not time dependent) and a short time after the wall temperature has been changed but well before the fluid has come to equilibrium with the new wall temperature.

11. (4.3) Benzine at a temperature of 100F flows in a tube 2-in. ID at a flow rate of 4.135 lb_m/min. Is the flow laminar or turbulent?

12. (4.3) What is the maximum mass flow rate possible with glycerine at 70F flowing in a 4-in. ID tube and still having laminar flow? What is the maximum flow rate for laminar flow with glycerine in this tube at 120F?

13. (4.3) Often a fluid may be difficult to work with for some reason or other. Liquid sodium at high temperature is such a case, but the operating advantage of low vapor pressure at high temperature makes it very attractive as a heat transfer fluid. At what temperature could water be used in order to model flow studies for sodium at 700F such as pressure drop and flow patterns for a given flow geometry and identical flow velocities?

14. (4.3) Air flows inside a tube with 1-in. ID with a bulk velocity of 35 ft/s at a temperature of 80F and a pressure of 50 psia. Determine the Reynolds number and the nature of the flow.

15. (4.4) Consider the three fluids air, water, and glycerine to flow parallel to a sharp edged flat plate (separately, not as a mixture) as indicated in Figure 4.5. The free stream velocity is 25 ft/s at a temperature of 100F. Determine the boundary layer thicknesses at locations 6 and 24 in. from the leading edge, arrange your results in a table, and comment.

16. (4.4) Compare the thermal and momentum boundary layer thicknesses for the three fluids and the conditions described in Problem 15 for the 24-in. location. Which fluid has the thickest, and which the thinnest thermal boundary layer?

17. (4.4) Estimate the force and power required to keep the wing section of an airplane in motion at 200 mph at an air temperature of 40F and atmospheric pressure. Consider the wing to be a flat plate 4 ft wide and 30 ft long.

18. (4.4) Determine the mean skin friction coefficient for the flow of air over the wing described in Problem 17 for the conditions given and compare with the value predicted from the curve in Figure 4.8.

19. (4.5) Air at 80F and 50 psia flows in a smooth 1-in. ID tube with a bulk velocity of 35 ft/s; the tube is 75 ft long. What is the pressure drop and power required to move the air through the tube?

20. (4.5) A small pump continually supplies 100 gpm of 50F water to an elevated storage tank located directly above the outlet of the pump. A vertical 1-in. schedule 40 galvanized iron pipe 80 ft long connects the pump to the top of the storage tank. Determine the pressure drop along the length of the pipe and the pumping power required.

21. (4.5) Helium gas flows in a smooth stainless-steel tube 0.430-in. ID having 13 in. heated electrically to an average temperature of 1549R. The gas temperature at the start of the heated part is 508R and at the end of the heated part is 822R at a mass flow rate of 0.007829 lb_m/s. The pressure at the start of the heated section is 120 psi. Determine the Reynolds number at the average bulk temperature, the friction factor, and pressure drop due to wall shear over the 13-in. length.

22. (4.5) Water at 400F and 2000 psi flows parallel to a bundle of tubes 1-in. OD that are spaced so that their centers form an isosceles triangle 1.25 in. on a side. The water flows in the space outside of the tube at a bulk velocity of 15 cm/s. Determine the flow Reynolds number and the pressure drop per foot length along the flow direction.

23. (4.5) In the laminar flow of a fluid in a smooth round duct, at what distance from the wall will the velocity be the same as the bulk velocity?

24. (4.5) Water at 70F flows in a smooth tube 2-in. ID at a flow rate of 500 lb_m/hr. What velocity would be indicated by a small impact tube placed at a radial location of 0.5 in. from the tube wall?

25. (4.6) A horizontal stainless steel pipe 1-in. OD has a surface temperature of 2040F and is in still air at 40F and 1 atm pressure. What surface tempera-

ture is required on a 2-in. OD horizontal cylinder in air under the same conditions to have the same Grashof number? Note that properties must be evaluated at the film temperature $T_f = (T_s + T_\infty)/2$.

26. (4.6) A 1-in. OD cylinder whose surface temperature is at 2060F has air at 1 atm and 40F flowing at a velocity of 50 ft/s over the outer surface in a direction perpendicular to the cylinder. What velocity would be required for air at 1 atm and 80F flowing over the inside of a 2-in. OD cylinder (whose surface temperature is at 280F) in a direction normal to the axis of the cylinder in order to have flow similarity for the two cylinders? Note properties must be evaluated at the film temperature $T_f = (T_s + T_\infty)/2$. Is free convection important in this physical situation? Why?

27. (4.6) A vertical pipe 2 ft long has water flowing upward at an average bulk temperature of 60F and at a bulk velocity of 2 ft/s. The pipe has an average inner surface temperature of 180F. Under these conditions are the free convection effects important to the heat transfer process? Discuss the aspects involved in this situation.

28. (4.6) Mercury at 300F flows in a square duct 2 in. on a side whose wall temperature is maintained at 500F with a unit area mass flow rate of 75,000 lb_m/hrft^2. Is it necessary to consider conduction of heat in the axial direction in assessing heat transfer for mercury in this flow situation?

29. (4.7) Estimate the heat transfer coefficient for the inside surface of a few foot lengths of the 1-in. ID smooth tube described in Problem 14 under the flow conditions given. Assume that the average bulk temperature of the air flowing in the tube is about 90F for a constant wall temperature of 100F.

30. (4.7) Liquid benzene at 60F flows in a 0.5-in.-ID tube that has an internal roughness of 0.000625 ft. Estimate the heat transfer coefficient for a bulk velocity of 7.5 ft/s.

References

1. Reynolds, O., "On the Extent and Action of the Heating Surface for Steam Boilers," *Proceedings of the Manchester Literary and Philosophical Society*, Vol. 14, 1874, p. 7.
2. Moody, L. F., "Friction Factors for Pipe Flow," *Transactions ASME*, Vol. 66, 1944, p. 671.
3. Deissler, R. G., "Analytical and Experimental Investigation of Adiabatic Turbulent Flow in Smooth Tubes," NACA TN 2138, Jan. 1950, pp. 24, 25.
4. Wolf, H., "The Experimental and Analytical Determination of the Heat Transfer Characteristics of Air and CO_2 in the Thermal Entrance Region of a Smooth Tube with Large Temperature Differences Between the Gas and the Tube Wall," Ph.D. Thesis, Purdue University, W. Lafayette, Indiana, Part I, 1958, p. 85.
5. Kreith, F., *Principles of Heat Transfer*, 3d Edition, Harper & Row, New York, 1973, p. 444.
6. Hall, N., *Thermodynamics of Fluid Flow*, Prentice-Hall, Englewood Cliffs, N.J., 1951, p. 46.
7. Zucrow, M. J., *Principles of Jet Propulsion and Gas Turbines*, Wiley, New York, 1948, p. 132.

8. Schlichting, H., *Boundary Layer Theory*, 6th Edition, McGraw-Hill, New York, 1968, p. 130.

9. Pearson, S. W., "The Effects of Axial Conduction and Arbitrary Flux on Heat Transfer at Low Peclet Numbers," Ph.D. Thesis, Mechanical Engineering Department, University of Arkansas, Fayetteville, AR, 1968.

10. Van Wylen, G. J. and Sontag, R. E., *Fundamentals of Classical Thermodynamics*, Wiley, New York, 1973, p. 607.

11. Eckert, E. R. G. and Drake, R. M., *Heat and Mass Transfer*, 2d Edition, McGraw-Hill, New York, 1959, p. 175.

12. Rouse, H., *Elementary Mechanics of Fluids*, Wiley, New York, 1946, p. 187.

Chapter 5
Free Convection

5.0 GENERAL REMARKS

In our discussion of the basic concepts of convection, we learned that the motion of the fluid in free convection was caused by buoyant forces. The most prevalent cause of such forces is the gravitational attractive force of the earth. There are other types of forces, such as centrifugal or inertial forces, which act in the same manner; therefore, the fluid responds in a similar way. As a matter of fact, we speak of centrifugal forces as so many g's in expressing magnitude, a direct analogy with gravitational forces. One example of such an area where free convection is important is in the interior cooling of gas turbine blades that spin at high rotational speed.

The forces that exist near the surface due to density differences caused by heat transfer make the fluid move along the surface. The resulting movement convects energy to or from the surface. We may infer that the buoyancy induced velocities extend into the fluid as far as the temperature changes do. Such a conclusion is valid for fluids whose Prandtl number is less than or equal to about 1. Figure 5.1 shows a qualitative sketch of the variation of the dimensionless temperature ratio $\theta/\theta_s = (T - T_\infty)/(T_s - T_\infty)$ and the dimensionless velocity ratio u/u_{max} with distance from the surface. When the Prandtl number is considerably greater than 1 (as for oils, for example), the greater diffusion of

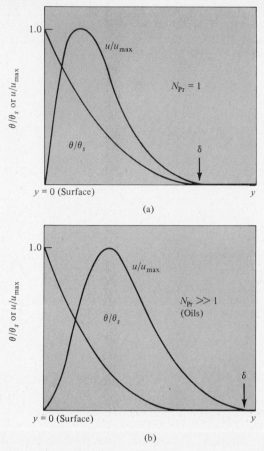

Figure 5.1 Dimensionless velocity and temperature profiles near the surface in free convection on a vertical flat plate; (a) $N_{Pr} = 1$ and (b) $N_{Pr} \gg 1$, oils.

momentum ($\nu \gg \alpha$) causes the viscous shear to extend past the influence of the temperature (energy) penetration.

The interaction layer where the velocity is greater than zero is called the boundary layer δ as indicated in Figures 5.1(a) and 5.1(b). The discussion so far and to follow describes the happenings near the surface of a vertical plate in a gravitational field because the events are easier to visualize; the general nature of the phenomena also applies to other geometries. The nature of the flow in the boundary layer will be either laminar (layerlike) or mildly turbulent. The nature or character of the free convection turbulent boundary layer is nowhere near as strong compared to what we can observe in forced flow. As a matter of fact, it would be more exact to call the flow wavy or tumbling rather than turbulent.

The flow in the boundary layer at the start, $x = 0$ (bottom for energy out of the plate, top for energy into the plate), is always laminar and as velocities along

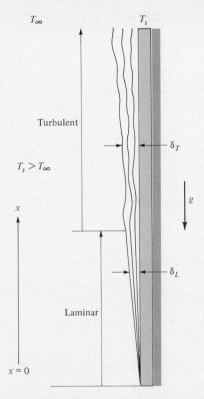

T_∞ T_s

Turbulent

δ_T

$T_s > T_\infty$

g

x

δ_L

Laminar

$x = 0$

Figure 5.2 Transition from laminar flow to turbulent flow in the boundary layer on a vertical flat plate with free convection on the surface.

the plate increase, there will be a transition location where the flow becomes unstable and starts to waver; Figure 5.2 illustrates the concepts. Clearly the heat transfer characteristics will be more pronounced in the turbulent portion of the boundary layer than in the laminar portion. Such differences are reflected in the average correlations that we will consider in the sections to follow.

We will first study the growth of the boundary layer under laminar and turbulent conditions, then we will consider design correlations for vertical, horizontal, and inclined flat surfaces; cylinders and spheres; and for enclosed spaces. The physical properties in the dimensionless parameters that comprise the correlations will be evaluated at the average film temperature $T_f = (T_s + T_\infty)/2$ unless specific instructions are given to use another temperature.

The physical situations we will study describe pure free convection; that means the fluid outside the boundary layer must be at rest, not in motion ($u_\infty = 0$). When a small finite u_∞ exists, Kreith (15) has shown that the effects of forced flow become important when the square of the Reynolds number is about the same magnitude as the Grashof number. Aspects of mixed free and forced convection will be discussed in Chapter 6.

5.1 BOUNDARY LAYER GROWTH

We have discussed the formulation of the flow pattern on a surface in free convection in our introductory comments and, therefore, have a good mental picture of the boundary layer thickness δ due to momentum diffusion. The heat transfer correlations will come from the concept of δ, but the aspects of fluid spatial relationships and mass flow involve the displacement thickness δ^*. The implications of δ^* are most important in free convection phenomena in enclosed spaces.

The concept of the displacement thickness arises from a calculation of the mass flow rate in the boundary layer. Figure 5.3 shows the relationship of the analytical formulation to the physical description of the velocity profile in the boundary layer. To determine the mass flow in the layer from $y = 0$ to δ, we apply the continuity Eq. 4.4 on a differential basis. Thus

$$dm = \rho u \, dA = \rho u (dyW) \tag{5.1}$$

where W is the width of the surface. In order to obtain the entire flow $\dot{m}$ in the boundary layer, we sum all of the incremental mass flow $d\dot{m}$ over the distance $y = 0$ to infinity; actually we need only sum to $y = \delta$ because $u = 0$ for $y > \delta$; thus

$$\dot{m} = W \int_0^\infty \rho u \, dy \tag{5.2}$$

since W is a constant. We note that the maximum velocity u_{max} is not a function of y; therefore, we can multiply and divide the right side of Eq. 5.2 by u_{max}; and for the case where the density ρ does not change a great deal with temperature

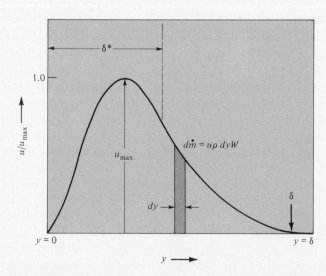

Figure 5.3 Formulation of mass flow rate in the free convection boundary layer on a vertical flat plate.

and can be assumed constant, we obtain

$$\dot{m} = u_{\max} \rho W \int_0^\infty \frac{u}{u_{\max}} dy \tag{5.3}$$

The integral in Eq. 5.3 represents a distance (check the dimensions) and is defined as the displacement thickness δ^*. Thus

$$\delta^* = \int_0^\infty \frac{u}{u_{\max}} dy \tag{5.4}$$

and

$$\dot{m} = u_{\max} \rho (W\delta^*) \tag{5.5}$$

The quantity $W\delta^*$ represents the flow area for $\dot{m}$ at the velocity $u_{\max}$ and density ρ so that δ^* represents a thickness wherein the major portion of the flow takes place.

In enclosed or restricted spaces it would be necessary to provide space sufficient for the largest amount of mass flow, thickest δ^* or δ, to be expected under the operating conditions envisioned. Such criteria must be applied, for example, to determine the size and number of tubes or flat channels to be used on the outside of large oil filled transformers to dissipate the joulian heat generated in the coils.

We will now direct our attention to the description and calculation of the momentum boundary layer thicknesses δ.

5.1.1 Laminar Boundary Layer

The determination of δ requires either a solution of the partial differential equation of motion or a somewhat simpler application of the integral method due to Von Karman using reliable estimates of the velocity and temperature profiles. Giedt (1) used Von Karman's method with the following assumptions for the velocity and temperature profiles:

$$\frac{\theta}{\theta_s} = \left(1 - \frac{y}{\delta}\right)^2 \tag{5.6}$$

$$\frac{u}{u_{\max}} = 6.75 \frac{y}{\delta} \left(1 - \frac{y}{\delta}\right)^2 \tag{5.7}$$

The analysis also assumed that $u_{\max}$ and δ varied as simple power functions of the distance x (see Figure 5.2) from the start of the boundary layer, so that for the laminar boundary layer δ_L, the thickness was formulated by Giedt as

$$\frac{\delta_L}{x} = 3.93(0.952 + N_{\text{Pr}})^{1/4} N_{\text{Pr}}^{-1/2} N_{\text{Gr}}^{-1/4} \tag{5.8}$$

where the Grashof number N_{Gr} is formulated with the length criterion x.

If we substitute Eq. 5.7 into 5.4 and evaluate the integral, the laminar displacement thickness can be related to the boundary layer thickness as

$$\delta_L^* = 0.562\delta_L \tag{5.9}$$

or slightly more than half of δ_L.

Calculation of the turbulent boundary layer thickness δ_T requires a different set of assumptions for the velocity and temperature profiles.

5.1.2 Turbulent Boundary Layer

The analysis performed by Eckert and Jackson (2) assumed that the velocity and temperature profiles had a structure similar to that in turbulent forced flow in that the variation with distance from the surface followed a $\frac{1}{7}$ power law; they assumed the following functions to hold.

$$\frac{\theta}{\theta_s} = 1 - \left(\frac{y}{\delta}\right)^{1/7} \tag{5.10}$$

$$\frac{u}{u_{max}} = 1.862\left(\frac{y}{\delta}\right)^{1/7}\left(1 - \frac{y}{\delta}\right)^4 \tag{5.11}$$

Eckert and Jackson's analysis also required that u_{max} and δ_T vary as simple power functions of the distance x and in addition the assumption was made that the boundary layer was turbulent from the start at $x = 0$. The latter assumption restricts the results to those cases where the extent of the initial laminar boundary layer is small compared to the length of the turbulent part. For the turbulent boundary layer they obtained

$$\frac{\delta_T}{x} = 0.565\left(1 + 0.494N_{Pr}^{2/3}\right)^{1/10}N_{Pr}^{-8/15}N_{Gr}^{-1/10} \tag{5.12}$$

a form similar to the result for the laminar case but with different constants and exponents. Again, the Grashof number is evaluated with the length dimension x as shown in Figure 5.2. The Grashof number, also the Nusselt number, calculated with x is often referred to as the local value of the parameter.

Also, by substituting Eq. 5.11 into Eq. 5.4 and evaluating the resulting integral permits us to relate the turbulent displacement thickness to the boundary layer thickness. Thus

$$\delta_T^* = 0.272\delta_T \tag{5.13}$$

The turbulent displacement thickness is a smaller fraction of δ_T than δ_L^* is of δ_L because the maximum velocity is much closer to the surface when the boundary layer is turbulent than when it is laminar.

Example 5.1

We would like to determine the boundary layer thickness variation with distance on a vertical flat plate in Wemco-C transformer oil and determine the approximate value of the Grashof number when the boundary layer changes from laminar to turbulent motion. The plate is 8 ft tall and 1 ft wide, and we want to make the calculation for two values of surface temperature 110 and 160F with the oil 10F higher in temperature than the surface for both cases.

SOLUTION
The boundary layer thicknesses at a given location x can be calculated from Eqs. 5.8 and 5.12 and properties for Wemco-C oil are given in Appendix C. Since these are repetitive calculations, a short computer program is the easiest way to accomplish the job. Table 5.1 presents the acronym symbols utilized in the program and Table 5.2 is the FORTRAN list of the program steps. Note that in this program the data is read in from cards according to statement 500 which allocates the first 24 columns on a card for the two listed quantities: surface temperature TS in columns 1 through 12, and temperature difference DT in columns 13 through 24. The programs we have considered so far have all had data internal to the program.

The results are presented numerically in Tables 5.3 and 5.4 and graphically in Figure 5.4. The figure shows that the laminar layer extends much farther down the plate for the cooler temperature of 110F as compared to the hotter 160F surface due primarily to viscosity effects. The transition occurs at a N_{Gr} of about 10^8, which is in agreement with average heat transfer results, a topic that we will discuss in more detail in Section 5.2.

We could also in principle determine the Grashof number at the transition location by noting that $\delta_L = \delta_T$ and solving the resulting expression for N_{Gr}. Therefore

$$3.93x(0.952 + N_{\text{Pr}})^{1/4} N_{\text{Pr}}^{-1/2} N_{\text{Gr}}^{-1/4}$$

$$= 0.565x(1 + 0.494N_{\text{Pr}}^{2/3})^{1/10} N_{\text{Pr}}^{-8/15} N_{\text{Gr}}^{-1/10}$$

Table 5.1   DICTIONARY OF TERMS EMPLOYED IN BOUNDARY LAYER PROGRAM LISTED IN TABLE 5.2

X =	distance from top of plate, ft
TF =	film temperature, F
TS =	surface temperature, F
DT =	fluid delta T, F
RO =	density, lb_m/ft^3
CP =	specific heat, $\text{Btu}/\text{lb}_m\text{F}$
ZU =	viscosity, lb_m/hrft
XK =	thermal conductivity, Btu/hrftF.
B =	expansion coefficient, $1/\text{F}$
PR =	Prandtl number
GR =	Grashof number
DLM =	laminar thickness δ_L, in.
DLMS =	laminar displacement thickness δ_L^*, in.
DTB =	turbulent thickness δ_T, in.
DTBS =	turbulent displacement thickness δ_T^*, in.

Table 5.2 FORTRAN LISTING FOR BOUNDARY LAYER THICKNESS
CALCULATION PROGRAM FOR 8-FT VERTICAL PLATE IN WEMCO-C
TRANSFORMER OIL (SEE APPENDIX C FOR PROPERTIES EQUATIONS)

```
C
C       INPUT DATA
C           TS = SURF TEMP, DEG F, XXX.
C           DT = FLUID DELTA T, DEG F, XX.
C
    10 READ 500, TS, DT
    15 PRINT 510, TS, DT
    20 X = 0.0
    21 X = X + 0.25
C
C       EVALUATE PROPERTIES AT THE FILM TEMPERATURE
C
    25 TF = TS + DT/2.
    30 RO = 56.57 − 0.02186*TF
    35 CP = 0.439
    40 ZU = 1650./((TF/10.)**1.935)
    45 XK = 0.07915 − 0.00002644*TF
    50 B = 0.0003856 + 0.0000001643*TF
    55 PR = CP*ZU/XK
    60 GR = (X*X*X*RO*RO*B*32.174*DT*3600.*3600.)/(ZU*ZU)
C
C       CALCULATE BL THICKNESSES
C
    64 PG = (PR**0.5)*(GR**0.25)
    65 DLM = (3.93*X*((0.952 + PR)**0.25)/PG)*12.
    70 DLMS = 0.562*DLM
    73 ZZ = (1. + 0.494*(PR**0.6667))**0.1
    74 PGG = (PR**0.533)*(GR**0.1)
    75 DTB = (0.565*X*ZZ/PGG)*12.
    80 DTBS = 0.272*DTB
    81 BB = B*10000.
    85 PRINT 520, X, DLM, DLMS, DTB, DTBS, PR, GR, RD
    90 IF(X − 8.24) 21, 95, 95
    95 GOTO 10
   500 FORMAT (2F12.6)
   510 FORMAT (1H1, 5X//////27X, 27HBOUNDARY LAYER CALCULATIONS///
       124X, 11HSURF TEMP = F5.0,    3X,   10HTEMP  DIF = F4.0//
       28X, 39H    X        DLM        DLMS        DTB    DTBS, 4X,
       321H PR      GR        RO//)
   520 FORMAT (8X, 5F8.3, F6.0, E12.3, F7.1)
   100 END
```

Since the location x is the same, we can solve for N_{Gr} as

$$N_{Gr}^{0.15} = \frac{3.93(0.952 + N_{Pr})^{1/4}N_{Pr}^{1/30}}{0.565\left(1 + 0.494N_{Pr}^{2/3}\right)^{1/10}}$$

For the plate temperature of 110F and oil temperature 120F, the Prandtl

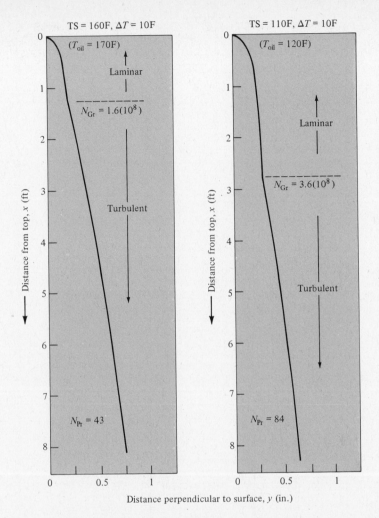

Figure 5.4   Variation of free convection boundary layer thickness with distance from the top of an 8-ft cooled vertical plate in Wemco-C oil.

number at 115F is 84 (see Table 5.3) so that the Grashof number then becomes

$$N_{\mathrm{Gr}} = \left[\frac{3.93(0.952 + 84)^{1/4} 84^{1/30}}{0.565 \left[1 + 0.494(84)^{2/3} \right]^{1/10}} \right]^{100/15}$$

$$= 3.8 \times 10^8$$

which agrees with the results given in Table 5.3. We should recognize that the laminar thickness δ_L in Tables 5.3 and 5.4 has meaning only as long as $\delta_L < \delta_T$ (also δ_T has meaning only when $\delta_T > \delta_L$).

Table 5.3 RESULTS OF BOUNDARY LAYER
THICKNESS CALCULATIONS FOR 8-FT
VERTICAL PLATE IN WEMCO-C OIL

SURF TEMP = 110. TEMP DIF = 10.

X	DLM	DLMS	DTB	DTBS	PR	GR	RO
0.250	0.159	0.090	0.056	0.015	84.	0.360E 06	54.1
0.500	0.189	0.106	0.091	0.025	84.	0.288E 07	54.1
0.750	0.210	0.118	0.121	0.033	84.	0.972E 07	54.1
1.000	0.225	0.127	0.148	0.040	84.	0.230E 08	54.1
1.250	0.238	0.134	0.173	0.047	84.	0.450E 08	54.1
1.500	0.249	0.140	0.197	0.053	84.	0.778E 08	54.1
1.750	0.259	0.146	0.219	0.060	84.	0.124E 09	54.1
2.000	0.268	0.151	0.241	0.065	84.	0.184E 09	54.1
2.250	0.276	0.155	0.261	0.071	84.	0.263E 09	54.1
2.500	0.283	0.159	0.281	0.076	84.	0.360E 09	54.1
2.750	0.290	0.163	0.301	0.082	84.	0.479E 09	54.1
3.000	0.296	0.167	0.320	0.087	84.	0.622E 09	54.1
3.250	0.302	0.170	0.338	0.092	84.	0.791E 09	54.1
3.500	0.308	0.173	0.356	0.097	84.	0.988E 09	54.1
3.750	0.313	0.176	0.374	0.102	84.	0.122E 10	54.1
4.000	0.319	0.179	0.391	0.106	84.	0.148E 10	54.1
4.250	0.323	0.182	0.408	0.111	84.	0.177E 10	54.1
4.500	0.328	0.184	0.424	0.115	84.	0.210E 10	54.1
4.750	0.332	0.187	0.441	0.120	84.	0.247E 10	54.1
5.000	0.337	0.189	0.457	0.124	84.	0.288E 10	54.1
5.250	0.341	0.192	0.473	0.129	84.	0.334E 10	54.1
5.500	0.345	0.194	0.488	0.133	84.	0.383E 10	54.1
5.750	0.349	0.196	0.504	0.137	84.	0.438E 10	54.1
6.000	0.352	0.198	0.519	0.141	84.	0.498E 10	54.1
6.250	0.356	0.200	0.534	0.145	84.	0.563E 10	54.1
6.500	0.360	0.202	0.549	0.149	84.	0.633E 10	54.1
6.750	0.363	0.204	0.564	0.153	84.	0.709E 10	54.1
7.000	0.366	0.206	0.578	0.157	84.	0.791E 10	54.1
7.250	0.370	0.208	0.593	0.161	84.	0.878E 10	54.1
7.500	0.373	0.209	0.607	0.165	84.	0.972E 10	54.1
7.750	0.376	0.211	0.621	0.169	84.	0.107E 11	54.1
8.000	0.379	0.213	0.635	0.173	84.	0.118E 11	54.1
8.250	0.382	0.215	0.649	0.176	84.	0.129E 11	54.1

The transition Grashof number calculated by equating δ_L and δ_T is indicated to be a function of the Prandtl number and the expression given above shows an increase in N_{Gr} with increasing N_{Pr}. The transition Grashof numbers range from about 10^6 at $N_{Pr} = 1$ to about 5×10^8 at a $N_{Pr} = 100$ which roughly agrees with the transition results indicated by average heat transfer measurements that we will discuss in the next section. As a point of interest at $N_{Pr} = 20$ the indicated transition Rayleigh number $(N_{Gr}N_{Pr})$ is very close to 10^9 which is the value indicated by heat transfer measurements; a result that is fortuitous considering the assumptions made in formulating the expression for the boundary layer thicknesses δ_L and δ_T. This concludes Example 5.1.

Table 5.4 RESULTS OF BOUNDARY LAYER THICKNESS CALCULATIONS FOR 8-FT VERTICAL PLATE IN WEMCO-C OIL

SURF TEMP = 160. TEMP DIF = 10.

X	DLM	DLMS	DTB	DTBS	PR	GR	RO
0.250	0.134	0.075	0.068	0.018	43.	0.143E 07	53.0
0.500	0.160	0.090	0.110	0.030	43.	0.114E 08	53.0
0.750	0.177	0.099	0.146	0.040	43.	0.385E 08	53.0
1.000	0.190	0.107	0.178	0.048	43.	0.913E 08	53.0
1.250	0.201	0.113	0.208	0.057	43.	0.178D 09	53.0
1.500	0.210	0.118	0.237	0.064	43.	0.308E 09	53.0
1.750	0.218	0.123	0.264	0.072	43.	0.489E 09	53.0
2.000	0.226	0.127	0.290	0.079	43.	0.730E 09	53.0
2.250	0.232	0.131	0.314	0.086	43.	0.104E 10	53.0
2.500	0.239	0.134	0.338	0.092	43.	0.143E 10	53.0
2.750	0.244	0.137	0.362	0.098	43.	0.190E 10	53.0
3.000	0.250	0.140	0.385	0.105	43.	0.246E 10	53.0
3.250	0.255	0.143	0.407	0.111	43.	0.313E 10	53.0
3.500	0.260	0.146	0.428	0.117	43.	0.391E 10	53.0
3.750	0.264	0.148	0.450	0.122	43.	0.481E 10	53.0
4.000	0.268	0.151	0.470	0.128	43.	0.584E 10	53.0
4.250	0.273	0.153	0.491	0.133	43.	0.701E 10	53.0
4.500	0.276	0.155	0.511	0.139	43.	0.832E 10	53.0
4.750	0.280	0.157	0.530	0.144	43.	0.978E 10	53.0
5.000	0.284	0.160	0.550	0.150	43.	0.114E 11	53.0
5.250	0.287	0.161	0.569	0.155	43.	0.132E 11	53.0
5.500	0.291	0.163	0.588	0.160	43.	0.152E 11	53.0
5.750	0.294	0.165	0.606	0.165	43.	0.174E 11	53.0
6.000	0.297	0.167	0.625	0.170	43.	0.197E 11	53.0
6.250	0.300	0.169	0.643	0.175	43.	0.223E 11	53.0
6.500	0.303	0.170	0.661	0.180	43.	0.251E 11	53.0
6.750	0.306	0.172	0.678	0.185	43.	0.281E 11	53.0
7.000	0.309	0.174	0.696	0.189	43.	0.313E 11	53.0
7.250	0.311	0.175	0.713	0.194	43.	0.348E 11	53.0
7.500	0.314	0.177	0.730	0.199	43.	0.385E 11	53.0
7.750	0.317	0.178	0.747	0.203	43.	0.425E 11	53.0
8.000	0.319	0.179	0.764	0.208	43.	0.467E 11	53.0
8.250	0.322	0.181	0.781	0.212	43.	0.513E 11	53.0

5.2 VERTICAL SURFACE CORRELATIONS

In order to calculate the heat transfer coefficient for a given convection situation, we recognize that the heat flux that passes by conduction through the very thin stagnant layer of fluid on the surface is the same heat flux that determines the convective coefficient in Newton's equation. Expressed mathematically, we have

$$-kA\left(\frac{dT}{dy}\right)_{y=0} = hA(T_s - T_\infty) \tag{5.14}$$

but the area A is the same in both expressions and since T_∞ is considered as a constant $dT = d(T - T_\infty) = d\theta$, also recalling $\theta_s = T_s - T_\infty$, we can write

$$-k\left(\frac{d\theta}{dy}\right)_{y=0} = h\theta_s \tag{5.15}$$

To proceed further requires knowledge of the temperature profile in order to evaluate $d\theta/dy$ at the wall. Some solutions to the equations of motion for a fluid in laminar flow are available in the literature but the $d\theta/dy$ cannot be simply evaluated. For the case we have considered in Section 5.1.1, we can differentiate Eq. 5.6 to obtain

$$\frac{d\theta}{dy} = \frac{-2\theta_s}{\delta} + \frac{2\theta_s y}{\delta^2} \tag{5.16}$$

so that at the surface

$$\left(\frac{d\theta}{dy}\right)_{y=0} = \frac{-2\theta_s}{\delta} \tag{5.17}$$

Combining Eqs. 5.17 and 5.15 yields the simple result $h = 2k/\delta$ which we can substitute into Eq. 5.8 and noting the variables $h_x x/k = \text{Nu}_x$, the local Nusselt number at the distance x from the start of the boundary layer formation, gives the local correlation

$$\text{Nu}_x = 0.509 N_{\text{Pr}}^{1/2}(0.952 + N_{\text{Pr}})^{-1/4} N_{\text{Gr}}^{1/4} \tag{5.18}$$

In engineering applications we are generally more interested in the heat transfer for the entire surface under consideration than in local values. We can obtain the average heat transfer coefficient from Eq. 5.18 by solving for the local coefficient, then integrating over the vertical length of the plate. Thus

$$h_x = f(\text{properties})x^{-1/4} \tag{5.19}$$

We are considering constant surface and ambient temperature so that the properties are essentially constant and the average coefficient is given by

$$\bar{h} = \frac{1}{L}\int_0^L h_x\, dx \tag{5.20}$$

$$\bar{h} = \frac{f(\text{properties})}{L}\int_0^L x^{-1/4}\, dx \tag{5.20a}$$

$$\bar{h} = \frac{f(\text{properties})}{L}\left(\frac{x^{3/4}}{3/4}\right)_0^L \tag{5.20b}$$

$$= \frac{f(\text{properties})L^{-1/4}}{3/4} \tag{5.20c}$$

Comparing Eq. 5.20c with Eq. 5.19, we see that $f(\text{properties})L^{-1/4} = h_L$ the coefficient at the location $x = L$, so that

$$\bar{h} = \tfrac{4}{3}h_L \tag{5.21}$$

The average Nusselt number over the length L is defined as $\mathrm{Nu} = \bar{h}L/k$ so that

$$\mathrm{Nu} = 0.679\left(\frac{N_{\mathrm{Pr}}}{0.952 + N_{\mathrm{Pr}}}\right)^{1/4}(N_{\mathrm{Pr}}N_{\mathrm{GrL}})^{1/4} \tag{5.22}$$

where we have split the $N_{\mathrm{Pr}}^{1/2}$ term and combined one term with the Grashof number. At values of N_{Pr} significantly over 1, the term $[N_{\mathrm{Pr}}/(0.952 + N_{\mathrm{Pr}})]^{1/4}$ is essentially 1; for example, for $N_{\mathrm{Pr}} = 10$ the term becomes 0.977 and even at $N_{\mathrm{Pr}} = 1$ the term is 0.846. The result of the theoretical analysis that produced Eq. 5.8 then predicts that for a vertical plate the relationship between the Nusselt number and the Prandtl and Grashof product should be of the form

$$\mathrm{Nu} = (\mathrm{constant})(N_{\mathrm{Pr}}N_{\mathrm{Gr}})^{1/4} \tag{5.23}$$

for the heat transfer situation where the free convection boundary layer is laminar.

For the case where the boundary layer is turbulent, differentiation of Eq. 5.10 leads to a unusable result. Eckert and Jackson (2) applied a modified form of Reynolds analogy and arrived at a result similar to Eq. 5.23 but with an exponent on the $N_{\mathrm{Pr}}N_{\mathrm{Gr}}$ product of $2/5$ instead of $\frac{1}{4}$. Eckert and Jackson verified their results with a heat transfer correlation based on the data reported by Jakob (4) and McAdams (3) as shown in Figure 5.5. The predicted slopes (1/4 and

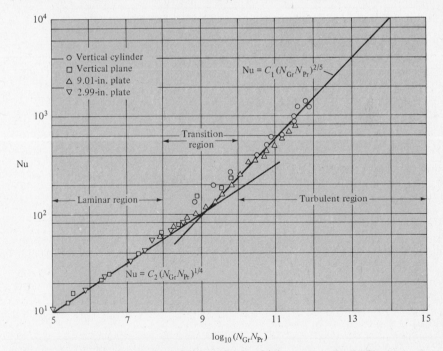

Figure 5.5 Relationship of average Nusselt number for vertical plates and cylinders with the Grashof-Prandtl number product based on the height L; according to data for air presented in McAdams (3) and Jakob (4) as given by Eckert and Jackson (2).

2/5) agree well with the laminar and turbulent regions. The transition from laminar to turbulent takes place at a value of $N_{Gr}N_{Pr}$ of about 10^9, a value that agrees, as we have seen from boundary layer thickness calculations. Jakob (4) further suggests that the simple relationship given by Eq. 5.23 would not hold over a very large range in the Rayleigh number ($N_{Ra} = N_{Gr}N_{Pr}$) and as we have seen, the exponent does change.

Churchill and Chu (5) present an extensive summary of experiments with air, water, ethanol, mercury, glycol, and oil (wide range in Prandtl numbers) for vertical plates and recommend the following correlations:

$N_{RaL} < 10^9$ laminar boundary layer, constant wall temperature

$$\mathrm{Nu} = 0.68 + 0.670 N_{RaL}^{1/4}\left[1 + \left(\frac{0.492}{N_{Pr}}\right)^{9/16}\right]^{-4/9} \tag{5.24}$$

$N_{RaL} < 10^9$ laminar boundary layer, uniform heat flux

$$\mathrm{Nu} = 0.68 + 0.670 N_{RaL}^{1/4}\left[1 + \left(\frac{0.437}{N_{Pr}}\right)^{9/16}\right]^{-4/9} \tag{5.25}$$

T_s at midpoint $(L/2)$

$N_{RaL} > 10^9$ turbulent (plus laminar) boundary layer,

constant wall temperature

$$\mathrm{Nu}^{1/2} = 0.825 + 0.387 N_{RaL}^{1/6}\left[1 + \left(\frac{0.492}{N_{Pr}}\right)^{9/16}\right]^{-8/27} \tag{5.26}$$

$N_{RaL} > 10^9$ turbulent (plus laminar) boundary layer, uniform heat flux

$$\mathrm{Nu}^{1/2} = 0.825 + 0.387 N_{RaL}^{1/6}\left[1 + \left(\frac{0.437}{N_{Pr}}\right)^{9/16}\right]^{-8/27} \tag{5.27}$$

T_s at midpoint $(L/2)$

Figure 5.6 shows how Eq. 5.26 agrees with the data surveyed by Churchill and Chu; agreement for Eqs. 5.24, 5.25, and 5.27 was equally as good. The length criteria in Nu and N_{Gr} is the height of the plate L and as we observed in Section 5.0, the properties are evaluated at $T_f = (T_s + T_\infty)/2$. The correlations presented for vertical plates hold also for vertical cylinders.

Example 5.2

A vertical cylinder is 4 in. in diameter and 10 in. in height. The surface temperature is maintained at 120F and the ambient air temperature is 40F at a pressure of 1 atm. We want to determine the amount of thermal energy lost by the cylindrical surface due to free convection effects.

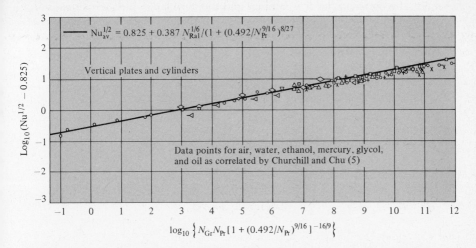

Figure 5.6 Free convection correlation for a vertical flat plate with predominantly turbulent boundary layer [Eq. (5.26)] due to Churchill and Chu (5) as given in Karlekar and Desmond (6). Reproduced by permission from *Engineering Heat Transfer* by B. V. Karlekar and R. M. Desmond, copyright © 1977, West Publishing Company.

SOLUTION

First, we must determine the Rayleigh number to see if we have a predominantly laminar or turbulent free convection boundary layer. After that condition is determined, we can select a correlation equation for the Nusselt number, then calculate the average heat transfer coefficient which permits calculation of the heat flux.

For the cylinder:

$$L = \tfrac{10}{12} \text{ ft} \qquad\qquad N_{Pr} = 0.709$$

$$D = 4 \text{ in.} \qquad\qquad \beta = \tfrac{1}{540} R$$

$$T_s = 120F \qquad\qquad \frac{g\beta}{\nu^2} = 2.09(10^6)\ 1/Rft^3$$

$$T_\infty = 40F \qquad\qquad k = 0.01516 \text{ Btu/hrftF}$$

$$T_f = 80F$$

$$N_{GrL} = L^3 \frac{\beta g}{\nu^2} \Delta T$$

$$= \left(\tfrac{10}{12}\right)^3 2.09(10^6)80 = 9.677(10^7)$$

$$N_{RaL} = N_{GrL} N_{Pr} = 9.677(0.709)10^7$$

$$= 6.861(10^7) < 10^9$$

Therefore, the boundary layer is laminar and the correlation of Eq. 5.24 applies.

$$
\mathrm{Nu} = 0.68 + 0.670 N_{\mathrm{Ra}L}^{1/4}\left[1 + \left(\frac{0.492}{N_{\mathrm{Pr}}}\right)^{9/16}\right]^{-4/9}
$$

$$
= 0.68 + 0.670(6.861)^{1/4}10^{7/4}\left[1 + \left(\frac{0.492}{0.709}\right)^{9/16}\right]^{-4/9}
$$

$$
= 0.68 + \frac{60.98}{1.303} = 47.5
$$

$$
\bar{h} = \frac{\mathrm{Nu}\,k}{L} = \frac{47.5(0.01516)}{10/12} = 0.864 \ \mathrm{Btu/hrft^2F}
$$

$$
q_c = \bar{h}A(T_s - T_\infty)
$$

$$
= 0.864\frac{4\pi 10}{144}(120 - 40)
$$

$$
= 60.3 \ \mathrm{Btu/hr}
$$

The quantity 60.3 Btu/hr is the energy carried away from the vertical surface by the heated air in the free convection boundary layer. We recognize that the surface will also radiate to the surroundings an amount depending on the nature of the surface, the geometrical orientation with respect to other surfaces nearby, and the temperature level of those other surfaces, aspects that we will explore in Chapter 8. This concludes Example 5.2.

For small angles of inclination θ from the vertical ($\theta = 0$ is the vertical), Kreith (15) recommends replacing the buoyant force $\beta g\,\Delta T$ in the Rayleigh number by $\beta g \cos\theta\,\Delta T$ to account for the decrease in buoyant force along the surface. The correction recommended by Kreith is probably valid for angles of inclination up to 20 or 30 degrees; for larger angles separation of the boundary layer at the top (of a heated plate) would result in different phenomena than those described by the vertical plate equations.

5.3 HORIZONTAL SURFACES CORRELATIONS

The boundary layer formation on horizontal plates is of a different nature than that formed on vertical plates and cannot be simply analyzed. The experimental work that has been done with horizontal surfaces has been correlated very well with the Grashof-Prandtl product form that was expressed previously as

$$
\mathrm{Nu} = CN_{\mathrm{Ra}}^m \tag{5.23}
$$

with properties evaluated at the film temperature $T_f = (T_s + T_\infty)/2$ and a characteristic length according to the geometry under consideration. The constants C and m vary with the Rayleigh number, but for most situations values can be selected which hold over a reasonable range.

We will distinguish between horizontal flat plates and horizontal cylinders; spheres, however, because of their complete symmetry, are in a class of their own.

5.3.1 Heated Plates Facing Up; Cooled Plates Facing Down

The correlation for heated plates facing up or cooled plates facing down given by McAdams (7) for square plates can be used for plates of any configuration according to Goldstein, Sparrow, and Jones (8) when the length dimension in the Nusselt and Rayleigh numbers is evaluated as

$$L^* = \frac{\text{surface area}}{\text{perimeter}} = \frac{A}{P} \tag{5.28}$$

A flat circular disk, for example, would have $L^* = D/4$ where D is disk diameter. The correlations recommended by McAdams are

$$\text{Nu} = 0.14 N_{\text{Ra}L^*}^{1/3} \qquad 2(10^7) < N_{\text{Ra}L^*} < 3(10^{10}) \tag{5.29}$$

$$\text{Nu} = 0.54 N_{\text{Ra}L^*}^{1/4} \qquad 10^5 < N_{\text{Ra}L^*} < 2(10^7) \tag{5.30}$$

where

$$L^* = \frac{A}{P} \qquad \text{Nu} = \frac{hL^*}{k_f}$$

$$N_{\text{Ra}L^*} = N_{\text{Gr}L^*} N_{\text{Pr}} \qquad \text{properties at } T_f = \frac{T_s + T_\infty}{2}$$

It is possible to select a narrow range of operation for a particular substance and make simple design equations of the form $\bar{h} = \text{constant}(\Delta T/L^*)^n$ where the constant has value and units depending on the particular system employed, an aspect that is explored in the home problems.

5.3.2 Heated Plates Facing Down; Cooled Plates Facing Up

The correlation presented by McAdams (7) for the subject physical arrangement was formulated from data on plates that had no restrictions to the flow of fluid around the edges of the plates. Restrictions of any kind would block the free convection flow of the fluid.

$$\text{Nu} = 0.27 N_{\text{Ra}L^*}^{1/4} \qquad 3(10^5) < N_{\text{Ra}L^*} < 3(10^{10}) \tag{5.31}$$

where

$$L^* = \frac{A}{P} \qquad \text{Nu} = \frac{\bar{h}L^*}{k_f}$$

$$N_{\text{Ra}L^*} = (L^*)^3 \left(\frac{\beta g}{\nu \alpha}\right)_f \Delta T \qquad \text{properties at } T_f = \frac{T_s + T_\infty}{2}$$

5.3.3 Horizontal Cylinders

The length criterion for the dimensionless parameters relating free convection for horizontal cylinders is the outer diameter D of the cylinder. McAdams (7) has collected data on air, hydrogen, carbon dioxide, water, aniline, glycerine, toluene, and carbon tetrachloride in free convection on horizontal cylinders of diameters ranging from 0.01 to 8.9 cm as shown in Figure 5.7. For $N_{Ra} > 1000$ the data are well correlated by a straight line up to 10^9, but below 1000 two equations are needed to maintain a prediction precision within the spread of the experimental data. The equations are

$$\text{Nu} = 1.08 N_{RaD}^{0.10} \qquad 10^{-4} < N_{RaD} < 1 \tag{5.32}$$

$$\text{Nu} = 1.08 N_{RaD}^{0.155} \qquad 1 < N_{RaD} < 10^3 \tag{5.33}$$

$$\text{Nu} = 0.537 N_{RaD}^{0.25} \qquad 10^3 < N_{RaD} < 10^9 \tag{5.34}$$

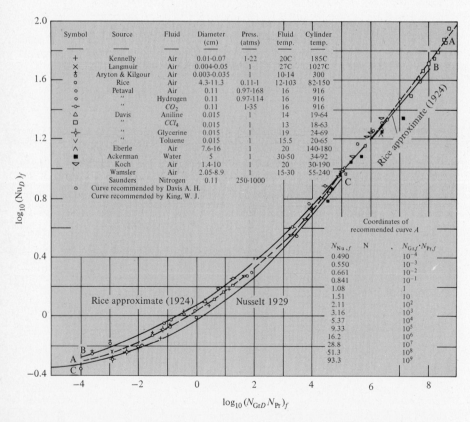

Symbol	Source	Fluid	Diameter (cm)	Press. (atms)	Fluid temp.	Cylinder temp.
+	Kennelly	Air	0.01-0.07	1-22	20C	185C
×	Langmuir	Air	0.004-0.05	1	27C	1027C
⚵	Aryton & Kilgour	Air	0.003-0.035	1	10-14	300
○	Rice	Air	4.3-11.3	0.11-1	12-103	82-150
○	Petaval	Air	0.11	0.97-168	16	916
○	"	Hydrogen	0.11	0.97-114	16	916
⟠	"	CO_2	0.11	1-35	16	916
△	Davis	Aniline	0.015	1	14	19-64
□	"	CCl_4	0.015	1	13	18-63
⟊	"	Glycerine	0.015	1	19	24-69
∨	"	Toluene	0.015	1	15.5	20-65
∧	Eberle	Air	7.6-16	1	20	140-180
■	Ackerman	Water	5	1	30-50	34-92
▽	Koch	Air	1.4-10	1	20	30-190
	Wamsler	Air	2.05-8.9	1	15-30	55-240
○	Saunders	Nitrogen	0.11	250-1000		

Curve recommended by Davis A. H.
Curve recommended by King, W. J.

Coordinates of recommended curve A

$N_{Nu,f}$	$N_{Gr,f} \cdot N_{Pr,f}$
0.490	10^{-4}
0.550	10^{-3}
0.661	10^{-2}
0.841	10^{-1}
1.08	1
1.51	10
2.11	10^2
3.16	10^3
5.37	10^4
9.33	10^5
16.2	10^6
28.8	10^7
51.3	10^8
93.3	10^9

Rice approximate (1924)

Nusselt 1929

Rice approximate (1924)

$\log_{10}(\text{Nu}_D)_f$

$\log_{10}(N_{GrD} N_{Pr})_f$

Figure 5.7 Free convection from single horizontal cylinders in liquids and gases according to McAdams (7) with the Grashof and Nusselt numbers based on the diameter D and properties at the film temperature. Reproduced with permission of McGraw-Hill Book Company.

where

$$\text{Nu} = \frac{\bar{h}D}{k_f}$$

$$N_{\text{Ra}D} = D^3 \left(\frac{\beta g}{\nu \alpha} \right)_f \Delta T \qquad \text{properties at } T_f = \frac{T_s + T_\infty}{2}$$

In the event that the reader requires more precision than can be obtained from Eqs. 5.32 and 5.33, the coordinates of the recommended curve (see Figure 5.7) can be plotted on graph paper with a suitable scale, or the region can be further split and additional equations formulated. As ΔT approaches zero, the Nusselt number approaches 0.45 as a limiting value.

Example 5.3

An electronics component module is in the shape of a vertical cylinder. Because of encapsulation and packaging all the energy dissipated by the module must be through the top of the cylinder which is a disk 6 in. in diameter. Estimate the power dissipation limitations in watts for the module for a top surface temperature of 150F and an ambient temperature of 90F.

SOLUTION
We observe that the least power dissipation will take place under free convection conditions, and therefore, heat flux dissipated from the top of the module (a horizontal, heated, circular disk facing up) will be the limiting case for the module.

$$T_\infty = 90F \qquad \left(\frac{g\beta}{\nu\alpha} \right)_f = \left(\frac{g\beta}{\nu^2} \right)_f N_{\text{Pr}f}$$

$$T_s = 150F \qquad = 2.280(10^6)0.703$$

$D = 6 \text{ in.} \qquad T_f = 120F \qquad = 1.603(10^6)\text{ft}^{-3}\text{R}^{-1}$

$$L^* = \frac{D}{4} = 1.5 \text{ in.} \qquad k_f = 0.01615 \text{ Btuhrft}^2\text{F}$$

The Rayleigh number determines the type of correlation equation. Therefore

$$N_{\text{Ra}L^*} = (L^*)^3 \left(\frac{g\beta}{\nu\alpha} \right)_f \Delta T = \left(\frac{1.5}{12} \right)^3 1.063(10^6)60$$

$$= 1.246 \times 10^5 \therefore \text{ Eq. 5.30 applies}$$

$$\text{Nu} = 0.54 N_{\text{Ra}L^*}^{0.25} = 0.54(124{,}600)^{0.25} = 10.14 = \frac{\bar{h}L^*}{k_f}$$

$$\therefore \bar{h} = \frac{0.01615(10.14)}{1.5/12} = 1.31 \text{ Btu/hrft}^2\text{F}$$

Limiting flux due to free convection, $q = \bar{h}A\,\Delta T$,

$$q = 1.31\frac{9\pi}{144}60 = 15.43\,\text{Btu/hr}$$

$$q = \frac{15.43}{3.412} = 4.52\,\text{W}$$

The limiting power that can be dissipated is about 4.5 W due to free convection. Depending on conditions at the surface and on the nature and temperature of other surfaces in the immediate vicinity, more or less energy than 4.5 W may be dissipated due to possible radiant energy exchange.

5.4 OTHER EXTERNAL CORRELATIONS

The physical phenomenon of free convection around spheres, blocks, and short cylinders involves a combination of buoyancy driven flow phenomena on the surfaces which results in a Nusselt number for the particular geometry-fluid combination which can be correlated with the Grashof-Prandtl product (the Rayleigh number), as we have seen previously. The following subsections present a few such correlations.

5.4.1 Spheres

A limited amount of data for free convection from spheres in air only has been reported by Yuge (9). The data were correlated on the basis of sphere diameter D with properties at the film temperature as

$$\text{Nu} = 2 + 0.43N_{\text{Ra}D}^{1/4} \qquad 1 < N_{\text{Ra}} < 10^5$$

$$\text{properties at } T_f = \frac{T_s + T_\infty}{2} \tag{5.35}$$

We can observe that as ΔT approaches zero, the Nusselt number approaches the value 2 which is the case for pure conduction from the spherical surface in a very large extent of fluid (see Problem 10 for details).

5.4.2 Rectangular Blocks

McAdams (7) presents a correlation for blocks which can also be used for short cylinders whose L/D is of the order of magnitude of unity that is also based on the Rayleigh number but with a particular characteristic length such that

$$\frac{1}{L^*} = \frac{1}{L_{\text{vert}}} + \frac{1}{L_{\text{horizontal}}} \tag{5.36}$$

The correlation is given as

$$\text{Nu} = 0.55N_{\text{Ra}L^*}^{1/4} \qquad 10^4 < N_{\text{Ra}} < 10^7 \tag{5.37}$$

$$\text{Nu} = 0.13N_{\text{Ra}L^*}^{1/3} \qquad 10^7 < N_{\text{Ra}} < 10^{12} \tag{5.38}$$

The reciprocal of the characteristic length is the sum of the reciprocal of the vertical height and the reciprocal of the longest horizontal dimension where the two dimensions are of the same order of magnitude (the concept would not apply to thin plates).

Example 5.4

We desire to determine the heat transfer coefficient for the surfaces of a short cylinder 4 in. tall and 6 in. in diameter whose surface is at 212F in air at 1 atm pressure and 20C under free convection conditions.

SOLUTION

We first determine the Rayleigh number to determine the nature of the free convection flow patterns and which of Eqs. 5.37 or 5.38 to use. The film temperature is given by $T_f = (212 + 68)/2 = 140F$ so that properties at 140F are

$$\left(\frac{g\beta}{\nu\alpha} \right)_f = 0.90936(10^6) \text{ ft}^{-3}\text{R}^{-1}$$

$$k_f = 0.01664 \text{ Btu/hrftF}$$

The characteristic length

$$\frac{1}{L^*} = \tfrac{1}{4} + \tfrac{1}{6} = \tfrac{5}{12} \text{ in.}^{-1}$$

$$L^* = 2.4 \text{ in.}$$

and the Rayleigh number

$$N_{\text{Ra}} = (L^*)^3 \frac{g\beta}{\nu\alpha} \Delta T$$

$$N_{\text{Ra}} = \left(\frac{2.4}{12} \right)^3 (0.90936)10^6(144)$$

$$= 1.048 \times 10^6$$

which requires use of Eq. 5.37 so that

$$\text{Nu} = 0.55 N_{\text{Ra}}^{1/4} = 0.55(1{,}048{,}000)^{1/4}$$

$$= 17.6$$

$$\bar{h} = \frac{17.6 k_f}{L^*} = \frac{17.6(0.01664)}{2.4/12}$$

$$= 1.46 \text{ Btu/hrft}^2\text{F}$$

a value that is in the range given in Table 1.1 in Chapter 1. This concludes the example.

5.5 CORRELATIONS FOR ENCLOSED SPACES

In the physical phenomena that we have discussed so far, the fluid extended indefinitely from the surface in question; actually, far enough so that the confining surface exerted no influence on the heat transfer happenings. In this section we will look at some correlation equations for energy exchange between surfaces that are close together so that the boundary layers interact. We consider the geometries shown in Figure 5.8; the plate separation distance is s and for the vertical orientation the height is L, the length of the plate. The Nusselt and Grashof or Rayleigh numbers are based on the plate spacing s as the length criterion, so that

$$\mathrm{Nu} = \frac{\bar{h}s}{k_f} \qquad N_{\mathrm{Gr}s} = s^3 \left(\frac{\beta g}{\nu^2} \right)_f \Delta T$$

$$N_{\mathrm{Ra}s} = s^3 \left(\frac{\beta g}{\nu \alpha} \right)_f \Delta T$$

As in previous correlations the subscript f indicates that the properties are evaluated at the average temperature between the surfaces $T_f = (T_1 + T_2)/2$ and in this case $\Delta T = T_1 - T_2$. For the configurations shown in Figure 5.8, we can define a heat transfer coefficient $\bar{h}$ to represent $\bar{h}_h$, $\bar{h}_v$, or $\bar{h}_\theta$ for the horizontal, vertical, or inclined case in terms of the temperatures T_1 and T_2 such that

$$q = \bar{h}A(T_1 - T_2) \tag{5.39}$$

where the area $A = LW$, the height times the enclosure width W. We observe that at T_1 approaching T_2 the buoyancy effects become negligible and only conduction remains as the mode of heat transfer. For conduction only across the

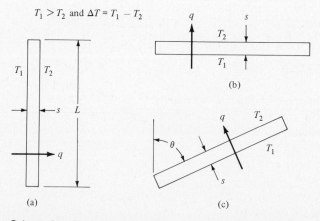

Figure 5.8 Orientation for free convection in enclosed spaces with separation distance s between surfaces. (a) Vertical. (b) Horizontal. (c) Inclined θ degrees from the vertical.

space s the flux is given by

$$q = k\frac{A}{s}(T_1 - T_2) \tag{5.40}$$

Equating 5.39 and 5.40 and rearranging terms shows that $\text{Nu} = hs/k = 1$ for pure conduction. Kraussold (16) has shown experimentally that pure conduction, that is, $\text{Nu} = 1$, exists at Rayleigh numbers less than 1000 in enclosed vertical, horizontal, and annular cylindrical spaces; the criteria may also be extended to spherical spaces.

The heat transfer coefficient $\bar{h}$ determined from the correlations presented in the subsections above and below accounts only for convective effects; no radiant contribution is included. The total flux is $q = q_c + q_r$ and we deal only with the determination of q_c.

5.5.1 Horizontal Enclosure Correlations

Experimental results have been analyzed by Jakob (4) to obtain the correlations applicable to the enclosure shown in Figure 5.8(b) as

$$\text{Nu}_h = 0.0766 N_{\text{Ra}s}^{1/3} \qquad N_{\text{Ra}s} > 400{,}000 \tag{5.41}$$

$$\text{Nu}_h = 0.213 N_{\text{Ra}s}^{1/4} \qquad 6000 < N_{\text{Ra}s} < 400{,}000$$

$$\text{properties at } T_f = \frac{T_1 + T_2}{2} \tag{5.42}$$

At values of N_{Gr} based on s less than 1000 the heat transfer is by conduction.

5.5.2 Vertical Enclosure Correlations

Jakob (4) includes an L/s term in the correlations for free convection in narrow vertical enclosures such that

$$\text{Nu}_v = 0.0732 N_{\text{Ra}s}^{1/3}\left(\frac{L}{s}\right)^{-1/9} \qquad 2(10^5) < N_{\text{Ra}s} < 10(10^6) \tag{5.43}$$

$$\text{Nu}_v = 0.197 N_{\text{Ra}s}^{1/4}\left(\frac{L}{s}\right)^{-1/9} \qquad 10^4 < N_{\text{Ra}s} < 2(10^5)$$

$$3 < \frac{L}{s} < 50 \qquad \text{properties at } T_f = \frac{T_1 + T_2}{2} \tag{5.44}$$

Jakob (4) indicates that the lower limit on L/s is about 3, and should not be exceeded because as the spacing increases, the physical situation approaches that for a vertical plate in an infinite medium and values less than 3 give unrealistically high Nusselt numbers with the correlation in Eq. 5.43.

5.5.3 Inclined Enclosures

Jakob suggests that a linear interpolation based on the angle θ from the vertical should adequately predict results for inclined closely spaced enclosures between flat surfaces. Such an interpolation can be expressed as

$$\text{Nu}_\theta = \text{Nu}_v + \frac{\theta}{90}(\text{Nu}_h - \text{Nu}_v) \tag{5.45}$$

where Nu_h and Nu_v are given by Eqs. 5.41 or 5.42 and 5.43 and 5.44, respectively.

5.5.4 Horizontal Cylindrical Annuli

Jakob (4) also reports on some early European experiments with oil and water in the annulus formed by two concentric horizontal cylinders, the inner cylinder being heated so that its surface temperature T_1 is greater than the inner surface temperature T_2 of the outer cylinder. The correlation given is

$$\text{Nu} = 0.11 N_{\text{Ra}s}^{0.29} \qquad 10^{3.8} < N_{\text{Ra}s} < 10^6 \tag{5.46}$$

$$\text{Nu} = 0.40 N_{\text{Ra}s}^{0.2} \qquad 10^6 < N_{\text{Ra}s} < 10^8 \tag{5.47}$$

$$s = r_2 - r_1$$

$$7 < N_{\text{Pr}} < 4000$$

$$\text{properties at } T_f = \frac{T_1 + T_2}{2}$$

The length criterion for Nu and $N_{\text{Ra}s}$ is the difference in radii of the concentric cylinders which is the spacing. Data for the gases, air, hydrogen, and carbon dioxide are also described in Reference 4 but not in a simple correlation.

Those circumstances where the surfaces at T_1 and T_2 are close together but do not form an enclosure involve a through-put of mass flow at low velocity dependent on the magnitude T_1 and T_2 compared to the ambient temperature T_∞ and involve mixed forced and free convection. Some aspects of mixed flow are covered in Chapter 6. For those cases where the surface temperatures are not the same, $T_1 \neq T_2$, it may be possible to use a superposition treatment of existing theoretical solutions in laminar flow. Such topics are discussed in graduate courses on convection and are beyond the scope of our objectives.

5.6 CONCLUDING REMARKS

We have described results and calculation techniques for the basic, most important geometries of vertical, horizontal, and inclined surfaces and for horizontal cylinders. The results in all cases were such that the Nusselt number was some function of the Grashof-Prandtl number product which is the Rayleigh number. The length criteria depends on the nature of the geometrical arrangement.

A very large body of information on free convection has accumulated in the heat transfer literature since about 1900, covering a myriad of geometrical arrangements and many different fluids (range in Prandtl number). Our purpose in this chapter was to become familiar with the dimensionless parameters that characterize free convection and to see what magnitude the free convection heat transfer coefficient takes. The reader with a particular application that cannot be modeled with the basic geometries we have discussed must peruse the literature for more apropos experimental or analytical results. It is impractical to catalog all experiments in an introductory text. The more important sources, especially the German literature, prior to 1945 are covered in Jakob (4), (10) and the period from 1945 to 1967 has been surveyed by Moore (11) and Moore and Wolf (12). More recent work can be found in the journals of professional societies that specialize in heat transfer topics. Two of the more well-known ones are *The Journal of Heat Transfer*, Series C of the *Transactions ASME* and the *International Journal of Heat and Mass Transfer*; the latter contains periodic surveys of the Russian literature on heat transfer topics.

PROBLEMS

A reference to the chapter section which deals with the problem topic has been included with the problems for Chapter 1 through 4 as an orientation aid to the reader. That reference is being omitted from the problems in this and subsequent chapters now that the reader has had some background and experience in assessing different problems. It is particularly important in convection situations to delineate the phenomena being considered and to match that phenomena with an applicable correlation equation. In subsequent chapters we will deal with combined phenomena (such as convection and radiation), a process that requires the reader to extend thinking over a broader range than has been required up to this point.

1. Compare the boundary layer thickness δ and the displacement thickness δ^* on a vertical flat plate in air and in water at surface temperatures of 200F and fluid temperatures of 160F at a location 1 ft from the bottom edge of the plate.
2. Determine the value of the Grashof and Rayleigh numbers at the transition from laminar to turbulent boundary layer for a vertical plate in water under the conditions of Problem 1 by the method discussed in Example 5.1
3. Show that the maximum velocity in the laminar boundary layer on a vertical plate under free convection conditions occurs at a distance $y = \delta/3$ from the surface.
4. Starting with the defining equation for δ^*, Eq. 5.4, insert the laminar velocity profile given by Eq. 5.7 and determine δ_L^* as a function of the boundary layer thickness δ_L.
5. Why is it not possible to evaluate Eq. 5.15 for the turbulent boundary layer and formulate a Nusselt number from the expression for boundary layer thickness as was done for the laminar boundary layer?
6. Determine the amount of heat transferred by free convection to a large vertical plate glass window that is 5 ft high and 7 ft wide from the air in the room when the surface of the glass is at 35F and the air is at 75F.

7. Eckert and Jackson (2) have given a value of 0.021 for C_1 and 0.555 for C_2 in the equations delineating the slopes in Figure 5.5. Check the magnitude of your Nusselt number calculated in Problem 6 by using the appropriate equation from Figure 5.5.

8. It is possible to formulate simple design equations for $\bar{h}$ in terms of a dimensional constant and $(\Delta T/L)^n$ for a narrow temperature range for a given fluid from a correlation equation of the type specified in Eq. 5.23. Formulate such an equation for air at a film temperature of 80F and 200F for the laminar boundary layer using a constant of 0.555 in Eq. 5.23.

9. A can of pears is suspended by pincers (all sides exposed) in still air from a conveyor belt and has just come out of a cooking bath at 212F. Estimate how long it would take to cool the contents to 100F in air at 70F. The can is 4 in. in diameter, 4.75 in. tall, and is completely filled with pears and water. Assume properties of pears are the same as water.

10. Show that for pure conduction from a spherical surface into a fluid ($\Delta T \to 0$) infinite in extent that the Nusselt number approaches the value 2. Do this by equating the convection Eq. 1.1 with the conduction Eq. 2.32 for a hollow sphere and take the limit as the outer radius approaches infinity.

11. A nickel wire 0.010 in. in diameter and 18 in. long is held horizontally in still air (at 80F) between two electrodes. Estimate the voltage drop across the wire required to produce a surface temperature of 1000F.

12. A horizontal tube 1-in. OD carries saturated Freon-12 liquid at a pressure of 23.849 psia. How many pounds per minute of Freon are evaporated in a 20-ft length of tube in still air at 120F?

13. A copper sphere 1 in. in diameter is used to determine a free convection heat transfer coefficient for that geometry in water. The sphere was very slowly immersed into the water and conditions allowed to stabilize so that free convection prevailed. The water temperature was 50F and the sphere took 15.4 s to cool from 75 to 65F. Determine the Nusselt number from the lumped parameter method and compare with the prediction from Yuge's equation.

14. Determine the maximum permissible velocity of air over a solid copper sphere 1.5 in. in diameter in order for free convection to be the dominant mechanism of heat transfer when the film temperature is 80F and there is a temperature difference of 50F between the surface and the undisturbed fluid. Note that the Reynolds number for a sphere is based on the diameter of the sphere.

15. Estimate the energy savings that could be achieved by filling the $3\frac{1}{2} \times 14\frac{1}{2}$ in. space between wall studs in the vertical wall of a residential house with glass wool of 4 lb_m/ft^3 density. Assume that the side (surface) of the space toward the inside of the house is at 65F, the surface toward the outside is at 15F, and the space is 8 ft tall in the vertical direction.

16. Water at the rate of 2 lb_m/s flows in a 0.5-in. OD horizontal copper tube 50 ft long. The tube passes through an unheated storage shed in which the air temperature is at 20F. If the water at the point of entrance into the storage

shed is at 140F, determine the drop in temperature of the water as it passes through the 50-ft length of tubing. Would it be worthwhile to insulate the tube?

17. Thermal conductivity for liquids is often measured (see Reference 17) by filling a spherical annulus with the fluid, assuming conduction to be the mode of heat transfer, and calculating k from Eq. 2.32 after precisely measuring: dimensions, ΔT, and heat flux (electrically). Determine the maximum ΔT for the conduction assumption to hold for k experiments with Freon-12 and with light oil (separately) in a spherical annulus having 0.12-in. spacing at an average temperature of 100F in the gap.

18. Determine the heat flux between two horizontal square plates 2 ft on a side and spaced 1 in. apart when the top plate is at 105F and the bottom plate at 95F with carbon dioxide in the enclosed space between the plates.

19. A block of ice $12 \times 12 \times 12$ in. is placed on a horizontal styrofoam pad (so that there is no heat transfer on the bottom) in a room where the relative humidity is 100 percent and the air temperature is 88F at atmospheric pressure. How long will it take for 10 percent of the block to melt if there is no motion of the air in the room?

20. A cylindrical tank containing liquid nitrogen at 1 atm pressure is 1 ft in diameter and 10 in. tall. The tank is well insulated on the sides and the top is covered with a loosely fitting insulating cover to allow gas to escape. The tank is filled to a depth of 4 in. with LN_2 and is suspended a distance of 57.5 in. from the floor in a moisture-free quiet room where the air temperature is 120F. Estimate the time required for the nitrogen to evaporate.

References

1. Giedt, W. H., *Principles of Engineering Heat Transfer*, Van Nostrand, New York, 1957, p. 212.
2. Eckert, E. R. G. and Jackson, T. W., "Analysis of Turbulent Free Convection Boundary Layer on a Flat Plate," NACA Tech. Report 1015, 1951.
3. McAdams, W. H., *Heat Transmission*, 2d Edition, McGraw-Hill, New York, 1942, p. 248.
4. Jakob, M., *Heat Transfer*, Vol. I, Wiley, New York, 1949, pp. 529, 535.
5. Churchill, S. W. and Chu, H. H. S., "Correlating Equations for Laminar and Turbulent Free Convection from a Vertical Plate," *International Journal of Heat and Mass Transfer*, Vol. 18, 1975, p. 1323.
6. Karlekar, B. V. and Desmond, R. M., *Engineering Heat Transfer*, West Publishing Co., New York, 1977, p. 438.
7. McAdams, W. H., *Heat Transmission*, 3d Edition, McGraw-Hill, New York, 1954, pp. 172, 176, 180.
8. Goldstein, R. J., Sparrow, E. M., and Jones, D. C., "Natural Convection Mass Transfer Adjacent to Horizontal Plates," *International Journal of Heat and Mass Transfer*, Vol. 16, 1973, p. 1025.
9. Yuge, T., "Experiments on Heat Transfer from Spheres Including Combined Natural and Forced Convection," *Transactions ASME, C, Journal of Heat Transfer*, Vol. 80, 1960, p. 214.

10. Jakob, M., *Heat Transfer*, Vol. II, Wiley, New York, 1957.
11. Moore, N. R., MSME Thesis, Mech. Engr. Dept., University of Arkansas, Fayetteville, AR, 1967.
12. Moore, N. R. and Wolf, H., "Heat Transfer Research with Application to Large Distribution Transformers: Part I-A, Survey and Critical Analysis of the Literature," University of Arkansas Engineering Experiment Station Research Report No. 10, College of Engineering, Fayetteville, AR, 1967.
13. Pera, L. and Gebhart, B., "Natural Convection Boundary Layer Flow over Horizontal and Slightly Inclined Surfaces," *International Journal of Heat and Mass Transfer*, Vol. 16, 1973, p. 1131.
14. Pera, L. and Gebhart, B., "On the Stability of Natural Convection Boundary Layer Flow over Horizontal and Slightly Inclined Surfaces," *International Journal of Heat and Mass Transfer*, Vol. 16, 1973, p. 1147.
15. Kreith, F., *Principles of Heat Transfer*, 3d Edition, Harper & Row, New York, 1973, pp. 396, 406.
16. Kraussold, H., *Forschung a.d. Gebiete d. Ingenieurwes*, Vol. 5, 1934, p. 186
17. Dyer, H. L. and Wolf, H., "Determination of Transport Properties of Transformer Oils," Engineering Experiment Station Research Report Series No. 12, University of Arkansas, Fayetteville, AR, Aug. 1968.

Chapter 6
Forced Convection

6.0 INTRODUCTORY REMARKS

The basic concepts involved in forced convection were discussed in the general overview of convection presented in Chapter 4. The two important basic phenomena influencing forced convection are the nature of the bulk flow (laminar or turbulent) and the character of the boundary layer. These two aspects and the recognition that the flow takes place either internally in ducts or externally over surfaces indicate a natural division and organization of the material that we will cover. It is even more important for forced convection, as compared to free convection, to emphasize that we are dealing here only with a simple survey of the fundamental phenomena. The recent space age (about 1950 to 1975) has given a tremendous impetus to work in forced convection, both theoretical and experimental, which cannot be covered in an introductory treatment. We will look at one exotic aspect only, that of high-speed flow, which can be treated within the scope of our understanding at this level.

We begin our description of forced convection with a look at entrance region flow, laminar and turbulent, in ducts followed by correlations for fully developed flow in ducts under various conditions. We will then consider external flow over plates, cylinders, and spheres and conclude with the effects of roughness, high-speed flow, and a consideration of the interaction between free and forced convection when the forced velocities are very low; a logical topic to conclude with because at that time we will probably be relatively short of breath.

In anticipation of the results, we will see that the Nusselt number can be functionally expressed as follows:

$$\text{Nu} = f\left(N_{\text{Re}}, N_{\text{Pr}}, \frac{D}{L}, \frac{T_s}{T_b} \right) \tag{6.1}$$

The flow conditions are reflected by the Reynolds number N_{Re}, the nature of the fluid by the Prandtl number, the status of the boundary layer by D/L (for entrance region cases), and the influence of physical property variations with temperature by T_s/T_b, the ratio of wall to bulk temperature. We have discussed properties in Section 4.1 and the Prandtl number in detail in Section 4.1.4 but it may be helpful to look at Figure 6.1 to associate the magnitude of the Prandtl number with a mental "physical feel" for the fluid. In general, low Prandtl number fluids make the best heat transfer fluids because of their superior thermal diffusivity and thermal conductivity.

In all of the correlations that we will encounter, some temperature must be specified at which the physical properties are to be evaluated. Clearly the temperature must be the same one that the investigator utilized to obtain the

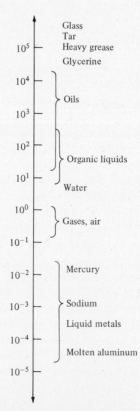

Figure 6.1 Scope of Prandtl number values for different substances.

least deviation of the data points from the line representing the correlation equation. That temperature is called the reference temperature T_R and generally lies somewhere between the bulk temperature T_b and the wall temperature T_s. The reference temperature can be expressed mathematically as

$$T_R = T_b + x(T_s - T_b) \tag{6.2}$$

When $x = 0$, the reference temperature is the bulk temperature; at $x = 0.5$, the reference temperature is the film temperature $T_f = (T_b + T_s)/2$; and at $x = 1$, the reference temperature is the wall temperature. The choice of the value of x is made by the experimentor in evaluating the data. Most heat transfer correlations for forced convection at low values of $|(T_s - T_b)|$ are correlated at the bulk temperature $(x = 0)$ and at high values of $|(T_s - T_b)|$, generally values of $x > 0.4$ are needed to get a reasonable correlation, or the term T_s/T_b to some power is included in the correlation equation. For liquids we will find the viscosity ratio μ_s/μ_b instead of T_s/T_b useful in achieving a good correlation. We now direct our attention to internal flow in the entrance region.

6.1 INTERNAL FLOW, LAMINAR

In the Reynolds number range below 2300, the flow of a fluid in a duct is laminar. When special pains are taken to avoid vibration or shock, flow conditions can be held laminar at values higher than 2300 up to about 30,000 or so, but under unstable conditions. In the laminar flow regime, the transfer of thermal energy in the fluid is by conduction and as the thermal boundary layer builds up in the flow direction, the resistance to heat transfer is increased. The Nusselt number (or heat transfer coefficient), therefore, starts out at a very high value at the location where energy exchange is initiated and then decreases to a constant value for the uniform heat flux case with constant properties or to a lower constant value for the constant wall temperature case. Specific qualitative details are explored in the following subsection.

6.1.1 Entrance Region, Constant Wall Temperature

For laminar flow in a round tube (with a fully developed velocity profile at the point where energy exchange with the fluid begins) the local and average analytical Nusselt numbers Nu_x and Nu_{av} have been theoretically determined by Sellars, Tribus, and Klein (1) in series form as a function of nondimensional distance $x^+ = x/(r_0 N_{\mathrm{Re}} N_{\mathrm{Pr}})$ and constants as

$$\mathrm{Nu}_x = \frac{\Sigma G_n \exp\left(-\lambda_n^2 x^+\right)}{2\Sigma\left(G_n/\lambda_n^2\right)\exp\left(-\lambda_n^2 x^+\right)} \tag{6.3}$$

$$\mathrm{Nu}_{av} = \frac{1}{2x^+}\ln\left[\frac{1}{8\Sigma\left(G_n/\lambda_n^2\right)\exp\left(-\lambda_n^2 x^+\right)}\right] \tag{6.4}$$

The constants λ_n^2 and G_n are given below in Table 6.1.

Table 6.1 INFINITE SERIES CONSTANTS
λ_n^2 AND G_n FOR USE IN EQS. 6.3
AND 6.4 FOR CONSTANT WALL
TEMPERATURE (CWT)

N	λ_n^2	G_n
0	7.312	0.749
1	44.62	0.544
2	113.8	0.463
3	215.2	0.414
4	348.5	0.382

For $n > 4$; $\lambda_n = 4n + \frac{8}{3}$
$G_n = 1.0127\lambda_n^{-1/3}$

Equations 6.3 and 6.4 are not too convenient to work with; therefore, Kays (2) has calculated some representative values that are presented in Table 6.2. For $x^+ > 0.1$, Table 6.2 shows that the local Nusselt number is essentially the same as the value for fully developed flow, thermally and hydraulically. We can therefore conclude that the entrance length for the flow situation described by Table 6.2 occurs at about the distance such that $x^+ = 0.1$ so that

$$x^+ \simeq 0.1 = x/r_0 N_{\text{Re}} N_{\text{Pr}} \tag{6.5}$$

or we can infer that

$$\left(\frac{x}{D}\right)_{\text{entrance}} \simeq 0.05 N_{\text{Re}} N_{\text{Pr}} \tag{6.6}$$

For air, for example, at a Reynolds number of 2300, the entrance length would be (for $N_{\text{Pr}} = 0.7$) about 80 diameters, whereas with an oil whose Prandtl number $N_{\text{Pr}} = 50$, the entrance length would be about 5750 diameters at the same Reynolds number.

The results for this theoretical analysis for the determination of Nusselt number hold only when the properties (such as viscosity and thermal conductivity) are essentially constant across the cross section. Such would be the case when the wall temperature and the bulk temperature are only a few degrees apart. When the viscosity varies sharply with temperature, as for organic liquids,

Table 6.2 NUSSELT NUMBERS FOR
LAMINAR FLOW IN CIRCULAR
TUBES, CWT

x^+	Nu_x	Nu_{av}
0	∞	∞
0.001	12.86	22.96
0.004	7.91	12.59
0.01	5.99	8.99
0.04	4.18	5.87
0.08	3.79	4.89
0.10	3.71	4.66
0.20	3.66	4.16
∞	3.66	3.66

a correction for viscosity effects as indicated in Section 6.1.3 must be included to avoid considerable error.

6.1.2 Entrance Region, Uniform Heat Flux

The boundary condition of uniform heat flux for laminar flow in the entrance region has been treated by Siegel, Sparrow, and Hallman (3) in an analysis similar to Reference 1. The local Nusselt number Nu_x given by Reference 3 can be written as

$$Nu_x = \left[\frac{1}{Nu_\infty} - \Sigma \frac{\exp(-\gamma_n^2 x^+)}{2 A_n \gamma_n^4} \right]^{-1} \tag{6.7}$$

The Nusselt number Nu_∞ far down the tube (at $x^+ = \infty$ in the mathematical solution) is $\frac{48}{11} = 4.364$ and a tabulation of the constants γ_n^2 and A_n is given in Table 6.3.

A calculation of representative values of local and average Nusselt numbers as a function of x^+ is given in Table 6.4. The values for average Nusselt number over the length of the tube x^+ were calculated from a graphical integration of the definition

$$Nu_{av} = \frac{1}{x^+} \int_0^{x^+} Nu_x \, dx^+ \tag{6.8}$$

for each of the x^+ values listed in the table.

The results in Table 6.4 show that we are within 5 percent of the fully developed Nusselt number at the location $x^+ = 0.1$ so that we can say (as was the case for CWT) the entrance region for uniform heat flux (UHF) can also be estimated by using Eq. 6.6. The results given in Table 6.4 are for constant properties also, so that the remarks made about viscosity effects in the previous section apply here in the same manner.

Figure 6.2 shows the variation of the local Nusselt numbers tabulated in Tables 6.2 and 6.4 with the dimensionless axial distance $x^+ = x/r_0 N_{Re} N_{Pr}$ for the two boundary conditions of CWT and UHF. The figure shows that the UHF results are slightly higher, but that both curves level out at about $x^+ \simeq 0.1$ as we observed above. If we took 5 percent as a criteria [$1.05(4.63) = 4.86$ and

Table 6.3 CONSTANTS γ_n^2 AND A_n FOR USE IN EQ. 6.7

n	γ_n^2	$A_n \times 10^3$
1	25.6796	7.631
2	83.8618	2.053
3	174.167	0.9026
4	296.536	0.4941
5	450.947	0.3068
6	637.387	0.2067
7	855.850	9.1476

Table 6.4 NUSSELT NUMBERS FOR
LAMINAR FLOW IN ROUND
TUBES, UHF

x^+	Nu_x	Nu_{av}
0	∞	∞
0.002	12	16.7
0.004	9.93	13.2
0.010	7.49	9.94
0.020	6.14	8.01
0.040	5.19	6.56
0.060	4.81	5.93
0.080	4.62	5.58
0.100	4.51	5.34
0.120	4.45	5.17
∞	4.36	4.36

1.05(3.66) = 3.84], we could say that the entrance region extended to about $x^+ = 0.07$ for CWT and only to $x^+ = 0.05$ for UHF. The determination of the entrance region, therefore, is not a precise matter but subject to engineering judgment for the particular application at hand.

6.1.3 Dimensionless Correlations

The analytical results presented above are in table form and are not very useful for design calculations or parametric studies with computer programs. Kays (4) has shown that the analytical solutions underestimate the Nusselt number for

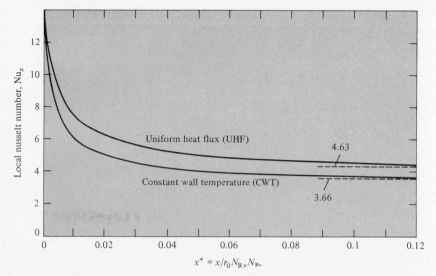

$$x^+ = x/r_0 N_{Re} N_{Pr}$$

Figure 6.2 Variation of theoretical local Nusselt number with dimensionless axial distance for laminar flow with CWT and UHF boundary conditions (from Tables 6.3 and 6.4).

laminar flow of gases by a considerable amount. For gases and liquids in laminar flow at a constant wall temperature, Kays (4) recommends that the average Nusselt number be calculated by the equation

$$\left[\begin{array}{l} \mathrm{Nu}_{av} = 3.66 + \dfrac{0.104\left[N_{Re} N_{Pr}(D/L) \right]}{1 + 0.016\left[N_{Re} N_{Pr}(D/L) \right]^{0.8}} \\[4mm] \text{(CWT)} \\ \text{properties at } \bar{T}_b \end{array} \right. \tag{6.9}$$

where L is the length of the circular tube and D the diameter.

The equation given by Kays (4) for the uniform heat flux case is for local Nusselt numbers at the location x from the start of heating, but can be integrated using the form given in Eq. 6.8 to obtain an expression for the average Nusselt number as follows:

$$\left[\begin{array}{l} \mathrm{Nu}_{av} = 4.36 + \dfrac{0.036 N_{Re} N_{Pr}}{(L/D)} \ln\left[\dfrac{(L/D)}{0.0011 N_{Re} N_{Pr}} + 1 \right] \\[4mm] \text{(UHF)} \\ \text{properties at } \bar{T}_b \end{array} \right. \tag{6.10}$$

Both Eqs. 6.9 and 6.10 represent the constant property case so that their application holds only for small differences between the wall and bulk temperatures (about 5 to 10F for liquids and about 50 to 75F for gases). Properties are evaluated at the bulk temperature.

When the temperature differences between the fluid and wall are large, viscosity effects must be considered and the effects of property variations may be estimated from

$$\frac{\mathrm{Nu}_{av}(\text{variable properties})}{\mathrm{Nu}_{av}(\text{constant properties})} = \left(\frac{\mu_b}{\mu_s} \right)^{0.14} \tag{6.11}$$

where μ_b and μ_s are the viscosities of the fluid evaluated at the bulk and wall temperatures, respectively. The results given by Eqs. 6.9 and 6.10 agree very well with experiments using air as the heat transfer fluid.

6.1.4 Noncircular Ducts

Analytical solutions to the energy equation have been obtained by Clark and Kays (4, 5) for rectangular and triangular tubes. The results are shown in Table 6.5 for conditions of fully developed flow that is downstream of the entrance region. The values in Table 6.5 can be used as average values for tube lengths that are at least twice as long as the entrance length. The Nusselt number and Reynolds number must be based on the hydraulic diameter as was discussed in Section 4.5.3; the criterion of $N_{Re} < 2300$ for laminar flow conditions also applies to noncircular ducts.

Table 6.5 NUSSELT NUMBERS FOR FULLY DEVELOPED
FLOW OUTSIDE THE ENTRANCE REGION IN
RECTANGULAR AND TRIANGULAR TUBES,
LAMINAR FLOW*

$\overline{Nu}(UHF)$	CROSS SECTION	(LENGTH/SIDE)	$\overline{Nu}(CWT)$
4.364		—	3.66
3.0		—	2.35
3.63		1.0	2.98
3.78		1.4	—
4.11		2.0	3.39
4.77		3.0	—
5.35		4.0	4.44
6.60		8.0	5.95
8.23		∞	7.54

*No variation of temperature around periphery.

Example 6.1

We would like to compare the average Nusselt number predicted by Eq. 6.10 for a Reynolds-Prandtl number product of 4000 in a tube of 40 cm heated length and 2-cm ID with the value given in Table 6.4.

SOLUTION
The length/diameter ratio for this case is $L/D = \frac{40}{2} = 20$ so that $N_{Re}N_{Pr}D/L = \frac{4000}{20} = 200$, and Eq. 6.10 becomes

$$Nu_{av} = 4.36 + 0.036(200)\ln[1/0.0011(200) + 1]$$

$$= 4.36 + 12.33 = 16.7$$

Now to compare with Table 6.4, we need to determine x^+. At $x = L = 40$ cm and $r_0 = D/2 = 1$ cm radius, the value for $x^+ = x/r_0 N_{Re}N_{Pr} = 40/1(4000) = 0.01$ and Table 6.4 shows a value for $Nu_{av} = 9.94$. We see therefore that the analytical prediction is lower than the correlation equation yields as we discussed above. We might further recognize that the Prandtl number of the fluid under consideration must be at least about 1.8 in order to have laminar flow. This concludes the example.

We conclude our discussion of laminar flow in ducts with a comment on the application of the heat transfer coefficient calculated from the average Nusselt number. In internal flow situations the surface and fluid temperatures generally vary in the flow direction so that the application of Eq. 4.1 to calculate heat flux must be based on average temperatures

$$q_c = h_{av}(\bar{T}_s - \bar{T}_b) \tag{6.12}$$

When the wall and fluid temperatures are essentially linear functions with axial distance, a simple numerical average can be used when $\Delta T_{in}/\Delta T_{out} \leq 2$. The quantities ΔT_{in} and ΔT_{out} are the entering and leaving differences in surface to fluid temperature as illustrated in Figure 6.3. When the ratio $\Delta T_{in}/\Delta T_{out}$ is greater than 2, the concept of the log mean temperature difference (LMTD) must be employed as we shall see in Chapter 7. With $\Delta T_{in}/\Delta T_{out} \leq 2$ the numerical averages referred to above can be expressed as

$$\bar{T}_s = (T_{s1} + T_{s2})/2 \tag{6.13}$$

$$\bar{T}_b = (T_{b1} + T_{b2})/2 \tag{6.14}$$

for use in Eq. 6.12. When working with fluids in the entrance region for uniform heat flux (UHF), the wall temperature is not a linear function of distance and an integrated average must be used where

$$\bar{T}_s = \frac{1}{L} \int_0^L T_s \, dx \tag{6.15}$$

can be evaluated numerically or graphically; with the constant wall temperature

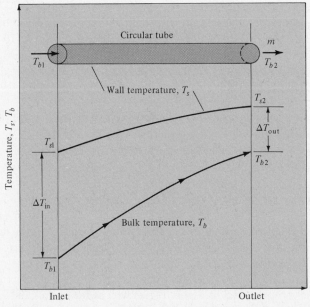

Figure 6.3 Variation of T_s and T_b along the tube length and definition of ΔT_{in} and ΔT_{out}.

(CWT) boundary condition there is, of course, no difficulty. The observations and comments just made apply also to turbulent flow heat transfer as well.

6.2 INTERNAL FLOW, TURBULENT

The range in Reynolds number from about 2000 to about 5000 is the transition region from laminar to turbulent flow under normal conditions of commercial pipe or tube roughness and the vibration of motor driven pumps. It is undesirable to operate heat transfer equipment in this region because of the unpredictable nature of the flow and the considerable difference in heat transfer coefficients for laminar and turbulent flow. It is generally desirable to operate at Reynolds numbers greater than 10,000 when using turbulent flow correlations.

We must first look at entrance region considerations, then fully developed flow, and last at the effect of variable properties; we will consider roughness effects and high-speed flow in separate sections.

6.2.1 Entrance Region, Turbulent Flow

The extent of the entrance region in turbulent flow is considerably less than in laminar flow because of the enhancing action of the turbulent fluctuations. Sparrow, Hallman, and Siegel (6) have also theoretically analyzed the turbulent flow entrance region problem and predict entrance lengths in the range from 12 to 14 diameters at Prandtl numbers near 1, based upon constant properties. Unfortunately, the eigenfunctions in their solution are strong functions of the Reynolds and Prandtl numbers so that their solution is cumbersome to use. For turbulent flow, the effect of wall boundary conditions (such as UHF or CWT) on the average Nusselt number is very small (Reference 7) for Prandtl numbers 0.7 and larger (gases and liquids). Accordingly, entrance length considerations are not strongly influenced by the wall boundary conditions; entrance lengths decrease as the Prandtl number increases.

Figure 6.4 presents some data for air obtained by Wolf and Lehman (8, 18) which shows that the local Nusselt number (the term corrected Nusselt number refers to corrections for small variations in the Reynolds number to obtain the nominal values shown on the figure) continues to decrease for the case where the air was heated under conditions of uniform heat flux at the wall. The continued decrease is due to the increasing thermal conductivity of the air as the bulk temperature increased along the duct. Consequently, the Nusselt number never really reached an asymptotic value and the entrance lengths indicated are unrealistically large. The data of Wolf (9, 18) show that the local heat transfer coefficient h_x reaches a constant value from which the thermal entrance length can be determined. Figure 6.5b shows the results of such measurements for the heating data for air shown in Figure 6.4 as well as for cooling data. The cooling data are based on the average bulk Reynolds number and the heating data on the initial Reynolds number at the location where heating starts. The parameters ϕ and β are parameters in the theoretical analysis of Deissler (10) with which the

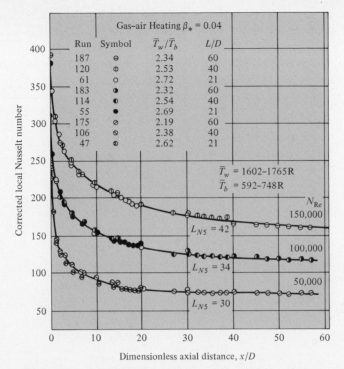

Figure 6.4   Variation of corrected local Nusselt number with x/D for air (Reference 8); uniform initial temperature profile, fully developed initial velocity profile, and uniform wall heat flux.

lines labeled theory (L_δ) in Figure 6.5b were calculated. The expression

$$\frac{L_{h1}}{D} = 2.4 N_{\mathrm{Re}b}^{0.2} \tag{6.16}$$

correlates entrance lengths for air in turbulent flow for heating and cooling within about ± 10 percent. The entrance length L_{h1} is defined as the length required to obtain an $h_x = 1.01 h_{\mathrm{asymptotic}}$. The bulk Reynolds numbers for the heating data shown in Figure 6.5 are less than the initial Reynolds numbers, an aspect that has the effect of shifting the curve up slightly so that the constant 2.1 is closer to 2.4 when the correlating equation is based on the bulk Reynolds number as given in Eq. 6.16. We can conclude from Figure 6.5b that entrance lengths for gases range from about 15 to 30 diameters. Data obtained by Hartnett (11) with water and oil shows entrance lengths from about 7 to 17 over the Reynolds number range from 10,000 to 46,000, decreasing with increasing N_{Re}; in general, the entrance length decreased as the Prandtl number increased at a given Reynolds number. For all practical purposes at locations beyond 60 diameters, the heat transfer coefficient can be considered to have its fully developed value for Reynolds numbers less than 500,000.

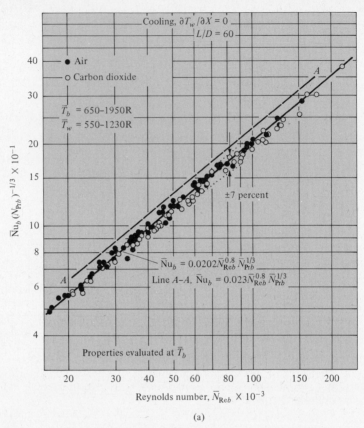

(a)

Figure 6.5a   Example of a forced convection correlation for the cooling of air and CO_2 in turbulent flow in a smooth round duct of $L/D = 60$ at constant wall temperature from Reference 39.

6.2.2 Low Temperature Difference Correlations

When the difference between the wall and the fluid temperature is not large, the Nusselt number can be expressed in the form indicated by Eq. 6.1 without the T_s/T_b term. Slightly different variations give better correlations for gases than for liquids.

In Section 4.7 we learned that Reynolds analogy predicted that the Stanton number $N_{St} = f/8$ as was given by Eq. 4.48 for constant properties. We can recognize that $N_{St} = h/c_p \rho u_b = Nu/N_{Re} N_{Pr}$ so that when we substitute Eq. 4.35 ($f = 0.190 N_{Re}^{-0.2}$) into Eq. 4.48 and remember that the Reynolds analogy was for $N_{Pr} = 1$ (or $\nu = \alpha$), we then find the Nusselt number should be of the form

$$Nu = 0.024 N_{Re}^{0.8} \tag{6.17}$$

Experiments conducted by Dittus and Boelter (12) with liquids showed that the correlation was improved by including the Prandtl number to a power and

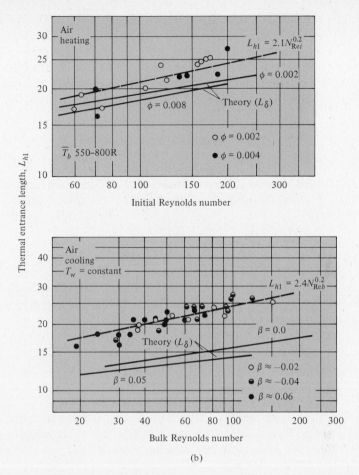

Figure 6.5b   Variation of the thermal entrance length for air with Reynolds number under conditions of heating and cooling the gas (References 8, 9).

reducing the constant to 0.023 so that

$$\begin{bmatrix} \mathrm{Nu} = 0.023 N_{\mathrm{Re}}^{0.8} N_{\mathrm{Pr}}^{0.4} \ (\text{heating}) \\ \mathrm{Nu} = 0.023 N_{\mathrm{Re}}^{0.8} N_{\mathrm{Pr}}^{1/3} \ (\text{cooling}) \\ \text{liquid properties at } T_b \end{bmatrix}$$

(6.17a)

(6.17b)

For gases the constant in Eq. 6.17 is somewhat lower and the exponent on the Prandtl number can be either 1/3 or 0.4, depending on the investigator. For example, Wolf (18) has correlated cooling experiments with air and CO_2 with a maximum deviation of ±7 percent (see Figure 6.5a) such that

$$\begin{bmatrix} \mathrm{Nu} = 0.0202 N_{\mathrm{Re}}^{0.8} N_{\mathrm{Pr}}^{1/3} \\ \text{properties at } T_b \end{bmatrix}$$

(6.17c)

over the Reynolds number range from about 15,000 to 220,000. The earlier experiments may not have been performed with tubes having a high degree of smoothness on the inside surface. When there is a fully developed velocity profile at the point where energy exchange is initiated and the tube length is less than about twice the entrance length, McAdams (13) recommends that an L/D correction be applied to Eq. 6.17 so that the Nusselt number is given by

$$
\left[
\begin{array}{l}
\text{Nu} = 0.023 N_{\text{Re}}^{0.8} N_{\text{Pr}}^{0.4} \left(1 + \dfrac{1.4}{L/D}\right) \\[2mm]
\text{properties at } T_b \text{ for liquids}
\end{array}
\right.
\qquad \bullet(6.18)
$$

For the case where there is a sharp contraction at the point where heating or cooling starts, the L/D correction takes the form

$$
\left[
\begin{array}{l}
\text{Nu} = 0.023 N_{\text{Re}}^{0.8} N_{\text{Pr}}^{0.4} \left[1 + \left(\dfrac{L}{D}\right)^{-0.7}\right] \\[2mm]
\text{properties at } T_b \text{ for liquids}
\end{array}
\right.
\qquad (6.19)
$$

For gases McEligot (37) and coauthors obtained experimental results with air, nitrogen, and helium and recommended the following correlation to account for the local dependence of the Nusselt number

$$
\left[
\begin{array}{l}
\text{Nu}_x = 0.021 N_{\text{Re}}^{0.8} N_{\text{Pr}}^{0.4} \sqrt{\dfrac{T_b}{T_s}} \left[1 + \left(\dfrac{x}{D}\right)^{-0.7}\right] \\[2mm]
\text{properties at } T_b \text{ (for gases)}
\end{array}
\right.
\qquad (6.20)
$$

The equation can be integrated for constant properties, and assuming that T_b/T_s does not vary with distance according to Eq. 6.8, with x as the variable rather than x^+ we obtain a correlation for average heat transfer as

$$
\left[
\begin{array}{l}
\text{Nu} = 0.021 N_{\text{Re}}^{0.8} N_{\text{Pr}}^{0.4} \sqrt{\dfrac{\overline{T_b}}{\overline{T_s}}} \left[1 + 3.33\left(\dfrac{L}{D}\right)^{-0.7}\right] \\[2mm]
\text{properties at } \overline{T_b} \text{ (for gases)}
\end{array}
\right.
\qquad (6.21)
$$

For values of T_s/T_b less than 1.5, the square root of the temperature ratio term can be left out of the correlation. The effect of T_s/T_b is discussed in the next section.

For other types of entry configurations, different forms of the L/D correction must be used; McAdams (13) and Rohsenow and Hartnett (14) catalog most of the experimental work reported in the open literature.

6.2.3 High Temperature Difference Correlations

When the temperature difference between the wall and the bulk of the fluid is large enough to cause significant changes in the viscosity of the fluid, heat transfer results can no longer be predicted or correlated with Eqs. 6.16 through 6.19. For liquids Sieder and Tate (15) recommend

$$\left[\begin{array}{l} \text{Nu} = 0.027 N_{\text{Re}}^{0.8} N_{\text{Pr}}^{1/3} \left(\dfrac{\mu_b}{\mu_s} \right)^{0.14} \\[2mm] \text{properties at } T_b \text{ except } \mu_s \text{ at } T_s \end{array} \right. \tag{6.22}$$

based on experiments with water, oils, and organic fluids whose viscosity changes rapidly with temperature. Experiments reported by Wolf, Gray, and Reese (16) with red and white fuming nitric acid were also correlated very well with Eq. 6.22.

For gases the viscosity variation with temperature is nearly linear on a log-log plot so that a relationship of the type $\mu = C_1 T^a$, where T is the absolute temperature, describes the temperature variation reasonably well as shown in Figure 6.6. Accordingly, the ratio of viscosities in Eq. 6.22 can then be written as

$$\frac{\mu_b}{\mu_s} = \left(\frac{T_b}{T_s} \right)^a \tag{6.23}$$

The value of a depends on the gas under consideration.

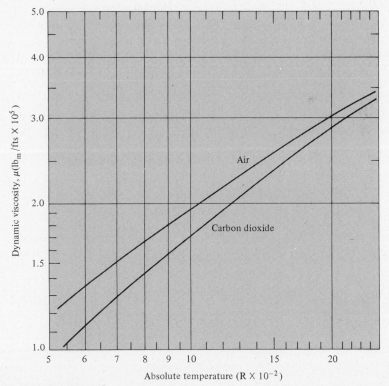

Figure 6.6 Relationship of dynamic viscosity for air and carbon dioxide with absolute temperature.

When the viscosity ratio is replaced by the temperature ratio, the following type of correlation equation results for gases:

$$\left[\begin{array}{l} \text{Nu} = CN_{\text{Re}}^{0.8}N_{\text{Pr}}^{0.4}\left(\dfrac{T_s}{T_b}\right)^n\left(\dfrac{L}{D}\right)^{-0.15} \\ \text{properties at } T_b \text{ or gases} \end{array} \right. \tag{6.24}$$

The exponent n and constant C depend on the magnitude of the wall to bulk temperature ratio T_s/T_b, and are given in the table below.

CONSTANTS C AND n FOR EQ. 6.24

TEMPERATURE RATIO	CONSTANT C	EXPONENT n
$T_s/T_b < 1$	0.038	0
$1 < T_s/T_b < 2.5$	0.038	-0.30
$T_s/T_b > 2.5$	0.048	-0.55

The values for C and n were obtained from the results of several investigators. Figure 6.7 shows data for air, CO_2, argon, helium, and hydrogen gas in the range of T_s/T_b from 0.45 to 9.7. Small changes in the value of a do not influence the value of n for the gases investigated. It has also been pointed out by Sleicher and

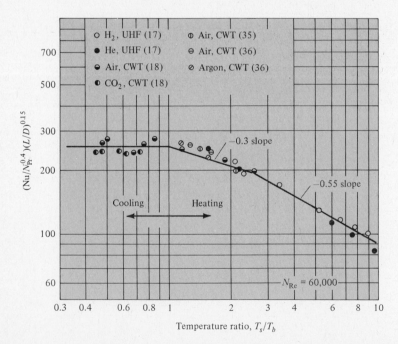

Figure 6.7 The relationship of $(\text{Nu}/N_{\text{Pr}}^{0.4})(L/D)^{0.15}$ with the wall to bulk temperature ratio for several gases at a Reynolds number of 60,000 under conditions of heating and cooling the gas.

Tribus (7) that the influence of wall boundary conditions on constant property heat transfer, $T_s/T_b = 1$, is small. The data shown in Figure 6.7 include constant wall temperature (CWT) and uniform heat flux (UHF) wall conditions at values of T_s/T_b as high as 2 and show no discernible effect.

The results of McEligot (37), McCarthy and Wolf (17), Wolf (18), Weiss (35), Prideaux (36), Taylor (38), and others all show a decrease in the Nusselt number when the gas is heated severely ($T_s/T_b \gtrsim 2$), but do not show an increase in Nusselt number when the gas is cooled severely ($T_s/T_b \lesssim 0.8$) as predicted by theory (Reference 20). When the gas is heated, the attendant acceleration appears to exert a damping effect on the turbulent fluctuations which would account for the decrease in heat transfer indicated by the higher value of n at the higher values of T_s/T_b. When the gas is cooled, the deceleration process apparently does not enhance the turbulent fluctuations and the heat transfer characteristics are not significantly influenced by T_s/T_b. Efforts by Seader and Wolf (19) to theoretically predict the influence of T_s/T_b for helium by using the variable property analysis by Deissler (20) were successful only for the heating case $T_s/T_b > 1$. Figure 6.8 shows that at a Reynolds number of 60,000 the exponent $n = -0.53$ for heating the gas is in good agreement with the results shown in Figure 6.7. When the gas is cooled, the theory does not agree with experiment and should not be used to predict heat transfer.

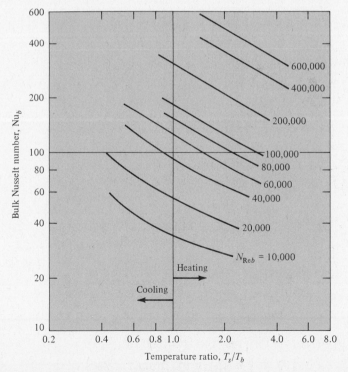

Figure 6.8 Theoretically computed Nusselt number for helium as a function of wall to bulk temperature ratio at different Reynolds numbers.

Example 6.2

A solar concentrator focuses sunlight on a bank of molybdenum alloy tubes that are 0.625-in. ID and 14 in. long. A pyrometer scan indicates an average surface temperature of 2980F for a tube and air enters the tube at 140F with a mass flow rate of 0.075 lb_m/s; we would like to predict the outlet temperature.

SOLUTION

Predicting the outlet temperature requires knowledge of h from the Nusselt number so that we can match the convective heat flux with the enthalpy rise for the air. Accordingly we estimate $\bar{T}_b$, then check our fluxes and reevaluate if need be; assume $\bar{T}_b = 600F$ which gives the following properties:

$$\mu_b = 0.07212 \ lb_m/fthr$$

$$k_b = 0.02655 \ Btu/hrftF$$

$$N_{Pr} = 0.681$$

Neglecting the temperature drop through the tube wall, we calculate the wall to bulk temperature ratio $\bar{T}_s/\bar{T}_b = 3440/1060 = 3.24$. Since we have an application with large temperature differences between the surface and the fluid, we select Eq. 6.24 to determine h. The constants $C = 0.048$ and $n = -0.55$ are for the range in $\bar{T}_s/\bar{T}_b > 2.5$ so that

$$Nu = 0.048 N_{Re}^{0.8} N_{Pr}^{0.4} \left(\frac{\bar{T}_s}{\bar{T}_b} \right)^{-0.55} \left(\frac{L}{D} \right)^{-0.15}$$

The Reynolds number is given by

$$N_{Re} = 48 \frac{\dot{m}}{\pi D \mu_b}; \quad (D \text{ in in.})$$

$$= 48 \frac{(0.0750)3600}{\pi(0.625)0.07212}$$

$$= 91{,}521 \text{ turbulent}$$

$$\therefore Nu = 0.048(91{,}521)^{0.8}(0.681)^{0.4}(3.24)^{-0.55} \left(\frac{14}{0.625} \right)^{-0.15}$$

$$= 125.9 = \frac{hD}{k_b}$$

so that

$$h = 125.9(0.02655)12/0.625$$

$$= 64.2 \ Btu/hrft^2F \ (364 \ W/m^2K)$$

The inside surface area of the tube is

$$A_i = \pi DL = \pi .625(14)/144 = 0.1909 \ ft^2$$

$$\therefore q = hA_i(\bar{T}_s - \bar{T}_b) = 64.2(0.1909)(2980 - 600)$$

$$= 29{,}168 \ Btu/hr \ (8548 \ W)$$

From the first law we can evaluate the temperature rise of the air for this heat flux neglecting changes in kinetic energy as

$$\Delta T = \frac{q}{\dot{m}\bar{c}_p}$$

For air at $\bar{T}_b = 600$, $\bar{c}_p = 0.2507$ Btu/lb$_m$R

$$\therefore \Delta T = 29{,}168/(0.075)3600(0.2507)$$

$$= 431 \quad \text{and} \quad T_{b2} = 140 + 431 = 571\text{F}$$

so that

$$\bar{T}_b = (140 + 571)/2 = 356\text{F}$$

This value shows our estimate to be much too high, therefore another iteration is required. Assume $\bar{T}_b = 340\text{F}$:

$$\mu_b = 0.05959 \text{ lb}_m/\text{fthr}$$

$$k_b = 0.02122 \text{ Btu/hrftF}$$

$$\bar{c}_p = 0.2439 \text{ Btu/lb}_m\text{F} \qquad N_{\text{Pr}} = 0.685$$

so that

$$N_{\text{Re}} = \frac{48(0.075)3600}{\pi(0.625)(0.05959)}$$

$$= 110{,}765$$

and

$$\text{Nu} = 0.048(110{,}765)^{0.8}(0.685)^{0.4}\left(\frac{3440}{800}\right)^{-0.55}(22.4)^{-0.15}$$

$$= 125.9 = \frac{hD}{k_b}$$

$$h = \frac{125.9(0.02122)12}{0.625} = 51.3 \text{ Btu/hrft}^2\text{F}$$

$$q = 51.3(0.1909)(3440 - 800) = 25{,}855 \text{ Btu/hr}$$

$$\Delta T = \frac{25{,}855}{0.075(3600)0.2439} = 393\text{F}$$

$$T_{b2} = 140 + 393 = 533\text{F}$$

Checking $\bar{T}_b = (140 + 533)/2 = 337\text{F}$.

The agreement is satisfactory so that the outlet temperature of the air from the tube is about 340F. This completes the example.

The extensive cooling experiments reported by Wolf (18) for air and carbon dioxide could be correlated within ±7 percent maximum deviation by the equation for average Nusselt number

$$\left[\begin{array}{l} \text{Nu} = 0.0202 N_{\text{Re}}^{0.8} N_{\text{Pr}}^{1/3} \\ \text{properties at } T_b \end{array}\right.$$

(6.25)

(repeat of 6.17b)

for essentially constant wall temperature boundary conditions and a tube length of 59.9 diameters; values of T_s/T_b as low as 0.44 were investigated. The value of the correlation constant 0.0202 is about 2 percent lower than that recommended for Eq. 6.17a for $T_s/T_b < 1$, due in part to the value of the Prandtl number and in part to the wall boundary condition. Equation 6.25 is also recommended for use for cases where gases are cooled at other than constant wall temperature.

6.2.4 Perspective on Low and High ΔT Convection

Because of the large number of correlation equations for turbulent flow in ducts, it is difficult to put all the information into perspective; compounding the difficulty is the choice of classification into low and high temperature difference categories. Unfortunately, the question of what is low temperature difference must be answered differently for gases than for liquids, and also differently for different liquids. We should note here that the remarks made in Section 6.1.3 as a rough guide for laminar flow (5 to 10F for liquids, 50 to 75F for gases) are not valid the turbulent region.

For gases, reference to Figure 6.7 shows that for $T_s/T_b \leqslant 1.5$ the influence of T_s/T_b is to decrease the ordinate about 12 percent or less, since $(1.5)^{-0.3} = 0.885$. The effect of T_s/T_b, in the range under discussion, is about the same as the experimental uncertainty in the data. Therefore, since T_s/T_b is proportional to temperature difference $T_s - T_b$, we conclude that heat transfer situations for gases in turbulent flow where $T_s/T_b \leqslant 1.5$ fall in the low temperature difference category. Accordingly, a simpler correlation equation that does not include the T_s/T_b term is permissible. The above information can be translated into an actual temperature difference when either the bulk or surface temperature is known.

For liquids the viscosity ratio term in Eq. 6.22 is the deciding criterion that indicates how much deviation from the constant property case exists. If we arbitrarily select 10 percent as the maximum deviation acceptable, the corresponding viscosity ratio can be determined by setting $(\mu_b/\mu_s)^{0.14} = 1.10$ and solving for μ_b/μ_s; therefore, $\mu_b/\mu_s \simeq 2$. Knowledge of the bulk or wall temperature and the fluid would permit translating the above information into a temperature difference. For example, water at 80F has $\mu_b = 2.08$ lb$_m$/hrft so that for a ratio of 2 the value of $\mu_s = 1.04$ lb$_m$/hrft, which corresponds to 150F or in this case a temperature difference of 70F. Note again that the temperature difference limit for a given percentage deviation from constant property conditions depends on the fluid under consideration.

To summarize, we observe that for $T_s/T_b \leqslant 1.5$ we have a 12 percent or less deviation from constant property correlations for gases, and for $\mu_b/\mu_s \leqslant 2$ we have a 10 percent or less deviation for liquids.

6.2.5 Correlation Techniques

We close this section on turbulent flow inside ducts with an observation on the technique of correlating data. The prime independent variable is the flow condition as indicated by the Reynolds number. Over a moderate range in Reynolds number most convection phenomena obey a simple power relationship of the kind $y = x^n$. After computing the N_u, N_{Pr}, N_{Re}, and any other pertinent dimensionless variables, one rearranges the form of Eq. 6.1 as

$$y = \frac{Nu}{f(N_{Pr}, D/L, T_s/T_b)}$$

(6.1)
(Repeat)

and

$$x = N_{Re}$$

and plots the value of y and x on log-log paper following the tenets given in Appendix A. The resulting slope of the straight line through the data determines the exponent on the Reynolds number. Such a plot may also reveal dependencies on other variables. To illustrate the concept, Figure 6.5(a) shows the data for air under cooling conditions together with cooling data for CO_2 as one would correlate them using the technique described above. Correlation aspects are explored in more detail in the problems at the end of the chapter.

6.3 EXTERNAL FLOW ON FLAT PLATES

In Section 4.7 we saw that Reynolds analogy ($\varepsilon_M = \varepsilon_H$, and $\nu = \alpha$ or $N_{Pr} = 1$) related the Stanton number to the friction factor for tube flow as was shown in Eq. 4.48 which stated that $N_{St} = f/8$. Von Karman (21) has applied Reynolds analogy to turbulent flow on flat plates, wherein he has defined a heat transfer number C_H, analogous to the Stanton number, as follows:

$$C_H = q/A\rho u_\infty c_p(T_s - T_\infty)$$

(6.26)

The heat transfer number C_H represents the ratio of the convected energy per unit area $[h = q/A(T_s - T_\infty)]$ in the direction perpendicular to the flow velocity, to the energy carried along by the fluid stream $(\rho u_\infty c_p)$ per unit cross-sectional area. By an integration of the simplified momentum and energy equations (Eqs. 4.12 and 4.14) for the value of Prandtl $N_{Pr} = 1$, Von Karman was able to show that $C_H = C_f/2$ where C_f is the local skin friction coefficient for flow on the flat plate that we have already seen in Eq. 4.17. The evaluation of C_f then depends on whether the boundary layer is laminar (i.e., $N_{Rex} < 5 \times 10^5$) or turbulent and the resulting correlation equation will then depend on how much of the plate has a laminar boundary layer and how much is covered by turbulent layer when the Reynolds number at the end of the plate is greater than 5×10^5.

6.3.1 Laminar Boundary Layer

In order to determine the energy requirement to maintain a flat plate at constant temperature $T_s > T_\infty$ in a fluid stream of velocity u_∞, we need to evaluate the average heat transfer number $\bar{C}_H$ over the surface as follows:

$$\bar{C}_H = \frac{1}{L} \int_0^L C_H \, dx$$

(6.27)

We consider here the case where the velocity u_∞ is such that the Reynolds number at the end of the plate is less than 5×10^5. We assume the Prandtl number $N_{Pr} = 1$ so that Von Karman's relationship $C_H = C_f/2$ can be used to evaluate Eq. 6.27 by substituting from Eq. 4.18. Thus

$$\bar{C}_H = \frac{1}{L} \int_0^L 0.332 N_{Rex}^{-0.5} \, dx \tag{6.28}$$

The local Reynolds number is given by $N_{Re} = \rho_\infty u_\infty x / \mu_\infty$; therefore, we note that x is the only variable with distance since our model shown in Figure 4.5 is for fixed free-stream temperature T_∞ and velocity u_∞. We can then write

$$\bar{C}_H = \frac{c_1}{L} \int_0^L x^{-0.5} \, dx \tag{6.29}$$

where $\quad c_1 = 0.332 \left(\frac{\rho_\infty u_\infty}{\mu_\infty} \right)^{-0.5}$

$$\tag{6.29a}$$

$$= 2c_1 L^{-0.5}$$

so that

$$\bar{C}_H = 0.664 N_{ReL}^{-0.5} \quad \text{for} \quad N_{ReL} < 5 \times 10^5 \tag{6.29b}$$

We can express the heat transfer number $\bar{C}_H$ in terms of an average Nusselt number by rewriting Eq. 6.26 in terms of the pertinent average quantities.

$$\bar{C}_H = \bar{q} / A \rho u_\infty c_p \left(\bar{T}_s - T_\infty \right) \tag{6.30}$$

but

$$\bar{h} = \bar{q} / A \left(\bar{T}_s - T_\infty \right) \tag{6.31}$$

and

$$\overline{Nu} = \bar{h} L / k \tag{6.32}$$

Therefore, substituting 6.31 and 6.32 into 6.30 yields

$$\bar{C}_H = \frac{\bar{h} L / k}{L \rho u_\infty / \mu} \frac{k}{c_p \mu}$$

$$\tag{6.33}$$

$$\bar{C}_H = \frac{\overline{Nu}}{N_{ReL} N_{Pr}}$$

and then replacing $\bar{C}_H$ in the final expression in 6.29b, we obtain the average Nusselt number for a flat plate at constant temperature with a laminar boundary layer in a fluid whose $N_{Pr} = 1$. Thus

$$\overline{Nu} = 0.664 N_{ReL}^{0.5} \qquad N_{ReL} < 5 \times 10^5$$

$$N_{Pr} = 1 \tag{6.34}$$

Experiments by Slack (22) have shown that Eq. 6.34 correlates with experiments very well when $N_{Pr}^{1/3}$ is included on the right-hand side so that the average

Nusselt number then becomes

$$\left[\begin{array}{ll} \overline{\mathrm{Nu}} = 0.664 N_{\mathrm{Re}L}^{0.5} N_{\mathrm{Pr}}^{1/3} & N_{\mathrm{Re}L} < 5 \times 10^5 \\ \text{properties at } T_\infty & 0.5 < N_{\mathrm{Pr}} < 10 \end{array} \right. \tag{6.35}$$

When the Reynolds number based on L, the plate length, is greater than 5×10^5, the latter portion of the boundary layer is turbulent and we must add a term to that effect in the definition of $\overline{C}_H$ as we show in the next section.

6.3.2 Turbulent Boundary Layer

The integral representing $\overline{C}_H$ in Eq. 6.27 must be split into two parts, one to cover the laminar boundary layer for $L < L_{cr}$ and the other part to cover the turbulent portion from L_{cr} to L. Thus

$$\overline{C}_H = \frac{1}{L} \left[\int_0^{L_{cr}} C_{H\,\mathrm{lam}} \, dx + \int_{L_{cr}}^L C_{H\,\mathrm{turb}} \, dx \right] \tag{6.36}$$

Again, we use Von Karman's result that $C_H = C_f/2$ for $N_{\mathrm{Pr}} = 1$ and the heat transfer numbers are given by

$$C_{H\,\mathrm{lam}} = 0.332 N_{\mathrm{Re}x}^{-0.5} \qquad (\text{Eq. } 4.18) \tag{6.36a}$$

$$C_{H\,\mathrm{turb}} = 0.0296 N_{\mathrm{Re}x}^{-0.2} \qquad (\text{Eq. } 4.19) \tag{6.36b}$$

The integration of Eq. 6.36 is more complicated but the procedure is the same. We let

$$c_2 = 0.332 \left(\frac{\rho u_\infty}{\mu} \right)^{-0.5} \tag{6.37}$$

$$c_3 = 0.0296 \left(\frac{\rho u_\infty}{\mu} \right)^{-0.2} \tag{6.37a}$$

in order to simplify the manipulation. The integral for $\overline{C}_H$ then becomes

$$\overline{C}_H = \frac{1}{L} \left[\int_0^{L_{cr}} c_2 x^{-0.5} \, dx + \int_{L_{cr}}^L c_3 x^{-0.2} \, dx \right] \tag{6.38}$$

$$= \frac{1}{L} \left[2c_2 L_{cr}^{0.5} + 1.25 c_3 \left(L^{0.8} - L_{cr}^{0.8} \right) \right] \tag{6.38a}$$

Replacing c_2 and c_3 by 6.37 and 6.37a together with some algebraic rearrangement gives

$$\overline{C}_H = 0.037 N_{\mathrm{Re}L}^{-0.2} - 0.001743 \frac{L_{cr}}{L} \tag{6.39}$$

or, using the results of Eq. 6.33, we can express 6.39 in terms of the average Nusselt number, and if we take Slack's (22) recommendation of including $N_{\mathrm{Pr}}^{1/3}$, we can write the average Nusselt number as

$$\left[\overline{Nu} = \left[0.037 N_{ReL}^{0.8} - 0.001743 \frac{L_{cr}}{L} N_{ReL} \right] N_{Pr}^{1/3} \right.$$
$$\left. \text{properties based on } T_\infty \right.$$

(6.40)

The transition location L_{cr} depends on the nature (physical properties) of the free stream fluid and is calculated from the value of the critical Reynolds number, 5×10^5, so that

$$L_{cr} = \frac{5(10^5)\nu}{u_\infty}$$

(6.41)

The results presented by Eq. 6.40 are in agreement with the survey reported by Whitaker (23).

Example 6.3

Atmospheric air flows over the top surface of a flat plate 3 ft long and 2 ft wide; the bottom surface is insulated. We would like to know how much energy must be supplied to the plate to keep it at 150F in an air stream of 70 ft/s and a temperature of 40F.

SOLUTION
First we must calculate the Reynolds number based on the length L of the plate to see if we have only laminar or laminar and turbulent boundary layers present as is shown in Figure 4.5. The physical properties for air at 40F are

$$\nu_\infty = 0.5301 \text{ ft}^2/\text{hr}$$
$$k_\infty = 0.01416 \text{ Btu/hrftF}$$
$$N_{Pr} = 0.714$$

Accordingly, the Reynolds number can be calculated as

$$N_{ReL} = \frac{u_\infty L}{\nu} = \frac{70(3)3600}{0.5301}$$
$$= 1.528(10^6)$$

This value is greater than $5(10^5)$ so a portion of the plate from the leading edge to L_{cr} has a laminar boundary layer and the rest of the plate has a turbulent boundary layer. The transition length is

$$L_{cr} = \frac{5(10^5)\nu}{u_\infty}$$

$$= \frac{5(10^5)0.5301}{(3600)70}$$

$$= 1.052 \text{ ft}$$

Therefore, about the first third of the plate has a laminar boundary layer. The average heat transfer coefficient for the plate can be obtained from the average Nusselt number given by Eq. 6.40.

$$\overline{Nu} = \left\{0.037\left[1.528(10^6)\right]^{0.8} - 0.001743(1.052/3)1.528(10^6)\right\}(0.714)^{1/3}$$

$$= (3277 - 932)(0.894) = 2096$$

and

$$\overline{h} = \frac{k}{L}\overline{Nu}$$

$$= \frac{0.01416}{3}2096$$

$$= 9.89 \text{ Btu/hrft}^2\text{F} \quad (56.2 \text{ W/m}^2\text{K})$$

and heat flux for the entire plate then becomes

$$\overline{q} = \overline{h}A(T_s - T_\infty)$$

$$= 9.89(6)(150 - 40)$$

$$= 6527 \text{ Btu/hr} \quad (1.91 \text{ kW})$$

About 2000 W are required to keep the plate at 150F. We might also ask what percentage of this flux is transferred through the laminar and through the turbulent boundary layer. To do so we must determine the average Nusselt number for the laminar portion.

$$\overline{Nu}_{lam} = 0.664\left(5(10^5)\right)^{0.5}(0.714)^{1/3}$$

$$= 419.6$$

$$\overline{h}_{lam} = \frac{k}{L_{cr}}\overline{Nu}_{lam}$$

$$= \frac{0.01416}{1.052}419.6$$

$$= 5.65 \text{ Btu/hrft}^2\text{F} \quad (32.1 \text{ W/m}^2\text{K})$$

$$\overline{q}_{lam} = 5.65(1.052)2(110)$$

$$= 1307 \text{ Btu/hr} \quad (0.383 \text{ kW})$$

This result shows that about 20 percent of the total flux is transmitted to the air in the first one-third length of the plate and 80 percent is transmitted by the last two-thirds of the plate length through the turbulent boundary layer. This is the end of the example.

The results we have shown above for the flat plate in incompressible flow were obtained for the very simple analogy developed by Von Karman and hold for values of the Prandtl number near 1. For applications where the fluid has a Prandtl number significantly larger than 1 (more than 10), it is necessary to consider other analogies as described in Reference 14 which consider the Prandtl number as a variable. In addition to the data of Slack (22) with air ($N_{Pr} = 0.7$), Whitaker (23) has summarized data for fluids with N_{Pr} to about 400 with the following equation,

$$\left[\begin{array}{l} \overline{Nu} = 0.036 N_{Pr}^{0.43} \left(N_{ReL}^{0.8} - 9200 \right) \left(\dfrac{\mu_\infty}{\mu_s} \right)^{0.25} \\ \text{properties at } T_\infty \text{ except } \mu_s \text{ at } T_s \end{array} \right. \tag{6.42}$$

when there is negligible free stream turbulence. When a significant amount of free stream turbulence exists, the Nusselt number at Reynolds numbers less than about 4×10^6 is higher than it would be with no free stream turbulence and Whitaker (23) recommends

$$\left[\begin{array}{l} \overline{Nu} = 0.036 N_{Pr}^{0.43} N_{ReL}^{0.8} \left(\dfrac{\mu_\infty}{\mu_s} \right)^{0.25} \\ \text{properties at } T_\infty \text{ except } \mu_s \text{ at } T_s \end{array} \right. \tag{6.43}$$

The lower limit on N_{ReL} for Eq. 6.43 is 2×10^5 and the upper limit on Reynolds number for 6.42 and 6.43 is about 10^7; the value of the viscosity ratio ranged from 0.26 to 3.5, covering cooling the fluid (energy *into* the surface) as well as heating the fluid (energy *out* of the surface).

Since the above sections dealt with incompressible flow on flat plates, it is natural to ask how heat transfer correlations are affected when the velocity is high enough to get into the compressible flow regime. We take a very brief, simple look at high speed flow in the next section.

6.3.3 High-Speed Flow

As the flow velocity becomes large, significant amounts of energy are taken up in the form of kinetic energy of translation, that is, $u_\infty^2 / 2 g_c J$, in Btu/lb$_m$ or cal/kg in consistent units. We must then distinguish between a temperature measurement that is made by a device moving along at the same speed as the fluid (static temperature) and a device that brings the fluid to rest without heat losses (total temperature), thus recovering the kinetic energy completely. Figure 6.9 shows how the concepts of static, adiabatic, and total temperature are visualized for high-speed flow on a flat plate. The free stream velocity u_∞ is large so that kinetic energy is significant. When the fluid is slowed to zero velocity at the surface through viscous action, the frictional effects due to viscosity then increase the internal energy of the fluid in the direction toward the plate; and, if the plate is insulated, the surface of the plate reaches a temperature somewhat less than the total temperature and is appropriately called the adiabatic wall temperature, T_{ad}. Of course, if the fluid is brought to rest without friction, there is no loss in kinetic energy through frictional dissipation and the resulting temperature is the total temperature T_0 (the subscript zero representing zero velocity). Special care must be exerted to achieve adiabatic reduction of fluid

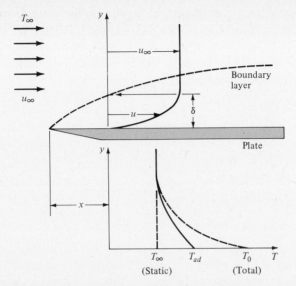

Figure 6.9 Temperature and velocity profiles on a flat plate in high-speed flow showing static, adiabatic, and total temperature.

velocity to zero; Figure 6.10 shows a schematic diagram of a total temperature probe. The probe is placed with its axis aligned with the fluid velocity vector and small bleed holes are placed at the measuring thermocouple to slowly bleed off the fluid brought to rest and to keep the thermocouple junction at the temperature of the fluid brought to rest, T_0, otherwise conduction effects would give an erroneous indication.

When the surface temperature is other than the adiabatic wall temperature, energy will enter or leave the surface depending on whether the surface temperature is less than or greater than the adiabatic wall temperature. Accordingly, we then define the heat transfer coefficient under such conditions with respect to the difference between the surface and the adiabatic wall temperature as the driving potential. Thus

$$q = \bar{h} A (T_s - T_{ad}) \tag{6.44}$$

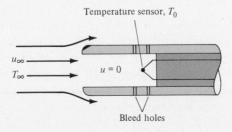

Figure 6.10 Cross section of a probe for measuring total temperature in a high-speed flow stream.

where $\bar{h}$ is the average coefficient over the length of the plate and, as was the case in low-speed flow, depends on whether the boundary layer is laminar or turbulent.

The adiabatic wall temperature can be related to the total temperature by means of the recovery factor r, defined as

$$r = \frac{T_{ad} - T_\infty}{T_0 - T_\infty} \tag{6.45}$$

Values of the recovery factor depend on the Prandtl number and whether the boundary layer is laminar or turbulent. Eckert (24) recommends the following relationships for gases:

$$\text{laminar boundary layer:} \qquad r = \left(N_{Pr}^* \right)^{1/2} \tag{6.46}$$

$$\text{turbulent boundary layer:} \qquad r = \left(N_{Pr}^* \right)^{1/3} \tag{6.47}$$

The procedure recommended by Eckert (24) for calculating heat transfer in high-speed flow makes use of the same form of the equations for local heat transfer as given in Chapter 4 but with a special reference temperature T^* to evaluate the properties.

For laminar boundary layer:

$$\text{Nu}_x^* = 0.332 N_{Rex}^{*1/2} N_{Pr}^{*1/3} \tag{6.48a}$$

$$N_{Rex}^* < 5 \times 10^5$$

For turbulent boundary layer:

$$\text{Nu}_x^* = 0.0296 N_{Rex}^{*0.8} N_{Pr}^{*1/3} \tag{6.48b}$$

$$5 \times 10^5 < N_{Rex}^* < 10^7$$

$$\text{Nu}_x^* = \frac{0.185 N_{Re}^* N_{Pr}^{*1/3}}{\left(\log N_{Rex} \right)^{2.584}} \tag{6.48c}$$

$$10^7 < N_{Rex}^* < 10^9$$

The reference temperature for evaluating the physical properties in Nu^*, N_{Re}^*, and N_{Pr}^* was given by Eckert as

$$T^* = T_\infty + 0.5(T_s - T_\infty) + 0.22(T_{ad} - T_\infty) \tag{6.49}$$

All that remains now is the formulation of an expression for the total temperature so that we can use Eqs. 6.45, 6.46, and 6.47 to determine the adiabatic wall temperature. To do so, we apply the first law of thermodynamics to a flow situation where the fluid with velocity u_∞ is brought to rest at zero velocity. We employ the letter i for enthalpy since the customary h would be a conflict with our designation for the heat transfer coefficient. Thus

$$i_\infty + \frac{u_\infty^2}{2g_c J} = i_0 \tag{6.50}$$

The conversion constant $J = 778$ ftlb$_f$/Btu or 4.186 cal/J is required to keep the units consistent. We can replace $i_0 - i_\infty$ in Eq. 6.50 with $c_p(T_0 - T_\infty)$ because the difference in T_0 and T_∞ is not large enough to cause appreciable variation in the specific heat so that

$$\frac{T_0}{T_\infty} = 1 + \frac{u_\infty^2}{2g_c c_p J T_\infty} \tag{6.51}$$

The ratio of total to static temperature can be expressed in terms of Mach number u_∞/a_∞ where the acoustic velocity at the static temperature is given by $a_\infty = (c_p g_c R T_\infty/c_v)^{1/2}$ and for perfect gas conditions $c_p - c_v = R/J$ so that

$$\frac{T_0}{T_\infty} = 1 + \frac{(\gamma - 1)M_\infty^2}{2} \tag{6.51a}$$

where γ is the specific heat ratio c_p/c_v, and R is the specific gas constant in the ideal gas equation of state, $Pv = RT$.

The above equations are applied in the same manner as we did previously for low-speed flow. The recovery factor is calculated by assuming a T^* to get the Prandtl number, then from the computed stagnation temperature and adiabatic wall temperature, T^* is calculated from Eq. 6.49 and compared to the assumed value for adjustment if necessary. The Reynolds number at the end of the plate determines whether only a laminar or both a laminar and turbulent boundary layer exists on the plate so that the average heat transfer coefficient can be evaluated by integration over the length of the plate;

$$\bar{h} = \frac{1}{L} \int_0^L h_x^* \, dx \tag{6.52}$$

using h_x^* from the appropriate equation for Nu$_x^*$ depending on the Reynolds number range. Example 6.4 illustrates the procedure.

Example 6.4

The leading edge of an aircraft wing traveling at a speed of Mach 1 has a de-icer strip that is 2.5 in. in length from the stagnation point (along the top and along the bottom) and extends the length of the wing. How much energy is required at 35,000-ft altitude to keep the surface of the de-icer strip at 150F when the free stream temperature is -60F and the pressure is 0.25 atm?

SOLUTION

To calculate the heat transfer involved, we must apply Eq. 6.44 which is based on the adiabatic wall temperature which in turn is calculated from the recovery factor and the stagnation temperature T_0. For air the specific heat ratio $\gamma = c_p/c_v = 1.4$ so that

$$T_0 = T_\infty \left[1 + (\gamma - 1)M^2/2\right]$$
$$= (-60 + 460)\left[1 + (1.4 - 1)(1)^2/2\right]$$
$$= 480\text{R} \equiv 20\text{F}$$

Since the adiabatic wall temperature depends on N_{Pr}^*, we must first estimate T^*; then calculate T_{ad} and then check our estimate. We do need to know whether we have a laminar or a combination laminar and turbulent boundary layer so that we can select the appropriate expression for recovery factor. To do so we calculate the Reynolds number based on free stream conditions to see where we are. Since $M = u_\infty/a_\infty = 1$, we calculate

$$a_\infty = (\gamma g_c RT)^{1/2}$$

$$= [1.4(32.174)53.34(400)]^{1/2}$$

$$= 980 \text{ ft/s}$$

Therefore

$$u_\infty = 1.0(980) = 980 \text{ ft/s}$$

Air properties at $T_\infty = -60F$:

$$\mu_\infty = 0.03513 \text{ lb}_m/\text{fthr}$$

$$\rho_\infty = P/RT_\infty$$

$$= \frac{0.25(14.7)144}{53.34(400)}$$

$$= 0.0248 \text{ lb}_m/\text{ft}^3$$

$$\nu_\infty = \frac{\mu_\infty}{\rho_\infty} = 1.4165 \text{ ft}^2/\text{hr}$$

The Reynolds number at the chordal end of the strip (in the flow direction), assuming the strip to be a flat plate:

$$N_{ReL} = \frac{Lu_\infty}{\nu_\infty} = \frac{2.5}{12} \frac{980(3600)}{1.4165}$$

$$= 5.19 \times 10^5$$

We have essentially all laminar boundary layer and therefore use $r = N_{Pr}^{*0.5}$. Since T_{ad} depends on N_{Pr}^*, we first estimate T^*, evaluate N_{Pr}^*, calculate T_{ad}, then check T^* and revise if needed. Generally T^* is a little higher than the average of T_∞ and T_s; therefore, assume $T^* = 60F$, which gives $N_{Pr} = 0.712$ and $r = N_{Pr}^{*0.5} = (0.712)^{0.5} = 0.8438$ so that

$$T_{ad} - T_\infty = r(T_0 - T_\infty)$$

$$= 0.8438[20 - (-60)] = 68F$$

giving

$$T_{ad} = 8F$$

Checking

$$T^* = T_\infty + 0.5(T_s - T_\infty) + 0.22(T_{ad} - T_\infty)$$

$$= -60 + 0.5(150 + 60) + 0.22(68)$$

$$= 60F$$

which agrees with our estimate so that no revision is needed. To calculate $\bar{h}$ we evaluate h_x^* from Nu_x^* and integrate over the streamwise length of the plate. Air properties at $T^* = 60\mathrm{F}$ are

$$\mu^* = 0.04339 \ \mathrm{lb_m/hrft}$$

$$\rho^* = \frac{0.25(14.7)144}{53.34(520)}$$

$$= 0.0191 \ \mathrm{lb_m/ft^3}$$

$$\nu^* = \frac{\mu^*}{\rho^*} = 2.274 \ \mathrm{ft^2/hr}$$

$$k^* = 0.01466 \ \mathrm{Btu/hrftF}$$

and, therefore, the Reynolds number is

$$N_{\mathrm{Re}L}^* = \frac{Lu_\infty}{\nu^*}$$

$$= \frac{2.5(980)3600}{12(2.274)}$$

$$= 3.232 \times 10^5 < 5 \times 10^5$$

so that we use Eq. 6.48 to obtain h_x^* from Nu_x^*. Thus

$$\bar{h} = \frac{1}{L} \int_0^L h_x^* \, dx$$

and

$$h_x^* = \frac{k^*}{x} \mathrm{Nu}_x^*$$

$$= \frac{k^*}{x} 0.332 \left(\frac{u_\infty}{\nu^*} \right)^{0.5} x^{0.5} N_{\mathrm{Pr}}^{*1/3}$$

$$= c_1 x^{-0.5}$$

where

$$c_1 = 0.332 k^* N_{\mathrm{Pr}}^{*1/3} \left(\frac{u_\infty}{\nu^*} \right)^{0.5}$$

Then

$$\bar{h} = \frac{1}{L} \int_0^L c_1 x^{-0.5} \, dx$$

$$= \frac{c_1}{L} (2x^{0.5})_0^L = 2c_1 L^{-0.5}$$

Substituting for c_1 from above gives

$$\bar{h} = 0.664 \frac{k^*}{L} N_{\mathrm{Pr}}^{*1/3} \left(\frac{Lu_\infty}{\nu^*} \right)^{0.5}$$

$$= 0.664 \frac{0.01466(12)}{2.5} (0.712)^{1/3} \left[3.232(10^5) \right]^{0.5}$$

$$= 23.7 \ \mathrm{Btu/hrft^2F}$$

We then calculate the heat transfer for a 1-ft length of strip that extends 2.5 in. on the top and bottom of the wing for the area $A = 2(2.5) \, 12/144 \, \text{ft}^2$ so that

$$q = \bar{h}A(T_s - T_{ad})$$

$$= 23.70 \frac{60}{144}(150 - 8)$$

$$= 1402 \, \text{Btu}/\text{hr ft length}$$

the amount of energy that must be supplied per foot length of the strip (about 400 W) to maintain the surface at 150F. The temperature of 150F is higher than would be needed to prevent ice formation, so that to conserve power a lower temperature would be more apropos. This is the end of the example.

We have looked at a brief introduction to heat transfer under conditions of compressible flow ($M > 0.2$) for the very simple boundary condition of constant wall temperature. When the wall temperature varies in the flow direction, Chapman and Rubesin (25) recommend a superposition technique to account for the variation in heat transfer from constant wall temperature conditions.

We shall look at some correlations of external flow over cylinders and spheres, then consider the influence of rough surfaces as compared to the smooth surfaces assumed in everything we have done so far, before concluding with the interaction between forced and free convection at low velocities.

6.4 OTHER EXTERNAL FLOW CORRELATIONS

The previous sections on external flow dealt with flow over flat surfaces where the boundary layer was either laminar or laminar and turbulent. In this section we will consider external flow normal to a cylinder, flow over spheres, and flow normal to noncircular cylinders.

6.4.1 Flow Normal to Single Cylinders

When there is curvature to a surface, there is the possibility that the flow streamlines can separate from the surface. When separation takes place, vortices and recirculatory flows result that cause local heat transfer to be much higher than if a normal boundary layer existed, and the resulting average heat transfer reflects such local effects. Figure 6.11 shows some sketches of what flow around circular cylinders or spheres looks like at different values of the Reynolds number. For very low flow velocities ($N_{Re} < 1$) the flow always adheres to the surface, is completely laminar in all aspects, and may be compared to potential flow where viscous effects are insignificant. When the Reynolds number is about 10, some weak eddies form at the back of the cylinder which die out within a short distance downstream; viscous effects in the fluid provide the damping forces. At Reynolds numbers of around 100, alternate vortexes shed first from one side, then the other, and a vortex street forms that is very strong and is not

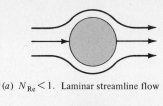

(a) $N_{Re} < 1$. Laminar streamline flow

(b) $N_{Re} \approx 10$. Weak eddies.

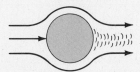

(c) $N_{Re} \approx 100$. Von Karman vortex street.

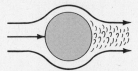

(d) $N_{Re} \approx 10^3$ to 10^5. Large turbulent wake with early separation.

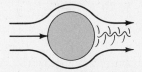

(e) $N_{Re} > 10^5$. Small turbulent wake with late separation.

Figure 6.11 Flow configuration normal to a cylinder in crossflow.

damped out until quite a distance downstream from the surface. At Reynolds numbers greater than 1000, turbulent wakes form which separate at locations between 80 and 110 degrees from the forward stagnation point, separation occurring later at larger values of Reynolds numbers. The inertia forces in these violent turbulent eddies cause very high heat transfer coefficients where separation takes place and at the rear of the cylinder or sphere. One would, therefore, expect the Nusselt number for such flow conditions to be a changing function of the Reynolds number for each progressively stronger set of eddies that form behind the cylinder as the Reynolds number increases.

The results of 13 different investigations of heat transfer to air flowing normal to a single cylinder have been correlated by McAdams (13) in terms of Nusselt number as a function of Reynolds number, as shown in Figure 6.12. The figure shows that at the higher Reynolds numbers, the slope of the correlation

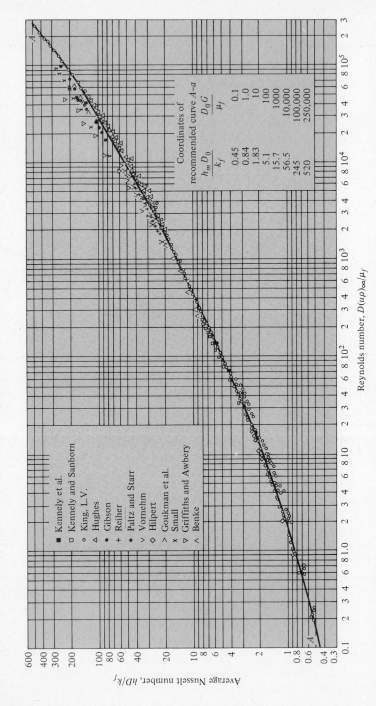

Figure 6.12 Nusselt number from 13 investigations of air flowing normal to a single cylinder as a function of Reynolds number as given by McAdams (13). Viscosity and thermal conductivity evaluated at $T_f = (T_s + T_\infty)/2$. Reproduced with permission of McGraw-Hill Book Co.

Coordinates of recommended curve A-a

$\dfrac{h_m D_0}{k_f}$	$\dfrac{D_0 G}{\mu_f}$
0.45	0.1
0.84	1.0
1.83	10
5.1	100
15.7	1000
56.5	10,000
245	100,000
520	250,000

Kennely et al.
Kennely and Sanborn
King, L.V.
Hughes
Gibson
Reiher
Paltz and Starr
Vornehm
Hilpert
Goukman et al.
Small
Griffiths and Awbery
Benke

Reynolds number, $D(u\rho)_\infty/\mu_f$

Average Nusselt number, hD/k_f

Table 6.6 VALUES OF THE CONSTANT C AND n
FOR USE IN EQ. 6.53

$N_{\mathrm{Re}f}$	n	C
1–4	0.330	0.989
4–40	0.385	0.911
40–4,000	0.466	0.683
4,000–40,000	0.618	0.193
40,000–250,000	0.805	0.0266

line is steeper than at lower values in accord with the description of physical phenomena described above. The data shown in Figure 6.12 are correlated by the equation

$$\left[\begin{array}{l} \mathrm{Nu}_D = CN_{\mathrm{Re}}^n N_{\mathrm{Pr}}^{1/3} \\ \text{properties at } T_f = (T_s + T_\infty)/2 \text{ except } \rho_\infty \end{array} \right. \tag{6.53}$$

The constants C and n vary throughout the range of Reynolds number as indicated in Table 6.6. The separate equations then represent straight line estimates of the curve $A - A$ shown in Figure 6.12. The data described by McAdams were obtained in experiments where the diameters of the cylinders ranged from 0.004 to about 6 in. and the surface temperature from 70 to 1840F, a considerable spread.

The phenomena of flow normal to cylinders can be used to determine gas or liquid velocity in the form of a hot wire anemometer. Instruments with wires small enough to measure large-scale turbulent fluctuations have been constructed.

For the flow of liquids normal to a single cylinder, Whitaker (23) has correlated the results of four different experimenters and recommends the equation

$$\left[\begin{array}{l} \mathrm{Nu} = \left(0.4 N_{\mathrm{Re}}^{1/2} + 0.06 N_{\mathrm{Re}}^{2/3} \right) N_{\mathrm{Pr}}^{0.4} \left(\dfrac{\mu_\infty}{\mu_s} \right)^{1/4} \\ \text{properties at } T_\infty \text{ except } \mu_s \text{ at } T_s \end{array} \right. \tag{6.54}$$

over the Reynolds number range from 4 to 10^5; the range in Prandtl number from 1 to 300; and the range in viscosity ratio from 0.25 to 5.2.

6.4.2 Flow Past Single Spheres

The physical phenomena of flow past spheres is similar to that for flow normal to single cylinders with the distinction that the happenings are three-dimensional rather than essentially two-dimensional for the cylinder. As the Reynolds number approaches zero, the Nusselt number approaches 2 when the surface temperature simultaneously approaches the fluid temperature. If the surface temperature remains constant, then free convection conditions prevail as the free stream velocity (N_{Re}) approaches zero. For the former case we can consider Eq. 2.32

and divide numerator and denominator by $r_1 r_2$ to obtain

$$q = \frac{4\pi k(T_1 - T_2)}{1/r_1 - 1/r_2} \qquad \text{(2.32)}$$
$$\text{(Repeat)}$$

If we note that the area of the inner surface is $A = 4\pi r_1^2$ and that we are applying this equation to a hollow sphere of fluid with a very large r_2, we can then define an h for an inner surface such that $h = q/(4\pi r_1^2)(T_1 - T_2)$ so that Eq. 2.32 becomes

$$\frac{2hr_1^2}{k} = \frac{2}{1/r_1 - 1/r_2} \qquad \text{(6.55)}$$

When we take the limit of Eq. 6.55 as $r_2 \to \infty$ and define the Nusselt number based on the diameter of the sphere with radius r_1, we obtain

$$\mathrm{Nu}_D = 2 \qquad \text{(6.56)}$$

for the sphere in an infinite fluid with conduction as the only mechanism of heat transfer for the case where T_s is very close to T_∞.

Since the phenomena of flow over spheres and cylinders are similar, Whitaker (23) proposed a similar correlation for spheres in the form

$$\boxed{\begin{array}{l} \mathrm{Nu}_D = 2 + \left(0.4 N_{\mathrm{Re}}^{1/2} + 0.06 N_{\mathrm{Re}}^{2/3}\right) N_{\mathrm{Pr}}^{0.4} \left(\dfrac{\mu_\infty}{\mu_s}\right)^{1/4} \\[2mm] \text{properties at } T_\infty \text{ except } \mu_s \text{ at } T_s \end{array}} \qquad \text{(6.57)}$$

Equation 6.57 correlates the data of three investigators working with air, water, and oil over the Prandtl number range from 0.7 to 380; the Reynolds number range from 3.5 to $7.6(10^4)$; and a viscosity ratio range from 1 to 3.2, that is, heating the fluid only.

6.4.3 Flow Normal to Tube Banks

After discussing flow normal to single cylinders, a logical extension of the topic is to consider the heat transfer situation for flow normal to multiple cylinders or tube banks. Strictly speaking the phenomenon we are discussing belongs to the heat exchanger (in particular crossflow heat exchangers) category of heat transfer, but when the surface temperature of the tubes is essentially constant, it is possible to treat the heat transfer problem as a simple iterative solution when the temperature of the fluid approaching the tubes is uniform and known.

There are two general types of orientation for the tubes, depending on whether heat transfer or pressure drop is the most important aspect. For minimum pressure drop, one would use an in-line arrangement and for maximum heat transfer, a staggered one. The in-line and staggered geometry is shown in Figure 6.13 where u_∞ and T_∞ are the flow velocity and temperature upstream from the tube bank and S_L is longitudinal spacing (in the flow direction)

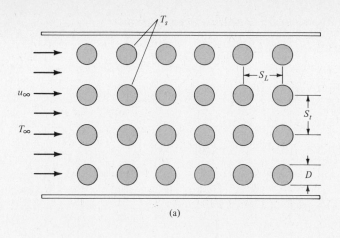

(a)

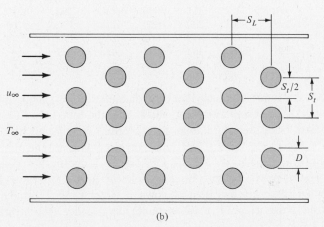

(b)

Figure 6.13 Arrangement of tube bank geometry in the (a) in-line and (b) staggered modes with flow normal to the tube axis illustrating longitudinal spacing, S_L, and transverse spacing, S_t.

between adjacent rows of tubes and S_t is the transverse spacing between tubes in a row (the rows are perpendicular to the flow direction).

The definitive experimental work in this area was done by Huge (29) and Pierson (30), and correlated by Grimison (28). McAdams (13) summarized Grimison's results in table form wherein the constants $C' = BN_{\text{Pr}}^{1/3}$ and n in Eq. 6.53 were tabulated as functions of S_t/D and S_L/D for air as the flowing fluid. The results presented by McAdams (13) are most useful because the Prandtl number term was included in the correlation so that

$$\left[\begin{array}{l} \text{Nu} = BN_{\text{Pr}}^{1/3}N_{\text{Re}}^{n} \\ \text{properties at } T_f = \left(T_s + \bar{T}_b\right)/2 \end{array}\right. \tag{6.58}$$

Table 6.7 CONSTANTS B AND n FROM GRIMISON'S (28) CORRELATION EVALUATED FROM MCADAMS (13) TO FIT EQUATION 6.58 FOR THE N_{REf} RANGE 2000 TO 40,000.

$(S_t/D) \rightarrow$	1.25		1.50		2.0		3.0	
$\downarrow(S_L/D)$	B	n	B	n	B	n	B	n
Staggered:								
0.60	—	—	—	—	—	—	.241	.636
0.90	—	—	—	—	.505	.571	.454	.581
1.00	—	—	.563	.558	—	—	—	—
1.125	—	—	—	—	.541	.565	.586	.560
1.25	.586	.556	.572	.554	.588	.556	.591	.562
1.50	.511	.568	.521	.562	.512	.568	.552	.568
2.00	.457	.572	.471	.568	.546	.556	.508	.570
3.00	.351	.592	.403	.580	.498	.562	.477	.574
In-Line:								
1.25	.394	.592	.311	.608	.113	.704	.0717	.752
1.50	.415	.586	.283	.620	.114	.702	.0768	.744
2.00	.473	.570	.339	.602	.259	.632	.224	.648
3.00	.328	.601	.404	.584	.423	.581	.324	.608

The Reynolds number is based on tube diameter and is calculated with the maximum mass flow per unit cross-sectional area in the tube bank. Table 6.7 has been constructed from the information given by McAdams for air by computing the constants B for an $N_{Pr} = 0.689$ (air) corresponding to the mean of the temperature range reported by Pierson (30) for the range in N_{Re} from 2000 to 40,000. The results predicted with Eq. 6.58 hold for 10 or more rows of tubes in the bank. When there are less than 10 rows, a factor from Table 6.8 must be applied to the Nusselt number to account for the decrease in turbulence (Reference 13).

The Reynolds number in Eq. 6.58 is calculated with the mass flow rate per minimum cross-sectional area or with the maximum velocity in the tube bank. The geometrical arrangement for in-line and staggered tubes is shown in Figure 6.14. The Reynolds number for heat transfer and pressure drop calculations is then given by:

$$N_{Re} = \frac{Du_{max}\rho}{\mu_f} = \frac{DG_{max}}{\mu_f} \tag{6.59}$$

The pressure drop Δp in lb_f/ft for flow across a bank of N tube rows can be

Table 6.8 RATIO OF Nu_N TO Nu_{10} FOR LESS THAN 10 ROWS DEEP.

$N \rightarrow$	1	2	3	4	5	6	7	8	9	10
Staggered	0.68	0.75	0.83	0.89	0.92	0.95	0.97	0.98	0.99	1
In-line	0.64	0.80	0.87	0.90	0.92	0.94	0.96	0.98	0.99	1

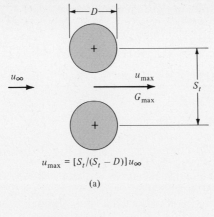

$$u_{max} = [S_t/(S_t - D)]u_\infty$$

(a)

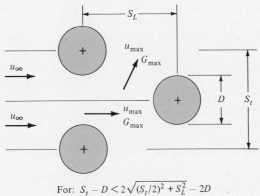

For: $S_t - D < 2\sqrt{(S_t/2)^2 + S_L^2} - 2D$

$$u_{max} = [S_t/(S_t - D)]u_\infty$$

For: $S_t - D > 2\sqrt{(S_t/2)^2 + S_L^2} - 2D$

$$u_\infty = (\sqrt{1 + 4(S_L/S_t)^2} - 2D/S_t]u_{max}$$

Figure 6.14 Visualization of maximum velocity location in (a) in-line and (b) staggered tube bank arrangements.

calculated according to Jakob (31) from

$$\Delta p = \left[\frac{f_t G_{max}^2 N}{\rho_\infty (2.09)10^8}\right]\left(\frac{\mu_s}{\mu_b}\right)^{0.14} (\text{lb}_f/\text{ft}^2) \tag{6.60}$$

where the quantities G_{max} and ρ_∞ must have the following units:

$G_{max} = \text{lb}_m/\text{hrft}^2$ mass flow at minimum

cross section (Figure 6.14)

$\rho_\infty = \text{lb}_m/\text{ft}^3$ density upstream from tube bank

where velocity is u_∞

and the empirical friction factor f_t for the tube banks is given by Jakob (31) as

$$f_t = N_{Re}^{-0.15} \left\{ 0.044 + \frac{0.08(S_L/D)}{[(S_t - D)/D]^a} \right\}$$

$$a = 0.43 + \frac{1.13D}{S_L}$$

(6.61)

for in-line tubes and

$$f_t = N_{Re}^{-0.16} \left\{ 0.25 + \frac{0.118}{[(S_t/D) - 1]^{1.08}} \right\}$$

(6.62)

for staggered tube arrangements.

Whitaker (23) has pointed out that flow through tube banks is very similar to flow through packed beds of particles if the ratio ε of void volume in the tube bank (bed) to total volume is less than 0.65 (the smaller the ratio ε, the tighter the bed is packed or the closer the tubes in the tube bank). Accordingly, Whitaker proposed a correlation equation for staggered tube banks similar to that for packed beds as

$$\left[\begin{array}{l} Nu^* = \left(0.5 N_{Re}^{*1/2} + 0.2 N_{Re}^{*2/3} \right) N_{Pr}^{1/3} \left(\frac{\mu_b}{\mu_s} \right)^{0.14} \\[2mm] \text{properties at } \overline{T}_b \end{array} \right.$$

(6.63)

where

$$Nu^* = \tfrac{3}{2} Nu \frac{\varepsilon}{1 - \varepsilon}$$

(6.64)

$$N_{Re}^* = \tfrac{3}{2} N_{Re}(1 - \varepsilon)$$

(6.65)

and

$$100 < N_{Re}^* < 30{,}000$$

$$0.7 < N_{Pr} < 30$$

$$\varepsilon \leqslant 0.65$$

The factor $\tfrac{3}{2}$ arises from Whitaker's definition of the characteristic dimension for flow and heat transfer in packed beds as $D_P = 6V_p/A_p = \tfrac{3}{2}D$ where V_p is volume of the packing particle or tube and A_p is the surface area of the particle or tube. Whitaker was not able to correlate data from in-line tube banks with an equation of the type given by Eq. 6.63.

Example 6.5

Cold air at 30F, 16 psia, and an average velocity of 3.75 ft/s enters a 10 by 20 in. rectangular duct that is fitted with 11 rows of staggered copper tubes 2 in. in diameter having the spacing shown in the figure for Example 6.5. The tubes contain saturated steam at 17.5 psia and are normal to the air flow in a direction parallel to the long side of the duct. We want to predict the outlet temperature of the air which is to be used for drying purposes by using Grimison's data and we also want to compare Nusselt numbers predicted by Grimison's data and Whitaker's equation.

SOLUTION

Grimison's results give the coefficient B and exponent n for use in Eq. 6.58 for specific combinations of S_L/D and S_t/D. From the figure for Example 6.5

$$\frac{S_L}{D} = \frac{2.5}{2} = 1.25$$

$$\frac{S_t}{D} = \frac{2.5}{2} = 1.25$$

Accordingly Table 6.7 gives $B = 0.586$ and $n = 0.556$ for staggered tube banks.

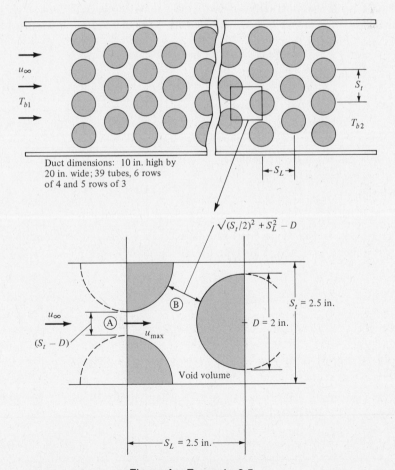

Figure for Example 6.5.

We next determine the physical properties at $T_f = (T_s + \bar{T}_b)/2$. Since we only know the inlet temperature $T_{b1} = 30F$ we must estimate $\bar{T}_b$ and later check our estimate and revise if need be; we estimate $\bar{T}_b = 100F$ as a convenient starting place. At 17.5 psia the saturation temperature is 220F and neglecting the very small temperature drop across the tube wall, we determine the reference temperature to be

$$T_f = (220 + 100)/2 = 160F$$

so that

$$\mu_f = 0.04959 \text{ lb}_m/\text{fthr}$$

$$k_f = 0.01712 \text{ Btu}/\text{hrftF}$$

$$N_{Pr_f} = 0.698$$

In order to determine the Reynolds number at the minimum cross-sectional area in the tube bank, we must determine the mass flow through the bank from the continuity equation $\dot{m} = \rho_\infty u_\infty A_\infty$; the density is calculated from the perfect gas law:

$$\rho_\infty = \frac{P}{RT} = \frac{16(144)}{53.34(490)}$$

$$= 0.08815 \text{ lb}_m/\text{ft}^3$$

$$A_\infty = \frac{10(20)}{144} = 1.3889 \text{ ft}^2$$

$$\therefore \dot{m} = 0.08815(3.75)1.3889$$

$$= 0.4591 \text{ lb}_m/s$$

or

$$\dot{m} = 1653 \text{ lb}_m/\text{hr}$$

Referring again to the figure for Example 6.5, we see that the minimum area occurs at either $\textcircled{A}$ or $\textcircled{B}$. Checking the dimensions:

At $\textcircled{A}$:

$$\frac{S_t - D}{2} = \frac{2.5 - 2}{2} = 0.25 \text{ in.}$$

At $\textcircled{B}$:

$$\sqrt{\left(\frac{S_t}{2}\right)^2 + S_L^2} - D = \sqrt{\left(\frac{2.5}{2}\right)^2 + (2.5)^2} - 2 = 0.795 \text{ in.}$$

Therefore, maximum velocity occurs at $\textcircled{A}$. There are four tubes at $\textcircled{A}$ so the flow area is $[10 - 4(2)]20/144 = 0.278 \text{ ft}^2$ and

$$G_{max} = \frac{\dot{m}}{A_{min}}$$

$$= \frac{1653}{0.278} = 5951 \text{ lb}_m/\text{hrft}^2$$

Therefore

$$N_{\text{Re}} = \frac{DG_{\text{max}}}{\mu_f}$$

$$= \frac{(2/12)5951}{0.04959}$$

$$= 20{,}000$$

From Table 6.7, $S_t/D = 1.25$ and $S_L/D = 1.25$, $B = 0.586$ and $n = 0.556$; then we compute the Nusselt number from Eq. 6.58:

$$\text{Nu}_f = BN_{\text{Pr}_f}^{1/3}N_{\text{Re}}^n$$

$$= 0.586(0.698)^{1/3}(20{,}000)^{0.556}$$

$$= 128$$

and

$$\bar{h} = \frac{\text{Nu}_f k_f}{D}$$

$$= \frac{128(0.01712)}{(2/12)}$$

$$= 13.15 \text{ Btu/hrft}^2\text{F}$$

To calculate the heat transferred to the air, we need to know the total area of the tube bundle. The 11 rows of staggered tubes contain 39 tubes; therefore,

$$A_s = \pi DLN = \pi \frac{2}{12} \frac{20}{12} 39$$

$$= 34.03 \text{ ft}^2$$

so that

$$q = \bar{h}A_s(T_s - \bar{T}_b)$$

$$= 13.15(34.03)(220 - 100)$$

$$= 53{,}705 \text{ Btu/hr}$$

Also

$$q = \dot{m}c_p(T_{b2} - T_{b1})$$

$$\therefore T_{b2} - T_{b1} = \frac{53{,}705}{1653(0.24)}$$

$$= 135\text{F}$$

and

$$\bar{T}_b = T_{b1} + \frac{T_{b2} - T_{b1}}{2}$$

$$= 30 + \frac{135}{2}$$

$$= 98\text{F}$$

This is close enough to our original assumption that we need not repeat the procedure.

We will now check Whitaker's correlation in order to compare Nusselt numbers. According to Whitaker we evaluate properties at $\bar{T}_b$; therefore

$$N_{Re} = \frac{DG_{max}}{\mu_b}$$

$$= \frac{(2/12)5951}{0.04594}$$

$$= 21,590$$

To calculate N_{Re}^* we need the void fraction $\varepsilon = V_{void}/V$ which we evaluate as follows (refer to figure for Example 6.5) for unit width:

$$V = (S_t S_L)1$$

$$V_{void} = \left(S_t S_L - \frac{\pi D^2}{4}\right)1$$

$$\therefore \varepsilon = 1 - \frac{\pi D^2}{4 S_t S_L}$$

$$= 1 - \frac{\pi}{4(S_t/D)(S_L/D)}$$

$$= 1 - \frac{\pi}{4(1.25)^2}$$

$$= 0.497$$

The value of $\varepsilon = 0.497$ is less than 0.65 so we are in the range of applicability of Whitaker's correlation.

$$N_{Re}^* = \tfrac{3}{2} N_{Re}(1 - \varepsilon)$$

$$= \tfrac{3}{2}(21,590)(1 - 0.497)$$

$$= 16,290$$

and

$$Nu^* = \left(0.5 N_{Re}^{*1/2} + 0.2 N_{Re}^{*2/3}\right) N_{Pr}^{1/3} \left(\frac{\mu_b}{\mu_s}\right)^{0.14}$$

$$= \left[0.5(16,290)^{1/2} + 0.2(16,290)^{2/3}\right](0.706)^{1/3}\left(\frac{4594}{4959}\right)^{0.14}$$

$$= 169$$

and

$$Nu = \tfrac{2}{3} Nu^* \frac{1 - \varepsilon}{\varepsilon}$$

$$= \tfrac{2}{3}(169)\frac{1 - 0.497}{0.497}$$

$$= 114$$

This value is somewhat less than the 128 predicted by Grimison's correlation by about 11 percent but that is within the precision of the data. This completes the example.

In closing our discussion of flow in tube banks, we should observe that such devices are really crossflow heat exchangers. The name crossflow is self-explanatory upon a little reflection on the physical nature of the flow directions. Where the surface temperature of the tubes does not change, as in Example 6.5, the simple treatment outlined is satisfactory when temperature differences are small, or more to the point when the ratio $\Delta T_{in}/\Delta T_{out} \lesssim 2$ as we discussed in the closing remarks of Section 6.1.4. The maximum temperature difference $\Delta T_{in} = T_s - T_{b1}$ and the minimum difference $\Delta T_{out} = T_s - T_{b2}$ for the configuration in Example 6.5. Using the results of Example 6.5 we find $\Delta T_{in} = 220F - 30F = 190F$ and $\Delta T_{out} = 220F - 165F = 55F$ so that the ratio $\Delta T_{in}/\Delta T_{out} = 190/55 = 3.5$, a value greater than 2. Accordingly we should treat the problem as a heat exchanger and use either effectiveness–NTU or the log-mean temperature difference to determine the heat transfer rate; topics that we will develop and discuss in detail in Chapter 7.

6.4.4 Flow Across Noncircular Cylinders

We conclude this section on other external flow correlations with a summary by Jakob (26) of Hilpert and Reiher's experiments with air; those experiments are summarized in Table 6.9. The length dimension to be used in the Nusselt and Reynolds numbers in Eq. 6.53 is the diameter of the cylinder perpendicular to the flow velocity. For example, for the elliptical cylinder in orientation (2) D = the minor diameter, whereas in orientation (10), D = the major diameter; and in (5) D is the distance across corners of the hexagonal cylinder, but in (4) D is the distance across flats; and similarly for the square cylinder. The surfaces of

Table 6.9 CORRELATION CONSTANTS FOR USE IN EQ. 6.53 ACCORDING TO JAKOB (26) FOR AIR FLOW OVER NONCIRCULAR CYLINDERS.

AIR FLOW ORIENTATION AND PROFILE	N_{Re} FROM	TO	n	C	OBSERVER
(1) → ◇	5,000	100,000	0.588	0.222	Hilpert
(2) → ⬭	2,500	15,000	0.612	0.224	Reiher
(3) → ◇	2,500	7,500	0.624	0.261	Reiher
(4) → ○	5,000	100,000	0.638	0.138	Hilpert
(5) → ○	5,000	19,500	0.638	0.144	Hilpert
(6) → □	5,000	100,000	0.675	0.092	Hilpert
(7) → □	2,500	8,000	0.699	0.160	Reiher
(8) → |	4,000	15,000	0.731	0.205	Reiher
(9) → ○	19,500	100,000	0.782	0.035	Hilpert
(10) → ◗	3,000	15,000	0.804	0.085	Reiher

the cylinders are considered to be at constant temperature T_s around the periphery and along the length.

6.5 SURFACE ROUGHNESS EFFECTS

In our discussion of the relationship between heat transfer and friction pressure drop in Section 4.7, we saw that $N_{St} = f/8$ for a smooth tube in turbulent flow and we observed that since roughness increased the friction factor in turbulent flow, there would also be an increase in the heat transfer number. After a brief discussion of roughness characterization, we will discuss the results of an extensive investigation of flow with heat transfer in round tubes with rough inner surfaces.

6.5.1 Roughness Characteristics

There are basically two types of roughness, irregular and regular. Irregular roughness might be described as natural roughness such as that produced by sand grains coated onto a surface. The sand grains could be all the same size, of a given size range, or completely random in size; we consider that the roughness sizes are a small fraction of the flow diameter. Other natural roughnesses (without regularity) are those that occur in manufacturing processes such as in casting (sand molds would give sand grain type of roughness) or in hot dip galvanizing. Irregular roughness can also be considered to be three-dimensional in nature.

Regular roughness is machine made such as threads cut into the surface or spiral wires attached to the surface in various configurations. Such surface roughness aspects are generally considered to be two-dimensional and it is, of course, easy to control the height of the roughness and also not too difficult to measure or characterize the roughness. Characterization of regular or two-dimensional roughness requires two parameters, the height to flow diameter ratio and the spacing in the flow direction.

Characterization of irregular or three-dimensional roughness requires only the height to flow diameter ratio but the height is not always simple to determine. One generally takes an average of the heights of the projections from the surface, either by direct measurement of the forming medium (sand, for example) or by using a stylus similar to a phonograph pickup to trace the surface irregularities and then amplify and do the averaging process electronically. We define ε as the average height of the roughness elements and ε/D the relative roughness ratio referred to the diameter of the flow passage. A particular duct with a specific value of ε/D then exhibits the friction factor behavior shown in the Moody diagram of Figure 4.10.

6.5.2 Turbulent Roughness Effects

For those fluids whose Prandtl number is near 1 (mainly gases) the results from Section 4.7 give a reasonably good estimate of the Stanton number for roughness conditions ε/D as $N_{St} = f/8$ where the friction factor is determined by the value of ε/D at the particular Reynolds number. For fluids whose N_{Pr} is greater than 1, the simple Reynolds analogy discussed in Section 4.7 does not adequately predict roughness effects. Early workers in the field proposed including the N_{Pr} to the $-2/3$ power so that $N_{\mathrm{St}} = (f/8)N_{\mathrm{Pr}}^{-2/3}$ to account for increases in the Prandtl number. The experiments by Dipprey and Sabersky (27) indicated an exponent of -0.44 as giving a much better account of the influence of the Prandtl number. Figure 6.15 shows the variation of the quantity $2C_H/C_F$ with the Prandtl number for flow in smooth tubes at a Reynolds number of 150,000 using the results of several investigators in addition to Dipprey and Sabersky's (27) data. We should note at this point that Sabersky's $C_F/2$ is the same as our $f/8$ since the friction factor in the Moody diagram is based on a slightly different relationship between the wall shear stress and the kinetic energy of the fluid; specifically

$$\tau_0 = C_F \frac{\rho u_b^2}{2g_c} \tag{6.66}$$

and

$$\tau_0 = \frac{f}{4} \frac{\rho u_b^2}{2g_c} \tag{6.67}$$

The figure shows that the $-2/3$ power considerably underestimates the data.

Dipprey and Sabersky (27) recommend correlation of the heat transfer number $C_H = q/A\rho u_b c_p (T_s - T_b) = h/c_p G = N_{\mathrm{St}}$ with the friction factor C_F (defined by Eq. 6.66) for fully turbulent flow in tubes with sand grain type of roughness as follows:

$$\left[\frac{C_F}{2C_H} = 1 + \sqrt{\frac{C_F}{2}} \left[5.19(\varepsilon^*)^{0.2} N_{\mathrm{Pr}}^{0.44} - 8.48 \right] \right. \tag{6.68}$$
$$\left. \text{properties at } T_b, \text{ and } \varepsilon^* > 50 \right.$$

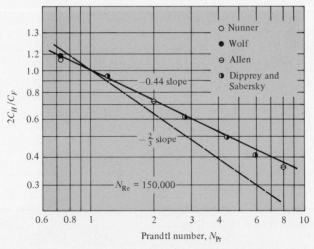

Figure 6.15   Variation of Stanton number to friction factor ratio with Prandtl number at 150,000 Reynolds number according to Sabersky (27), including data from Wolf (18).

where the nondimensional roughness height ε^* is given by

$$\varepsilon^* = \frac{\varepsilon u_b \sqrt{C_F/2}}{\nu} \tag{6.69}$$

From Eqs. 6.66 and 6.67, we note that $C_F = f/4$ and f is the friction factor given in Figure 4.10. Equation 6.68 holds for the fully rough region or $\varepsilon^* > 50$; for values less than 50, the Prandtl number effects cannot be accounted for by a simple power law and the experimental curves given in Reference 27 must be used.

Since roughness increases the Stanton number C_H, it is possible to use a shorter length of tube to accomplish the same heating effect as compared to a smooth tube. The increased roughness, however, results in an increase in pumping power even though the length is shorter when operating in the fully rough region. Dipprey and Sabersky (27) have found a limited range of ε^* in the transition region $(4 < \varepsilon^* < 50)$ where favorable ratios of rough to smooth tube performance with regard to heat transfer and pumping power were found; the computation procedure is, however, complicated. Example 6.6 illustrates the concepts under discussion for the fully rough region.

Example 6.6

Water at an average bulk temperature of 100F and a mass flow rate of 2.4 $\text{lb}_\text{m}/\text{s}$ flows in a round tube 40 in. long with sand grain type roughness on the internal surface such that the average roughness height $\varepsilon = 0.016$ in. The nominal inside diameter of the tube is 0.80 in. We would like to determine the Stanton number and heat transfer coefficient for the rough tube, the length of smooth tube that would be required to give the same fluid outlet temperature, and the ratio of pumping power of the rough tube compared to the smooth tube.

SOLUTION
We note that the average bulk temperature of the fluid in the tube is known or given so that we can base properties for Reynolds and Prandtl number on that temperature. The pertinent properties of water at 100F are

$$\mu = 1.65 \text{ lb}_\text{m}/\text{fthr}$$
$$N_{\text{Pr}} = 4.52 \quad \text{and} \quad c_p = 0.997 \text{ Btu/lb}_\text{m}\text{F}$$
$$\nu = 0.0266 \text{ ft}^2/\text{hr}$$
$$\rho = 61.99 \text{ lb}_\text{m}/\text{ft}^3$$

Figure for Example 6.6.

The relative roughness

$$\frac{\varepsilon}{D} = \frac{0.016}{0.8} = 0.02$$

The Reynolds number

$$N_{\text{Re}} = \frac{Du_b\rho}{\mu} = \frac{4\dot{m}}{\pi D\mu}$$

so that

$$N_{\text{Re}} = \frac{48(2.40)3600}{\pi(0.80)(1.65)}$$

$$= 100{,}007 = 10^5$$

From Figure 4.10, we read $f = 0.049$ at $N_{\text{Re}} = 10^5$ and $\varepsilon/D = 0.02$, therefore

$$C_F = \frac{f}{4} = \frac{0.049}{4} = 0.0123$$

We calculate the bulk velocity from the continuity equation and then compute the nondimensional roughness height ε^*.

$$A = \frac{\pi D^2}{4} = \frac{\pi(0.8)^2}{4(144)} = 0.00349 \text{ ft}^2$$

$$u_b = \frac{\dot{m}}{\rho A} = \frac{2.40}{61.99(0.00349)} = 11.09 \text{ ft/s}$$

and

$$\varepsilon^* = \frac{\varepsilon u_b\sqrt{C_F/2}}{\nu}$$

$$= \frac{0.016(11.09)\sqrt{0.0123/2}}{12(0.0266/3600)}$$

$$= 156.9 \text{ dimensionless}$$

We can now evaluate C_H and h from Eq. 6.68:

$$\frac{C_F}{2C_H} = 1 + \sqrt{\frac{C_F}{2}}\left[5.19(\varepsilon^*)^{0.2}N_{\text{Pr}}^{0.44} - 8.48\right]$$

$$= 1 + \sqrt{0.00612}\left[5.19(156.9)^{0.2}(4.52)^{0.44} - 8.48\right]$$

$$= 2.504$$

We must recognize that the ratio we just calculated is for the rough inner surface condition so that we really should distinguish by writing $C_F/2C_H = C_{FR}/2C_{HR} = 2.504$ and for a smooth inner surface, we would use the designation C_{FS} and

C_{HS}. Accordingly

$$C_{HR} = \frac{C_{FR}/2}{2.504}$$

$$= \frac{0.0123/2}{2.504} = 0.00246$$

For the heat transfer coefficient h we then obtain

$$h_R = c_p G C_{HR}$$

where

$$G = \frac{\dot{m}}{A} = \frac{2.40(3600)}{0.00349}$$

$$= 2.476(10^6)\ \text{lb}_\text{m}/\text{hrft}^2$$

Therefore

$$h_R = 0.997(2.476)10^6(0.00246)$$

$$= 6072\ \text{Btu}/\text{hrft}^2\text{F}$$

for the rough surfaced tube. To compare with the smooth surface condition, we use Dipprey and Sabersky's recommendation

$$C_{HS} = \frac{C_{FS}}{2} N_\text{Pr}^{-0.44}$$

At a $N_\text{Re} = 10^5$ Figure 4.10 gives $f = 0.018$; therefore,

$$\frac{C_{FS}}{2} = \frac{f}{8} = \frac{0.018}{8} = 0.00225$$

so that

$$C_{HS} = (0.00225)(4.52)^{-0.44}$$

$$= 0.00116\ \text{dimensionless}$$

and

$$h_S = c_p G C_{HS}$$

$$= 0.997(2.476)10^6(0.00116)$$

$$= 2864\ \text{Btu}/\text{hrft}^2\text{F}$$

The ratio $C_{HR}/C_{HS} = h_R/h_S = 2.12$ then shows a 112 percent increase in the heat transfer coefficient and Stanton number. We naturally are now curious to see what the pumping power requirements are for the rough tube and the length of an equivalent smooth tube which will transfer the same amount of heat flux at the same surface to average bulk temperature difference. Thus

$$q_R = h_R A_R \left(T_s - \bar{T}_b\right) = h_S A_S \left(T_s - \bar{T}_b\right) = q_s$$

$$h_R(\pi D L_R) = h_S(\pi D L_S)$$

or

$$L_S = \frac{h_R}{h_S} L_R$$

$$= 2.12 L_R$$

We therefore need a little more than twice the length ($L_S = 2.12(40) = 85$ in.) of rough tube to obtain the same outlet temperature in a smooth tube.

The pumping power requirement for the rough tube in horsepower is

$$\text{HP}_R = \frac{\dot{m}\,\Delta P_R}{\rho 550}$$

$$= \frac{\dot{m} f_R \left(u_b^2 / 2g_c \right) \left(L_R / D \right)}{550}$$

After substitution for ΔP_R from Eq. 4.24 and similarly for the smooth tube

$$\text{HP}_S = \frac{\dot{m} f_S \left(u_b^2 / 2g_c \right) \left(L_S / D \right)}{550}$$

Now since we have the same $\dot{m}$ and average bulk temperature, the bulk velocity u_b will be the same in both tubes so the ratio of pumping powers then is

$$\frac{\text{HP}_R}{\text{HP}_S} = \frac{f_R L_R}{f_S L_S}$$

$$= \frac{0.0123}{0.0045(2.12)}$$

$$= 1.29$$

Therefore, we would need 29 percent more power to pump the fluid through the rough tube than through the smooth tube, but the rough tube is half as long so there are less capital costs and higher operating costs. The choice is dictated by economics and space requirements. This completes the example.

Most random type roughnesses can be estimated with Sabersky's method. Special two-dimensional roughnesses create particular flow patterns and each requires a distinctive correlation for the particular set of experimental parameters investigated. Specific details for such cases must be sought in the literature.

6.6 COMBINED FREE AND FORCED CONVECTION

At very low velocities in forced convection, the geometrical orientation of the heat transfer surface with respect to the gravity vector determines whether there is enhancement or interference between free and forced convection phenomena. For example, in a vertical tube where $T_s > T_b$ the free convection boundary layer will move in an upward direction so that forced flow at low velocities in the upward direction will enhance the heat transfer, whereas forced flow downward at low velocities will decrease the heat transfer. In a horizontal tube direction is immaterial and low forced velocities result in a spiral circulation pattern that enhances heat transfer.

Metais and Eckert (32) report the results of a survey of data in the free, mixed, and forced flow regions for horizontal and vertical tubes. Figure 6.16 shows the results of a number of investigators for flow through vertical tubes; the figure includes both experiments where the flow enhanced the heat transfer and experiments with interference (opposing flow). The specific investigators listed in Figure 6.16 are discussed by Metais in Reference 33. The figure shows a definite region of mixed flow between the forced convection region and the free convection region. To use Figure 6.16 one calculates the Reynolds number N_{Re} and the (Grashof)(Prandtl)(D/L) product, enters the figure with those values, and the intersection indicates the region of operation that determines the correlation equation from the appropriate section in this chapter or Chapter 5. It is interesting to note that at a Reynolds number of 10^4 or 10,000 at a $N_{Gr}N_{Pr}(D/L)$ value of 10^9, we would be operating in the free convection region with a turbulent boundary layer and the influence of the forced velocity would be less than 10 percent according to Metais and Eckert (32).

Figure 6.17 shows similar information for horizontal tubes with flow on the inside. No definite mixed convection region was definable in this case and

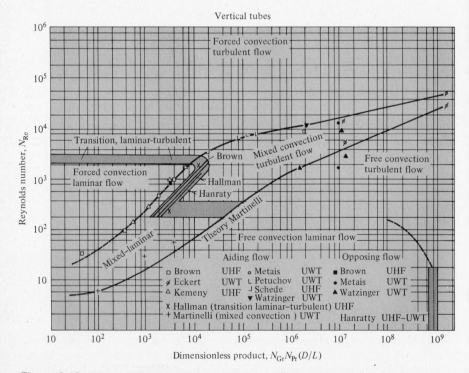

Figure 6.16 Interaction of free and forced convection phenomena in vertical tubes as given by Metais and Eckert (32) for the range $0.01 < N_{Pr} < 1$. Reproduced with permission of the American Society of Mechanical Engineers. (UHF = uniform wall flux. UWT = uniform wall temperature.)

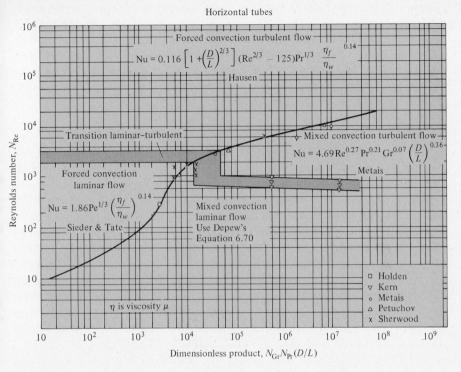

Figure 6.17 Interaction of free and forced convection phenomena in horizontal tubes as given by Metais and Eckert (32) for the range $0.01 < N_{Pr} < 1$. Reproduced with permission of the American Society of Mechanical Engineers.

the correlation equations recommended by Metais and Eckert are given on the figure. More recent information obtained by Depew and August (34) show that the average Nusselt number for the mixed convection laminar flow region is best correlated as

$$\left[\text{Nu} = 1.75 \left[N_{Gz} + 0.12 \left(N_{Gz} N_{Gr}^{1/3} N_{Pr}^{0.36} \right)^{0.88} \right]^{1/3} \left(\frac{\mu_b}{\mu_s} \right)^{0.14} \right.$$

$$\left. \text{properties at } \overline{T}_b \text{ except } \mu_s \text{ at } \overline{T}_s \right. \tag{6.70}$$

where

N_{Gz} = Graetz number

$\quad = \dfrac{\dot{m} c_p}{kL}$

$\quad = \dfrac{\pi}{4} N_{Re} N_{Pr} \dfrac{D}{L}$

N_{Gr} = Grashof number

$\quad = \dfrac{D^3 \rho^2 \beta g \, \Delta t}{\mu^2}$

Depew and August maintain that D/L is not an important parameter for mixed convection phenomena with L/D ratios less than 50. Metais and Eckert (32) have also correlated the data shown in Figure 6.17 with $N_{Gr}N_{Pr}$ as the ordinate and obtain similar results. The range in the parameter $N_{Pr}(D/L)$ for both Figures 6.16 and 6.17 extends from 0.01 to 1, which precludes application of these results to liquid metals.

6.7 CLOSURE

In this chapter we have briefly discussed the basic aspects of internal and external flow configurations with the simplest of boundary conditions; constant wall temperature or uniform heat flux. In all the flow situations that we examined, no variation in either wall temperature or heat flux was considered in the direction perpendicular to the flow (around the periphery). Often situations are encountered in engineering practice which cannot adequately be modeled by the simple phenomena described in the preceding sections. In that case engineers must peruse the heat transfer literature (Reference 14 is a good place to start) to see if their particular situation, or one sufficiently alike, has been analyzed. If not, either analysis or experiment must be performed depending on the importance of the application. The space age has given a tremendous impetus to research and engineering experimentation in the field of heat transfer and there are several professional journals devoted exclusively to the topic.

PROBLEMS

1. Determine the local Nusselt number at a point 10 in. from the entrance for water flowing in a 0.5-in. ID tube at a mass flow rate of 1.687 lb_m/min under constant wall temperature conditions. Use Eq. 6.3 and check your resulting magnitude with Table 6.1 and Figure 6.2. Assume the water bulk temperature to be 100F.

2. The tube in Problem 1 is maintained at a temperature of 210F. Determine the temperature rise of the water in passing through the tube.

3. A light oil flows through a 1-in. ID tube at the flow rate of 35 lb_m/min and is at a temperature of 150F. Determine the entrance length in feet under these conditions and discuss the significance of the result.

4. Consider the information presented in Table 6.4 and make a qualitative sketch (no computations) showing the relationship of bulk temperature and wall temperature with x^+ for the uniform heat flux (UHF) boundary condition. Give a brief discussion of the reasoning that led you to your result.

5. Consider the information given in Table 6.2 for the constant wall temperature (CWT) boundary condition and make a qualitative sketch (no computations) of the relationship between unit area heat flux q/A and the fluid bulk temperature as a function of dimensionless distance x^+. Give a brief discussion of the reasoning that led you to your result.

6. A process requires a small amount of warm air at 120F and essentially atmospheric pressure. The air is to be heated from ambient conditions of 60F at uniform heat flux in a tube 1.5 in. in diameter and 24 in. long. What average surface temperature is required for 2-cfm flow of air into the tube? Estimate the wall temperature at the outlet.

7. Determine the average Nusselt number for the flow conditions and geometry of Problem 1 with the UHF boundary condition instead of CWT using the correlation given in Eq. 6.10 and compare the result with Table 6.4.

8. Determine the average Nusselt number for the flow situation in Problem 1 from the correlation equation recommended by Kays in Eq. 6.9 for CWT and compare with the theoretical result given in Table 6.2 and the result for the UHF boundary condition calculated in Problem 7.

9. What mass rate of flow for air is required to achieve an average Nusselt number of 9 (experimental correlation) at $\bar{T}_b = 100F$ in a 1-in. ID smooth round tube 30 in. long at UHF boundary conditions?

10. Air at a bulk temperature of 200F enters a smooth tube of 1.5-in. ID at a mass flow rate of 0.25 lb_m/s. Determine the thermal entrance length under these conditions.

11. The table gives some data obtained from experiments where air and CO_2 where heated in a smooth round tube having a 0.569-in. ID and a heated length of 34.7 in. The ratio T_s/T_b in each case is less than 1.2 so the data can be considered as low temperature difference data. Calculate the heat transfer and flow parameters and plot the result on log-log paper to obtain a correlation of the type recommended by McEligot in Eq. 6.21.

GAS	$\bar{T}_b$ (R)	$\bar{T}_s$ (R)	$\dot{m}$ (lb_m/s)	q (Btu/s)
Air	564.5	642.4	0.0267	0.382
Air	561.4	634.8	0.0421	0.549
Air	560.1	642.3	0.0590	0.807
Air	558.5	646.1	0.0729	1.015
CO_2	562.9	640.7	0.0328	0.377
CO_2	557.8	631.7	0.0518	0.531
CO_2	554.9	641.3	0.0733	0.778
CO_2	551.8	638.4	0.0911	0.952

Use 1×2 cycle log-log paper and note that the ordinate and abscissa values printed on the sheet can only be multiplied or divided by factors of 10.

12. Since the tube in which the data for Problem 11 were obtained is quite long, the results should not be greatly influenced by the entrance region. Plot the data as a simple Dittus-Boelter type of correlation on log-log paper and compare your results with Eq. 6.17 for heating.

13. The table shows the results of local heat transfer experiments for air heated in a smooth tube for runs 175 and 183 as reported in Reference 18. Compare

the data with McEligot's correlation, using the $\sqrt{T_s/T_b}$ term in the correlation.

	RUN 175				RUN 183			
x/D	T_s	T_b	N_u	N_{Re}	T_s	T_b	N_u	N_{Re}
2.49	1340	557	148.2	70,600	1368	548	237.6	141,300
7.47	1531	601	117.3	66,300	1532	587	203.2	134,500
14.92	1663	666	101.4	62,500	1653	646	176.2	126,200
29.8	1817	797	84.93	55,400	1813	762	147.2	113,000
44.7	1930	927	76.11	49,800	1929	879	131	102,400
49.6	1958	970	74.59	48,200	1962	917	127.2	99,300
54.7	1978	1015	73.82	46,600	1982	957	125.6	96,200

Could you use this correlation to predict cooling experiments where T_s/T_b < 1? Discuss.

14. A proposed energy conversion unit utilizing a solar powered Brayton Cycle focuses sunlight onto several tubes from a large concentrating mirror. The tubes are made from a special high-temperature alloy and have smaller cylindrical mirrors in back so that the solar flux is fairly well distributed around the circumference. A pyrometer indicates an average surface temperature on a tube to be 2740F. Air leaves the compressor and enters the heater tubes at 620F with a flow rate of 0.283 lb_m/s. Estimate the turbine inlet temperature for a tube length of 12 in. and 0.4-in. ID.

15. Pressurized water in a reactor such as the Shippingport Unit enters the steam generation tubes at about 540F and 2000 psi. Saturated steam at 600 psi is generated on the outside of the tubes. If the steam generator tubes are 1-in. ID and the bulk velocity of the pressurized water 17.5 ft/s, estimate the heat transfer coefficient on the water side and the unit area flux through the tube wall.

16. The bed of a large earth moving truck is fabricated with square channels in the floor through which the hot diesel exhaust gases are vented in order to keep moist payload from freezing during extremely cold weather. The channels are 4 in. square and the exhaust gas is diluted with air so that it enters the channels at about 800F. Assuming an average temperature of the hot gases of 600F at 15 psia and a velocity of 30 ft/s with properties essentially those for air, determine the average heat flux per unit area available to melt ice.

17. A synthesis plant produces n-butyl alcohol at 200F and it must be cooled to 100F in order to safely package the fluid. Would it be possible to cool the alcohol in a single 2-in. ID tube at a flow velocity of 10 ft/s with $\bar{T}_s = 60F$?

18. A high-temperature, gas-cooled reactor (HTGR) has coolant channels in the core which are 0.825 in. in diameter and 31 in. long. Helium gas flows through the channels with a flow rate of 0.0176 lb_m/s, an inlet temperature of 636F, and an outlet temperature of 1377F. Estimate the average surface temperature of the coolant channel wall under these conditions. Start with a simple Dittus-Boelter correlation, then refine with a more precise equation.

19. Beginning with Eq. 6.27 insert Eq. 4.18 and perform the necessary steps to obtain Eq. 6.34 for the case where the Prandtl number is 1.

20. Sheets of carbon steel (1.5 percent) which are 2 by 6 ft and $\frac{1}{4}$ in. thick come from an annealing furnace at a temperature of 600F and are cooled by a crossflow of air at 100F which flows parallel to the short side at a velocity of 30 ft/s. The sheets move slowly on rollers spaced far enough apart so that air contacts both sides of the sheet in the same manner. Estimate how long it would take to cool the sheet to 200F by forced convection alone.

21. Consider the top surface of a four-door sedan to be a flat plate with typical dimensions of 132 by 168 cm. When the car is traveling at 100 km/hr, what percent of the roof is covered with a laminar boundary layer when the air temperature is 20C?

22. A small aircraft is flying at 10,000 ft on an NACA standard day at a speed of 150 mph. The wing has a chord of 5 ft and contains a section filled with fuel. In order to determine the temperature history of the fuel we must estimate the heat transfer coefficient under these conditions.

23. The struts of a hydroplane boat serve not only to support the craft, but the rear struts also support the bearing for the propellor driveshaft and thus contain lubrication lines. The hydrofoil struts are 18 in. wide and about $1\frac{1}{2}$ in. thick at the thickest point. Estimate the heat transfer coefficient between the strut and the sea at a craft velocity of 40 mph when the sea temperature is 4.5C.

24. The bottom of a pleasure boat can be considered to be a flat plate 18 ft long by 8 ft wide. How much energy is required to keep the bottom surface temperature at 50F in 40F water at a speed of 15 knots?

25. A cruise missile is designed to travel at Mach 0.8 an very low altitude where there can be considerable moisture in the atmosphere. The wings on the missile are 2 ft wide and 4 ft in length. At an air temperature of 20F how much energy needs to be supplied to the wing to prevent ice formation?

26. The straightening vanes in a low-speed wind tunnel create quite a bit of free stream turbulence as the boundary layer peels off the trailing edge. Estimate the power required to maintain the top surface of a flat plate 15 in. (in the flow direction) by 10 in. at a temperature of 400F when the air temperature is 60F (at 1 atm) with a free stream velocity of 200 ft/s.

27. Compare the predicted values of Nusselt number for air flow normal to a cylinder at Reynolds numbers of 4 and 100,000, based on free stream temperature from McAdams' and Whitaker's correlations at a surface temperature of 1000F and an air temperature of 100F.

28. An electric dryer operates with a set of 20-gauge (Brown & Sharpe) nichrome wires 18 in. long as the heat source. Determine the current in amperes for each wire to remain at 1200F in an 80F air stream at 1 atm which is moving at a velocity of 10 ft/s normal to the wires.

29. A constantan wire (no. 26 Brown & Sharpe gauge) is used as a resistance thermometer element in a flowing air stream. What is the time constant for

this wire in an air stream that changes temperature suddenly from 500 to 300F at a velocity of 40 ft/s normal to the wire? Air pressure is 1 atm.

30. Saturated steam at 20-psia pressure enters a 2-in. schedule 40 pipe that is 50 ft long. The pipe is positioned 40 ft from the ground between two buildings. How many pounds per hour of condensate leave the pipe when a 21-mph breeze at 92F of atmospheric air flows normal to the pipe?

31. Estimate how far solidified lead shot $\frac{3}{32}$ in. in diameter must fall to cool from 320 to 120F in atmospheric air if the rate of fall is 160 ft/s.

32. Steel ball bearings ($k = 20$ Btu/hrftF and $\alpha = 0.35$ ft^2/hr) that are 0.25 in. in diameter come out of a heat treat furnace at 600F and are dropped into a light oil quench bath. The bath is 3 ft deep and the bearings take 2.3 s to reach the bottom. Estimate the temperature of the bearings when they hit the bottom of the oil bath whose temperature is 100F.

33. A tube bank contains eight rows of six 1-in. pipes in an in-line arrangement. The longitudinal and transverse spacing is 1.644 in. and the duct size is 10×18 in. with the six pipes along the 10-in. dimension. The air velocity upstream to the tube bank is 5 ft/s at a temperature of 20F and atmospheric pressure. Determine the surface temperature of the pipes required to raise the air temperature at the exit of the tube bank to 70F.

34. Calculate the pressure drop across the eight rows of tubes in Problem 33. What kind of measuring instrument would you use to measure the pressure difference you determined?

35. Six rows of four 2-in.-OD tubes in a staggered arrangement are housed in a 11.35 by 18 in. duct. The tubes extend along the 18-in. dimension of the duct and contain saturated steam on the inside at 12 psia. Water enters the duct at 40F and a velocity of 0.75 ft/s. The transverse and longitudinal spacing is 2.5 in. Determine the temperature of the water downstream from the tube bank.

36. Evaluate the Nusselt number for the flow situation described in Problem 35 using Grimison's correlation which was based on air data. Compare your result with the Nusselt number from Whitaker's correlation.

37. Determine the pressure drop across the tube bank in Example 6.5 for the flow situation described.

38. A hexagonal brass bar, 1 in. across flats and 12 in. long, is removed from an annealing furnace at 800F and placed upright on an end on a transite block to be cooled by a fan. The fan blows atmospheric air at 25 ft/s and 60F over the brass bar. How long will it take to cool to 200F?

39. Water flows in a 2-in. ID tube that has a sand grain type of roughness on the inside surface with an average roughness height of 0.036 in. The tube is 36 in. long and the bulk velocity of the water at the entrance is 10 ft/s at a temperature of 60F; pressure is 160 psia. What average wall temperature is required to obtain an outlet temperature of 100F? Is such a wall temperature feasible?

40. Determine the pressure drop and pumping power required to maintain the flow conditions for the rough tube described in Problem 39.

41. Freon-12 at an average bulk temperature of 60F and 500 psia flows in a 1-in. ID round tube with rough inner surface. The tube is 24 in. long and has an average roughness height of 0.015 in.; the mass rate of flow of the Freon-12 in the tube is 5.3 lb_m/s. The tube is electrically heated such that the average surface temperature is 160F. What rise in temperature of the Freon-12 can be expected under these conditions as it passes through the tube?

42. Air at a bulk temperature of 100F and atmospheric pressure flows inside a 6-in. horizontal tube at a velocity of 1.4 ft/s. The tube is 36 in. long and the inside wall temperature is 900F. Is the flow laminar or turbulent? Determine the applicable flow regime and compare the resulting Nusselt number with what you would predict for laminar or turbulent forced convection only.

43. Water flows upward in a vertical pipe 6 in. in diameter and 6 ft tall at a flow rate of 1300 lb_m/hr. The water is at an average bulk temperature of 200F and a pressure of 500 psia, the inside wall temperature of the pipe is maintained at 40F. Determine the flow regime for this situation.

References

1. Sellars, J. R., Tribus, M., and Klein, J. S., "Heat Transfer to Laminar Flow in a Round Tube or Flat Conduit—The Graetz Problem Extended," *Transactions ASME*, Vol. 78, 1956, p. 441.

2. Kays, W. M., *Convective Heat and Mass Transfer*, McGraw-Hill, New York, 1966, p. 126.

3. Siegel, R., Sparrow, E. M., and Hallman, T. M., "Steady Laminar Heat Transfer in a Circular Tube with Prescribed Wall Heat Flux," *Applied Scientific Research*, Section A, Vol. 7, 1958, p. 386.

4. Kays, W. M., "Numerical Solutions for Laminar Flow Heat Transfer in Circular Tubes," *Transactions ASME*, Vol. 77, 1955, p. 1265.

5. Clark, S. H. and Kays, W. M., "Laminar Flow Forced Convection in Rectangular Tubes," *Transactions ASME*, Vol. 75, 1953, p. 859.

6. Sparrow, E. M., Hallman, T. M., and Siegel, R., "Turbulent Heat Transfer in the Thermal Entrance Region of a Pipe with Uniform Heat Flux," *Applied Scientific Research*, Section A, Vol. 7, 1957, p. 37.

7. Sleicher, C. A. and Tribus, M., "Heat Transfer in a Pipe with Turbulent Flow and Arbitrary Wall Temperature Distribution," *Transactions ASME*, Vol. 79, 1957, p. 789.

8. Wolf, H. and Lehman, J. H., "The Determination of Thermal Entrance Lengths for Gases in Turbulent Flow in Smooth Round Ducts," *Jet Propulsion*, Aug., 1957, p. 897.

9. Wolf, H., "Heating and Cooling Air and Carbon Dioxide in the Thermal Entrance Region of a Circular Duct with Large Gas to Wall Temperature Differences," *ASME Journal of Heat Transfer*, Vol. 81, 1959, p. 267.

10. Deissler, R. G., "Analysis of Turbulent Heat Transfer and Flow in the Entrance Regions of Smooth Passages," NACA TN 3016, Oct. 1953.

11. Hartnett, J. P., "Experimental Determination of the Thermal Entrance Length for the Flow of Water and Oil in Circular Pipes," *Transactions ASME*, Vol. 77, 1955, p. 1211.

12. Dittus, F. W. and Boelter, L. M. K., *Univ. of Calif. Berkeley, Publications Engineering*, Vol. 2, 1930, p. 443.
13. McAdams, W. H., *Heat Transmission*, 3d Edition, McGraw-Hill, New York, 1954, pp. 225, 273.
14. Rohsenow, W. M. and Hartnett, J. P., Eds., *Handbook of Heat Transfer*, McGraw-Hill, New York, 1972, p. 7–1.
15. Sieder, E. N. and Tate, G. E., "Heat Transfer and Pressure Drops of Liquids in Tubes," *Industrial and Engineering Chemistry*, Vol. 28, 1936, p. 1429.
16. Wolf, H., Gray, F. L., and Reese, B. A., "Heat Transfer and Friction Characteristics of Red and White Fuming Nitric Acid," *Jet Propulsion*, Vol. 26, Nov. 1956, p. 979.
17. McCarthy, J. R. and Wolf, H., "The Heat Transfer Characteristics of Gaseous Hydrogen and Helium," Research Report RR-60-12, Rocketdyne Division of North American Aviation, Inc., Canoga Park, Calif., 1960.
18. Wolf, H., Ph.D. Thesis, Part I, Purdue University, W. Lafayette, Ind., March 1958.
19. Seader, J. D., and Wolf, H., "Theoretical Analysis of Heat Transfer to Gases in Smooth Round Tubes Under Conditions of Turbulent Flow and High Heat Flux," *ARS Journal*, May 1961, p. 650.
20. Deissler, R. G. and Eian, C. S., "Analytical and Experimental Investigations of Fully Developed Turbulent Flow of Air in a Smooth Tube with Heat Transfer and Variable Fluid Properties," NACA TN 2629, Feb. 1952.
21. Von Karman, T. H., "The Analogy Between Fluid Friction and Heat Transfer," *Transactions ASME*, Vol. 61, Nov. 1939.
22. Slack, E. G., "Experimental Investigation of Heat Transfer Through Laminar and Turbulent Boundary Layers on Flat Plates," NACA TN 2686, April 1952.
23. Whitaker, S., "Forced Convection Heat Transfer Correlations for Flow in Pipes, Past Flat Plates, Single Cylinders, Single Spheres, and for Flow in Packed Beds and Tube Bundles," *AIChE Journal*, Vol. 18, No. 2, March 1972, p. 361.
24. Eckert, E. R. G., "Engineering Relations for Heat Transfer and Friction in High-Velocity Laminar and Turbulent Boundary-Layer Flow over Surfaces with Constant Pressure and Temperature," *Transactions ASME*, Vol. 78, 1956, p. 1273.
25. Chapman, D. R. and Rubesin, M. W., "Temperature and Velocity Profiles in the Compressible Laminar Boundary Layer with Arbitrary Distribution of Surface Temperature," *Journal Aeronautical Science*, Vol. 16, No. 9, Sept. 1949, pp. 547–565.
26. Jakob, M., *Heat Transfer*, Vol. I, Wiley, New York, 1949, p. 562.
27. Dipprey, D. F. and Sabersky, R. H., "Heat and Momentum Transfer in Smooth and Rough Tubes at Various Prandtl Numbers," *International Journal of Heat and Mass Transfer*, Vol. 6, 1963, p. 329.
28. Grimison, E. D., "Correlation and Utilization of New Data on Flow Resistance and Heat Transfer for Cross Flow of Gases over Tube Banks," *ASME Transactions*, Vol. 59, 1937, p. 583.
29. Huge, E. C., "Experimental Investigation of Equipment Size on Convective Heat Transfer and Flow Resistance in Cross Flow of Gases over Tube Banks," *ASME Transactions*, Vol. 59, 1937, p. 573.
30. Pierson, O. L., "Experimental Investigation of Tube Arrangement on Convective Heat Transfer and Flow Resistance in Cross Flow of Gases over Tube Banks," *ASME Transactions*, Vol. 59, 1937, p. 563.
31. Jakob, M., "Discussion of Grimison, Huge, and Pierson Papers," *ASME Transactions*, Vol. 60, 1938, p. 384.

32. Metais, B. and Eckert, E. R. G., "Forced, Mixed, and Free Convection Regimes," *Transactions ASME*, Series C., Vol. 86, 1964, p. 295.
33. Metais, B., "Criteria for Mixed Convection," HTL Tech. Dept. 51, Heat Transfer Lab, University of Minnesota, Minneapolis, Minn., 1963.
34. Depew, C. A. and August, S. E., "Heat Transfer Due to Combined Free and Forced Convection in a Horizontal and Isothermal Tube," *Transactions ASME*, Series C, Vol. 93, 1971, p. 380.
35. Weiss, T. G., MSME Thesis, Mechanical Engineering Department, Fayetteville, University of Arkansas, 1963.
36. Prideaux, J. P., Ph.D. Thesis, Mechanical Engineering Department, Fayetteville, University of Arkansas, 1966.
37. McEligot, D. M., Magee, P. H., and Leppert, G., "Effect of Large Temperature Gradients on Convective Heat Transfer: The Downstream Region," *Transactions ASME*, Series C, Vol. 87, 1965, p. 67.
38. Taylor, M. F., "Experimental Local Heat Transfer Data for Precooled Hydrogen and Helium at Surface Temperatures up to 5300 R," NASA TND 2595, 1965.
39. Wolf, H., Ph.D. Thesis, Purdue University, West Lafayette, Ind., 1958.

Chapter 7
Heat Exchangers

7.0 INTRODUCTION

In our discussions so far we have considered only heat exchange between a surface and a fluid in contact with that surface. The fluid was either large in extent (external to the surface) or flowed inside the surface in a channel. In the latter case our basic definition of the heat transfer coefficient was simple and straightforward for constant surface temperature as expressed by Eq. 1.1 where

$$q = hA(T_s - T_\infty) \tag{7.1}$$

For the internal flow situation the fluid temperature changed along the length of the duct so that we were required to modify our defining equation for the heat transfer coefficient in terms of the bulk temperature T_b such that

$$q_x = h_x A(T_s - T_b) \tag{7.2}$$

and in this case the heat transfer coefficient is the local value at the particular point in the duct where the surface temperature is T_s and the fluid bulk temperature is T_b. For the entire channel we used the average concept where

$$q = \bar{h}A(\bar{T}_s - \bar{T}_b) \tag{7.3}$$

and $\bar{T}_b$ was a simple numerical average of the inlet and exit bulk temperatures of the fluid. The average surface temperature we defined as an integrated average over the length of the tube.

In considering heat exchangers we extend the above ideas and concepts to the situation where energy is exchanged between two fluids through a wall rather than between a fluid and a surface. Consequently, the resistance to the flow of energy comprises two convective resistances separated by a conduction resistance. The concept is shown in Figure 7.1 together with an electrical analog of the thermal circuit. On a local basis, we then define an overall coefficient U (Btu/hrft^{2}F, W/m^{2}K) that relates the local heat flux to the difference in temperature between the two fluids such that

$$q = UA(T_{b1} - T_{b2})$$ (7.4)

In terms of the thermal resistance concept we can write Eq. 7.4 as

$$q = \frac{T_{b1} - T_{b2}}{1/UA} = \frac{T_{b1} - T_{b2}}{R_{total}}$$ (7.5)

where the total resistance R_{total} in the thermal circuit consists of the two convection resistances and the wall resistance mentioned above; thus

$$R_{total} = R_{c1} + R_k + R_{c2}$$ (7.6)

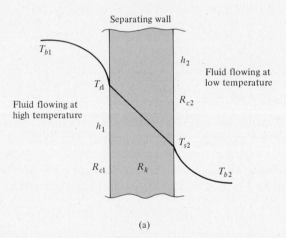

(a)

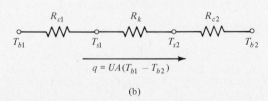

(b)

Figure 7.1 Physical arrangement (a) and electrical analogy (b) of heat transfer between two fluids separated by a wall as is embodied in the concept of the overall coefficient U.

In terms of the definitions of the resistances from Chapter 1 we can express the overall coefficient then as

$$UA = \frac{1}{(1/h_1 A) + (\Delta x/kA) + (1/h_2 A)} \qquad (7.7)$$

For the simple plane wall exchanger shown in Figure 7.1 the area in contact with each fluid is the same, but in other types of exchangers such as the concentric tube or the shell and tube types, the areas are not the same. Consequently, the UA product is generally given or the overall coefficient must be specified for either inside or outside area.

The question that we need to address then is: what temperature difference must be used with the overall coefficient for the *entire* exchanger similar to Eq. 7.4 such that

$$q = UA \text{ (temperature difference)} \qquad (7.8)$$

The answer requires a mathematical analysis of the heat exchanger. A grouping of the results of such an analysis in the form of Eq. 7.8 will show the required form the temperature difference must take.

7.1 ANALYSIS OF SIMPLE HEAT EXCHANGERS, LMTD

A simple concentric tube heat exchanger is illustrated in Figure 7.2. The high-temperature fluid flows in the central tube and the cold fluid in the annular space surrounding the inner tube. The hot fluid enters at T_{h1} and leaves at T_{h2}; corresponding temperatures for the cold fluid are T_{c1} entering and T_{c2} leaving. The exchanger has a length L and a corresponding area A. We will analyze a small infinitesimal section, dA, of the heat exchanger, then sum over the entire length of the exchanger in an integration process. We consider the following conditions to hold for the analysis:

1. Steady flow for both hot and cold fluids; the mass flow rates $\dot{m}_h$ and $\dot{m}_c$ are constants.
2. Specific heats of the fluids are not functions of temperature and may be considered constants.
3. Fully developed flow exists in the passages and heat transfer coefficients and the overall coefficient U are not functions of distance along the exchanger.

The last assumption is the most serious and requires that the exchanger be fairly long, since we know that the heat transfer coefficient is always higher in the entrance region than in fully developed flow.

We begin the analysis by using the first law of thermodynamics to relate the infinitesimal heat fluxes to the infinitesimal temperature changes. Thus

$$dq_h = -\dot{m}_h c_{ph} \, dT_h \qquad (7.9)$$

$$dq_c = \dot{m}_c c_{pc} \, dT_c \qquad (7.9a)$$

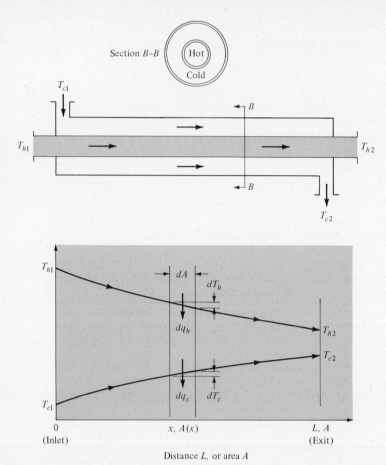

Figure 7.2 Diagram and temperature profile of a simple concentric tube heat exchanger operating in the parallel flow mode.

We note that dq_h must have a minus sign since heat energy leaving a control volume is negative. Solving for dT_h and dT_c yields

$$dT_h = -\frac{1}{\dot{m}_h c_{ph}} dq_h \tag{7.10}$$

$$dT_c = \frac{1}{\dot{m}_c c_{pc}} dq_c \tag{7.10a}$$

We recognize that the magnitudes of dq_h and dq_c are the same and if we designate that magnitude as dq, we can subtract Eq. 7.10a from 7.10 to obtain

$$d(T_h - T_c) = -\left[\frac{1}{\dot{m}_h c_{ph}} + \frac{1}{\dot{m}_c c_{pc}}\right] dq \tag{7.11}$$

by recognizing that the differential operator is a linear operator so that $dT_h -$

$dT_c = d(T_h - T_c)$. We can express the local heat exchanger flux $dq = U\,dA(T_h - T_c)$ according to the definition of the overall coefficient given in Eq. 7.4 and for simplification in manipulation call the bracketed term $[f(\dot{m})]$. Equation 7.11 then becomes

$$d(T_h - T_c) = -[f(\dot{m})]U\,dA(T_h - T_c) \tag{7.12}$$

Observing the conditions in assumptions (1), (2), and (3) we can separate variables and integrate over the length of the exchanger.

$$\int \frac{d(T_h - T_c)}{T_h - T_c} = -[f(\dot{m})]U\int dA \tag{7.13}$$

The boundary conditions for the integration of Eq. 7.13 over the length of the exchanger are:

at the entrance: $A = 0$ and $T_h - T_c = T_{h1} - T_{c1}$

at the exit: $A = A$ and $T_h - T_c = T_{h2} - T_{c2}$

The area of the exchanger is either the outside or the inside area of the inner tube, depending on which area the overall coefficient is based in Eq. 7.7. The integration can be performed easily by noting that the variable of integration is $\theta = T_h - T_c$. Accordingly, the $\int d\theta/\theta$ is equal to $\ln\theta$, where ln is the natural logarithm to the base e. Thus

$$\ln\left[\frac{T_{h2} - T_{c2}}{T_{h1} - T_{c1}}\right] = -[f(\dot{m})]UA \tag{7.14}$$

We are making progress! If we relate the mass flow term to the total heat exchanger flux we can then recast the result in the form of Eq. 7.8 and answer our original question. Applying the first law to the hot and cold fluids flowing through the entire exchanger we can write

$$q_h = \dot{m}_h c_{ph}(T_{h1} - T_{h2}) \tag{7.15}$$

$$q_c = \dot{m}_c c_{pc}(T_{c2} - T_{c1}) \tag{7.15a}$$

for the magnitude of the heat fluxes. We note that the heat exchanger heat flux $q_{HX} = q_h = q_c$. Solving Eqs. 7.15 and 7.15a for the reciprocals of the $\dot{m}c_p$ terms, adding the reciprocals, and using q_{HX} for q_h and q_c gives

$$[f(\dot{m})] = [(T_{h1} - T_{h2}) + (T_{c2} - T_{c1})]/q_{HX} \tag{7.16}$$

Substituting Eq. 7.16 into Eq. 7.14 and rearranging the result in terms of q_{HX} yields the form we seek. Thus

$$q_{HX} = UA\left\{\frac{(T_{h1} - T_{h2}) + (T_{c2} - T_{c1})}{-\ln[(T_{h2} - T_{c2})/(T_{h1} - T_{c1})]}\right\} \tag{7.17}$$

Comparing Eq. 7.17 with Eq. 7.8 we note that the bracketed term is the form we must use for the temperature difference when we use the overall coefficient U.

The bracketed term in Eq. 7.17 is called the log mean temperature difference, or LMTD in acronym terminology.

We can manipulate the LMTD term and put it into a simpler form that is easier to relate to the heat exchanger. If we regroup the numerator, we obtain

$$\text{LMTD} = \frac{(T_{h1} - T_{c1}) - (T_{h2} - T_{c2})}{\ln[(T_{h1} - T_{c1})/(T_{h2} - T_{c2})]} \tag{7.18}$$

where we have recognized that $\ln(\alpha/\beta) = -\ln(\beta/\alpha)$. Now, referring back to our sketch in Figure 7.2, we can see that $T_{h1} - T_{c1} = \Delta T_{\text{in}}$, the temperature difference at the inlet to the HX (acronym for heat exchanger) and $T_{h2} - T_{c2} = \Delta T_{\text{out}}$, the temperature difference at the exit of the HX. Accordingly, we can express the LMTD in these terms as

$$\text{LMTD} = \frac{\Delta T_{\text{in}} - \Delta T_{\text{out}}}{\ln(\Delta T_{\text{in}}/\Delta T_{\text{out}})} \tag{7.19}$$

We note and define the inlet to the HX as the location of the *inlet of the hot fluid*. For the parallel flow exchanger the definition is self-evident, but for the counterflow HX the inlet would be ambiguous as would be the case for crossflow exchangers. We will discuss crossflow exchangers in more detail in a later section. It is possible that in a counterflow HX that the inlet ΔT could be less than the exit ΔT, depending on the flow rates and thermal capacities of the fluids. Equation 7.19 still gives a positive value for LMTD, because the natural logarithm of a number less than 1 is negative. Accordingly, the LMTD is defined by some engineers in terms of the maximum and minimum ΔT's associated with the exchanger, so that

$$\text{LMTD} = \frac{\Delta T_{\text{max}} - \Delta T_{\text{min}}}{\ln(\Delta T_{\text{max}}/\Delta T_{\text{min}})} \tag{7.20}$$

Either form for LMTD can be used in calculating heat exchanger flux. The above concepts are illustrated in the following examples.

Example 7.1

A parallel flow HX has hot fluid entering at 250F and leaving at 150F while the cold fluid enters at 80F and leaves at 120F. Calculate the average temperature difference and the LMTD.

SOLUTION
We sketch the temperature profiles of the parallel flow HX so that we can visualize the temperature differences involved. From the figure we note:

$$\Delta T_{\text{max}} = 250F - 80F = 170F$$

$$\Delta T_{\text{min}} = 150F - 120F = 30F$$

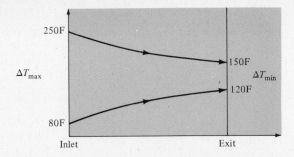

Figure for Example 7.1.

so that

$$\text{LMTD} = \frac{170 - 30}{\ln(170/30)}$$

$$\text{LMTD} = 80.7\text{F}$$

Considering average temperature

$$\bar{T}_h = (250\text{F} + 150\text{F})/2 = 200\text{F}$$

$$\bar{T}_c = (80\text{F} + 120\text{F})/2 = 100\text{F}$$

and

$$\overline{\Delta T} = \bar{T}_h - \bar{T}_c = 100\text{F}$$

or

$$\overline{\Delta T} = (170\text{F} + 30\text{F})/2 = 100\text{F}$$

The average temperature difference is considerably larger than the LMTD in this case. It is only when the ratio $\Delta T_{max}/\Delta T_{min} \leqslant 2$ that the average temperature difference approaches the LMTD.

Example 7.2

A counterflow heat exchanger has the hot fluid entering at 250F and leaving at 100F, while the cold fluid enters at 60F and leaves at 230F. Evaluate the LMTD for this HX and compare with the average temperature difference.

SOLUTION
The figure shows the temperature happenings in the HX. The figure shows that the maximum temperature difference is at the exit of the HX in this case and the minimum is at the entrance.

$$\Delta T_{max} = 100\text{F} - 60\text{F} = 40\text{F}$$

$$\Delta T_{min} = 250\text{F} - 230\text{F} = 20\text{F}$$

Therefore

$$\text{LMTD} = \frac{40 - 20}{\ln(40/20)} = 28.9\text{F}$$

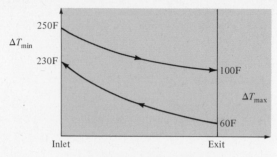

Figure for Example 7.2.

and the average temperature difference

$$\bar{T}_h = (250\text{F} + 100\text{F})/2 = 175\text{F}$$

$$\bar{T}_c = (60\text{F} + 230\text{F})/2 = 145\text{F}$$

and

$$\Delta\bar{T} = 175\text{F} - 145\text{F} = 30\text{F}$$

or

$$\Delta\bar{T} = (40\text{F} + 20\text{F})/2 = 30\text{F}$$

In this case we note that $\Delta T_{max}/\Delta T_{min} = 2$ and LMTD and $\Delta\bar{T}$ are within about 1 degree. Accordingly, when the ratio of maximum to minimum ΔT is less than 2, the average temperature difference is essentially the same as the LMTD. This concludes the two examples.

So far we have discussed only two types of simple heat exchangers; the concentric tube parallel and counterflow types. To gain some perspective we will look at several different types of exchangers, then consider an alternate method of analysis before concluding this chapter with some design considerations.

7.2 TYPES OF HEAT EXCHANGERS

There are basically two general categories of heat exchangers; one in which the flow directions are parallel (either in the same direction or in opposite directions) and the other in which the flow directions are at an angle near 90 degrees, called crossflow exchangers. In the latter category, there are numerous variations depending on the manufacturer. Kays and London (1) give the most extensive compilation of compact crossflow type of heat exchangers in the literature at the present time.

7.2.1 Concentric Tube Exchangers

This simple type of exchanger was shown in Figure 7.2 and described in the analysis leading to the concept of the LMTD. Flow of the fluids can either be in the same direction (parallel flow) or in opposite directions (counterflow), depending on the desired exit temperatures. Various combinations can be used in series or in parallel circuits.

7.2.2 Crossflow Heat Exchangers

Crossflow heat exchangers are more economical in space utilization than the concentric tube type. In crossflow exchangers the fluid velocity directions are generally at right angles to each other, and the flow can either be mixed or unmixed. By mixed flow we mean that no baffles exist to prevent the fluid from mixing within itself due to the existing turbulent fluctuations. Figures 7.3(a) and

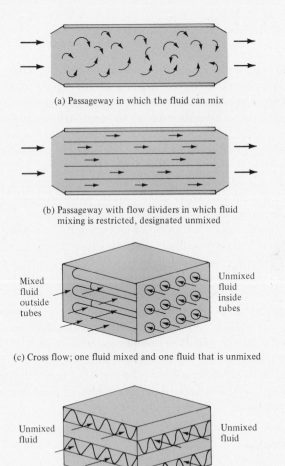

(a) Passageway in which the fluid can mix

(b) Passageway with flow dividers in which fluid mixing is restricted, designated unmixed

Mixed fluid outside tubes

Unmixed fluid inside tubes

(c) Cross flow; one fluid mixed and one fluid that is unmixed

Unmixed fluid

Unmixed fluid

(d) Cross flow; both of the fluids are unmixed

Figure 7.3 Concept of mixed and unmixed fluid flow in crossflow heat exchangers.

7.3(b) compare the aspects of mixed and unmixed flow. Aircraft oil coolers are an example of an application of crossflow heat exchangers that require minimum space configurations.

There are several variations in use according to whether the fluids are mixed or unmixed. Figure 7.3(c) shows the orientation for one fluid mixed and one fluid unmixed, while Figure 7.3(d) shows a configuration where both fluids are unmixed. Generally speaking, the mixed fluid case has the least pressure drop associated with the flow, because there is less surface for fluid shear to take place. But less surface is not as good from the heat transfer standpoint. Exchangers of the type shown in Figure 7.3(d) are very efficient in transferring energy, because of the large amount of surface area in a compact volume.

7.2.3 Shell and Tube Heat Exchangers

When large volumes of fluid are to be handled such as in the chemical process industry, the food industry, or the petroleum industry, aspects of pumping power (pressure drop) are more important than space considerations. It is desirable to keep equipment size reasonable from the standpoint of capital costs. The shell and tube exchangers evolved to meet such requirements. The fluid flowing outside of the tubes (see Figure 7.4) is mixed and the baffles are installed to promote mixing in addition to keeping the velocity across the tubes high and nearly at right angles. Figure 7.4(a) shows an arrangement with one pass for the fluid inside the tubes through the exchanger and one pass on the shell side as compared to Figure 7.4(b) where there are two passes through the exchanger on the tube side. Many other configurations can be achieved by combining and modifying the types shown in Figure 7.4 and using them in series or parallel flow circuits. Figure 7.5 shows an illustration of a typical heat exchanger currently being manufactured for industrial use.

The shell and tube units sketched in Figure 7.4 have plate baffles to direct the flow on the shell side. Recent developments by Small and Gentry (8) in regard to replacing the plates with rods inserted between the tubes have shown a decrease in pressure drop and an increase in heat transfer over the plate baffles for a given tube bundle and flow conditions. Such units have been called rod baffled heat exchangers in contrast to plate baffle types.

Our purpose is to understand the fundamental operating principles and to become familiar with the more common types of heat exchangers used in engineering practice.

7.2.4 Other Types

We conclude our brief description of heat exchanger types with a mention of other configurations in use. The heat pipe exchanger utilizes a phase change principle that we will discuss in Chapter 9. At this point we can describe the heat pipe simply as a closed tube with a particular fluid inside, whose saturation temperature is in the range of temperature to be expected in the exchanger. The

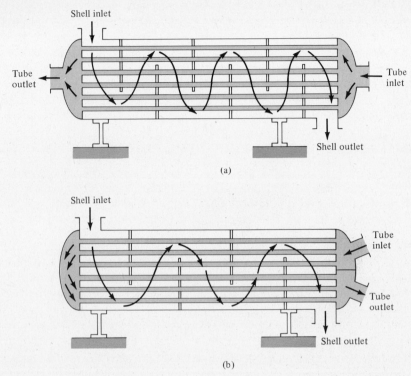

Figure 7.4   Examples of shell and tube heat exchangers. (a) One shell pass; one-tube pass. (b) One shell pass; two-tube pass.

fluid is vaporized at the hot end of the pipe and condensed at the cooled end of the pipe. The condensed liquid is returned to the hot end (inside the pipe) by either capillary action or gravity or both.

Regenerative heat exchangers in fossil electric power plants are massive drums that store energy from hot stack gases in the metal vanes of the drum and are rotated slowly into the path of incoming cooler combustion air, thereby transferring the energy to the air and heating it.

Other types of heat exchangers that are perhaps more well known are the finned tube cores that are employed in automobile radiators. The name radiator is, of course, a misnomer; convector would be more exact terminology, since very little energy is exchanged by radiation from an automotive radiator. Finned tube exchangers are also used in industrial space heaters and baseboard heaters for residential applications.

We now look at applying the LMTD method to crossflow exchangers by introducing a correction factor for use with Eq. 7.17; then after a brief discussion of fouling factors we will continue our analysis with the effectiveness–NTU* method.

*NTU = number of transfer units.

Figure 7.5 Tube bundle of a large shell and tube heat exchanger for use in a Sulfino processing plant, 3553 tubes $\frac{3}{4}$ in. OD and 20 ft long. Photo courtesy of the Shell and Tub Company, Tulsa, Oklahoma.

7.2.5 Temperature Difference for Crossflow HX

The concept of the log mean temperature difference developed for the concentric tube heat exchanger was relatively simple to formulate. For crossflow HX the mathematical procedure becomes more difficult and complicated. Consequently, a simple correction procedure applied to the LMTD as it would be calculated for a counterflow heat exchanger has been adopted wherein the heat exchanger flux is written as

$$q = UA_o(F \cdot \text{LMTD}) \tag{7.21}$$

The correction factor F is obtained from charts like those shown in Figure 7.6 for crossflow heat exchangers with fluids mixed or unmixed, and the area A_o is the outside area of the tubes (Reference 2). There are many different variations of fluid flow arrangements in crossflow and in shell and tube heat exchangers and each arrangement requires a particular correlation (graph) to determine F. Figure 7.6 is an example of how the method is applied to three cases; more extensive tables and charts are given in References 2 and 3.

Example 7.3

A crossflow heat exchanger having both fluids unmixed is to be used for heating 0.1 lb_m/s water from 50 to 180F by hot exhaust gases at 430F that are cooled to 220F. The exchanger has a $U_o = 20$ Btu/hrft^{2}F based on the outside tube area under such service, and we want to know how many square feet of heating surface are required.

SOLUTION
We diagram the heat exchanger as a counterflow HX with the desired temperatures, evaluate the LMTD, obtain F, and calculate the area from Eq. 7.21. Thus

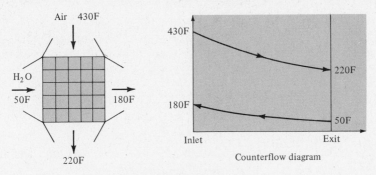

Figure for Example 7.3

From the counterflow diagram we note
$$\Delta T_{\text{max}} = 430 - 180 = 250\text{F}$$
$$\Delta T_{\text{min}} = 220 - 50 = 170\text{F}$$

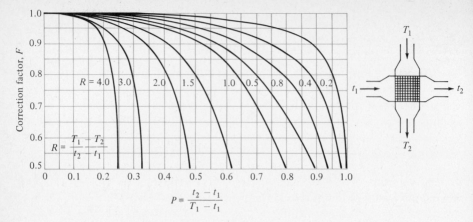

(a) Cross flow HX; both fluids unmixed

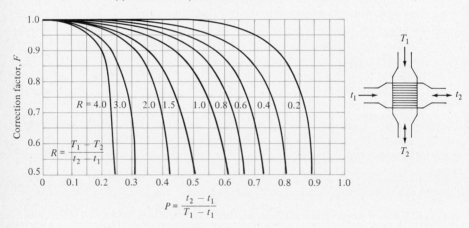

(b) Cross flow HX; one fluid mixed, one unmixed

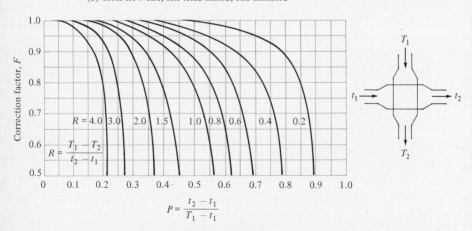

(c) Cross flow HX; both fluids mixed

Figure 7.6 Correction factor F for crossflow heat exchangers to obtain mean temperature difference from LMTD calculated for counterflow conditions. Reproduced from Reference 3 with the permission of the American Society of Mechanical Engineers.

Therefore

$$\text{LMTD} = \frac{250 - 170}{\ln(250/170)}$$

$$= 207.5\text{F}$$

Evaluating the parameters R and P,

$$R = \frac{430 - 220}{180 - 50} = 1.6 \quad \text{and} \quad P = \frac{180 - 50}{430 - 50} = 0.34$$

so that we read $F = 0.93$ from Figure 7.6(a). The heat exchanger flux is determined by the enthalpy rise of the water

$$q_{\text{HX}} = \dot{m}c_p(T_{co} - T_{ci})$$

$$= 360(1)(180 - 50) = 46,800 \text{ Btu/hr}$$

so that the area becomes

$$A_o = \frac{q}{U_o(F \cdot \text{LMTD})}$$

$$= \frac{46,800}{20(0.93)207(0.5)}$$

$$= 12.1 \text{ ft}^2$$

We would use about 12.5 ft^2 to account for the uncertainty in the correction factor.

7.3 FOULING FACTORS

During the course of operation of heat exchangers, corrosion and/or scale deposits add an additional resistance to the heat transfer surface. The result is a decrease in the performance or effectiveness of the heat exchanger. The amount of surface fouling that occurs depends on the type of fluid, the velocities involved, and the length of service.

For tubular heat exchangers TEMA (2) has recommended basing the overall coefficient on the outside area of the tube. We would, therefore, write Eq. 7.7 for such an application as

$$U_o A_o = \frac{1}{(1/h_i A_i) + R_k + (1/h_o A_o)} \tag{7.22}$$

for the clean condition. The wall resistance for a tube we recall from Chapter 3 as $R_k = \ln(r_o/r_i)/2\pi Lk$. The overall coefficient, based on the outside area, is then given by

$$U_{o(\text{clean})} = \frac{1}{(A_o/h_i A_i) + A_o R_k + (1/h_o)} \tag{7.23}$$

again for the clean condition. When the exchanger surface becomes fouled by dirt and scale an additional resistance term $\mathcal{R}_F = \mathcal{R}_i + \mathcal{R}_o$, the sum of internal

Table 7.1 REPRESENTATIVE FOULING RESISTANCES, $\mathcal{R}_i$ OR $\mathcal{R}_o$, SELECTED FROM REFERENCE 2

FLUID	RESISTANCE, $\mathcal{R}$ (hrFft^2/Btu)
Seawater	0.0005
Well water	0.001
Hard water	0.003
Mississippi River water	0.003
Chicago Sanitary Canal	0.008
Distilled water	0.0005
Transformer oil	0.001
Fuel oil	0.005
Quench oil	0.004
Exhaust steam	0.001
Compressed air	0.002
Refrigerant liquids	0.001
Hydraulic fluid	0.001
Heavy fuel oils	0.005
Asphalt and residuum	0.010
Kerosene	0.001
Natural gas	0.001
Vegetable oils	0.003
Engine exhaust gas	0.010
Manufacturer's gas	0.010

and external (to the tube) fouling resistances, must be added to the denominator to account for the decrease in overall coefficient. (A representative sample of values for $\mathcal{R}_i$ and $\mathcal{R}_o$ is given in Table 7.1.) The decreased overall coefficient in the fouled condition is then given by

$$U_{o(\text{fouled})} = \frac{1}{(A_o/h_i A_i) + A_o R_k + \mathcal{R}_F + (1/h_o)} \tag{7.24}$$

or by combining Eqs. 7.23 and 7.24 we can express the fouling resistance as

$$\mathcal{R}_F = \frac{1}{U_{o(\text{fouled})}} - \frac{1}{U_{o(\text{clean})}} \tag{7.25}$$

In the above expressions we should be careful to distinguish between R_k, hrF/Btu, K/W, and $\mathcal{R}_F$, hrFft^2/Btu, Km^2/W. Since the overall coefficient is based on the outside area of the tube, all other resistances must also be based on the outside area of the tube. For example, since $A_o = \pi D_o L N$, where N is the number of tubes, the unit area wall resistance would be given by

$$\mathcal{R}_k = A_o R_k = \frac{D_o N \ln(r_o/r_i)}{2k} \tag{7.26}$$

and

$$\mathcal{R}_i = \frac{A_o R_i}{A_i} \tag{7.27}$$

$$\mathcal{R}_o = A_o R_o \tag{7.27a}$$

It is possible to gain experience in HX performance by selectively cleaning a heat exchanger and using Eqs. 7.25 and 7.27 to evaluate either $\mathcal{R}_i$ or $\mathcal{R}_o$, or both. Example 7.4 illustrates the above concepts.

Example 7.4

In Example 7.3 the cooling water is taken from the Chicago Sanitary Canal. We would like to estimate the value of the overall coefficient for the HX after it has been in operation for some time and to obtain an indication of the effect on HX performance at the same mass flow rates.

SOLUTION
From Table 7.1 we obtain the fouling resistances $\mathcal{R} = 0.008$ hrFft^2/Btu for Chicago Sanitary Canal water and $\mathcal{R} = 0.010$ hrFft^2/Btu for exhaust gas. Therefore

$$\mathcal{R}_F = 0.010 + 0.008 = 0.018 \text{ hrFft}^2/\text{Btu}$$

and from Eq. 7.25 we can estimate the overall coefficient in the fouled condition, so that

$$\frac{1}{U_{o(\text{fouled})}} = \mathcal{R}_F + \frac{1}{U_{o(\text{clean})}}$$

$$= 0.018 + \tfrac{1}{20}$$

$$= 0.068 \text{ hrft}^2\text{F}/\text{Btu}$$

and

$$U_{o(\text{fouled})} = 14.7 \text{ Btu}/\text{hrft}^2\text{F}$$

a 26.5 percent reduction in the overall coefficient. We recognize that the inlet temperatures to the exchanger will not vary, but the exit temperatures will change. The exit water temperature will decrease and the exit gas temperature will increase, because less energy is exchanged. Performance prediction is not easy; we would have to guesstimate an exit temperature; calculate the other exit temperature from first law; determine the corresponding LMTD and F; evaluate the HX flux; and then check our guesstimate with the first law, adjust, and repeat probably several times to get agreement. Such a procedure is tedious and time consuming and can be avoided by using the effectiveness–NTU concept discussed in the next section. This concludes the example and leads us into the next topic.

7.4 EFFECTIVENESS–NTU METHOD OF ANALYSIS

In the development of the LMTD approach we obtained a result that involved all four inlet and exit temperatures of the heat exchanger. A knowledge of LMTD and the heat exchanger flux lets us determine the surface area required when the overall coefficient is known or can be predicted. As we have seen in

Example 7.4, when the exit temperatures are not known, it is a tedious procedure to predict performance with the LMTD method from the overall coefficient and the known area. The effectiveness–NTU method presents a more direct approach.

7.4.1 Definitions

We define the effectiveness $\mathcal{E}$ of a heat exchanger as the ratio of the actual heat transferred to the maximum possible heat transfer, if the fluid with the minimum thermal capacitance were to go through the maximum temperature difference in the exchanger. Thus

$$\mathcal{E} = \frac{q_{HX}}{q_{\text{max possible}}} \tag{7.28}$$

The exchanger heat flux is equal to the enthalpy change of either of the fluids going through the exchanger. For liquids there is generally negligible change in kinetic energy, but for gases the change in kinetic energy may be significant for large flow rates and large changes in temperature. Generally changes in kinetic and potential energy are neglected. The first law gives for such an assumption

$$q_{HX} = \dot{m}_h c_{ph}(T_{h1} - T_{h2}) \tag{7.29}$$

$$q_{HX} = \dot{m}_c c_{pc}(T_{c2} - T_{c1}) \tag{7.29a}$$

The maximum possible heat flux must be the enthalpy change of the fluid with the smallest thermal capacitance in passing through the maximum temperature difference in the heat exchanger as dictated by the first law. Accordingly

$$q_{\text{max possible}} = \left(\dot{m}c_p\right)_{\text{min}}(T_{h1} - T_{c1}) \tag{7.30}$$

Either the cold fluid or the hot fluid may have the minimum capacitance; that depends on the values of flow rate $\dot{m}$, and the specific heat c_p.

At this point it is convenient to make a slight change in nomenclature. We will be dealing with parallel flow (PF), counterflow (CF), and crossflow heat exchangers, and the change in nomenclature will minimize the number of equations that we will have to deal with. We have designated the inlet temperatures in PF as T_{h1} and T_{c1}, so that for CF we would have to designate the inlet for the cold fluid as T_{c2} which would be confusing with the outlet designation of the hot fluid as T_{h2}. We recall that the *inlet* to the exchanger is always at the inlet of the hot fluid, and crossflow exchangers are analyzed on the basis of counterflow nomenclature. Accordingly, we will use the subscript i for inlet and e for exit as we show in Figure 7.7.

The effectiveness for a PF exchanger with the hot fluid having minimum capacitance is then given by

$$\mathcal{E}_{HFM} = \frac{q_{HX}}{q_{\text{max}}} = \frac{(\dot{m}_h c_{ph})(T_{hi} - T_{he})}{(\dot{m}_h c_{ph})(T_{hi} - T_{ci})} = \frac{T_{hi} - T_{he}}{T_{hi} - T_{ci}} \tag{7.31}$$

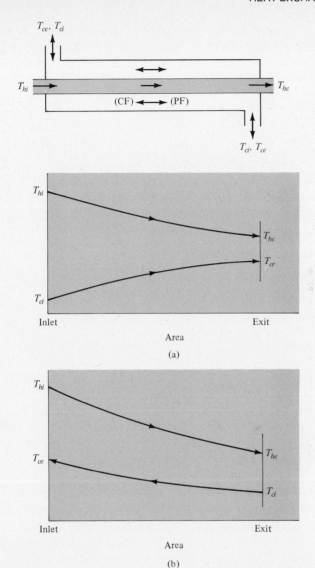

Figure 7.7   Inlet and exit temperature designations for parallel and counterflow, concentric tube heat exchangers as employed in the effectiveness definition. (a) Parallel flow (PF) temperature profiles. (b) Counterflow (CF) temperature profiles.

For the counterflow exchanger, when the hot fluid is the minimum capacitance fluid, reference to Figure 7.7 shows that

$$\mathcal{E}_{\text{HFM}} = \frac{q_{\text{HX}}}{q_{\text{max}}} = \frac{T_{hi} - T_{he}}{T_{hi} - T_{ci}} \tag{7.32}$$

which is exactly the same expression we obtained above. For the case where the cold fluid is the minimum capacitance we get a similar result for both parallel and counterflow exchangers. Thus

$$\mathcal{E}_{\text{CFM}} = \frac{T_{ce} - T_{ci}}{T_{hi} - T_{ci}} \tag{7.33}$$

$$\mathcal{E}_{\text{HFM}} = \frac{T_{hi} - T_{he}}{T_{hi} - T_{ci}} \tag{7.34}$$

The subscripts CFM stand for cold fluid minimum, and HFM stand for hot fluid minimum. Most importantly we note that each expression for $\mathcal{E}$ contains only three temperatures, two of which are generally known (inlet of both fluids)! Now if we can obtain an expression for $\mathcal{E}$ in terms of U, A, and the flow rates, we can then solve for the outlet temperatures; that is the thrust of the $\mathcal{E}$–NTU method.

7.4.2 Formulation of the Method

After integration of Eq. 7.12 to obtain Eq. 7.14 we take a slightly different interpretation of the mass flow term $[f(\dot{m})]$. From our definition in Eq. 7.11 we recall that

$$[f(\dot{m})] = \left[\frac{1}{\dot{m}_c c_{pc}} + \frac{1}{\dot{m}_h c_{ph}} \right] \tag{7.35}$$

for the parallel flow HX. It is vital to note that the analysis to follow is specifically for the PF case. The $\dot{m}c_p$ products are the thermal capacitances of the hot and cold fluids; we designate $C = \dot{m}c_p$ either in Btu/hrF or W/K. One of the capacitance terms will be smaller than the other, so we can write Eq. 7.35 in terms of C_{min} and C_{max} as

$$[f(\dot{m})] = \left[\frac{1}{C_{\text{min}}} + \frac{1}{C_{\text{max}}} \right] \tag{7.36}$$

We define Z as the ratio of minimum to maximum capacitance. Thus

$$Z = \frac{C_{\text{min}}}{C_{\text{max}}} \tag{7.37}$$

so that the mass flow function can be written as

$$[f(\dot{m})] = \frac{1 + Z}{C_{\text{min}}} \tag{7.38}$$

We substitute Eq. 7.38 into Eq. 7.14 to proceed with the analysis,

$$\ln\left[\frac{T_{he} - T_{ce}}{T_{hi} - T_{ci}} \right] = -\left[\frac{UA}{C_{\text{min}}} \right](1 + Z) \tag{7.39}$$

With the assumptions utilized in obtaining Eq. 7.14 (Eq. 7.39) we recognize that the term $[UA/C_{\text{min}}]$ will be a constant for the HX operating under steady flow conditions. We define NTU $= UA/C_{\text{min}}$, the number of transfer units; the dimensionless term NTU is indicative of the physical and thermal size of the exchanger. Note the temperature ratio on the left side of Eq. 7.39 is almost like an effectiveness. The denominator is correct, but the numerator is not, and requires some manipulation. First, however, we take the antilogarithm of Eq. 7.39 to obtain

$$\frac{T_{he} - T_{ce}}{T_{hi} - T_{ci}} = \exp[-\text{NTU}(1 + Z)] \tag{7.40}$$

Substitution for the temperature ratio can be achieved by first considering the first law for the exchanger; from Eqs. 7.15 and 7.15a and the definition of Z we can write for $Z = C_c/C_h$, the cold fluid minimum

$$Z(T_{ce} - T_{ci}) = T_{hi} - T_{he} \tag{7.41}$$

Now by expanding Eq. 7.41, adding T_{ce} and $-T_{ci}$ to both sides, and rearranging,

$$T_{he} - T_{ce} = (T_{hi} - T_{ci}) + (1 + Z)(T_{ci} - T_{ce}) \tag{7.42}$$

Dividing through by $T_{hi} - T_{ci}$ gives the temperature ratio on the left side of Eq. 7.40 that we seek; therefore

$$\frac{T_{he} - T_{ce}}{T_{hi} - T_{ci}} = 1 + (1 + Z)\left[\frac{T_{ci} - T_{ce}}{T_{hi} - T_{ci}}\right] \tag{7.43}$$

where we recognize the bracketed term on the right as $-\mathcal{E}_{\text{CFM}}$. Thus

$$\frac{T_{he} - T_{ce}}{T_{hi} - T_{ci}} = 1 - \mathcal{E}_{\text{CFM}}(1 + Z) \tag{7.44}$$

and finally substituting back into Eq. 7.40 and solving for the effectiveness term give

$$\mathcal{E}_{\text{CFM}} = \frac{1 - \exp[-\text{NTU}(1 + Z)]}{1 + Z} \tag{7.45}$$

The analysis just completed was for the parallel flow HX and in our evaluation of Eq. 7.36 we did not distinguish which fluid was minimum capacitance, so that we could just as well have rearranged Eq. 7.41 to give us $\mathcal{E}_{\text{HFM}}$. Our result then would be exactly the same as Eq. 7.45.

For the counterflow HX we would be required to start with an analysis similar to the original LMTD analysis, and obtain an expression like Eq. 7.14, and proceed like we have done above. In this instance, however, the algebra is much more involved and cumbersome, but we would obtain a single expression for $\mathcal{E}$ as we did for the parallel flow case. The results can be summarized as follows:

$$\mathcal{E}_{\text{PF}} = \frac{1 - \exp[-\text{NTU}(1 + Z)]}{(1 + Z)} \tag{7.45}$$
$$\text{(Repeat)}$$

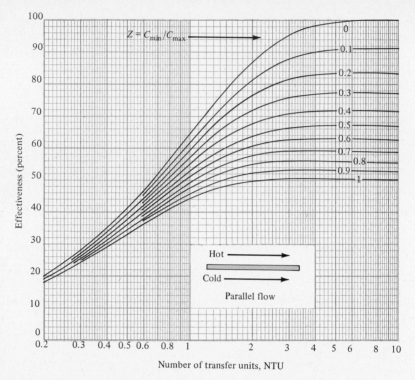

Figure 7.8   Effectiveness as a function of NTU for the simple concentric tube heat exchanger in parallel flow.

for the parallel flow concentric tube HX and

$$\mathscr{E}_{CF} = \frac{1 - \exp[NTU(1 - Z)]}{Z - \exp[NTU(1 - Z)]}$$
(7.46)

for the counterflow concentric tube HX.

A close examination of $\mathscr{E}_{CF}$ shows that for equal capacitances ($Z = 1$) the effectiveness is indeterminate as given by Eq. 7.46. That case is treated in one of the home problems. Some typical values of Eqs. 7.45 and 7.46 have been plotted in Figures 7.8 and 7.9, which can be conveniently read to two significant figures for use in applications.

Example 7.5

The heat exchanger described in Example 7.1 is used to recover waste heat from the hot fluid. Using the effectiveness–NTU method, estimate the size (square feet) of the heat exchanger required to heat 100 gpm of water with the temperatures given if the HX has an overall coefficient $U = 35$ Btu/hrft²F.

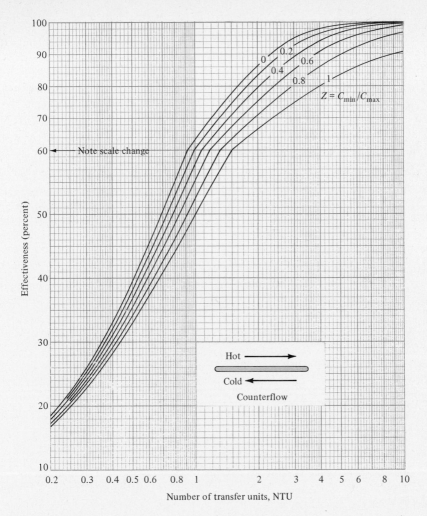

Figure 7.9   Effectiveness as a function of NTU for the simple concentric tube heat exchanger in counterflow.

SOLUTION

The temperature change of the fluids shows which fluid has the minimum capacitance. The hot fluid drops 100F and the cold fluid increases 40F so that an energy balance on the HX shows

$$C_h(T_{hi} - T_{he}) = C_c(T_{ce} - T_{ci})$$
$$C_h(100) = C_c(40)$$

so that for $Z = C_{min}/C_{max}$ in this case the hot fluid must be the minimum capacitance fluid for Z to be less than 1; therefore $Z = 0.4$. The effectiveness for the exchanger with hot fluid minimum is given by Eq. 7.34

$$\mathcal{E}_{\text{HFM}} = \frac{T_{hi} - T_{he}}{T_{hi} - T_{ci}}$$

and referring back to the figure in Example 7.1 which shows the temperatures, we obtain

$$\mathcal{E}_{\text{HFM}} = \frac{250 - 150}{250 - 80}$$

$$= 0.588 \quad \text{or} \quad 58.8 \text{ percent}$$

For the parallel flow exchanger we enter Figure 7.8 and read NTU = 1.24 at $Z = 0.4$ and $\mathcal{E}_{\text{HFM}} = 58.8$ percent. The water stream has the maximum capacitance and we can determine the minimum capacitance from Z. Therefore

$$C_{max} = 100(62.4)\frac{231}{1728}60(1)$$

$$= 50{,}050 \text{ Btu/hrF}$$

where we have used the conversion factor of 231 in.^{3}/gal to obtain mass of water per gallon. Then

$$C_{min} = ZC_{max}$$

$$= 0.4(50{,}050)$$

$$= 20{,}020 \text{ Btu/hrF}$$

and from the definition of NTU = UA/C_{min} we calculate the area as

$$A = C_{min}(\text{NTU})/U$$

$$= 20{,}020(1.24)/35$$

$$= 709.3 \text{ ft}^2$$

The NTU method must be consistent with the LMTD method of analysis so we can check our result by using the value of LMTD calculated in Example 7.1. The heat flux for the exchanger is

$$q_{\text{HX}} = \dot{m}_h c_{ph}(T_{hi} - T_{he})$$

$$= 20{,}020(250 - 150)$$

$$= 2{,}002{,}000 \text{ Btu/hr}$$

There is a slight discrepancy because only two significant figures can be read from Figure 7.8; the third figure is estimated. This concludes the example.

7.4.3 Effectiveness for Crossflow HX

For the crossflow exchanger the analysis is much more complicated than we have just performed for the simple concentric tube exchangers. Stevens (4) and co-authors present the results of several analyses and tables of results based on numerical analysis of the unmixed flow exchanger. We will briefly discuss the simple one-pass crossflow exchanger as an example of this type. Figure 7.10

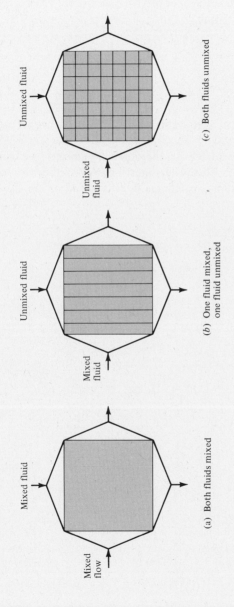

(a) Both fluids mixed

(b) One fluid mixed, one fluid unmixed

(c) Both fluids unmixed

Figure 7.10   Flow variations for the simple, one-pass, crossflow heat exchanger.

illustrates the three types of flow variations for the simple one-pass crossflow exchanger.

For the case where both fluids are mixed (case (a), Figure 7.10) the effectiveness is given by

$$\mathcal{E}_{(a)} = \frac{NTU}{(NTU/A) + (ZNTU/B) - 1} \tag{7.47}$$

where

$$A = 1 - \exp(-NTU) \tag{7.47a}$$

$$B = 1 - \exp(-ZNTU) \tag{7.47b}$$

The quantities $Z = C_{min}/C_{max}$ and $NTU = UA/C_{min}$ are as we have defined before.

When one fluid is mixed and the other is unmixed (see case (b) in Figure 7.10) the expression for effectiveness must be expressed in two forms depending on whether the minimum capacitance fluid is mixed or unmixed. Thus, for the *minimum fluid mixed*

$$Z = \frac{C_{min}}{C_{max}} = \frac{C_{mixed}}{C_{unmixed}}$$

$$\mathcal{E}_{(b)} = 1 - \exp\{-[1 - \exp(-ZNTU)]/Z\} \tag{7.48}$$

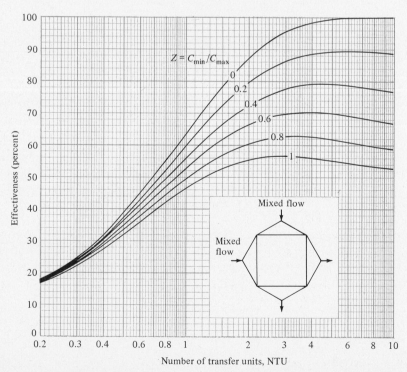

Figure 7.11 Effectiveness as a function of NTU for the crossflow heat exchanger with both fluids mixed.

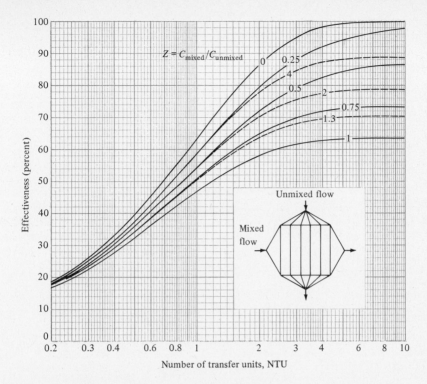

Figure 7.12 Effectiveness as a function of NTU for the crossflow heat exchanger with the minimum capacitance fluid mixed.

and when the *minimum fluid is unmixed*

$$Z = \frac{C_{min}}{C_{max}} = \frac{C_{unmixed}}{C_{mixed}}$$

$$\mathcal{E}_{(b)} = \left[1 - \exp\{ -Z[1 - \exp(-NTU)]\} \right] / Z \qquad (7.49)$$

where NTU is as before. In both expressions Z must be evaluated as stated according to whether the minimum capacitance fluid is mixed or unmixed. Equations 7.47, 7.48, and 7.49 are plotted in Figures 7.11 and 7.12, respectively, for ease in application.

When both fluids are unmixed, the effectiveness cannot be given in simple closed form. A series solution has been worked out by Mason (5) and evaluated numerically by Stevens (4); the results are plotted in Figure 7.13. The form for effectiveness $\mathcal{E}$ depends on whether the hot or the cold fluid is the minimum capacitance fluid as was shown in Eqs. 7.33 and 7.34.

Example 7.6

In Example 7.4 we noted how tedious it would be to estimate the performance of the crossflow heat exchanger under the estimated fouling conditions. The effectiveness–NTU method is a straightforward approach to such situations;

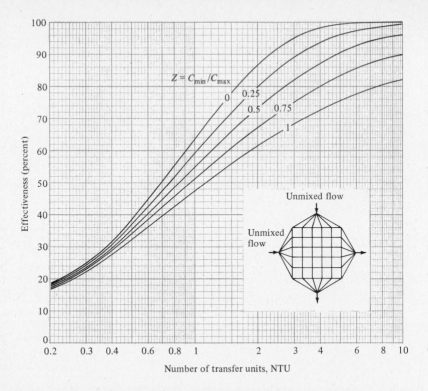

Figure 7.13 Effectiveness as a function of NTU for the crossflow heat exchanger with both fluids unmixed.

therefore, we would like to predict the fluid exit temperatures under the fouled condition where $U_o = 14.7$ Btu/hrft^{2}F.

SOLUTION

We must determine the capacitance ratio Z and the number of transfer units NTU so that we can determine the exchanger effectiveness from Figure 7.13. Accordingly, the energy balance (first law) for the exchanger requires

$$C_{H_2O}(180F - 50F) = C_{gas}(430F - 220F)$$

in the unfouled condition. Since we are considering the same mass flows in the fouled as in the clean condition, the capacitance ratio Z will not change for constant properties. We see that

$$130F\ C_{H_2O} = 210F\ C_{gas}$$

so that the exhaust gas must be the minimum capacitance fluid and

$$Z = \frac{C_{min}}{C_{max}} = \frac{C_{gas}}{C_{H_2O}} = \frac{130}{210} = 0.619$$

The area of the HX is 12.1 ft^2 and we can obtain $C_{min} = ZC_{max} = ZC_{H_2O}$ from

Example 7.3; therefore

$$C_{min} = 0.619(360)(1) = 223 \text{ Btu/hrF}$$

Then the number of transfer units becomes

$$\text{NTU} = \frac{UA}{C_{min}} = \frac{14.7(12.1)}{223}$$

$$= 0.798$$

Entering Figure 7.13 we read $\mathcal{E} = 0.473$ from which we can determine the hot fluid outlet temperature with the aid of Eq. 7.34:

$$\mathcal{E} = \frac{T_{hi} - T_{he}}{T_{hi} - T_{ci}} = 0.473$$

or

$$430\text{F} - T_{he} = 0.473(430\text{F} - 50\text{F})$$

$$T_{he} = 250\text{F}$$

and since $T_{ce} - T_{ci} = Z(T_{hi} - T_{he})$ we can determine the exit cold fluid temperature as

$$T_{ce} - 50\text{F} = 0.619(430 - 250)$$

$$T_{ce} = 161\text{F}$$

Accordingly, we note that the fouling deposits have decreased the exit water temperature from 180 to 161F and increased the exit gas temperature from 220 to 250F when the mass flows of both streams are the same as they were in the clean condition. The pressure drops through the exchanger are higher when the surfaces become coated with deposits; the amount of increase depends on the nature of the deposit. More pumping power (costs) is required to maintain the same flow or alternatively even more pumping is required to increase the overall coefficient to maintain performance. This completes the example.

7.5 HEAT EXCHANGER RATING AND DESIGN

In the above sections our discussion has centered about the analysis of an existing heat exchanger; such a process is generally called rating the equipment as to its performance. Such a determination or prediction of performance (temperature and heat flux or heat duty) can be made when the operating flow rates and inlet temperatures are known or specified. It may be worthwhile to recapitulate the most important assumptions made in the rating analysis.

1. The overall coefficient U is a constant and not a function of position or temperature.
2. Physical properties of the fluids are constants and not functions of temperature.
3. Steady flow conditions exist for both fluids flowing through the exchanger.

From the information we studied and discussed in Chapter 6, it is apparent that the first assumption is the weakest of the three. The overall coefficient U represents an average of the conditions that exist in the exchanger. For short exchangers where there is a considerable portion of the flow passage in the entrance region, there may be large variations in U along the length. For liquids the constant property assumption is reasonable for specific heat and thermal conductivity; viscosity, however, is very temperature sensitive and contributes to the variation of heat transfer coefficients along the length. For gases, properties can vary if the change in temperature is very large, especially for substances like carbon dioxide.

It is possible to make a pseudovariable properties analysis of a heat exchanger by applying the above three assumptions to a very short section, ΔA, of the HX and calculating the outlet temperature at the end of the short section for use as the inlet temperature of the next ΔA, reevaluating all parameters at the new temperature, and so stepping along the exchanger. Such a procedure can be programmed for computer application to any degree of refinement that is economically practical for use with a given class of heat exchanger. Such programs, when combined with empirical experience factors, can also be useful tools in the design of new heat exchangers.

We will conclude our simple overview of heat exchangers with a brief comment on the considerations necessary when a new piece of equipment is to be designed and constructed for a given specific purpose.

7.5.1 Approach to Design

It is essential in an introductory text to include the topic of design by virtue of its shear importance even though we cannot make a pretense at completeness of presentation. Our purpose is to convey some sense of the enormity of the task.

The most important aspects of design are summarized by Kern (6) in a book that has long been the standard in the field. The aspects of heat exchanger design considerations and techniques have been summarized by Taborek (7) in an extensively detailed article. He identifies 10 steps to be considered after a preliminary subjective choice of physical arrangements has been made. They are:

1. Refinement of overall design geometry wherein actual dimensions are assigned as much as possible.
2. Determination of thermal resistances from selected or mandated flow rates. The possibility of switching fluids (shell side or tube side) allocation can be finalized at this point.
3. Ascertain the amount of overdesign desired or required; usually held to less than 10 percent.
4. Check pressure drop utilization to make sure most pressure drop is consumed in areas of highest heat transfer.
5. Analyze flow velocities to make sure velocities are high enough to minimize fouling, but not large enough to cause erosion, vibration, or excessive pressure drop.

6. Analyze the shell-side flow distribution with regard to crossflow, bypass streams, and leakage streams (tube to hole and baffle to shell).
7. Detailed analysis of baffle design, spacing, and baffle flow patterns.
8. Check out factors that modify LMTD to obtain effective temperature difference.
9. Review heat transfer coefficients and recheck resistances.
10. Final check on tube vibration analysis to verify the integrity of the tube bank.

The above short summary of Taborek's procedure should give the reader an idea of the magnitude of the design problem. Most manufacturers combine analysis with engineering experience and judgment into a detailed computer routine based on the above ideas in order to start the design process. With computer flexibility, design changes can easily be explored in order to determine the best arrangement for a particular set of specifications.

7.6 CLOSURE

Chapters 5, 6, and 7 have covered convection phenomena for a single-component substance without change of phase. Fundamentally, mixtures of substances can be treated with the same correlation equations as dictated by the geometrical arrangement as single substances, the only difficulty being the estimation or determination of the transport properties of the mixture. Reid and Sherwood (9) give some methods for determining properties of mixtures that are based on kinetic theory. Specific heat and density for gases can be determined from the properties of ideal gas mixtures (Reference 10) if the temperature and pressure are such that ideal gas behavior can be assumed; liquid mixtures are difficult to treat.

We will consider some basic aspects of heat transfer with phase change, boiling, condensing, freezing, and melting after we consider radiation phenomena in the next chapter.

PROBLEMS

1. A plane steel tank wall $\frac{1}{4}$ in. thick with $k = 25$ Btu/hrftF has air at 20F blowing on one side such that $h = 28$ Btu/hrft^{2}F on the surface, and hot oil at 130F on the other side with an $h = 10$ Btu/hrft^{2}F due to free convection. Determine the overall coefficient and the unit area heat flux through the tank wall.
2. Hot water at a bulk temperature of 200F flows inside a 1-in. ID red brass tube that has a 1.5875-mm wall thickness. The water flows at such a rate that the bulk Reynolds number is 56,500. The 1-in. tube is concentric with a 2-in. ID tube and light oil at the bulk Reynolds number of 96,000 and a bulk temperature of 80F flows in a annular space. Determine the overall coefficient based on the inside area A_i of the 1-in. tube, and the flux per unit inside area of the 1-in. tube.

3. The high-temperature fluid enters a parallel flow HX at a temperature of 623F and leaves at a temperature of 322F. The cold fluid enters at 76F and leaves at 252F. Sketch the temperature profiles for the HX and compare the LMTD with the average temperature difference.

4. The HX described in Problem 3 is operated in the counterflow mode so that the hot fluid that enters at 623F now leaves at 280F and the cold fluid which enters at 76F now leaves at 400F. Sketch the temperature profile for this case and compare the LMTD with the average temperature difference.

5. The flow conditions in the HX of Problem 4 are changed such that the hot fluid which enters at 623F now leaves at 276F which results in a corresponding change in the cold fluid so that the exit temperature of the cold fluid is 423F (inlet is still 76F). Determine the LMTD under these conditions.

6. A crossflow exchanger has both fluids unmixed and a ratio of thermal capacitance of the hot fluid to the cold fluid of 1.88. The hot fluid enters at 265F and leaves at 183F while the cold fluid enters at 45F and leaves at 199F. Determine the mean temperature difference to use in evaluating the HX heat flux.

7. Examine Eq. 7.46 for the effectiveness of the counterflow heat exchanger and note that $\mathcal{E}$ is indeterminate when the capacitance ratio is unity. Apply L'Hôspital's rule and determine the limit of the equation as the capacitance ratio approaches 1.

8. Calculate the heat flux for the exchanger analyzed in Example 7.6 under the conditions determined there by a first-law analysis, and also by using the LMTD method. Compare your results and comment.

9. The effectiveness for the crossflow HX with one fluid mixed and one fluid unmixed is given by Eqs. 7.48 and 7.49 and plotted in Figure 7.12. Note that the parameter $Z = C_{mixed}/C_{unmixed}$ is the only one given on the figure. For the value of NTU $= 3$ and a value of the ratio $C_{unmixed}/C_{mixed} = 0.25$, calculate $\mathcal{E}$ from the appropriate equation and check against the value read from the figure.

10. A crossflow heat recovery unit is being considered which comprises a number of tubes that have engine exhaust flowing normal to the tubes in mixed flow and water inside the tubes in unmixed flow. The hot exhaust gas enters at 1200F at a flow rate of 2000 lb_m/hr, and the water enters at 40F at 1800 lb_m/hr. Under these conditions the exchanger has an overall heat transfer coefficient of 55.7 Btu/hrft^{2}F for the tube area of 27.5 ft^2. At what pressure must the water be kept to prevent boiling in the tubes?

11. Determine the LMTD and the heat flux for the exchanger in Problem 7 and check your results with a first-law analysis of the cold fluid.

12. Steam at 250 psia condenses on the outside of a 20-mm OD copper tube whose wall thickness is 1 mm and length is 1.5 m. Air at 14.7 psia and 45F enters the tube with a bulk velocity of 40 ft/s. Assuming a heat transfer coefficient of 5000 Btu/hrft^{2}F on the condensing side of the copper tube, determine the outlet air temperature.

13. Calculate the heat flux through the wall of the copper tube in Problem 12 by (a) Newton's equation (Eq. 7.3), (b) using the LMTD concept and Eq. 7.17, and (c) from a first-law analysis of the cold fluid. Briefly discuss your results and using the most reliable value, determine the amount of steam condensed per hour.

14. Show that Eqs. 7.45 and 7.46 reduce to the same expression as the ratio of minimum to maximum capacitance approaches zero. Determine the resulting expression and explain the physical significance involved.

15. Determine the effects of fouling on the performance of the copper heat exchanger tube described in Problem 12. Assume that the air came from a compressed air source and determine the outlet temperature of the air under fouled conditions.

16. Hot oil enters a crossflow heat exchanger in which both fluids are mixed at a temperature of 475F and a flow rate of 1600 lb_m/hr. The hot oil is used to heat 2000 lb_m/hr of water entering at 50F. What area is required for the HX to obtain an outlet water temperature of 150F with an overall coefficient of 35.8 Btu/hrft^{2}F? Check your result by computing the exchanger flux with the LMTD method and compare with first-law analysis.

17. A compact crossflow HX with both fluids unmixed is used to heat air in order to keep cargo space above freezing in a jet freighter aircraft by taking hot lubricating oil at 200F to heat 0.5 lb_m/s of air from -20 to 80F. How much oil flow is required to accomplish the purpose?

18. A concentric tube HX is used to heat oil with hot water. The water enters at 220F and leaves at 170F while the oil enters at 90F and leaves at 140F. Determine the effectiveness of the exchanger and whether counterflow or parallel flow would be the smallest exchanger, assuming flow rates the same.

19. Hot water enters a concentric tube HX at a flow rate of 1000 lb_m/hr while 2000 lb_m/hr of kerosene ($c_p = 0.5$ Btu/lb_mF) enters the HX at 20F. With flow conditions such that the overall coefficient is 50 Btu/hrft^{2}F what is the smallest (area) HX that will produce the maximum effectiveness in parallel flow and what are the associated outlet temperatures? What outlet temperatures would be possible if that exchanger were run in the counterflow mode with the same mass flow rates?

20. Compare values of the LMTD for the parallel flow and counterflow heat exchangers discussed in Problem 19 for the flow conditions given therein.

References

1. Kays, W. M. and London, A. L., *Compact Heat Exchangers*, McGraw-Hill, New York, 1964.
2. Tubular Exchanger Manufacturers Association, *TEMA Standards*, 6th Edition, 707 Westchester Ave., White Plains, NY., 10604, 1978.
3. Bowman, R. A., Mueller, A. C., and Nagle, W. H., "Mean Temperature Difference in Design," *Transactions ASME*, Vol. 62, 1940, p. 288.

4. Stevens, R. A., Fernandez, J., and Woolf, J. R., "Mean Temperature Difference in One, Two, and Three Pass Cross Flow Heat Exchangers," *Transactions ASME*, Vol. 79, 1957, p. 287.
5. Mason, J. L., "Heat Transfer in Cross Flow," *Proceedings of the Second U.S. National Congress of Applied Mechanics, ASME*, New York, 1955, p. 801.
6. Kern, D. Q., *Process Heat Transfer*, McGraw-Hill, New York, 1950.
7. Taborek, J., "Evolution of Heat Exchanger Design Techniques," *Heat Transfer Engineering*, Vol. 1, No. 1, 1979.
8. Small, W. R. and Gentry, C. C., Phillips Petroleum Company, Bartlesville, Okla., personal communication, January, 1980.
9. Reid, R. C. and Sherwood, T. K., *The Properties of Gases and Liquids*, McGraw-Hill, New York, 1958.
10. Van Wylen, G. J. and Sonntag, R. E., *Fundamentals of Classical Thermodynamics*, 2d Edition, Wiley, New York, 1973, p. 431.

Part III
RADIATION AND PHASE CHANGE

Chapter 8
Radiation

8.0 GENERAL REMARKS

In the broad sense when we speak of energy radiation we would have to include all wavelengths in the discussion. The basic nature, composition, and speed of propagation of all radiation is the same; the only distinguishing feature is frequency (or wavelength). Radiation is propagated at the speed of light in a vacuum c, and the relationship between wavelength λ, velocity c, and frequency f is

$$\lambda f = c \tag{8.1}$$

Values for the velocity of light are:

$$c = 2.99776 \times 10^{10} \text{ cm/s}$$

$$c = 9.83517 \times 10^{8} \text{ ft/s}$$

Since wavelengths are very small in the region of the spectrum that we are interested in, the unit for describing wavelengths is the micron, μ or the angstrom, Å. The relationship between μ and Å is as follows:

$$1\,\mu \equiv 10^{-6}\,\text{m} \equiv 10^{-4}\,\text{cm} \equiv 10^{-3}\,\text{mm} \equiv 10^{4}\,\text{Å}$$

$$1\,\text{Å} \equiv 10^{-4}\,\mu \equiv 10^{-10}\,\text{m} \equiv 10^{-8}\,\text{cm} \tag{8.2}$$

The angstrom is the smaller unit, since 10,000 Å are required to make up a distance of 1 μ so that a four-digit whole number will conveniently describe radiation in the visible region with reasonable accuracy. For example, the NBS standard of length is the radiation from krypton 86 isotope (orange) which has a wavelength of $\lambda = 0.60578021$ μ or we could say that the wavelength was 6058 Å if we did not need all the precision of the NBS measurement. Such precision is useful when we recognize that krypton 86 has 1650763.73 waves in 1 m. Counting interference fringes electronically can give a very precise measurement of length, but that is not the topic we want to delve into here.

The electromagnetic spectrum or range in wavelength for radiation is shown in part in Figure 8.1. The thermal range lies between about 0.1 μ to 100 μ; it is in this range that significant thermal energy (due to the absolute temperature of the surface) can be detected or transmitted. Of course, as the wavelength gets shorter, the radiation carries more and more energy so that rather than depositing the energy content (of the radiation) on the surface, a significant portion penetrates the surface and even passes through the solid. Radiation between 0.1 and 100 μ essentially interacts at the surface and does not penetrate opaque solids or liquids; interaction with gases is possible and we will discuss that topic separately. Some solids transmit radiation in selected wavelength regions. Air, with no water vapor or CO_2, is essentially transparent to radiation at temperatures below about 8000 to 9000R.

The visible portion of the spectrum (radiation to which the human eye is sensitive) lies in the thermal range, and is probably due to our evolution in a sea of radiation (of all kinds) from the sun. The visible band (about 0.4 to 0.7 μ) lies around the maximum monochromatic energy emission of the sun. The discovery of thermal radiation by Sir William Herschel (1) in about 1800 occurred while he was investigating the energy content of the visible spectrum. He found significant amounts of energy where he could not see any light dispersed by his prism. His conclusion that thermal radiation was just like visible radiation and also obeyed the laws of geometric optics was prophetic in his time.

A real explanation of radiation phenomena had to await Planck's (4) quantum theory, which states that the energy of radiation is proportional to the ratio of c/λ and the constant of proportionality is a universal constant called Planck's constant h so that

$$E_0 = \frac{hc}{\lambda_0}$$

where

$$h = 6.6256(10^{-27}) \text{ erg/s}$$
$$h = 1.7433(10^{-40}) \text{ Btu/hr}$$

(8.3)

The quantity E_0 represents the energy content of a quantum or packet of radiation of wavelength λ_0 and has proved the key to understanding surface (molecular) interaction effects. In our study we will be concerned with macroscopic effects and we will treat radiant energy as a continuous phenomenon

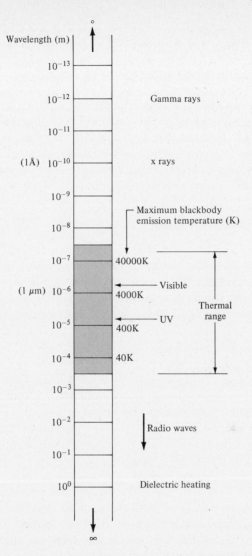

Figure 8.1 Portion of the radiation spectrum showing the region of importance for thermal energy.

according to the Stefan-Boltzmann law. We will not be concerned with the duality nature of radiation where in some aspects the radiation behaves as a continuous wave (diffraction, reflection phenomena) or as particle propagation (quanta and the photoelectric effect). According to the Stefan-Boltzmann law the maximum total energy emitted from a surface to the hemisphere above it depends on the fourth power of the absolute temperature; therefore

$$E_b = \sigma T^4$$

where

$$\sigma = 5.675 \times 10^{-8} \text{ W/m}^2\text{K}^4$$
$$\sigma = 0.1714 \times 10^{-8} \text{ Btu/hrft}^2\text{R}^4$$

(8.4)

and E_b is called the blackbody emissive power.

We are then concerned about the fraction of the maximum which real surfaces emit; how the radiation is distributed over the hemisphere; and the geometrical effects that determine exchange between two surfaces. Before we can proceed further we must establish a number of concepts and definitions to give us a vocabulary and a basis for discussion.

8.1 RADIATION DEFINITIONS

There are a number of loosely related concepts that must all be known in order to understand a discussion or development of radiation heat transfer topics. Some of the nomenclature has meaning in the everyday sense of the word, but some terms will be entirely new to the reader. These concepts have been very briefly introduced in Section 1.3, and we will elaborate on them in more detail here.

8.1.1 Propagation Concepts

We mentioned that the maximum amount of energy a surface could emit was given by the Stefan-Boltzmann law. We define

a. Emissive power

E, Btu/hrft^2 or W/m^2

E_λ, Btu/hrμft^2 or W/μm^2

as that radiant energy leaving the surface, in all directions, due only to the absolute thermodynamic temperature level of the surface. The quantity E we noted as q_{rad}/A in Section 1.3 and is the total energy summed over all wavelengths. The monochromatic emissive power E_λ is the energy contained in an infinitesimally small wavelength band centered about the particular wavelength λ being considered. We can at this stage observe that

$$E = \int_0^\infty E_\lambda \, d\lambda$$

(8.5)

and note that $E_\lambda = f(\lambda)$ where the functional relationship $f(\lambda)$ is called the spectral distribution function which we will discuss in some considerable detail later on. We also do not know the surface requirements that will produce the maximum emissive power at a given absolute temperature level. Again we will discuss the concept of a black surface or maximum emitter later. We now consider energy that strikes the surface and define

b. Irradiation

G, Btu/hrft^2 or W/m^2

G_λ, Btu/hrμft^2 or W/μm^2

as the combination, from all sources and directions, of radiant energy that strikes the surface. The concept requires that irradiation has meaning only in a plane just imperceptibly above the surface in question. The determination of G is made at the surface. The everyday meaning of the word *irradiation* is correct when we include the possibility of several sources. Energy leaving a surface is more complex; we define

c. Radiosity

J, Btu/hrft^2 or W/m^2

J_λ, Btu/hrμft^2 or W/μm^2

as the sum total of all kinds of radiation leaving the surface in all directions. We have already considered one kind, the emissive power. Of the irradiation that strikes the surface a fraction is reflected, and if the surface is transparent, radiation could also be transmitted from some source (or sources) on the other side of the surface. The radiosity then is the sum of the reflected energy, the transmitted energy, and the energy emitted due to the absolute temperature level of the surface. Thus

$$J = \rho G + \tau(\underline{G}) + E \tag{8.6}$$

We note that the irradiation $(\underline{G})$ is not from the same sources as G and we emphasize that fact by the sub bar and the parenthesis. If we are considering a plate, $(\underline{G})$ is the irradiation on the other side.

We have delineated incoming, G, outgoing, J, and emitted radiation, E as propagation concepts. Since radiation is a phenomenon that takes place in three-dimensional space, the concept of intensity is required and applies to the propagation concepts just discussed. We define

d. Intensity

I, Btu/hrft^2_p Ω

I_λ, Btu/hrμft^2_p Ω

as the amount of radiant flux Btu/hr per steradian of solid angle Ω per unit of projected area in the direction of propagation. The monochromatic intensity I_λ is the same concept with the radiant energy at a particular wavelength λ only. The solid angle concept is illustrated in Figure 8.2 and compared with that of the plane angle. The plane angle in radians is defined as arc length over radius, whereas the solid angle is defined as surface area of the sphere divided by the

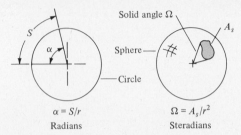

$$\alpha = S/r$$
Radians

$$\Omega = A_s/r^2$$
Steradians

Figure 8.2   Sketch illustrating the concept of plane and solid angles.

square of the radius. We may note that a circle contains 2π radians, (rad) and a sphere contains 4π steradians (sr). The reader will recognize the prefix "stereo" more readily in reference to audio equipment, but the three-dimensional or space aspect is similar.

The intensity of radiant energy is based on flux leaving the surface per unit of projected area as shown in Figure 8.3. As an angle ϕ between the normal to the surface and the direction of I approaches 90 degrees (or $\pi/2$) the flux approaches zero, and so does the projected area; consequently, it is possible for the intensity to remain at a finite or even a constant value at $\phi = \pi/2$. We will say more about that possibility when we discuss diffuse surface characteristics.

It would be worthwhile for the reader to review Section 1.3 in connection with the above, especially the concept of radiosity. The intensity concept was not discussed in Chapter 1. We now examine some aspects of surface behavior toward radiation.

8.1.2 Surface Characteristics

All surfaces emit with an intensity that is a function of the angle ϕ from the normal. Such emission can be calculated according to Fresnel's law (Reference 3) and a qualitative sketch of such calculations is shown in Figure 8.4. Nonconductors show a relative insensitivity with polar angle, whereas conductors

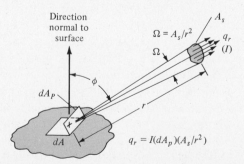

Figure 8.3   Geometrical aspects involved in the concepts of radiation intensity I.

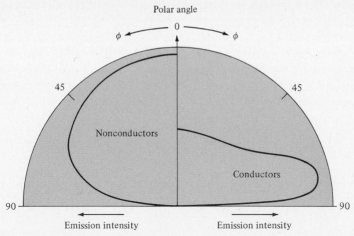

Figure 8.4 Relationship of radiant intensity due to emission with polar angle for conductors and nonconductors.

(metals) show a strong dependence on polar angles. Those substances whose intensity of radiation is not a function of ϕ we call diffuse. Surfaces that have a roughness whose dimensions are large compared to the wavelength of the radiation in question also show an intensity of reflected radiation that is not a function of ϕ, the polar angle. We define a diffuse surface as one whose intensity of emitted and reflected radiation is not a function of polar angle ϕ nor of azimuthal angle θ. Fortunately, a large number of real surfaces behave in a very nearly diffuse manner.

To complete the discussion we observe that surfaces which reflect incoming radiation at the same angle to the normal as the irradiation are called specular (mirrorlike) surfaces. Real surfaces reflect with varying degrees of a specular component. Concrete block surfaces painted with flat paint or untreated are good examples of diffuse surfaces, while vinyl floor tiles are examples of surfaces that have a strong specular component. In the discussion in the sections to follow we will treat exchange between diffuse surfaces only; the mathematical complexities in calculating exchange for nondiffuse surfaces are extremely difficult.

We have encountered the concepts of absorptance, reflectance, and transmittance in Chapter 1 where we learned that $\alpha + \rho + \tau = 1$ (Eq. 1.6) for the total hemispherical properties. The properties α, ρ, and τ have directional dependence when the incoming radiation is specular, but we will consider only aspects of diffuse radiation and surface properties that are summed over the entire hemisphere above the surface. Such a view is consistent with our definition of irradiation and is inherent in the statement energy absorbed $= \alpha G$. The absorptance α then is the fraction of energy G coming into the surface from all directions which is absorbed. We can also define monochromatic absorptance α_λ as the fraction of the monochromatic irradiation G_λ which is absorbed. Also ρ or ρ_λ is the fraction of G or G_λ reflected to the hemisphere (all directions) above the surface, and for a diffuse surface the intensity of that reflected radiation is not a function of direction.

We observed that the emissive power E is the radiant energy that leaves a surface due to its absolute temperature level and that the maximum emissive power is given by the Stefan-Boltzmann law in Eq. 8.4. We define the hemispherical emittance of a surface as the ratio of the actual emitted energy to the maximum possible emitted energy at the absolute temperature of the surface; thus

$$\varepsilon = \frac{E}{E_b} = \frac{E}{\sigma T^4} \tag{8.7}$$

If the radiation is monochromatic, we would write

$$\varepsilon_\lambda = \frac{E_\lambda}{E_{b\lambda}} \tag{8.8}$$

and if ε_λ and α_λ do not vary with λ we say the surface is gray. At this point in our discussion it might be of interest to look into the origin of the term *gray*. When a surface absorbs all incident radiation, there can be nothing reflected for the eye to see. Such surfaces are, therefore, called *black* surfaces or perfect absorbers. In a later section we will see that perfect absorbers are also perfect emitters. An opaque surface that absorbs no radiation must then reflect all radiation. When the eye sees all radiation to which it is sensitive (in relatively equal energy amounts); the description of the visual sensation is called *white*. Therefore, surfaces that absorb and reflect in between these limits and are insensitive to wavelength are called gray surfaces, because a combination of black and white color gives gray.

Values of surface properties have been obtained by a large number of investigators under varying conditions. Reference 2 presents the most complete survey of surface radiation properties published prior to 1960. Some values for emittance of different substances are given in Appendix C. The reader will note that the terms *emittance* and *emissivity* are used in the literature. An effort has been made to establish the term emissivity (also absorptivity, reflectivity, and transmissivity) as being a true property of the material or surface and not subject to surface conditions. The ending "ance" was proposed to apply to data for materials with a particular surface condition, dirt, oxide layer, roughness, or whatever. Unfortunately, the effort has not been successful, and the two endings (ance) and (ivity) continue to be used almost interchangeably.

8.2 KIRCHHOFF'S LAW

In the sections on exchange between gray surfaces we will want to know the relationship between emittance and absorptance for a surface. Accordingly, we consider a large evacuated cavity whose inside walls are at a uniform absolute temperature T as shown in Figure 8.5. The cavity is filled with radiant energy and at any location in the cavity the irradiation G must be the same. If it were not, it would be possible to connect a heat engine between locations of high and

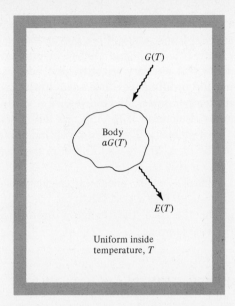

Figure 8.5 Large cavity enclosure whose inside walls are at uniform temperature T.

low irradiation and extract work continuously from a single reservoir in violation of the Kelvin-Planck statement of the second law of thermodynamics. We then place an arbitrary body in the enclosure and let it come to equilibrium, which means that it will reach the temperature of the enclosure. The cavity is evacuated; therefore, radiation is the only means of energy transport between the cavity walls and the surface of the body so that at equilibrium the energy absorbed from the irradiation is radiated from the surface of the body as emissive power. When the surface of the body is black, $\alpha = 1$, so that the energy balance terms become

$$\text{energy absorbed} = \alpha G(T) = G(T)$$
$$\text{energy emitted} = E(T) = E_b(T)$$
(8.9)

At equilibrium the two quantities are equal; therefore, we note that the irradiation $G(T) = E_b(T)$ in the enclosure is the maximum possible at the temperature T. When the surface of the body is nonblack, the energy balance is written as

$$\text{energy absorbed} = \alpha G(T) = \alpha E_b(T)$$
$$\text{energy emitted} = E(T) = \varepsilon E_b(T)$$
(8.10)

so that at equilibrium $\alpha = \varepsilon$. We must emphasize that the absorptance of a surface equals the emittance of the surface only when the source temperature of the irradiation is the same as the surface temperature. The foregoing is Kirchhoff's law. Fortunately, most substances have α that is relatively insensitive to wavelength so that a change in the temperature of the irradiating source (or irradiation from several different source temperatures) does not change the absorptance; consequently, the emittance remains the same. We reinforce our definition of a gray surface then as one whose α and ε are not functions of wavelength. Changing the temperature of the gray surface has a minimal effect on α and ε unless there is a very large change in temperature; we must, of course, exclude surfaces that are very selective absorbers in a particular wavelength band. At very large differences (like solar irradiation) in temperatures compared to ambient temperatures, the absorptance is vastly different from the emittance. For example, Reference 2 (pp. 23 and 249) lists for clean, smooth 24ST aluminum the absorptance at $T = 75F$ to solar radiation as $\alpha(T_{solar}) = 0.45$ and the emittance at 75F as $\varepsilon(75F) = 0.09$ so that in this case α is five times as large as ε.

We will use Kirchhoff's law in enclosure calculations where the surface temperatures are near enough that $\alpha = \varepsilon$. When the reader encounters situations where that may not be the case, the semigray approximation described by Wiebelt (3) is a reasonable approach.

8.3 SPECTRAL DISTRIBUTION, PLANCK'S LAW

A knowledge of the spectral distribution of radiant energy is essential in calculating energy transmission through transparent substances. The variation of black emissive power with wavelength was determined by Planck (4) to be

$$E_{b\lambda} = \frac{c_1\lambda^{-5}}{\exp(c_2/\lambda T) - 1} \tag{8.11}$$

where the constants c_1 and c_2 are given by Love (5) as

$$c_1 = 3.7404(10^4)\ W\mu^4/cm^2$$
$$c_1 = 1.1870(10^8)\ Btu\mu^4/hrft^2$$
$$c_2 = 1.4387(10^4)\ \mu K$$
$$c_2 = 2.5896(10^4)\ \mu R$$

Figure 8.6 shows the variation of $E_{b\lambda}$ with wavelength at three particular temperatures. The dashed line in the figure represents the locus of the maximum emissive power and we note the maximum shifts toward lower wavelengths as the surface temperature increases. The location of the maximum emissive power at a given temperature can be calculated from Wien's law, which states

$$\lambda_{max}T = 5215.6\ \mu R$$
$$\lambda_{max}T = 2897.6\ \mu K \tag{8.12}$$

Planck's law is deceptively simple in form; however, even simple hand calculations with the equation are tedious at best. It is more convenient to work with the distribution equation in table form in terms of the variable λT. We can recast

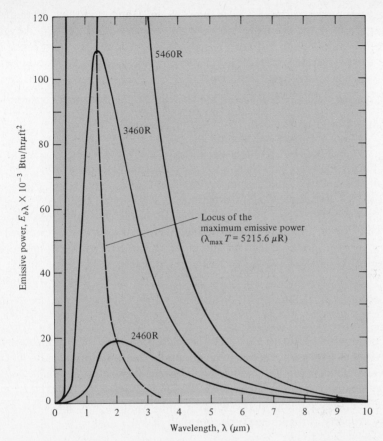

Figure 8.6 Monochromatic emissive power as a function of wavelength with the absolute temperature as the parameter according to Planck's law.

Eq. 8.5 in such terms by noting that $d\lambda = d(\lambda T)/T$ and multiplying and dividing by T^4 so that

$$E_b = T^4 \int_0^\infty \frac{E_{b\lambda T}}{T^5} d(\lambda T) \tag{8.13}$$

$$= T^4 c_1 \int_0^\infty \frac{(\lambda T)^{-5}}{\exp(c_2/\lambda T) - 1} d(\lambda T) \tag{8.14}$$

If we make the change of variable $z = c_2/\lambda T$ or $\lambda T = c_2 z^{-1}$, then $d(\lambda T) = -c_2 z^{-2} dz$ and substituting into Eq. 8.14 we get

$$E_b = -\frac{T^4 c_1}{c_2^4} \int_\infty^0 \frac{z^3}{e^z - 1} dz$$

$$= T^4 \frac{c_1}{c_2^4} \int_0^\infty \frac{z^3}{e^z - 1} dz \tag{8.15}$$

The definite integral in Eq. 8.15 has the value $\pi^4/15$ so that the black emissive power is given by

$$E_b = \frac{c_1\pi^4}{15c_2^4} T^4 \tag{8.16}$$

and comparing with Eq. 8.4 we can see that the Stefan-Boltzman constant is

$$\sigma = \frac{c_1\pi^4}{15c_2^4} \tag{8.17}$$

Values for σ computed with Planck's first and second constants agree very well with values calculated from careful measurements of E_b.

Table 8.1 lists the radiation functions $E_{b\lambda}/\sigma T^5$ and $E_b(0-\lambda T)/\sigma T^4$ as functions of λT. The values in Table 8.1 were originally calculated by Dunkle (6) from Planck's law, and were reevaluated with a digital computer by Wiebelt (3). The first column in the table is the integrand in Eq. 8.14 and is the monochromatic emissive power at λT,

$$\frac{E_{b\lambda}}{\sigma T^5} = \frac{c_1(\lambda T)^{-5}}{\sigma[\exp(c_2/\lambda T)-1]} \tag{8.18}$$

which is useful in calculating the total surface properties from monochromatic measurements. The second column is more useful to us; it represents the fraction of the energy quantity σT^4 which lies between zero and the value of λ in question. Figure 8.7 shows graphically the representation of $E_b(0 \to \lambda T)$. To obtain the energy content in the window from λ_1 to λ_2, we make use of the linearity property of the integral operator such that

$$\frac{E_b(\lambda_1 \to \lambda_2)}{\sigma T^4} = \int_{\lambda_1 T}^{\lambda_2 T} \frac{E_{b\lambda}}{\sigma T^5} d(\lambda T)$$

$$= \int_0^{\lambda_2 T} \frac{E_{b\lambda}}{\sigma T^5} d(\lambda T) - \int_0^{\lambda_1 T} \frac{E_{b\lambda T}}{\sigma T^5} d(\lambda T)$$

$$= \frac{E_b(0 \to \lambda_2 T)}{\sigma T^4} - \frac{E_b(0 \to \lambda_1 T)}{\sigma T^4} \tag{8.19}$$

where the last two terms on the right side of Eq. 8.19 are tabulated in Table 8.1. Application of the radiation functions is best described by Example 8.1.

Example 8.1

A sample of plate glass transmits radiant energy in two windows of the spectrum. In the wavelength range from 0.2 to 1 μ, 90 percent of the irradiation is transmitted; and from 4 to 10 μ, 40 percent is transmitted. We would like to know what fraction of radiation from a black emitter at 5000F will pass through the plate glass and the amount of that flux in Btu/hrft^2.

Table 8.1 BLACKBODY RADIATION FUNCTIONS

$\lambda T \,(\mu\mathrm{R})$	$\dfrac{E_{b\lambda} \times 10^5}{\sigma T^5}$	$\dfrac{E_b(0-\lambda T)}{\sigma T^4}$	$\lambda T \,(\mu\mathrm{R})$	$\dfrac{E_{b\lambda} \times 10^5}{\sigma T^5}$	$\dfrac{E_b(0-\lambda T)}{\sigma T^4}$
1,000.0	0.000039	0.0000	10,400.0	5.142725	0.7183
1,200.0	0.001191	0.0000	10,600.0	4.921745	0.7284
1,400.0	0.012008	0.0000	10,800.0	4.710716	0.7380
1,600.0	0.062118	0.0000	11,000.0	4.509291	0.7472
1,800.0	0.208018	0.0003	11,200.0	4.317109	0.7561
2,000.0	0.517405	0.0010	11,400.0	4.133804	0.7645
2,200.0	1.041926	0.0025	11,600.0	3.959010	0.7726
2,400.0	1.797651	0.0053	11,800.0	3.792363	0.7803
2,600.0	2.761875	0.0098	12,000.0	3.633505	0.7878
2,800.0	3.882650	0.0164	12,200.0	3.482084	0.7949
3,000.0	5.093279	0.0254	12,400.0	3.337758	0.8017
3,200.0	6.325614	0.0368	12,600.0	3.200195	0.8082
3,400.0	7.519353	0.0507	12,800.0	3.069073	0.8145
3,600.0	8.626936	0.0668	13,000.0	2.944084	0.8205
3,800.0	9.614973	0.0851	13,200.0	2.824930	0.8263
4,000.0	10.463377	0.1052	13,400.0	2.711325	0.8318
4,200.0	11.163315	0.1269	13,600.0	2.602997	0.8371
4,400.0	11.714711	0.1498	13,800.0	2.499685	0.8422
4,600.0	12.123821	0.1736	14,000.0	2.401139	0.8471
4,800.0	12.401105	0.1982	14,200.0	2.307123	0.8518
5,000.0	12.559492	0.2232	14,400.0	2.217411	0.8564
5,200.0	12.613057	0.2483	14,600.0	2.131788	0.8607
5,400.0	12.576066	0.2735	14,800.0	2.050049	0.8649
5,600.0	12.462308	0.2986	15,000.0	1.972000	0.8689
5,800.0	12.284687	0.3234	16,000.0	1.630989	0.8869
6,000.0	12.054971	0.3477	17,000.0	1.358304	0.9018
6,200.0	11.783688	0.3715	18,000.0	1.138794	0.9142
6,400.0	11.480102	0.3948	19,000.0	0.960883	0.9247
6,600.0	11.152254	0.4174	20,000.0	0.815714	0.9335
6,800.0	10.807041	0.4394	21,000.0	0.696480	0.9411
7,000.0	10.450309	0.4607	22,000.0	0.597925	0.9475
7,200.0	10.086964	0.4812	23,000.0	0.515964	0.9531
7,400.0	9.721078	0.5010	24,000.0	0.447405	0.9579
7,600.0	9.355994	0.5201	25,000.0	0.389739	0.9621
7,800.0	8.994419	0.5384	26,000.0	0.340978	0.9657
8,000.0	8.638524	0.5561	27,000.0	0.299540	0.9689
8,200.0	8.290014	0.5730	28,000.0	0.264157	0.9717
8,400.0	7.950202	0.5892	29,000.0	0.233807	0.9742
8,600.0	7.620072	0.6048	30,000.0	0.207663	0.9764
8,800.0	7.300336	0.6197	40,000.0	0.074178	0.9891
9,000.0	6.991475	0.6340	50,000.0	0.032617	0.9941
9,200.0	6.693786	0.6477	60,000.0	0.016479	0.9965
9,400.0	6.407408	0.6608	70,000.0	0.009192	0.9977
9,600.0	6.132361	0.6733	80,000.0	0.005521	0.9984
9,800.0	5.868560	0.6853	90,000.0	0.003512	0.9989
10,000.0	5.615844	0.6968	100,000.0	0.002339	0.9991
10,200.0	5.373989	0.7078			

SOURCE: Calculated by J. A. Wiebelt *Engineering Radiation Heat Transfer*, Holt, Rinehart and Winston, New York, 1966, and reproduced with the permission of the author, Dr. J. A. Wiebelt and the publisher, Holt, Reinhart and Winston.

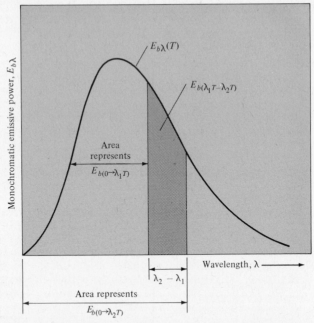

Figure 8.7 Representation of the energy content in the window λ_1 to λ_2 of a black radiator at temperature T.

SOLUTION
The fraction transmitted through the first window is given by Eq. 8.19 as

$$\frac{E_b(\lambda_1 T \to \lambda_2 T)}{\sigma T^4} = \frac{E_b(0 \to \lambda_2 T)}{\sigma T^4} - \frac{E_b(0 \to \lambda_1 T)}{\sigma T^4}$$

and the fraction transmitted through the second window by

$$\frac{E_b(\lambda_3 T \to \lambda_4 T)}{\sigma T^4} = \frac{E_b(0 \to \lambda_4 T)}{\sigma T^4} - \frac{E_b(0 \to \lambda_3 T)}{\sigma T^4}$$

The fraction of σT^4 that passes through the plate glass is then the sum of the fractions that pass through the two windows. The λT products for the source temperature $T = 5000F = 5460R$ are

$$\lambda_1 T = 0.2(5460) = 1092 \ \mu R$$

$$\lambda_2 T = 1(5460) = 5460 \ \mu R$$

$$\lambda_3 T = 4(5460) = 21{,}840 \ \mu R$$

$$\lambda_4 T = 10(5460) = 54{,}600 \ \mu R$$

and from Table 8.1 we obtain

$$\frac{E_b(0 \to \lambda_1 T)}{\sigma T^4} = 0.0000$$

$$\frac{E_b(0 \to \lambda_2 T)}{\sigma T^4} = 0.2813$$

Therefore, we see that 28.13 percent of σT^4 lies in the window from 0.2 to 1 μ and 90 percent of this amount of radiant energy passes through the glass. Therefore, $0.90(0.2813)\sigma T^4 = 0.2532\sigma T^4$ is transmitted. For the second window we obtain from Table 8.1

$$\frac{E_b(0 \to \lambda_3 T)}{\sigma T^4} = 0.9465$$

$$\frac{E_b(0 \to \lambda_4 T)}{\sigma T^4} = 0.9952$$

so that $(0.9952 - 0.9465) = 0.0487$ or only 4.87 percent of σT^4 lies in the range from 4 to 10 μ and 40 percent of this amount or $0.4(0.0487)\sigma T^4 = 0.0195\sigma T^4$ is transmitted. The total fraction of σT^4 which passes through the plate glass then is $(0.2532 + 0.0195) = 0.2727$ or 27.3 percent. The energy amount is given by

$$\begin{aligned}
G_{\text{trans}} &= 0.2727(0.1714)(5460)^4 10^{-8} \\
&= 415{,}400 \text{ Btu/hrft}^2 \\
&= 1{,}310{,}400 \text{ W/m}^2
\end{aligned}$$

If we had information on the reflectivity of the glass, we could determine the amount of energy reflected and also the amount absorbed.

8.4 DIFFUSE INTENSITY AND EMISSIVE POWER

When we consider net interchange between black surfaces that are diffuse emitters we will need the relationship between E_b and I in order to calculate net flux; we establish that relationship here. The emissive power has been defined as the energy in Btu/hrft^2 emitted by the surface and radiated to the hemisphere above it. To relate to the intensity, we must sum the radiant flux due to the intensity over the entire hemisphere above the surface as shown in Figure 8.8. The amount of flux d^2q coming out of the differential surface area of the hemisphere dA_2 can be expressed as

$$d^2q = I_b \, dA_s \cos \phi \, d\Omega \tag{8.20}$$

where we recognize that $dA_s \cos \phi$ is the projected area of dA_s in the I direction, and $d\Omega$ is the solid angle dA_2 subtends to the center of the base circle of the hemisphere. It may be helpful to refer back to Figure 8.3 with regard to the physical meaning of the intensity. We must consider the flux a second-order

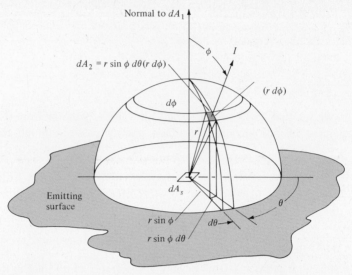

Figure 8.8 Geometrical considerations required in the formulation of the relationship between diffuse intensity and the emissive power of a surface.

differential because two integrations will be required to obtain the entire flux through the hemisphere. The differential area $dA_2 = (r \sin \phi \, d\theta)(r \, d\phi)$; accordingly, the solid angle is given by

$$d\Omega = \frac{dA_2}{r^2} = \sin \phi \, d\phi \, d\theta \tag{8.21}$$

The range on the variable θ is from zero to 2π and on ϕ from 0 to $\pi/2$ so that integration of Eq. 8.20 yields

$$q = I_b \, dA_s \int_0^{2\pi} \left(\int_0^{\pi/2} \cos \phi \sin \phi \, d\phi \right) d\theta \tag{8.22}$$

since for a diffuse surface I is not a function of direction ϕ or θ and the surface element dA_s is also independent of ϕ and θ. The integral over θ is very simple, and the integral over ϕ is easy to visualize if we let $z = \sin \phi$ so that $dz = \cos \phi \, d\phi$. We note that the corresponding limits on z are 0 and 1 when ϕ is 0 and $\pi/2$, respectively. Equation 8.22 then becomes

$$q = 2\pi I_b \, dA_s \int_0^1 z \, dz$$

$$= 2\pi I_b \, dA_s \left[\frac{z^2}{2} \right]_0^1$$

$$= \pi I_b \, dA_s \tag{8.23}$$

We can recognize that the unit area flux q/dA_s is the emissive power of the surface E_b so that

$$E_b = \pi I_b \tag{8.24}$$

for a diffuse black surface. We observe that intensity could be due to some other aspect than emission, such as reflection or transmission, for example. We could then express radiosity $J = \pi I_J$ in terms of a diffuse intensity I_J that included reflected, emitted, and transmitted radiation flux.

We will use the concept just developed in the next section where we will look at net exchange between black surfaces.

8.5 BLACK DIFFUSE SURFACE EXCHANGE

In this section we consider exchange between black and diffuse surfaces. The diffuse condition is essential or else the mathematics becomes extremely complicated. In Chapter 1 we introduced the concept of the radiation interchange factor F_{12} by simply defining it as the fraction of black diffuse radiation that left surface A_1 and struck surface A_2. We will show in this section that F_{12} depends only on the geometrical orientation of the surfaces when those surfaces are diffuse radiators.

8.5.1 Interchange Factor Formulation

In this section we emphasize the physical interrelationship of the concept of intensity and net flux exchange. We will trace the development of the interchange factor from a very small area to a finite large area, and from that concept to the interchange factor applicable between finite large areas. We might note at this point that the reader may find several terms in the technical literature such as angle factor, shape factor, or configuration factor to describe the phenomena. The term *interchange factor* seems to have the best connotation and the least conflict with other ideas for our purpose.

The physical orientation of the diffuse surfaces A_1 and A_2 is shown in Figure 8.9. We should note that the normals n_1 and n_2 to dA_1 and dA_2, respectively, do not lie in the same plane, and that an observer on either A_1 or A_2 can see all of the other area. The quantity r is the distance from dA_1 to dA_2 and its formulation depends on the coordinate system employed. We consider only energy emitted by the surfaces in order to formulate the F_{12} concept. The infinitesimal flux from dA_1 which arrives at dA_2 can be expressed as

$$d^2q_{dA_1 \rightarrow dA_2} = I_{b1}\cos \phi_1 \, dA_1 \, d\Omega_2 \qquad (8.25)$$

The solid angle $d\Omega_2$ is given by $(dA_2\cos \phi_2)/r^2$ and noting that $I_{b1} = E_{b1}/\pi$, the infinitesimal flux becomes

$$d^2q_{dA_1 \rightarrow dA_2} = (E_{b1} \, dA_1)\cos \phi_1\cos \phi_2 \, dA_2/\pi r^2 \qquad (8.26)$$

If we reflect a moment on the result we just obtained, we will see the concept of the interchange factor emerge. Physically $d^2q_{dA_1 \rightarrow dA_2}$ is the amount of energy in watts or Btu/hr from dA_1 which strikes dA_2, and $E_{b1} \, dA_1$ is the total amount of energy that leaves the element of area dA_1 and goes to the hemisphere above

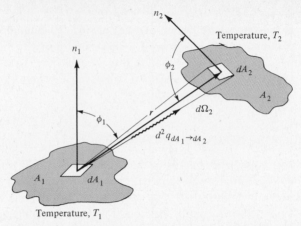

Figure 8.9 Geometric aspects for formulating the radiant interchange factor F_{12} between black diffuse surfaces A_1 and A_2 which are at constant temperature T_1 and T_2.

dA_1. The ratio $dq_{dA_1 \to dA_2}/(E_{b1}dA_1)$ then represents the fraction of black diffuse radiation that leaves dA_1 and strikes dA_2 which is exactly our definition of the interchange factor. In this case, since the exchange is between infinitesimal areas, we define the infinitesimal interchange factor as

$$dF_{dA_1 \to dA_2} = \cos \phi_1 \cos \phi_2 \, dA_2/\pi r^2 \tag{8.27}$$

Were we to consider the other direction from dA_2 to dA_1 and formulate $d^2q_{dA_2 \to dA_1}$ and note that $E_{b2} = \pi I_{b2}$, we could obtain

$$dF_{dA_2 \to dA_1} = \cos \phi_2 \cos \phi_1 \, dA_1/\pi r^2 \tag{8.28}$$

Since the expressions for the dF interchange factors are symmetrical, we can see that multiplying Eq. 8.27 by dA_1 and 8.28 by dA_2 makes the two equations equal so that

$$dA_1 \, dF_{dA_1 \to dA_2} = dA_2 \, dF_{dA_2 \to dA_1} \tag{8.29}$$

which is the reciprocity relationship for infinitesimal interchange factors. The results given in Eqs. 8.27 and 8.28 can be used to approximate interchange factors between small areas.

To obtain the fraction of black diffuse radiation leaving dA_1 and striking the entire area A_2 requires that we sum all the infinitesimal dF's over the area A_2 and obtain

$$F_{dA_1 \to A_2} = \int_{A_2} \frac{\cos \phi_1 \cos \phi_2}{\pi r^2} \, dA_2 \tag{8.30}$$

At this point we have to recognize that evaluating the double integral in Eq. 8.30 is a very difficult task for other than very simple geometries. Hamilton and

Morgan (7) present analytical solutions of Eqs. 8.30 and 8.34 for several geometries as well as tables of values. We will use Eq. 8.30 as a basis for an approximation technique that we will describe in a later section for determining the interchange factor from a very small area ΔA_1 to a larger area A_2. Before proceeding we will consider an example for a simple geometry where the mathematics is easy to follow.

Example 8.2

Consider the geometrical arrangement shown in the accompanying figure. We want to determine the interchange factor from the small area dA_1 to the disk of radius R_1 spaced a distance H above dA_1 in a plane parallel to the plane of dA_1. The small area dA_1 is positioned such that the normal to dA_1 passes through the center of the disk.

SOLUTION
We observe that the normals n_1 and n_2 are parallel; therefore, the angles ϕ_1 and ϕ_2 are equal. Consequently, $\cos \phi_1 = \cos \phi_2 = H/r$ so that Eq. 8.30 becomes

$$F_{dA_1 \to A_2} = \int_{A_2} \frac{H^2}{\pi r^4} dA_2$$

The distance r between dA_1 and dA_2 is $r = \sqrt{H^2 + R^2}$ and the area $dA_2 = R\, d\theta\, dR$ so that the integral over A_2 ranges over the variables R and the angle θ in the plane of A_2; we note that θ ranges from 0 to 2π and R ranges from 0 to R_1. The integral over the area A_2 can then be expressed as the definite double

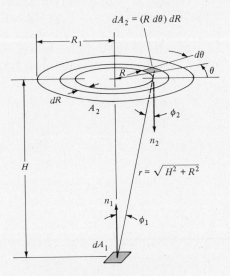

Figure for Example 8.2.

integral

$$F_{dA_1 \to A_2} = \int_0^{2\pi} \int_0^{R_1} \frac{RH^2 \, dR}{\pi (H^2 + R^2)^2} \, d\theta$$

None of the variables in the integrand are functions of θ so $\int_0^{2\pi} d\theta = 2\pi$ and the double integral reduces to

$$F_{dA_1 \to A_2} = H^2 \int_0^{R_1} (H^2 + R^2)^{-2} 2R \, dR$$

which is easy to integrate because it is like $\int z^n \, dz$. Therefore

$$F_{dA_1 \to A_2} = H^2 \left[\frac{-1}{H^2 + R^2} \right]_0^{R_1}$$

$$= \frac{R_1^2}{H^2 + R_1^2}$$

$$= \frac{1}{1 + (H/R_1)^2} \tag{8.31}$$

We observe that $F_{dA_1 \to A_2}$ is always less than 1, and as H approaches zero $F_{dA_1 \to A_2}$ approaches 1 which is in accord with our understanding and definition of the interchange factor.

Table 8.2 gives the solutions to Eq. 8.30 for four representative configurations. In case II the normal to dA_1 is parallel to the plane of A_2 and also parallel to the short edge; whereas in case I the normal to dA_1 is perpendicular to the plane of A_2 and passes through the upper left corner of A_2. Case III is the orientation worked out in Example 8.2 with the normal to dA_1 perpendicular to the plane of A_2 and passing through the center of the disk. In case IV the orientation is identical to case III except there is a hole in the center of the disk, and the radiation from dA_1 is to the washer-shaped area of A_2.

8.5.2 Formulation of F_{12} for Finite Areas

The amount of radiant flux from dA_1 to dA_2 is given by Eq. 8.25, and the orientation of dA_1 and dA_2 is shown in Figure 8.9. The entire amount of radiant flux from all of A_1 to all of A_2 requires a summation of all infinitesimal fluxes over both areas; thus

$$q_{A_1 \to A_2} = \int_{A_1} \int_{A_2} d^2 q_{dA_1 \to dA_2} \tag{8.32}$$

Substituting for the infinitesimal flux from Eq. 8.26 and noting that E_{b1} is a constant, since the temperature is assumed constant over the area A_1 we obtain

$$q_{A_1 \to A_2} = \frac{1}{A_1} (E_{b1} A_1) \int_{A_1} \int_{A_2} \frac{\cos \phi_1 \cos \phi_2}{\pi r^2} \, dA_1 \, dA_2 \tag{8.33}$$

Table 8.2 INTERCHANGE FACTOR EQUATION $F_{dA_1 \to A_2}$ FOR RECTANGULAR AND CIRCULAR AREAS AS ILLUSTRATED

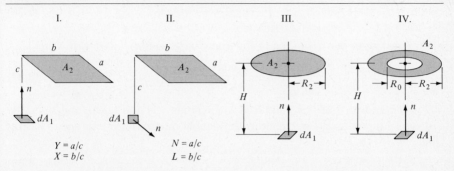

I.	II.	III.	IV.

$Y = a/c$
$X = b/c$

$N = a/c$
$L = b/c$

I. Area dA_1 to rectangle; normal to dA_1 perpendicular to plane of rectangle as shown.

$$F_{12} = \frac{1}{2\pi} \left[\frac{Y}{\sqrt{1+Y^2}} \tan^{-1}\left(\frac{X}{\sqrt{1+Y^2}}\right) + \frac{X}{\sqrt{1+X^2}} \arctan\left(\frac{Y}{\sqrt{1+X^2}}\right) \right]$$

II. Area dA_1 to rectangle; normal to dA_1 parallel to plane of rectangle as shown.

$$F_{12} = \frac{1}{2\pi} \left[\arctan\frac{1}{L} - \frac{L}{\sqrt{N^2+L^2}} \arctan\left(\frac{1}{\sqrt{N^2+L^2}}\right) \right]$$

III. Area dA_1 to disk; normal to dA_1 perpendicular to and passing through center of disk.

$$F_{12} = \frac{R_2^2}{H^2 + R_2^2}$$

IV. Area dA_1 to washer; normal to dA_1 perpendicular to and passing through center of hole.

$$F_{12} = \frac{R_2^2}{H^2 + R_2^2} - \frac{R_0^2}{H^2 + R_0^2}$$

The right side of Eq. 8.33 has been multiplied and divided by A_1 in order to obtain the physical representation of F_{12} as the fraction of the black diffuse radiation leaving A_1 and striking A_2, or

$$F_{12} = \frac{q_{A_1 \to A_2}}{E_{b1} A_1}$$

$$= \frac{1}{A_1} \int_{A_1} \int_{A_2} \frac{\cos\phi_1 \cos\phi_2}{\pi r^2} dA_1 \, dA_2 \tag{8.34}$$

The result shows that F_{12} depends only on the geometrical orientation of A_1 and A_2. We can in a similar manner formulate the infinitesimal flux from $dA_2 \to dA_1$ such that

$$d^2 q_{dA_2 \to dA_1} = E_{b2} \, dA_2 \cos\phi_2 \cos\phi_1 \, dA_1 / \pi r^2 \tag{8.35}$$

so that

$$q_{A_2 \to A_1} = \frac{1}{A_2} (E_{b2} A_2) \int_{A_2} \int_{A_1} \frac{\cos \phi_2 \cos \phi_1}{\pi r} dA_2 \, dA_1 \qquad (8.36)$$

and according to our definition of the interchange factor F_{21} we can write

$$F_{21} = \frac{q_{A_2 \to A_1}}{E_{b2} A_2} = \frac{1}{A_2} \int_{A_2} \int_{A_1} \frac{\cos \phi_2 \cos \phi_1}{\pi r^2} dA_2 \, dA_1 \qquad (8.37)$$

The integrands in Eqs. 8.33 and 8.37 are symmetrical and equal, and since the order of integration doesn't matter, the double integrals are also equal. Therefore,

$$A_1 F_{12} = A_2 F_{21} \qquad (8.38)$$

which is the reciprocity theorem that we arrived at by intuitive reasoning in Chapter 1. The reciprocity theorem is very useful especially when F_{21} is easier to obtain than F_{12}; we will make extensive use of the theorem in the sections to follow.

Closed form solutions to Eq. 8.34 are very difficult to obtain. Hamilton and Morgan (7) catalog known solutions and solutions they were able to work out. Wiebelt (3) presents an abbreviated summary of Hamilton and Morgan's work which is perhaps more accessible. Two of the most prevalent geometrical arrangements are the directly opposed rectangles with parallel sides and the adjacent rectangles with one mutual side whose planes are perpendicular. Figures 8.10 and 8.11 present F_{12} as a function of the pertinent geometrical parameters for those two arrangements.

The interchange factor for the orientation of directly opposed unequal disks whose planes are parallel and whose centers lie on a common line perpendicular to the planes of the disks is given in Figure 8.12. Hamilton and Morgan (7) present the closed form solution for F_{12} for such a geometry to be

$$F_{12} = \left(z - \sqrt{z^2 - 4E^2 B^2} \right) / 2 \qquad (8.39)$$

where

$$z = 1 + (1 + E^2) B^2$$

$$B = \frac{H}{r_1}$$

$$E = \frac{r_2}{H}$$

H = separation distance between disks

The parameters for Eq. 8.39 are illustrated in the small sketch in Figure 8.12. For the case where the disks are equal in size, $r_1 = r_2$, the interchange factor becomes

$$F_{12} = \left(B^2 + 2 - B\sqrt{B^2 + 4} \right) / 2 \qquad (8.39a)$$

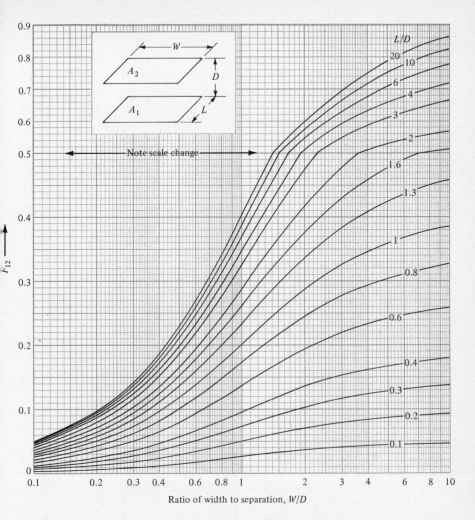

Figure 8.10 Variation of F_{12} with pertinent geometrical parameters for directly opposed equal rectangles whose sides and planes are parallel.

The disk equation is deceptively simple, compared to the opposed and adjacent rectangle equations. As we can see in Table 8.2 the equations are complicated and cumbersome to use in hand calculations. The interchange factor for case I in Table 8.2 is shown in Figure 8.13 as a function of the parameters Y and X. We will, in a later section, show how we can obtain other interchange factors for rectangles which are not directly opposed and nonadjacent rectangles by a method called interchange factor algebra.

We next consider some aspects of net radiant energy exchange for black diffuse surfaces before looking at the interchange factor algebra method.

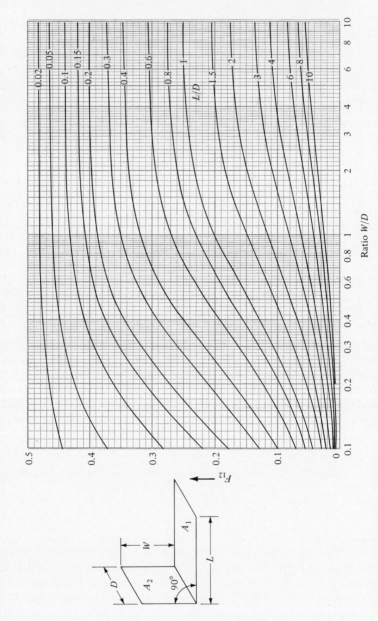

Figure 8.11 Variation of F_{12} with geometrical parameters for mutually perpendicular rectangles with one common side.

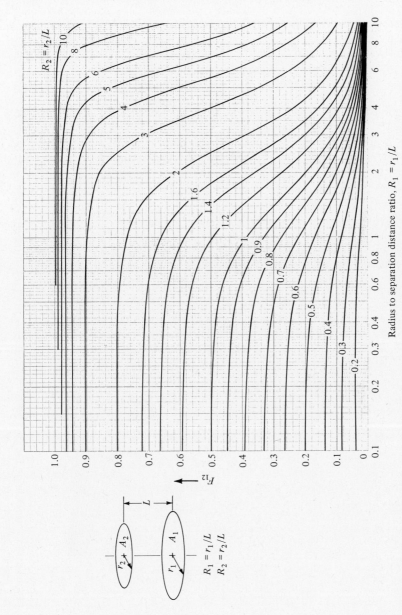

Figure 8.12 Variation of the interchange factor F_{12} for unequal, directly opposed disks whose planes are parallel with geometrical parameters R_1 and R_2.

353

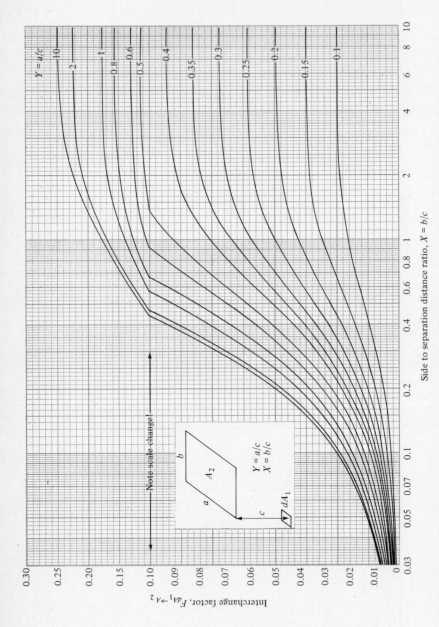

Figure 8.13 Interchange factor from an infinitesimal area dA_1 located under one corner of a finite rectangle A_2, planes of both areas parallel.

8.5.3 Net Exchange for Two Black Surfaces

Referring to Figure 8.9 we can note that the net exchange of radiant energy between the two surfaces A_1 and A_2 is

$$q_{12\text{net}} = q_{A_1 \to A_2} - q_{A_2 \to A_1}$$
$$= E_{b1} A_1 F_{12} - E_{b2} A_2 F_{21}$$
$$= A_1 F_{12}(E_{b1} - E_{b2})$$
$$= A_2 F_{21}(E_{b1} - E_{b2}) \tag{8.40}$$

We can, of course, substitute σT^4 for the emissive power to make an actual calculation. We note that both A_1 and A_2 also emit to the surroundings (which we must consider to be at zero Rankine to permit an unambiguous assessment) so that $q_{12\text{net}}$ represents an exchange quantity between the two surfaces. Were we to inquire how much energy it would take to maintain surface A_1 at temperature T_1 we would write

$$q_{\text{net1}} = A_1 E_{b1} - A_2 F_{21} E_{b2} \tag{8.41}$$

We must carefully note the difference between $q_{12\text{net}}$ and q_{net1}.

As was mentioned in Chapter 1, we can write Eq. 8.40 in the form of a potential difference $E_{b1} - E_{b2}$ divided by a resistance $1/A_1 F_{12}$ in order to visualize the heat flux transport in terms of the electrical analogy. The resistance $1/A_1 F_{12}$ is called a space resistance because its magnitude is determined by the spatial orientation of the surfaces A_1 and A_2.

8.5.4 Net Exchange with Refractory Surface

A refractory surface is a surface that receives radiant energy and reaches a temperature that is just high enough to reradiate all the energy received. Consequently, the surface has no net heat flux associated with it. Such surfaces are also sometimes called no-flux or reradiating surfaces, but since the application is mostly in furnace calculations where such surfaces are made of refractory materials, the name refractory surface is used more often. Figure 8.14 shows a generalized orientation of two black surfaces A_1 and A_2 and a refractory surface A_R. As shown in the diagram there is a net exchange $q_{12\text{net}}$ between A_1 and A_2 and also net radiant energy from A_1 to A_R and then to A_2 so that the net exchange between A_1 and A_2 is enhanced by the presence of A_R. The total flux $q_{12R\text{net}}$ can be determined from the electrical analog circuit diagram shown in Figure 8.14(b) as

$$q_{12R\text{net}} = \frac{E_{b1} - E_{b2}}{R_{\text{eq}}} \tag{8.42}$$

The reciprocal of the equivalent resistance for parallel resistances is the sum

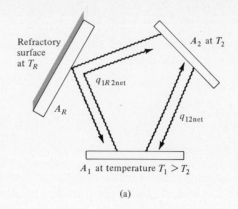

(a)

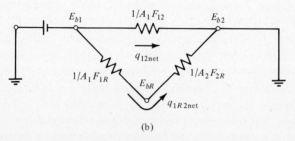

(b)

Figure 8.14 Radiant exchange between two black surfaces in the presence of a refractory surface. (a) Generalized orientation of two black surfaces in the presence of a refactory surface. (b) Equivalent circuit diagram for radiant exchange between surfaces shown in (a).

divided by the product so that

$$\frac{1}{R_{eq}} = \frac{(1/A_1 F_{12}) + (1/A_1 F_{1R}) + (1/A_2 F_{2R})}{(1/A_1 F_{12})[(1/A_1 F_{1R}) + (1/A_2 F_{2R})]} \tag{8.43}$$

If we call the bracketed term in the denominator B, multiply numerator and denominator by $A_1 F_{12}$, and factor out A_1 we obtain

$$\frac{1}{R_{eq}} = A_1 \left[F_{12} + \frac{1}{A_1 B} \right] \tag{8.44}$$

Replacing B by the bracketed term and inserting in Eq. 8.42 gives

$$q_{net12R} = A_1 \left[F_{12} + \frac{A_2 F_{1R} F_{2R}}{A_2 F_{2R} + A_1 F_{1R}} \right] (E_{b1} - E_{b2}) \tag{8.45}$$

Equation 8.45 gives the net exchange between A_1 and A_2 in the presence of the refractory surface A_R when the surroundings do not contribute to the radiation exchange. When the surfaces A_1, A_2, and A_R form an enclosure where A_1 and A_2 are flat nonreentrant surfaces (no radiation leaving A_1 strikes A_1 directly), we can write

$$F_{1R} = 1 - F_{12}$$

$$F_{2R} = 1 - F_{21}$$

which we can insert in Eq. 8.45. Some algebraic manipulation yields

$$q_{net12R} = A_1 \left[\frac{A_2 - A_1 F_{12}^2}{A_1 + A_2 - 2A_1 F_{12}} \right] (E_{b1} - E_{b2}) \tag{8.46}$$

for the net exchange between A_1 and A_2 when the surfaces are connected by refractory walls to form an enclosure. Example 8.3 illustrates the ideas.

Example 8.3

Consider the orientation of surfaces A_1, A_2, and A_R in the figure. Surface A_1 is at 3600R, surface A_2 is at 2000R, and both surfaces are black. The surfaces A_1 and A_2 are 10 by 10 ft and are separated by a distance of 10 ft so that the dimensions of A_R are the same as A_1 and A_2. We would like to compare the net exchange without and with the refractory surface, and the flux required to maintain A_1 at 3600R under those conditions.

SOLUTION
Consider first the net flux without the refractory surface present. For the directly opposed rectangles Figure 8.10 gives $F_{12} = 0.198$ and the black emissive powers are

$$E_{b1} = \sigma T_1^4 = 0.1714(10^{-8})(3600)^4 = 287,900 \text{ Btu/hrft}^2$$

$$E_{b2} = \sigma T_2^4 = 0.1714(10^{-8})(2000)^4 = 27,400 \text{ Btu/hrft}^2$$

Therefore:

$$q_{net12} = A_1 F_{12}(E_{b1} - E_{b2})$$

$$= 100(0.198)(287,900 - 27,400)$$

$$= 5.21 \times 10^6 \text{ Btu/hr}$$

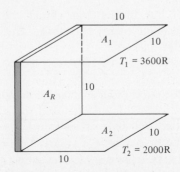

Figure for Example 8.3.

The energy required to keep A_1 at 3600R is given by

$$q_{net1} = A_1 E_{b1} - A_2 F_{21} E_{b2}$$

and in this case $F_{21} = 0.198$ because of symmetry so that

$$q_{net1} = 100(287,900) - 100(0.198)27,400$$

$$= 28.24 \times 10^6 \text{ Btu/hr}$$

so that we can see that surface A_1 loses considerable energy to the surroundings compared to the net amount it transmits to A_2.

With the refractory surface in place we see that $F_{1R} = F_{2R}$ because of the symmetrical orientation and Figure 8.11 gives $F_{1R} = 0.20$ so that

$$F_{12} + \frac{A_2 F_{1R} F_{2R}}{A_2 F_{2R} + A_1 F_{1R}} = 0.198 + \frac{100(0.2)(0.2)}{100(0.2) + 100(0.2)}$$

$$= 0.298$$

and

$$q_{net12R} = 100(0.298)(287,900 - 27,400)$$

$$= 7.76 \times 10^6 \text{ Btu/hr}$$

as compared to 5.21×10^6 Btu/hr without the refractory surface present.

The net flux to maintain surface A_1 at 3600R with the refractory surface present is

$$q_{net1R} = A_1 E_{b1} - A_2 F_{21} E_{b2} - A_R F_{R2} E_{bR}$$

$$= A_1 E_{b1} - A_2 F_{21} E_{b2} - A_2 F_{2R} E_{bR}$$

Reference to Figure 8.14(b) shows that when the space resistances $1/A_1 F_{1R}$ and $1/A_2 F_{2R}$ are equal, as they are in this example, then the emissive power of the refractory surface lies just halfway between E_{b1} and E_{b2}. Therefore

$$E_{bR} = (E_{b1} + E_{b2})/2 = 157,655 \text{ Btu/hrft}^2$$

and

$$q_{net1R} = 100(287,900) - 100(0.198)(27,400) - 100(0.2)157,655$$

$$= 25.1 \times 10^6 \text{ Btu/hr}$$

which is some 3.1×10^6 Btu/hr less than without the refractory because the refractory surface radiates some energy back to A_1. This concludes the example.

8.5.5 Enclosures

We close this section with a short discussion on reentrant surfaces and enclosures. All the surfaces we have analyzed and discussed so far have been flat so that none of the radiation that left the surface struck the surface itself directly (nonreentrant). For surfaces that have a concave curvature a fraction of the radiant energy that leaves the surface strikes the surface directly; that fraction is designated as F_{11} or F_{ii} in the general sense.

The definition of the interchange factor requires that all the fractions of the black diffuse radiation leaving a surface A_1 and striking other surfaces that form an enclosure above the surface add up to unity. We, therefore, can write $\Sigma F_{1j} = 1$ where j ranges from 2 to the nth surface in the enclosure. When the surface A_1 is reentrant (not flat) then the summation must include surface A_1 and the index j must range from 1 to the nth surface. So that for nonreentrant surfaces we would write

$$F_{12} + F_{13} + F_{14} + \cdots + F_{1n} = 1 \tag{8.47}$$

and for reentrant surfaces

$$F_{11} + F_{12} + F_{13} + F_{14} + \cdots + F_{1n} = 1 \tag{8.47a}$$

For example, if we were to consider a right circular cylinder and to call the cylindrical surface A_1, one flat end surface A_2, and the other flat end surface A_3, we could determine the reentrant fraction F_{11} from $F_{11} + F_{12} + F_{13} = 1$. We note that $F_{12} = F_{13}$ by symmetry and from the observation that $F_{31} + F_{32} = 1$ together with ordinary reciprocity, $A_3 F_{31} = A_1 F_{13}$, we can determine F_{13} because F_{32} can be obtained from the chart in Figure 8.12.

8.6 INTERCHANGE FACTOR ALGEBRA

It is possible to evaluate interchange factors for a particular geometric orientation without solving Eq. 8.34 if the configuration has a basic common orientation with a known interchange factor (like Figures 8.10 through 8.13, for example) by a straightforward algebraic procedure. For instance, the interchange factor for rectangles whose planes are perpendicular and having two sides parallel but *not* having a common side, can be expressed in an algebraic relationship in terms of interchange factors for mutually perpendicular rectangles *with* a common side as given by Figure 8.11. The method relies on the fact that the geometry under consideration can be extended to form two areas A_N and A_M that in themselves have a known interchange factor F_{NM}. The area A_N contains one of the areas A_n and the area A_M contains the other area A_m for which we desire to evaluate the interchange factor F_{nm}.

8.6.1 Formulation of the Method

From the definition of F_{NM} we can express the radiant heat flux from A_N to A_M as

$$q_{N \to M} = E_{bN} A_N F_{NM} \tag{8.48}$$

In this discussion no other source can contribute to $q_{N \to M}$; therefore, we consider A_N and A_M isolated in surroundings at zero Rankine. The flux $q_{N \to M}$ will then be the irradiation G_{NM} on surface A_M and come only from surface A_N. If we choose the temperature of surface A_N such that the emissive power $E_{bN} = 1$ in whatever units we are working with, then

$$q_{N \to M} = G_{NM} = A_N F_{NM} \tag{8.49}$$

The innovative key to the method is the judicious subdivision of A_N and A_M into the minimum number of subareas for which the interchange factors can be resolved into known factors except for F_{nm}. After subdivision we note that the total irradiation G_{NM} is the sum of the individual irradiations G_{ij} between the subdivided areas according to the law of conservation of energy; therefore,

$$G_{NM} = \sum_{j}^{M} \sum_{i}^{N} G_{ij} \tag{8.50}$$

The objective is to arrange the subdivision so that all G_{ij} can be expressed in terms of known AF products except $G_{nm} = A_n F_{nm}$ where F_{nm} is the unknown interchange factor. The procedure is probably best understood by working out a few specific cases.

8.6.2 Rectangles with Two Sides Parallel and Planes Perpendicular

In Section 8.5.2 we saw that integration of Eq. 8.32 gave an expression for the interchange factor for rectangles whose planes are perpendicular and having one common edge. Numerical results of that integration were presented graphically in Figure 8.11. In this section we will show how interchange factors for rectangles of the subject orientation can be obtained when one rectangle is offset and/or skewed. We will consider the offset case first, then the skewed.

Consider the orientation of areas A_1 and A_3 shown in Figure 8.15(a). We would like to determine F_{13} for the offset area A_1 to area A_3 with the

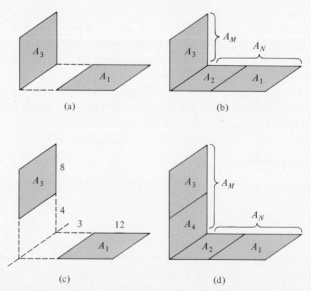

Figure 8.15 Orientation of rectangles whose planes are perpendicular and which have two sides parallel but one or both planes are offset from the line of intersection of the planes.

interchange factor algebra or geometric flux algebra method. According to the discussion above we form $A_N = A_1 + A_2$ as shown in Figure 8.15(b) and A_M simply remains as A_3; the interchange factor F_{NM} then is of a *known* category. The sum of the individual irradiations, or geometrical fluxes, then gives

$$G_{NM} = \sum_{j=3}^{3} \sum_{i=1}^{2} G_{ij} = G_{13} + G_{23} \tag{8.51}$$

We solve the result of the double sum for G_{13}, substitute the $A_i F_{ij}$ products, look up the known interchange factors F_{NM} and F_{23}, and solve for the unknown F_{13}. Thus

$$G_{13} = G_{NM} - G_{23}$$

$$A_1 F_{13} = A_N F_{NM} - A_2 F_{23} \tag{8.52}$$

$$F_{13} = \frac{A_N}{A_1} F_{NM} - \frac{A_2}{A_1} F_{23}$$

The case just illustrated involved only three areas; other cases would involve more areas so that a word about notation would be helpful. Often there will be combinations of subareas (always for A_N and A_M) so that it is necessary to distinguish the area to which the subscripts refer. We do so by grouping with parentheses as follows

$$G_{12} = G_{A_1 \rightarrow A_2}$$

$$G_{(12)3} = G_{(A_1 + A_2) \rightarrow A_3}$$

$$G_{(12)(34)} = G_{(A_1 + A_2) \rightarrow (A_3 + A_4)} \tag{8.53}$$

$$G_{(12)^2} = G_{(A_1 + A_2) \rightarrow (A_1^* + A_2^*)}$$

the convention being to indicate contiguous areas by enclosing the subscripts for those areas in parentheses. The last designation $G_{(12)^2}$ is used when the areas in question are corresponding or identical but belong in different planes; directly opposed rectangles fall into such a category. It might be helpful to consider an extension to the above case with actual numbers in order to consolidate the technique. Reentrant areas are not considered in this method.

Example 8.4

Consider the arrangement of areas shown in Figure 8.15(c); area A_1 is 10 by 12 m, area A_3 is 8 by 10 m, the planes of the two areas are perpendicular, and the sides are aligned as shown with A_3 being offset 4 m from the intersection line and A_1 offset 3 m from the intersection line. We want to determine a value for F_{13}.

SOLUTION

We recognize that interchange factors for mutually perpendicular rectangles with a common edge are known; therefore, we construct $A_M = A_3 + A_4$ and $A_N = A_1 + A_2$ as shown in Figure 8.15(d). We write the geometric flux summation as

$$G_{NM} = \sum_{j=3}^{4} \sum_{i=1}^{2} G_{ij} = G_{13} + G_{14} + G_{23} + G_{24}$$

and note that G_{24} and G_{NM} are knowns (really $A_2 F_{24}$ and $A_N F_{NM}$ are known) and that G_{14} and G_{23} have been worked out in the illustration above. Were that not the case, we would have to apply the technique to those areas and determine those interchange factors separately. Using the nomenclature explained above we write G_{NM} as $G_{(12)(34)} = A_{12} F_{(12)(34)}$ and after substituting the other AF products we can solve for the unknown $A_1 F_{13}$ as follows:

$$A_1 F_{13} = A_{12} F_{12(34)} - A_1 F_{14} - A_2 F_{23} - A_2 F_{24}$$

The factor F_{24} is known and F_{14} and F_{23} can be expressed as shown in the development of Eq. 8.52 with the appropriate subscripts. Thus

$$A_1 F_{14} = A_{12} F_{12(4)} - A_2 F_{24}$$

$$A_2 F_{23} = A_2 F_{2(34)} - A_2 F_{24}$$

so that the final expression for F_{13} becomes

$$F_{13} = \frac{A_{12}}{A_1} \left(F_{12(34)} - F_{(12)4} \right) - \frac{A_2}{A_1} \left(F_{2(34)} - F_{24} \right) \tag{8.54}$$

The individual interchange factors $F_{12(34)}$, $F_{12(4)}$, $F_{2(34)}$, and F_{24} can be obtained from Figure 8.11 as shown in the following table:

FACTOR	W/D	L/D	VALUE
$F_{12(34)}$	$12/10 = 1.2$	$15/10 = 1.5$	0.160
$F_{12(4)}$	$4/10 = 0.4$	$15/10 = 1.5$	0.090
$F_{2(34)}$	$12/10 = 1.2$	$3/10 = 0.3$	0.355
F_{24}	$4/10 = 0.4$	$3/10 = 0.3$	0.291

The area ratios are $A_{12}/A_1 = 150/120 = 1.25$ and $A_2/A_1 = 30/120 = 0.25$ so that

$$F_{13} = (1.25)(0.160 - 0.090) - (0.25)(0.355 - 0.291) = 0.071$$

According to the physical meaning of F_{13} there is 7.1 percent of the black diffuse radiation leaving surface A_1 which strikes A_3 for the given orientation. We note that the final calculation of F_{13} involves differences between F's which are close

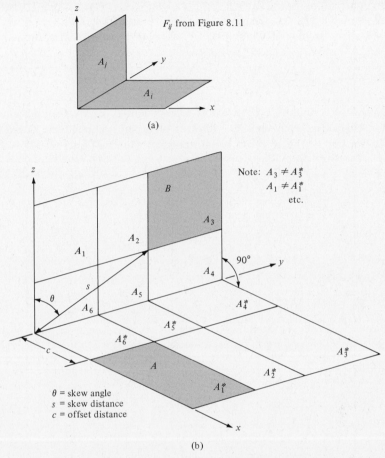

Figure 8.16 Orientation for unequal rectangles with planes perpendicular, two sides parallel, and with and without a common edge. (a) Mutually perpendicular rectangles with a common edge. (b) General orientation for skewed unequal rectangles.

in value. In such cases it may be necessary to compute the individual F's more precisely in order to ensure the desired accuracy. This completes the example.

The general case of skewed arbitrary orientation for rectangles with two sides parallel and planes perpendicular is shown in Figure 8.16. The expression for $F_{AB} = F_{13}^*$ was determined using interchange factor algebra (Reference 7) to be

$$2A_1^*F_{13}^* = 2A_AF_{AB} = 2G_{13}^* = G_{(123456)^2} + G_{(456)^2}$$
$$+ G_{(25)^2} + G_{(5)^2} + X_I - X_{II} - X_{III} \tag{8.55}$$

where

$$X_\mathrm{I} = G_{1256(5\overset{**}{6})} + G_{2345(4\overset{**}{5})} + G_{56(1\,2\,5\,6)}^{****} + G_{45(2\,3\,4\,5)}^{****}$$

$$X_\mathrm{II} = G_{(2345)^2} + G_{(1256)^2} + G_{456(1\,2\,3\,4\,5\,6)}^{******} + G_{123456(4\,5\,6)}^{***}$$

$$X_\mathrm{III} = G_{(25)\overset{*}{5}} + G_{5(2\overset{**}{5})} + G_{(45)^2} + G_{(56)^2}$$

The expression contains 16 AF products. Consequently, extreme caution must be taken in evaluating each term to sufficient precision such that the cumulative error in the sum of the 16 terms is not excessive. Generally one would evaluate

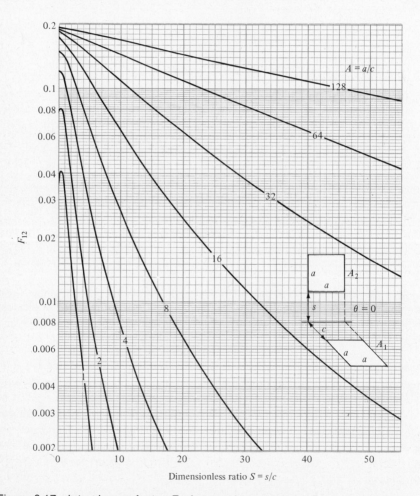

Figure 8.17 Interchange factor F_{12} for equal squares of side a whose planes are perpendicular, two sides parallel, having no common edge, and with skew angle $\theta = 0$ (see Figure 8.16), according to Mujahid and Lee (8).

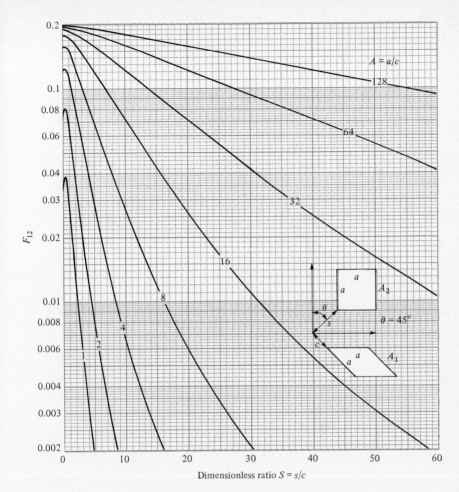

Figure 8.18   Interchange factor F_{12} for equal squares of side *a* whose planes are perpendicular, two sides parallel, having no common edge, and with skew angle $\theta = 45°$ (see Figure 8.16).

the component terms of F_{AB} with a double precision computer routine. Such an evaluation has been made by Mujahid and Lee (8) for the special case of equal squares whose planes are mutually perpendicular, and two sides parallel but with no common edge; a summary of their results is presented in Figures 8.17 and 8.18 for skew angles of $\theta = 0$ and 45 degrees. The parameters on the figures are the dimensionless distance $S = s/c$, which is the ratio of the skew distance s of square area A_2 from the origin to the offset distance c of square area A_1 from the intersection of the planes of the two areas. The dimensionless parameter $A = a/c$ is the ratio of the side of the square a to the offset distance c. We compare the use of the figures with the results of the interchange factor algebra method in Example 8.5.

Example 8.5

Consider the 10×10 squares whose planes are mutually perpendicular and offset as shown in the figure below, and also in Figure 8.15(d).

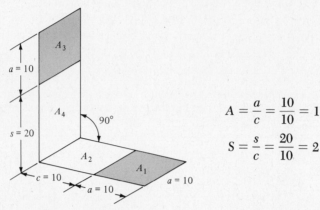

$$A = \frac{a}{c} = \frac{10}{10} = 1$$

$$S = \frac{s}{c} = \frac{20}{10} = 2$$

Figure for Example 8.5.

From Figure 8.17, read $F_{13} = 0.016$.

We desire to compare the value of F_{13} read from Figure 8.17 with the value calculated from Eq. 8.54 which was formulated by the interchange factor algebra method in Example 8.4.

SOLUTION

To apply Equation 8.54 we must evaluate the individual interchange factors involved from Figure 8.11. The tabulation below lists the specific factors with their pertinent parameters in the order that the factors appear in Eq. 8.54:

FACTOR	W/D	L/D	VALUE
$F_{12(34)}$	$30/10 = 3$	$20/10 = 2$	0.162
$F_{12(4)}$	$20/10 = 2$	$20/10 = 2$	0.149
$F_{2(34)}$	$30/10 = 3$	$10/10 = 1$	0.242
F_{24}	$20/10 = 2$	$10/10 = 1$	0.233

The area ratios are $A_{12}/A_1 = 200/100 = 2$ and $A_2/A_1 = 100/100 = 1$ so that substitution into Eq. 8.54 yields

$$F_{13} = 2(0.162 - 0.149) - (0.242 - 0.233) = 0.017$$

a value that compares favorably with 0.016 read from the chart. Generally a precision greater than two significant figures cannot be expected unless the F's are calculated from the defining equations. This completes the example.

8.6.3 Rectangles with Sides and Planes Parallel

The application of interchange factor algebra permits evaluation of F_{12} for rectangles whose sides and planes are parallel but not directly opposed in terms of interchange factors for directly opposed rectangles as given in Figure 8.10. The general orientation for skewed equal rectangles is shown in Figure 8.19. The planes of areas A and B are parallel and separated by the distance c. The skew

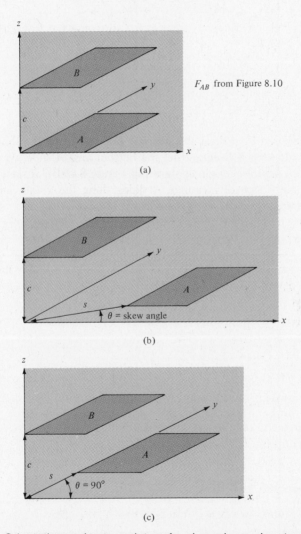

Figure 8.19 Orientation and nomenclature for skewed, equal rectangles with corresponding sides and planes parallel. (a) Directly opposed equal rectangles. (b) General orientation for skewed, equal rectangles with corresponding sides and planes parallel. (c) Orientation for skew angle $\theta = 90°$.

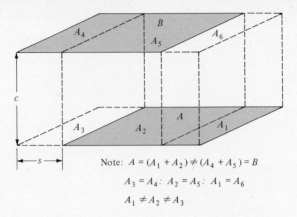

Note: $A = (A_1 + A_2) \neq (A_4 + A_5) = B$

$A_3 = A_4 : \ A_2 = A_5 : \ A_1 = A_6$

$A_1 \neq A_2 \neq A_3$

Figure 8.20   Orientation of Figure 8.19(b) for zero skew angle in expanded detail showing directly opposed subareas.

angle θ is measured from the x axis which is oriented along the short side. If we consider the top area B fixed and let the bottom area A slide, we measure the skew distance s from the origin as shown in Figure 8.19(b); at skew angle $\theta = 0$ degrees, the short side of rectangle A slides along the x axis, and at $\theta = 90$ degrees, the long side slides along the y axis [Figure 8.19(c)].

We continue our illustration of the flux algebra method with a little more involved case. Consider the orientation in Figure 8.19(c) with area A moved such that s is less than the length of the rectangle. To obtain F_{AB} we complete the geometry to form directly opposed rectangles A_{123} and A_{456} with the subareas as shown in the detailed sketch of Figure 8.20 and apply Eq. 8.50 as follows:

$$G_{123(456)} = \sum_{j=4}^{6} \sum_{i=1}^{3} G_{ij}$$

$$= \ \ G_{14} + G_{15} + G_{16}$$
$$+ G_{24} + G_{25} + G_{26}$$
$$+ G_{34} + G_{35} + G_{36} \tag{8.56}$$

We are interested in obtaining $F_{AB} = F_{12(45)}$ which we recognize is contained in $G_{12(45)}$. Applying Eq. 8.50 again we see that

$$G_{12(45)} = G_{14} + G_{15} + G_{24} + G_{25}$$

and combining with Eq. 8.56 yields

$$G_{12(45)} = G_{123(456)} - G_{16} - G_{26} - G_{34} - G_{35} - G_{36} \tag{8.57}$$

where we note that $G_{123(456)}$, G_{16}, and G_{34} are directly opposed equal rectangles for which F_{ij} is known. We need only work with G_{26}, G_{35}, and G_{36}. For G_{26} we write

$$G_{12(56)} = G_{16} + G_{15} + G_{26} + G_{25}$$

but we note that $A_2 F_{26} = A_5 F_{51}$ due to symmetry or identical geometric orientation and applying the reciprocity theorem $A_5 F_{51} = A_1 F_{15}$ so that $A_2 F_{26} = G_{26} = G_{15} = A_1 F_{15}$. Therefore,

$$2G_{26} = G_{12(56)} - G_{16} - G_{25}$$

which are all directly opposed orientations. Similarly we can show that

$$2G_{35} = G_{23(45)} - G_{25} - G_{34}$$

are also all directly opposed known orientations. For G_{36} we must go back and work with Eq. 8.56, recognizing that $A_1 F_{14} = G_{14} = G_{36} = A_6 F_{63} = A_3 F_{36} = G_{36}$ due to the symmetry and reciprocity. Substituting G_{26} and G_{35} into Eq. 8.56, collecting and canceling identical terms gives

$$2G_{36} = G_{123(456)} - G_{12(56)} - G_{23(45)} + G_{25}$$

again all directly opposed known orientations. When we substitute the above expressions for G_{26}, G_{35}, and G_{36} into Eq. 8.57 we obtain

$$G_{12(45)} = \tfrac{1}{2}\left[G_{123(456)} - G_{16} - G_{34} + G_{25}\right] \tag{8.57a}$$

a surprisingly simple result. The reader will recognize that as the number of subareas increases, the complexity of the method also increases. It is, however, still simple algebra as compared to the formidable calculus problems involved in solving Eq. 8.34 in closed form.

The most general orientation for parallel skewed unequal rectangles is the case where area A in Figure 8.19(b) is slid completely out from under area B as shown in Figure 8.21. Hamilton and Morgan (7) have obtained an expression for F_{AB} using interchange factor algebra, symmetry, and reciprocity as follows:

$$4A_A F_{AB} = 4A_3^* F_{37}^* = G_{(123456789)^2} + G_{(456)^2} + G_{(258)^2} - X_{\mathrm{II}} + X_{\mathrm{IV}} - X_{\mathrm{VI}} \tag{8.58}$$

where

$$X_{\mathrm{II}} = G_{(25)^2} + G_{(45)^2} + G_{(58)^2} + G_{(56)^2}$$

$$X_{\mathrm{IV}} = G_{(1256)^2} + G_{(2345)^2} + G_{(4589)^2} + G_{(5678)^2}$$

$$X_{\mathrm{VI}} = G_{(125678)^2} + G_{(234589)^2} + G_{(456789)^2} + G_{(123456)^2}$$

The reader will observe that there is a fascinating geometric symmetry and pattern to the X_{II}, X_{IV}, and X_{VI} terms. As was the case with Eq. 8.55, the 15 individual G's must be evaluated with extreme precision in order to minimize error in F_{AB}. Computer evaluation using double precision functions is the best way to make calculations with Eq. 8.58. Wills and Wang (9) and Huang and Jing (10) have made such computations for skewed 4:1 rectangles and for skewed squares; representative results of those calculations are shown in Figures 8.22 and 8.23. Example 8.6 shows how symmetry can be utilized to simplify the procedure and again illustrates the use of the charts.

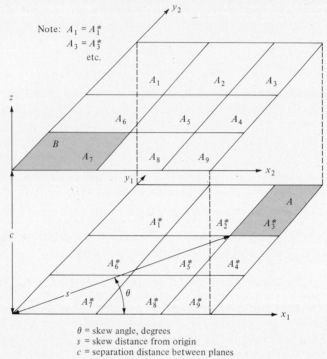

Note: $A_1 = A_1^*$
$A_3 = A_3^*$
etc.

θ = skew angle, degrees
s = skew distance from origin
c = separation distance between planes

Figure 8.21 Arbitrary, unequal rectangles A and B with planes and sides parallel in skewed orientation.

Example 8.6

Consider the skewed squares A_1 and A_3^* in the figure. The squares have sides a and their parallel planes are separated by a distance $c = a$. We want to determine F_{13}^* by using the flux algebra method and check the value obtained from Figure 8.23.

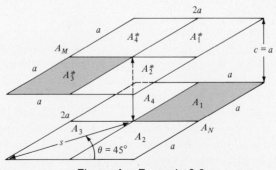

Figure for Example 8.6.

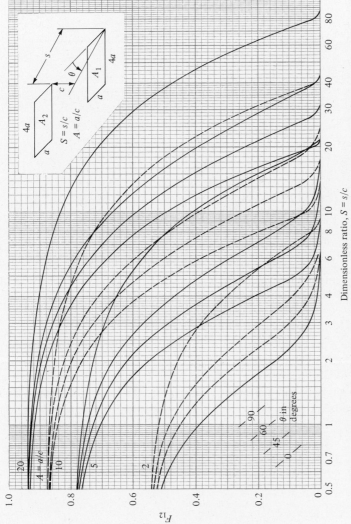

Figure 8.22 Interchange factor F_{12} for skewed, equal, $4:1$ rectangles whose sides and planes are parallel according to Wills and Wang (9).

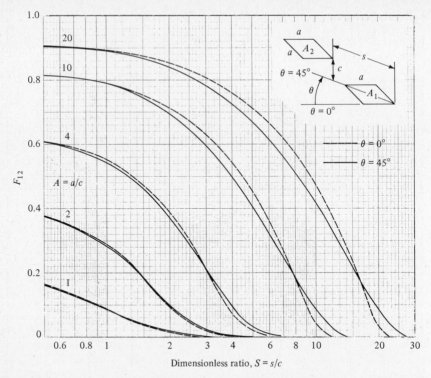

Figure 8.23 Interchange factor F_{12} for skewed equal squares whose sides and planes are parallel, according to Jinq and Huang (10).

SOLUTION

In this case s is the diagonal of the square of side a so that $s = a\sqrt{2}$ or $S = s/c = 1.414$. The parameter $A = a/c = 1$ for the geometry shown; accordingly, from Figure 8.23 we read $F_{13}^* = 0.042$. To apply the flux algebra method we again begin with Eq. 8.50 where for A_M we take the upper square, $2a$ units ' on a side, and A_N, the lower square directly opposed to A_M, also $2a$ units on a side. Thus

$$G_{NM} = \sum_{j=1}^{4} \sum_{i=1}^{4} G_{ij}$$

$$
\begin{aligned}
G_{(1234)^2} = \ & G_{1^2} + G_{12}^* + G_{13}^* + G_{14}^* \\
& + G_{21}^* + G_{2^2} + G_{23}^* + G_{24}^* \\
& + G_{31}^* + G_{32}^* + G_{3^2} + G_{34}^* \\
& + G_{41}^* + G_{42}^* + G_{43}^* + G_{4^2}
\end{aligned}
$$

Since all the subareas are equal, there is diagonal symmetry ($G_{13}^* = G_{24}^*$) and symmetry along the sides so that

$$G_{13}^* = G_{31}^* = G_{42}^* = G_{24}^*$$

and

$$G_{12}^* = G_{14}^* = G_{21}^* = G_{23}^*$$
$$= G_{32}^* = G_{34}^* = G_{41}^* = G_{43}^*$$

After collecting the symmetry terms and solving for G_{13}^*, we obtain

$$G_{13}^* = \tfrac{1}{4} G_{(1234)^2} - G_{(12)^2} + G_{(1)^2}$$

whereupon noting $A_{1234} = 4A_1$ and $A_{12} = 2A_1$, we obtain for the interchange factor

$$F_{13}^* = F_{(1234)^2} - 2F_{(12)^2} + F_{1^2}$$

the interchange factors needed can be obtained from Figure 8.10 and are summarized in the following table:

FACTOR	W/D	L/D	VALUE
$F_{(1234)^2}$	2	2	0.415
$F_{(12)^2}$	2	1	0.286
F_{1^2}	1	1	0.200

Therefore $F_{13}^* = 0.415 - 2(0.286) + 0.20 = 0.043$, a value that checks closely with what we could read from Figure 8.23. In this case the precision read from the charts was sufficient, but then there were only three terms in the final summation for G_{NM}. This completes the example.

8.6.4 Special Reciprocity

In the course of applying the ideas developed for geometric flux algebra, or interchange factor algebra, simplification of the results requires recognition of special and geometric symmetry. For the directly opposed rectangle orientation, it is easy to see the geometric symmetry. For example, if we look at Figure 8.21 and note that $A_1 = A_1^*$ and $A_7 = A_7^*$, we can observe that $A_1 F_{17}^* = A_1^* F_{17}^*$, which we designate special reciprocity because the geometry involved is identical. Contrast that observation with ordinary reciprocity, as was shown by Eq. 8.38, where we would write $A_1 F_{17}^* = A_7^* F_{71}^*$ which only involves the same areas.

For other geometries where corresponding areas are not equal, special reciprocity is not intuitively apparent and must be shown by the defining equation of the interchange factor together with the property of multiple integrals that the order of integration is immaterial. Consider the orientation shown in Figure 8.24; the areas A_1, A_2, A_3, and A_4 are all different sizes, but they do have a common edge. The defining expression for the interchange factor is given by Eq. 8.34 which we write here as the AF product for areas A_1 and A_3 as

$$A_1 F_{13} = \int_{A_3} \int_{A_1} (\text{integrand}) \, dA_1 \, dA_3 \qquad (8.59)$$

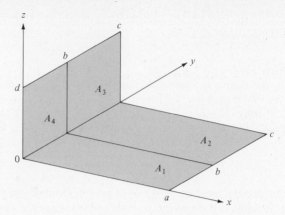

Figure 8.24 Geometric orientation of multiple adjoining rectangles for special reciprocity.

and for areas A_2 and A_4 as

$$A_2 F_{24} = \int_{A_4} \int_{A_2} (\text{integrand}) \, dA_2 \, dA_4 \tag{8.59a}$$

The integrand in both Eqs. 8.59 and 8.59a is the same ($\cos \phi_1 \cos \phi_2 / \pi r^2$) so that the integrals are functions only of the limits. If the limits are the same, the integrals are the same. We note that dA_1 and dA_2 lie in the x, y plane; therefore, $dA_1 \equiv dA_2 \equiv dx \, dy$ and $dA_3 \equiv dA_4 \equiv dz \, dy$ because they lie in the y, z plane. The concept of special reciprocity is easier to see if we write the integrals in expanded operational notation and include the limits. Thus

$$A_1 F_{13} = \underbrace{\int_0^d dz \int_b^c dy}_{\text{over } A_3} \underbrace{\int_0^a dx \int_0^b dy}_{\text{over } A_1} (\text{integrand}) \tag{8.60}$$

$$A_2 F_{24} = \underbrace{\int_0^d dz \int_0^b dy}_{\text{over } A_4} \underbrace{\int_0^a dx \int_b^c dy}_{\text{over } A_2} (\text{integrand}) \tag{8.60a}$$

Inspection of Eqs. 8.60 and 8.60a shows that on A_3 the limits on $\int dy$ extend from b to c which are identical to the limits on $\int dy$ on A_2; similarly on A_1 and A_4. Therefore, the quadruple integrals have the same value and $A_1 F_{13} = A_2 F_{24}$, or in flux terms $G_{13} = G_{24}$ for the orientation of Figure 8.24. The result just demonstrated can also be shown to be so if the areas were offset and also separated, as is the orientation for A_1^* and A_3 in Figure 8.16. In that case $A_1^* F_{13}^* = A_3^* F_{31}^*$, whereas ordinary reciprocity would yield $A_1^* F_{13}^* = A_3 F_{31}^*$.

The technique can also be applied to the directly opposed parallel rectangle orientation to substantiate visually observed symmetry. The results of special reciprocity together with ordinary reciprocity permitted expressing Eqs. 8.55

and 8.58 in only 16 and 15 terms instead of the 36 and 81 terms obtained from the geometric flux double sum.

8.7 GRAY SURFACE EXCHANGE

In Chapter 1 we noted that the maximum amount of energy that can be emitted by a surface at absolute temperature T is given by the Stefan-Boltzmann law and in Eq. 8.7 we defined the total hemispherical emittance ε as

$$\varepsilon = \frac{E}{E_b} = \frac{E}{\sigma T^4}$$

(8.7)
(Repeat)

where by the term *total* we mean over all wavelengths. All real surfaces have ε and α less than unity and the emittance and absorptance are in general functions of wavelength. Fortunately, most engineering materials have a continuous (albeit erratic) variation of α_λ or ε_λ with wavelength such that a value of ε as defined by Eq. 8.7 and not dependent on wavelength can be determined. We now explore those aspects.

8.7.1 Gray Surface Characteristics

A typical variation of emissive power with λ for a real surface is illustrated in Figure 8.25. The shaded area in the figure represents the emissive power of the

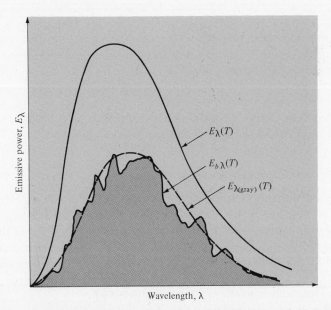

Figure 8.25   Illustration of a typical variation of monochromatic emissive power $E_\lambda(T)$ for a real surface at temperature T with wavelength compared to $E_{b\lambda}(T)$ and equivalent $E_{\lambda(\text{gray})}(T)$.

surface given by Eq. 8.10 as

$$E = \int_0^\infty E_\lambda \, d\lambda \tag{8.61}$$

and since $E_\lambda = \varepsilon_\lambda E_{b\lambda}$ by definition we can write

$$E = \int_0^\infty \varepsilon_\lambda E_{b\lambda} \, d\lambda \tag{8.61a}$$

so that substitution in Eq. 8.7 above gives

$$\varepsilon = \int_0^\infty \varepsilon_\lambda \frac{E_{b\lambda}}{\sigma T^4} d\lambda \tag{8.62}$$

for the total (over all wavelengths) emittance, which is also total hemispherical emittance when ε_λ is hemispherical. Most measurements are made with receivers directed normal to the surface, and if the surface is essentially diffuse, the measurement approximates the hemispherical value. When the variable in Eq. 8.62 is changed to λT, the term $E_{b\lambda}/\sigma T^5$ can be obtained from Table 8.1 and Eq. 8.62 can be solved by approximate means if the variation of ε_λ with λ is known.

Hottel and Sarofim (11) define a gray surface as one whose absorptivity is not a function of wavelength. For surfaces where the monochromatic absorptivity is insensitive to surface temperature changes (within moderate limits this is so for most diffuse real surfaces), the total absorptivity is also constant. Consequently, when $\alpha = \varepsilon$ at equilibrium, then $\alpha = \varepsilon$ also when the change in irradiating surface temperature does not influence α. As a result of the above argument, one must conclude that surfaces can be considered gray because their monochromatic absorptance is not influenced by changes in irradiation intensity and not because of Kirchhoff's law.

A large number of measurements of monochromatic surface properties have been compiled by Gubareff and coauthors (2) and current information is analyzed, compiled, and reported in CINDAS (Center for Information and Numerical Data Analysis and Synthesis) publications (12).

8.7.2 Gray Surface Resistance

Since a gray surface emits less energy than a black surface at a given temperature, we can model a gray surface with a black surface that has a fictitious layer distributed over the surface which adds a resistance in the path of the radiant energy and reduces the amount to that emitted by the gray surface. Knowing the magnitude of the resistance permits constructing an electrical analog for the radiation transport process which is easier to solve for the desired flux than tracing multiple reflections from all surfaces involved.

To proceed requires determination of the relationship between radiant flux, black emissive power, and radiosity. When the irradiation and temperature on a surface are uniform, then the radiosity is also uniform and we can write for the flux

$$q = (J - G)A \tag{8.63}$$

For an opaque surface the radiosity is comprised only of reflected and emitted radiation; thus

$$J = \rho G + \varepsilon E_b \tag{8.64}$$

Solving for G and substituting into Eq. 8.63 gives

$$\frac{q}{A} = J - \frac{J}{\rho} + \frac{\varepsilon E_b}{\rho}$$

$$= \frac{J(\rho - 1)}{\rho} + \frac{\varepsilon E_b}{\rho} \tag{8.65}$$

For opaque surfaces $\alpha + \rho = 1$ and assuming Kirchhoff's law to hold so that $\alpha = \varepsilon$, then $\varepsilon = 1 - \rho$. Substituting into Eq. 8.65 and rearranging yields

$$q = \frac{E_b - J}{\rho/\varepsilon A} \tag{8.66}$$

$$= \frac{E_b - J}{R_\varepsilon} \tag{8.66a}$$

In the light of the above introductory remarks, we see that the radiant flux q is the potential difference $E_b - J$ divided by a surface resistance R_ε, which we must consider to be uniformly distributed over the black surface according to the uniform J and G requirement. In the next few sections we will apply the above result to analyzing different gray surface exchange problems using the electrical analog. It would be instructive to list the assumptions involved in the surface resistance model:

1. Uniform diffuse irradiation and radiosity.
2. Gray surface conditions, $\alpha \neq f(\lambda)$.
3. Kirchhoff's law holds, $\alpha = \varepsilon$.
4. Opaque surface, $\alpha + \rho = 1$.

The first assumption restricts the geometry in the strictest sense to infinite parallel planes, concentric spheres, and infinite concentric cylinders for which $F_{12} = 1$. For other geometries the results represent an approximation that improves as F_{12} approaches 1. We can observe that finite parallel rectangles with F_{12} near one would have J and G fairly uniform, whereas mutually perpendicular rectangles with a common edge would have very nonuniform J and G, values being highest in the corner and decreasing toward the edge. Note for the mutually perpendicular rectangles, F_{12} is always less than 0.5.

8.7.3 Net Exchange for Two Gray Surfaces

For parallel planes large in extent and close together, concentric spheres and long cylinders, the analogy shown in Figure 8.26 holds exactly; for other

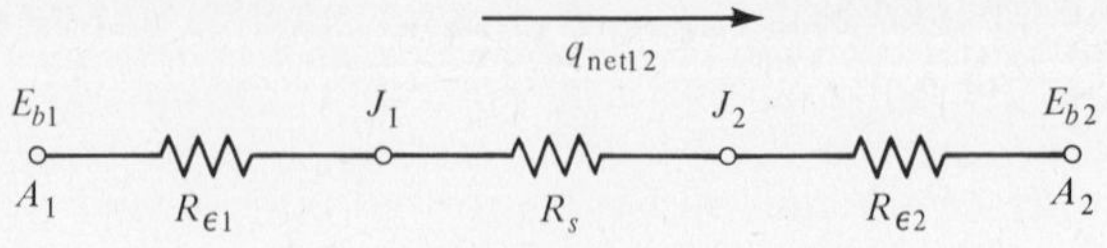

$$R_{\epsilon 1} = \rho_1/\epsilon_1 A_1$$
$$R_{\epsilon 2} = \rho_2/\epsilon_2 A_2$$
$$R_s = 1/A_1 F_{12}$$

Figure 8.26 Electrical analogy for radiant exchange between two gray surfaces that have uniform irradiation and radiosity.

two-surface orientations it is an approximation. The net flux between the surfaces can be written as

$$q_{net} = \frac{E_{b1} - E_{b2}}{R_{eq}}$$

$$= \frac{E_{b1} - E_{b2}}{R_{\epsilon 1} + R_s + R_{\epsilon 2}} \tag{8.67}$$

The term R_s is the space resistance $1/A_1 F_{12}$. Substituting the quantities for the resistance terms gives

$$q_{net12} = \frac{E_{b1} - E_{b2}}{\rho_1/\epsilon_1 A_1 + 1/A_1 F_{12} + \rho_2/\epsilon_2 A_2} \tag{8.68}$$

Actual evaluation then requires $E_b = \sigma T^4$ and values for ϵ and ρ. The term q_{net12} represents the energy required to maintain surface A_1 at temperature T_1 when $F_{12} = 1$. When Eq. 8.68 is used as an approximation for situations where $F_{12} < 1$, the flux q_{net} represents only the net exchange between A_1 and A_2, and the energy required to maintain A_1 at T_1 then depends on exchange with other surrounding temperature potentials.

Example 8.7

A Dewar flask is fabricated from two concentric spheres of polished stainless steel with a separation distance of $\frac{1}{4}$ in. between the shells. We would like to determine the boil-off rate when the flask is filled with liquid nitrogen at 1 atm pressure and the space between the shells is evacuated to a very low pressure to eliminate convection. The OD of the inner shell is 12 in. and the temperature of the outer shell is 70F.

SOLUTION
Since the space between the shells is highly evacuated, there can be no convection; consequently, the boil-off is due to radiation heat flux into the inner shell. The physical situation is a two-surface problem; we let A_1 be the outer surface of the inner shell and A_2 the inner surface of the outer shell; we also neglect the opening to fill the Dewar. The emittance of polished stainless steel is

$\varepsilon = 0.074$ for both surfaces. The radiant flux is given by Eq. 8.68,

$$q = \frac{E_{b1} - E_{b2}}{R_{1\varepsilon} + R_s + R_{2\varepsilon}}$$

where

$$R_{1\varepsilon} = \frac{\rho_1}{\varepsilon_1 A_1} \qquad R_s = \frac{1}{A_1 F_{12}}$$

$$R_{2\varepsilon} = \frac{\rho_2}{\varepsilon_2 A_2}$$

From the dimensions stated we calculate

$$A_1 = \pi D^2 = \pi \text{ ft}^2$$

$$A_2 = \pi \left(\frac{12.5}{12} \right)^2 = 3.049 \text{ ft}^2$$

$$R_{1\varepsilon} = \frac{0.926}{\pi(0.074)} = 3.983 \text{ ft}^{-2}$$

$$R_s = \frac{1}{\pi(1)} = 0.3183 \text{ ft}^{-2}$$

$$R_{2\varepsilon} = \frac{0.926}{0.074(3.049)} = 4.104 \text{ ft}^{-2}$$

The temperature of LN_2 at 1 atm pressure is 139R so that the black emissive powers are

$$E_{b1} = \sigma T_1^4 = 0.1714(1.39)^4 = 0.64 \text{ Btu/hrft}^2$$

$$E_{b2} = \sigma T_2^4 = 0.1714(5.3)^4 = 135.2 \text{ Btu/hrft}^2$$

and the radiant flux is

$$q = \frac{0.64 - 135.2}{3.983 + 0.3183 + 4.104} = -16 \text{ Btu/hr}$$

into the inner shell. The heat of vaporization for LN_2 at constant pressure is $h_{fg} = 85.67$ Btu/lb$_m$, therefore, the boil-off rate due to radiant flux is

$$\dot{m} = \frac{q}{h_{fg}} = \frac{16}{85.67} = 0.19 \text{ lb}_m/\text{hr}$$

The rate changes as the liquid level in the Dewar drops. The area of contact between the liquid nitrogen and the inner shell becomes less, thus increasing the equivalent resistance.

8.7.4 Small Gray Body in a Large Enclosure

The case of a small gray body with no reentrant portions on the surface can be treated with Eq. 8.68 to obtain a conveniently simple result. For such a physical arrangement the area A_1 is much less than A_2 so that $A_1/A_2 \cong 0$ and $F_{12} = 1$. Dividing both sides of Eq. 8.68 by A_1 yields

$$\frac{q_{net}}{A_1} = \frac{E_{b1} - E_{b2}}{(\rho_1/\varepsilon_1) + 1}$$

but since $\rho_1 = 1 - \varepsilon_1$ and assuming that the temperature range is such that $\alpha = \varepsilon$, then

$$\frac{\rho_1}{\varepsilon_1} + 1 = \frac{1}{\varepsilon_1} - 1 + 1 = \frac{1}{\varepsilon_1}$$

and setting $E_b = \sigma T^4$ we can express the net flux for the small gray body A_1 as

$$q_{net} = \sigma \varepsilon_1 A_1 (T_1^4 - T_2^4) \tag{8.69}$$

The flux q_{net} in this case is the energy required to maintain surface A_1 at the temperature T_1 under the assumptions given.

Example 8.8

A thermocouple is inserted into a high-temperature air stream flowing in a 4-in. diameter duct. The thermocouple indicates a temperature of 550F while the wall temperature of the duct is 200F. We would like to determine the actual gas temperature and an expression for the error in the measurement. The conditions are shown in the figure.

SOLUTION
The thermocouple can be considered as a small gray body in a large enclosure. When the junction of the couple reaches an equilibrium state the energy convected to the junction is radiated to the wall of the duct, thus $q_{ct} = q_{Rw}$ so that

$$hA_t(T_g - T_t) = \sigma \varepsilon_t A_t (T_t^4 - T_s^4)$$

which gives for the gas temperature

$$T_g = T_t + [\sigma \varepsilon_t (T_t^4 - T_s^4)/h]$$

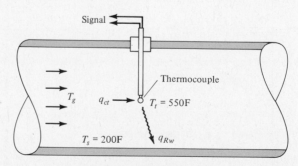

Figure for Example 8.8.

The term in brackets indicates the amount of error in the gas temperature measurement due to radiation to the duct wall. For the temperatures given with $\varepsilon_t = 0.9$ and $h = 12$ Btu/hrft^{2}F we calculate

$$T_g = 550 + 0.1714(0.9)(10.1^4 - 6.6^4)/12 = 659F$$

showing that the thermocouple indicates about 109F low. Increasing the heat transfer coefficient and decreasing the emittance decreases the measurement error. It is also possible to add radiation shields to decrease the error, a topic we will discuss later on in Section 8.8.3.

8.7.5 Two Gray Surfaces with a Refractory Surface

The physical situation described in this section is the same as that shown in Figure 8.14(a) except, of course, the surfaces are gray instead of black which means that in addition to the radiation exchange flux vectors shown there are also multiple reflections taking place. Some of the energy that leaves surface A_1 returns to A_1 from A_2 and from the refractory surface A_R and leaves again; aspects that are embodied in the concept of radiosity.

In the application of the surface resistance principle to the subject arrangement we must note that our result will be an approximation because with the refractory surface present F_{12} will be less than one. Figure 8.27 shows the generalized orientation for gray surfaces A_1, A_2, and A_R. The refractory surface A_R behaves as though it were black because there is no flux passing through its surface resistance. Therefore, J_R equals E_{bR} and A_R reaches a temperature T_R such that the surface reradiates all the energy it receives. The additional resistances $1/A_1F_{1R}$ and $1/A_2F_{2R}$ complicate the equivalent resistance needed to determine the next flux q_{net12R}. The resistance of the parallel branch is the product over the sum so that

$$R_{eq} = \frac{\rho_1}{\varepsilon_1 A_1} + \frac{1/A_1F_{12}(1/A_1F_{1R} + 1/A_2F_{2R})}{1/A_1F_{12} + 1/A_1F_{1R} + 1/A_2F_{2R}} + \frac{\rho_2}{\varepsilon_2 A_2} \tag{8.70}$$

The expression for the equivalent resistance can be algebraically rearranged and some terms combined to obtain the net flux from A_1 to A_2 in the presence of A_R as follows:

$$q_{net12R} = \frac{A_1(E_{b1} - E_{b2})}{\dfrac{1}{\left[F_{12} + \dfrac{F_{1R}A_2F_{2R}}{A_1F_{1R} + A_2F_{2R}}\right]} + \dfrac{\rho_1\varepsilon_2A_2 + \rho_2\varepsilon_1A_1}{\varepsilon_1\varepsilon_2A_2}} \tag{8.71}$$

We note that as ε_1 and ε_2 approach the value 1 then ρ_1 and ρ_2 approach zero and Eq. 8.71 reduces to Eq. 8.45 for two black surfaces in the presence of a refractory surface.

The concepts embodied in Eqs. 8.71, 8.68, and 8.45 have also been expressed in terms of the modified interchange factors; $\bar{F}$, $\mathcal{F}$, and $\bar{\mathcal{F}}$. The

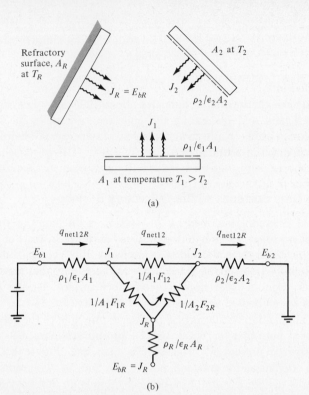

(a)

(b)

Figure 8.27   Radiant exchange between two gray surfaces in the presence of a refractory surface. (a) Generalized orientation for two gray surfaces in the presence of a refactory surface. (b) Equivalent circuit diagram for radiant exchange between the surfaces shown in (a).

modified factors are useful when a number of calculations must be made for a given geometry with the emittance as a parameter. The factors relate to the fluxes as follows; from Eq. 8.45 we write

$$q_{\text{net}12R} = A_1 \bar{F}_{12}(E_{b1} - E_{b2}) \tag{8.72}$$

where

$$\bar{F}_{12} = F_{12} + \frac{A_2 F_{2R} F_{1R}}{A_2 F_{2R} + A_1 F_{1R}} \tag{8.73}$$

From Eq. 8.68 we write

$$q_{\text{net}12} = A_1 \mathscr{F}_{12}(E_{b1} - E_{b2}) \tag{8.74}$$

where

$$\mathscr{F}_{12} = \frac{\varepsilon_1 \varepsilon_2 A_2 F_{12}}{\rho_1 \varepsilon_2 A_2 F_{12} + \varepsilon_1 \varepsilon_2 A_2 + \rho_2 A_1 \varepsilon_1 F_{12}} \tag{8.75}$$

and from Eq. 8.71 we define

$$q_{\text{net}12R} = A_1 \bar{\mathscr{F}}_{12}(E_{b1} - E_{b2}) \tag{8.76}$$

where

$$\overline{\overline{\mathscr{F}}}_{12} = \frac{\varepsilon_1\varepsilon_2 A_2 \overline{F}_{12}}{\rho_1\varepsilon_2 A_2 \overline{F}_{12} + \varepsilon_1\varepsilon_2 A_2 + \rho_2 A_1\varepsilon_1 \overline{F}_{12}} \tag{8.77}$$

The modified interchange factors hold only for the physical conditions defined for the two surfaces and $\mathscr{F}_{12}$ and $\overline{\overline{\mathscr{F}}}_{12}$ are approximations when F_{12} is less than one.

When more than two surfaces plus refractory are present, the modified factors loose their significance and it is necessary to determine the individual surface radiosities; the particular surface flux can then be calculated from Eq. 8.66 as is shown in the next section.

8.7.6 Network Analysis

When there are three or more active surfaces involved in the physical arrangement, the multisurface problem can be modeled with dc circuits such as those shown in Figure 8.28 for three and four surfaces. For the case of N surfaces there

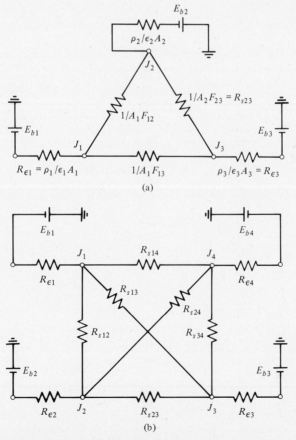

Figure 8.28 Electrical analog circuits for (a) three gray surface and (b) four gray surface radiant exchange problems.

will be N surface resistances and when the surfaces are all nonreentrant ($F_{ii} = 0$) each surface will have $N - 1$ space resistances associated with it; consequently, there will be $N(N - 1)/2$ unique space resistances in the diagram, because each space resistance serves two surfaces. For example, in Figure 8.28(b) we count six space resistances and $N(N - 1)/2 = 4(4 - 1)/2 = 6$ as we have noted above. Solution of the network involves writing an energy balance at each node [summation of currents into a junction (node) is zero] and obtaining a series of N equations for N radiosities which can be solved by algebraic or matrix means. We explore these aspects in Example 8.9.

Example 8.9

Two square gray surfaces 2 by 2 ft are oriented as shown in the figure. Surface A_1 is to be maintained at 1000R and A_2 at 2000R in a large surroundings whose walls are at 500R. The surface properties are $\varepsilon_1 = 0.5$ and $\varepsilon_2 = 0.3$. We would like to know the energy supply requirements to maintain the surfaces at the temperatures indicated.

SOLUTION

We construct the electrical analogy for this orientation and note that it is essentially identical to the diagram shown in Figure 8.28(a). Since area A_3 is very large $\rho_3/\varepsilon_3 A_3$ is essentially zero so that $J_3 = E_{b3}$ and we need only two equations for J_1 and J_2. First we determine all resistances and emissive powers as follows:

$$E_{b1} = \sigma T_1^4 = 0.1714(10^4) = 1714 \text{ Btu/hrft}^2$$

$$E_{b2} = \sigma T_2^4 = 0.1714(20^4) = 27{,}424 \text{ Btu/hrft}^2$$

$$E_{b3} = \sigma T_3^4 = 0.1714(5^4) = 107.1 \text{ Btu/hrft}^2$$

$$R_{\varepsilon1} = \frac{\rho_1}{\varepsilon_1 A_1} = \frac{0.5}{0.5(4)} = 0.250 \text{ ft}^{-2}$$

$$R_{\varepsilon2} = \frac{\rho_2}{\varepsilon_2 A_2} = \frac{0.7}{13(4)} = 0.583 \text{ ft}^{-2}$$

$$R_{s12} = \frac{1}{A_1 F_{12}} = \frac{1}{4(0.2)} = 1.25 \text{ ft}^{-2}$$

$$R_{s13} = \frac{1}{A_1 F_{13}} = \frac{1}{4(0.8)} = 0.3125 \text{ ft}^{-2}$$

$$R_{s21} = \text{same as } R_{s12}$$

$$R_{s23} = \frac{1}{A_2 F_{23}} = \frac{1}{4(0.8)} = 0.3125 \text{ ft}^{-2}$$

To determine the equations in J_1 and J_2 we sum the *currents* around the nodal

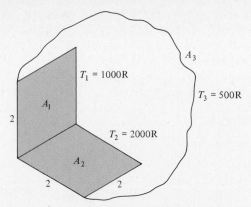

Figure for Example 8.9.

junctions [Figure 8.28(a)] representing J_1 and J_2. At node J_1:

$$\frac{E_{b1} - J_1}{0.25} + \frac{J_2 - J_1}{1.25} + \frac{J_3 - J_1}{0.3125} = 0$$

At node J_2:

$$\frac{E_{b2} - J_2}{0.583} + \frac{J_3 - J_2}{0.3125} + \frac{J_1 - J_2}{1.25} = 0$$

At node J_3:

$J_3 = E_{b3}$ since A_3 is very large

Substituting values for E_{b1}, E_{b2}, and E_{b3} and collecting like terms yields

$$7198.8 + 0.80J_2 - 8J_1 = 0$$
$$47{,}355.4 + 0.80J_1 - 5.714J_2 = 0$$

which are easily solved for the radiosities as

$$J_1 = 1753.16 \ \text{Btu/hrft}^2$$
$$J_2 = 8533.06 \ \text{Btu/hrft}^2$$

The net flux at surface A_1 is the *current* flowing between E_{b1} and J_1; therefore,

$$q_{\text{net1}} = \frac{E_{b1} - J_1}{0.25} = \frac{1714 - 1753.16}{0.25}$$
$$= -156.6 \ \text{Btu/hr}$$

which means that 156.6 Btu/hr must be removed from surface A_1 to maintain the temperature at 1000R. At surface A_2

$$q_{\text{net2}} = \frac{E_{b2} - J_2}{0.583} = \frac{27{,}424 - 8533}{0.5833}$$
$$= 32{,}384 \ \text{Btu/hr}$$

Accordingly, 32,384 Btu/hr must be supplied to surface A_2 to maintain the temperature at 2000R. Of this energy most goes to the surroundings A_3 and some goes to area A_1. We can compute the latter two amounts separately and the sum should agree with q_{net2} as a check. The net amount that leaves A_2 and goes to the surroundings is

$$_2q_{3net} = \frac{J_2 - J_3}{R_{s23}} = \frac{8533.06 - 107.1}{0.3125}$$

$$= 26{,}963 \text{ Btu/hr}$$

and from A_2 to surface A_1

$$_2q_{1net} = \frac{J_2 - J_1}{R_{s12}} = \frac{8533.06 - 1753.16}{1.25}$$

$$= 5423.9 \text{ Btu/hr}$$

The sum of the latter two net fluxes $_2q_{3net} + _2q_{1net} = 32{,}386 \text{ Btu/hr}$ agrees very closely with q_{net2} calculated above and serves as a check on our calculation of J_1 and J_2. Network analysis for more than three surfaces proceeds in similar manner except that the amount of work to determine the radiosities increases. This completes the example.

8.8 RADIATION SHIELDING

It is possible to decrease the amount of radiant energy exchange between two surfaces by placing a thin, opaque plate between the surfaces. Doing so introduces additional surface and space resistances into the thermal circuit which decrease the radiant flux. Several such shields are very effective as was demonstrated in the early part of the space age where it was necessary to store liquid oxygen, hydrogen, and nitrogen in the hot desert climate of southern California. Such tanks were fabricated with double walls and layers of crinkled aluminized mylar between them. A low vacuum was continuously maintained between the double walls to eliminate convection and the radiation shields effectively reduced boil-off to a few percent per month.

8.8.1 Black Shields

If we consider two large parallel black planes A_1 and A_2 with a black opaque surface A_3 of negligible mass between the surfaces as shown in Figure 8.29(a), when the planes are large in extent and the separation distance small, F_{12} is essentially 1 so that the unit area heat flux without shields is

$$\frac{q_{ns}}{A_1} = E_{b1} - E_{b2} \tag{8.78}$$

Adding one shield A_3 adds an additional space resistance $1/A_3F_{32}$ so that the flux becomes for $F_{32} = 1$,

$$\frac{q_{1s}}{A_1} = \frac{E_{b1} - E_{b2}}{2} \tag{8.79}$$

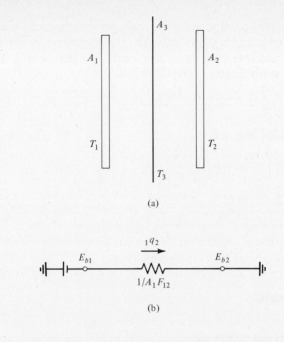

(a)

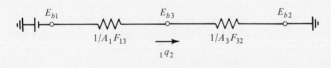

(b)

(c)

Figure 8.29   Orientation of black parallel planes with and without one shield and corresponding electrical analog circuits. (a) Black parallel planes with shield A_3 of negligible mass in between A_1 and A_2. (b) Network for surfaces A_1 and A_2 without the shield in place. (c) Network for surfaces A_1 and A_2 with one shield in place.

and we see that the flux has been halved. With two shields we would have one-third the flux and with three shields, one-fourth the flux so that we can generalize and say that N black shields between two black planes gives $1/(N + 1)$ of the flux without shields. The temperature of the shield depends on the emissive power the shield requires to reradiate all the energy received.

Example 8.10

Determine the temperature T_3 of the shield between A_1 and A_2 in Figure 8.29 when surface A_1 is at 1500F and surface A_2 is at 500F.

SOLUTION

The circuit diagram in Figure 8.29(c) shows that E_{b3} lies halfway between E_{b1} and E_{b2} because the two space resistances are the same ($F_{12} = F_{23} = 1$ and

$A_1 = A_2$). Therefore,

$$E_{b1} = \sigma T_1^4 = 0.1714(19.6)^4 = 25{,}295 \text{ Btu/hrft}^2$$

$$E_{b2} = \sigma T_2^4 = 0.1714(9.6)^4 = 1455.8 \text{ Btu/hrft}^2$$

$$E_{b3} = \frac{E_{b1} + E_{b2}}{2}$$

$$= 13{,}375.4 \text{ Btu/hrft}^2$$

$$T_3 = \left(\frac{E_{b3}}{\sigma} \right)^{0.25} = 100 \left(\frac{13{,}375.4}{0.1714} \right)^{0.25}$$

$$= 1671R \text{ or } 1211F$$

The shield temperature is closer to the higher temperature surface than to the lower temperature surface due to the fourth power influence in the Stefan-Boltzmann law. This concludes the example.

8.8.2 Gray Shields

When the surfaces and the shields are gray, the surface resistances $\rho/\varepsilon A$ must be added to the thermal circuit and those resistances serve to further decrease the radiant flux. When the surface emittances (and consequently the surface resistances) are not equal, no general expression or conclusion can be stated with regard to the amount of decrease. Each case must be calculated individually as

$$q_{\text{with shields}} = (E_{b1} - E_{b2})/R_{\text{eq}} \tag{8.80}$$

where E_{b1} and E_{b2} are the black emissive powers of the surfaces that have the interstitial shields and R_{eq} is the equivalent resistance in the thermal circuit. When the shields all have the same emissivity, some simplification results; for example, for one gray shield between two gray surfaces, as shown in Figure 8.30, the heat flux with no shields is

$$q_{ns} = \frac{E_{b1} - E_{b2}}{2R_\varepsilon + R_s} \tag{8.81}$$

With one shield in place the resistances shown in Figure 8.30(c) then comprise the equivalent resistance and the flux becomes

$$q_{1s} = \frac{E_{b1} - E_{b2}}{4R_\varepsilon + 2R_s}$$

$$= \frac{E_{b1} - E_{b2}}{2(2R_\varepsilon + R_s)} \tag{8.82}$$

which is half of the amount without the shield present. Adding another shield would add two more surface resistances and one more space resistance so that the equivalent resistance would be $3(2R_\varepsilon + R_s)$ which would make the heat flux for two shields be one-third of the heat flux with no shields. For N shields we

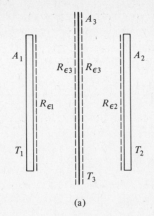

(a)

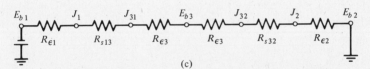

(b)

(c)

Figure 8.30 Orientation of gray parallel planes with gray shield and corresponding electrical analog circuit. (a) Gray parallel planes with gray shield A_3 of negligible mass in between A_1 and A_2. (b) Network for gray surfaces A_1 and A_2 without the shield in place. (c) Network for two gray surfaces A_1 and A_2 with the gray shield A_3 in place.

obtain the same reduction in heat flux $1/(N+1)$ from the case with no shields as we did when all surfaces were black. When the surface emittances are different no particular generalization can be made as we have noted above.

Example 8.11

An aluminum shield made of thin commercial sheet ($\varepsilon = 0.09$) is placed between a large cast-iron plate at 1500F ($\varepsilon = 0.70$) and a wrought iron plate at 500F ($\varepsilon = 0.91$). We would like to determine the reduction in heat flux caused by the shield and the temperature of the shield at equilibrium.

SOLUTION
The emittance of the aluminium commercial sheet is about $\varepsilon = 0.09$ in clean condition and we assume it remains so. In Figure 8.30 we let A_1 be the cast-iron

plate and A_2 the wrought iron plate. We assume the plates are large enough so that F_{13} and F_{32} are essentially 1 and we can make the calculation per unit surface area. From Example 8.9 we see that

$$E_{b1} = 25{,}295 \text{ Btu/hrft}^2$$

$$E_{b2} = 1455.8 \text{ Btu/hrft}^2$$

and the pertinent resistances for the case with no shields are shown in Figure 8.30(b). Thus on a unit area basis we obtain:

$$R_{\varepsilon1} = \frac{\rho_1}{\varepsilon_1} = \frac{0.3}{0.7} = 0.429$$

$$R_{\varepsilon2} = \frac{\rho_2}{\varepsilon_2} = \frac{0.09}{0.91} = 0.099$$

$$R_{s12} = \frac{1}{F_{12}} = \frac{1}{1} = 1$$

The equivalent resistance for the circuit in Figure 8.30(b) is the sum of the individual resistances; therefore,

$$R_{eq} = 0.429 + 1 + 0.099 = 1.527$$

We should also note that on a unit area basis all resistances are dimensionless. The heat flux with no shields then is

$$\frac{q_{ns}}{A} = \frac{E_{b1} - E_{b2}}{R_{eq}}$$

$$= \frac{25{,}295 - 1455.8}{1.527}$$

$$= 15{,}608 \text{ Btu/hrft}^2$$

With the shield in place there is an additional space resistance and two more surface resistances as shown in Figure 8.30(c).

$$R_{\varepsilon3} = \frac{\rho_3}{\varepsilon_3} = \frac{0.91}{0.09} = 10.11$$

$$R_{s13} = R_{s32} = \frac{1}{1} = 1$$

and the equivalent resistance is the sum

$$R_{eq} = 0.429 + 1 + 10.11 + 10.11 + 1 + 0.099 = 22.75$$

We can recognize immediately that the shield will be very effective because $R_{\varepsilon3}$ is very high compared to the equivalent resistance with no shields. Thus

$$\frac{q_{1s}}{A_1} = \frac{25{,}295 - 1455.8}{22.75}$$

$$= 1048 \text{ Btu/hrft}^2$$

which is about 6.7 percent of the flux with no shields. The shield temperature is determined from E_{b3} which is given by

$$E_{b3} = E_{b1} - (E_{b1} - E_{b2})(R_{\varepsilon 1} + R_{s13} + R_{\varepsilon 3})/R_{eq}$$

$$= 25{,}295 - (25{,}295 - 1455.8)(11.539)/22.75$$

$$= 13{,}203.5 = \sigma T_3^4$$

$$\therefore T_3 = (13{,}203.5/0.1714)^{0.25}(10^2) = 1666R$$

The potential E_{b3} in this example is very close to the value calculated in Example 8.9 because $R_{\varepsilon 3}$ is high and the same on both sides of the shield; therefore, E_{b3} is located essentially halfway between E_{b1} and E_{b2}. Since E_{b3} is nearly the same in both examples, the shield temperatures are also nearly the same. Were the surface resistances of the shield not the same, the value of E_{b3} would shift away from the median position and the shield temperature would be higher or lower than what we have calculated. We conclude the example with the final observation that the shield temperature is uncomfortably close to the melt point of pure aluminum; therefore, that aspect would need careful consideration for the alloy comprising the commercial sheet.

8.8.3 Shielding in Temperature Measurement

A very important application of shielding arises in the measurement of gas temperature. We have seen in Example 8.7 that the thermocouple indicates a temperature that is 109F less than the actual gas temperature. This error can be considerably reduced by placing a cylindrical shield over the thermocouple (or other transducer) in such a manner that the gas flow over the thermocouple is not impeded, but radiation from the couple to the wall is partially blocked. Example 8.12 considers such an application.

Example 8.12

A cylindrical stainless-steel shield $\frac{3}{4}$ in. in diameter and 4 in. long is attached to the thermocouple of Example 8.7 and aligned in the flow direction as shown in the figure. The thermocouple temperature is 550F and the wall temperature is 200F. We want to determine the true gas temperature under these conditions.

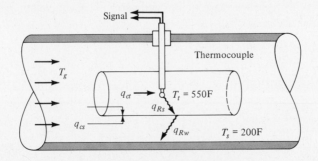

Figure for Example 8.12.

SOLUTION

The presence of the shield requires an additional energy balance in order to determine the shield temperature. An energy balance on the thermocouple requires the convected heat flux from the gas q_{ct} to be equal to the radiant flux to the shield q_{Rs}. Thus

$$q_{ct} = q_{Rs}$$

or

$$h_t A_t(T_g - T_t) = (E_{bt} - E_{bs})/R_{eq}$$

$$h_t(T_g - T_t) = \sigma(T_t^4 - T_s^4)/A_t R_{eq} \tag{8.83}$$

The energy balance for the shield requires that the sum of the radiant input from the thermocouple q_{Rs} plus the convective input q_{cs} be equal to the radiation flux from the outer surface of the shield q_{Rw} to the duct wall.

$$q_{cs} + q_{Rs} = q_{Rw}$$

or

$$2h_s A_s(T_g - T_s) + \sigma(T_t^4 - T_s^4)/R_{eq} = \sigma \varepsilon_s A_s(T_s^4 - T_w^4) \tag{8.84}$$

where $2A_s$ represents the inner and outer surface area of the shield assuming a thin walled cylinder. The convection coefficients would normally be calculated from geometry and flow data as discussed in Chapter 6. For this example we will assume $h_t = 12$ Btu/hrft^{2}F for the thermocouple and $h_s = 15$ Btu/hrft^{2}F for the surfaces of the shield, and for emittances $\varepsilon_t = 0.9$ and $\varepsilon_s = 0.5$ for the thermocouple and shield, respectively.

We note that F_{ts} is very close to 1 (it could be made equal to 1 with suitable baffles at the ends which minimize flow restriction) so that the radiation between the thermocouple and the shield is a two-surface problem with an analogous circuit as shown in Figure 8.26. Accordingly, the equivalent resistance is given by

$$R_{eq} = \frac{\rho_t}{\varepsilon_t A_t} + \frac{1}{A_t F_{ts}} + \frac{\rho_s}{\varepsilon_s A_s}$$

$$A_t R_{eq} = \frac{\rho_t}{\varepsilon_t} + \frac{1}{F_{ts}} + \frac{\rho_s A_t}{\varepsilon_s A_s}$$

We assume the ratio of shield inner area A_s to thermocouple area to be $A_s/A_t = 50$, as representative of the geometry. The equivalent resistance area product then becomes

$$A_t R_{eq} = \frac{0.1}{0.9} + \frac{1}{1} + \frac{0.5}{0.5(50)} = 1.131$$

and with $T_t = 550$F $\equiv 1010$R we can write Eq. 8.83 as

$$T_g = 550 + 0.0126\left(10.1^4 - \left(\frac{T_s}{100}\right)^4\right)$$

$$T_g = 681.3 - 0.0126\left(\frac{T_s}{100}\right)^4 \tag{8.85}$$

Substituting Eq. 8.83 into Eq. 8.84 and rearranging gives

$$T_g - T_s + \left(T_g - T_t\right)\frac{h_t A_t}{2h_s A_s} = \left(T_s^4 - T_w^4\right)\frac{\sigma \varepsilon_s}{2h_s}$$

and substituting the pertinent data yields the working equation

$$1.008 T_g - T_s - 0.008 T_t = 0.00286\left(\frac{T_s}{100}\right)^4 - 5.42 \tag{8.86}$$

We now have two equations in T_g and T_s, but the fourth power term prevents an explicit solution. We must assume a value for T_s, calculate T_g from Eq. 8.85, then see if Eq. 8.86 is satisfied. Accordingly, we assume $T_s = 995R = 535F$ for the shield temperature and from Eq. 8.85 we calculate the gas temperature as

$$T_g = 681.3 - 0.0126(9.95)^4 = 557.8F$$

and checking with Eq. 8.86 gives

$$1.008 T_g - T_s - 0.008 T_t = 562.3 - 535 - 4.4 = 22.86 \text{ on left side,}$$

$$0.00286\left(\frac{T_s}{100}\right)^4 - 5.42 = 0.00286(9.95)^4 - 5.42 = 22.61 \text{ on right}$$

which very nearly shows equality; it looks like the shield temperature is a little bit low. The gas temperature in this case is only 7 degrees above the thermocouple temperature due to the presence of the shield, as compared to 109F without the shield present.

It would also be of interest to consider the gas temperature known as $T_g = 659F$ from Example 8.7 and the wall temperature $T_w = 200F$ and use Eqs. 8.83 and 8.84 to see what values of thermocouple temperature and shield temperature would result under the conditions given, but since both T_t and T_s appear to the fourth power, the trial-and-error procedure is very sensitive to the mathematics and the calculations are best performed on a computer. The above examples have demonstrated the effectiveness of shields. An inert gas also can act as a shield, since the gas absorbs energy and reradiates what it receives, thereby adding a resistance to the circuit. The gas is not as effective as the opaque shield, because it transmits as well as reradiates. We now direct our attention to the subject of radiation to and from gases.

8.9 GAS RADIATION

In the radiation phenomena we have worked with so far the energy distribution has been over the entire spectrum, according to Planck's law for black surfaces, and in a continuous but irregular relationship for real surfaces. For gray surfaces we assumed that ε was not a function of wavelength; therefore, the energy distribution was proportional to the black Planckian relationship at, of course, a lower level for the same temperature.

For gases the spectral distribution of radiation is different due to the nature of the interaction of radiant quanta with gas molecules. We will look briefly at some fundamental concepts of the interaction of radiation with gas molecules and then see how we can relate Beer's law to an engineering approach to calculating radiant exchange between an idealized gray gas and a surface.

8.9.1 Fundamental Concepts

We recall from physics that the interaction of radiation quanta or photons of energy content hf, where h is Planck's constant and f the frequency, is with the electrons in orbit around the nucleus. If the *character* of the radiation is of the correct nature (proper value of f), the quanta will knock an electron into the next higher orbit, the radiation is considered absorbed, and the electron is excited into a higher energy state in a shell that it normally does not occupy. Conversely, if one of the excited electrons drops to a lower energy orbit a quantum of energy hf will be emitted and the frequency depends on which energy level the electron was in. We are not here specifically concerned with exactly what frequencies result, but to note that there will be certain bands of frequency (or wavelength if you prefer) where the radiation will be observable. Further we also note that only the phenomena of absorption and emission have significance; radiant energy that is not absorbed passes through the gas (transmitted); accordingly, the phenomenon of reflection has no significance. Radiation in the thermal region of the spectrum does not have enough energy to penetrate the force field around the nucleus, but it can be deflected or scattered from its direction. We do not consider any interactions with solid particles that may be in the gas such as dust or other fine particles.

The spectral distribution relationship of $E_{g\lambda}$ as a function of wavelength is then a series of peaks located at particular windows $\Delta\lambda_n$ depending on the type of gas under consideration. Figure 8.31 shows a typical spectral distribution for a gas at temperature T_g compared to the Planckian blackbody curve $E_{b\lambda}$ also at temperature $T = T_g$. The amount of energy emitted or absorbed depends on the number of gas molecules present, therefore, the gas pressure p and the optical depth L become the important parameters; the product pL is called the optical thickness of the gas. If the radiating gas is in a mixture of other gases, then the partial pressure is indicative of the number of radiating molecules present. Later we will see that the product pL in atm-ft or atm-m is the correlating parameter for gas emissivity. As the pL product increases, the peaks shown in the $\Delta\lambda$ windows of Figure 8.31 become stronger (larger), and when the gas is very dense the peaks broaden slightly and approach the Planckian curve. We can define a gas emittance ε_g as the energy emitted by the gas divided by the energy of a Planckian emitter at the gas temperature; thus

$$\varepsilon_g = \frac{E_g}{E_b} = \frac{E_g}{\sigma T^4} \tag{8.87}$$

The nature of the phenomena shows that even when the gas is dense and thick enough to be black, the emittance is still significantly less than one. Further, since the absorption or emission peaks are located in a particular wavelength

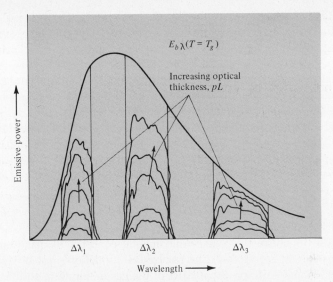

Figure 8.31   Typical spectral distribution of gas radiation at temperature T_g as a function of wavelength with optical thickness as a parameter.

region, when incoming radiation is from a source that is different from the gas temperature, the absorptance cannot equal the emittance; therefore, we cannot use Kirchhoff's law with gases.

Let's consider for a moment parallel or specular radiation of intensity $I_{\lambda,0}$ at location $x = 0$, a portion of which passes through the cylindrical gas volume shown in Figure 8.32. The amount of radiation absorbed is proportional to the local intensity $I_{\lambda,x}$ and the proportionality constant m_λ is called the monochromatic absorption coefficient. Thus

$$dI_{\lambda,x} = -m_\lambda I_{\lambda,x}\,dx \qquad (8.88)$$

The absorption coefficient m_λ has the units (length)$^{-1}$. Separating variables in Eq. 8.88 and integrating over the limits $x = 0$ where $I_{\lambda,x} = I_{\lambda,0}$ and $x = L$ where $I_{\lambda,x} = I_{\lambda,L}$ yields

$$\ln\frac{I_{\lambda,L}}{I_{\lambda,0}} = -m_\lambda L \qquad (8.89)$$

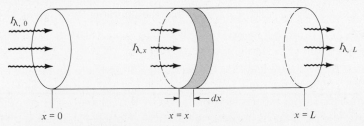

Figure 8.32   Radiation intensity nomenclature for absorption and transmission in a parallel beam of radiation passing through a gas.

Exponentiating both sides gives

$$\frac{I_{\lambda,L}}{I_{\lambda,0}} = e^{-m_\lambda L} = \tau_{g\lambda} \tag{8.90}$$

The ratio of the energy coming out of the gas volume $I_{\lambda,L}$ to the energy entering the gas volume $I_{\lambda,0}$ is exactly what we understand by the concept of transmittance $\tau_{g\lambda}$. The amount of radiant energy absorbed in the length L is $I_{\lambda,0} - I_{\lambda,L}$ and the ratio of energy absorbed to incident energy gives the absorptance of the gas, $\alpha_{g,\lambda}$. Thus

$$(I_{\lambda,0} - I_{\lambda,L})/I_{\lambda,0} = 1 - e^{-m_\lambda L} = 1 - \tau_{g\lambda} = \alpha_{g\lambda} \tag{8.91}$$

or $\alpha_{g\lambda} + \tau_{g\lambda} = 1$ since reflectance for a gas is zero. The above concepts were formulated on a monochromatic basis so that we would have to sum over all wavelengths to obtain total quantities such that $\alpha_g + \tau_g = 1$.

In order to utilize the above fundamental concepts to make engineering calculations several aspects must be resolved. Information must be available on the absorption and emission capabilities of the gas in question, and the relationship between the two since Kirchhoff's law cannot be applied. Also, since the radiation situations we are likely to encounter do not deal with specular radiation in nice parallel bundles, we must relate the radiation from geometric volumes to the particular spot of interest on the bounding surface of the volume to an equivalent specular beam length L as described above. Such a procedure is necessary, because the experimental determination of emission data for gases can only be obtained from specular radiation.

Accordingly, with effective beam length information and information for relating emissivity and absorptivity for a gas, we can calculate the net unit area heat flux to a black surface exposed to that gas as

$$\frac{q_{net}}{A} = \sigma\left(\varepsilon_g T_g^4 - \alpha_g T_s^4\right) \tag{8.92}$$

where ε_g is the gray gas emissivity at the gas temperature T_g and α_g is the gray gas absorptivity of the gas at temperature T_g for radiation from a black source at temperature T_s; in later discussion we will use the functional notation $\varepsilon_g(T_g)$ and $\alpha_g(T_g, T_s)$ to indicate these ideas. When the surface in contact with the gas is gray, the multiple reflections complicate the exchange process so that we cannot simply use the surface emissivity ε_s. Jakob (14) recommends for most cases an effective emissivity $\varepsilon_e = (\varepsilon_s + 1)/2$ so that the unit area flux is given by

$$\frac{q_{net}}{A} = \varepsilon_e \sigma\left(\varepsilon_g T_g^4 - \alpha_g T_s^4\right) \tag{8.93}$$

The effective beam length is the radius of a hemisphere containing the same gas which produces the same unit area heat flux at the center of the base circle due to radiation from the gas as is produced by the volume under consideration. Effective beam lengths have been calculated by Hottel (11) and Jakob (14) and others. Some representative values are given in Table 8.3. For shapes not listed in the table the mean beam length can be estimated to be $L = 3.5V/A_s$ where V

Table 8.3 MEAN EFFECTIVE BEAM LENGTHS

VOLUME	CHARACTERISTICS DIMENSION	BEAM LENGTH, L
1. Sphere	Diameter D	$0.63D$
2. Infinite cylinder	Diameter D	$0.94D$
3. Right circular cylinder, height = diameter		
a. Radiating to center of base	Diameter D	$0.70D$
b. Radiating to whole surface	Diameter D	$0.60D$
4. Right circular cylinder, height = diameter/2		
a. Radiating to center of base	Diameter D	$0.43D$
b. Radiating to cylinder surface only	Diameter D	$0.46D$
c. Radiating to entire surface	Diameter D	$0.45D$
5. Right circular cylinder, height = 2 diameters		
a. Radiating to center of base	Diameter D	$0.60D$
b. Radiating to cylindrical surface	Diameter D	$0.76D$
c. Radiating to entire surface	Diameter D	$0.73D$
6. Semi-infinite cylinder		
a. Radiating to center of base	Diameter D	$0.90D$
b. Radiating to entire base	Diameter D	$0.65D$
7. Rectangular parallelepiped		
a. Cube	Edge E	$0.60E$
b. $1 \times 1 \times 4$ to 1×4 face		$0.82S$
c. $1 \times 1 \times 4$ to 1×1 face	Shortest	$0.71S$
d. $1 \times 1 \times 4$ to all faces	side S	$0.81S$
8. Infinite parallel planes	Separation Distance S	$1.76S$
9. Infinite half-circular cylinder radiating to dA in middle of flat side	Radius, R	$1.26R$

is volume of the gas and A_s the bounding surface area. At very low gas pressures the values in Table 8.3 must be multiplied by a factor of 1.14 so that $L_0 = 1.14L$ because very little radiation is reabsorbed by a very dilute gas. The values for L_0 must be used with H_2O gas to evaluate the size influence on the number of radiating molecules with regard to emissivity, the topic to which we now direct our attention.

8.9.2 Gray Gas Properties

As we have discussed above, the radiant emission from a gas depends on the number of emitting molecules present which is proportional to the $p_g L$ product in ft atm or cm atm. When the gas is water vapor, the partial pressure must be very low in order to avoid the complication of reabsorption; consequently, the data for water was extrapolated to zero partial pressure to present the emissivity data and a separate correction is made to account for the influence of partial pressure p_w and total pressure P. Data for other gases are presented at a total pressure of 1 atm with separate charts to correct for the effect of total pressure. Since water vapor and carbon dioxide are the chief products of combustion for most fuels, we will restrict our discussion to those two gases. The principles we will work out apply equally to other gases except that mutual absorption information may be lacking. Hottel and Sarofim (11) present emissivity data for SO_2, CO, NH_3, HCl, NO, NO_2, and CH_4 in addition to CO_2 and H_2O. Jakob (14) presents an expanded version of the original CO_2 and H_2O emissivity data of Hottel which is reproduced in Figures 8.33 and 8.34. The two figures are at a total pressure of 1 atm and various beam lengths. The data for CO_2 (Figure 8.33) are at the actual partial pressures obtained in the experiments and a total pressure of 1 atm; corrections for total pressure can be obtained from Figure 8.35. The data for H_2O vapor have been extrapolated to zero partial pressure (at 1 atm total pressure) because reabsorption is particularly severe for the H_2O molecule; corrections for total and partial pressure are then obtained from Figure 8.36 at the pertinent values of $p_w L$.

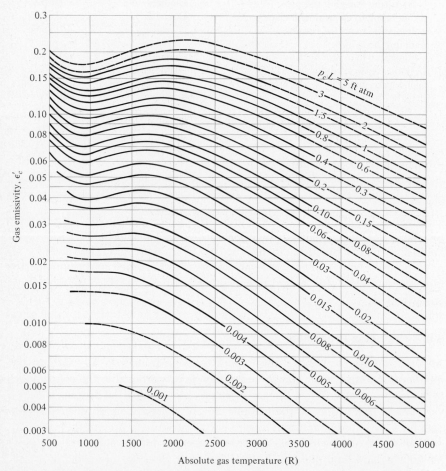

Figure 8.33 Emissivity data for CO_2 at a total pressure of 1 atm for different values of optical thickness, $p_c L$, according to Jakob (14). Reproduced with permission of John Wiley & Sons, Inc.

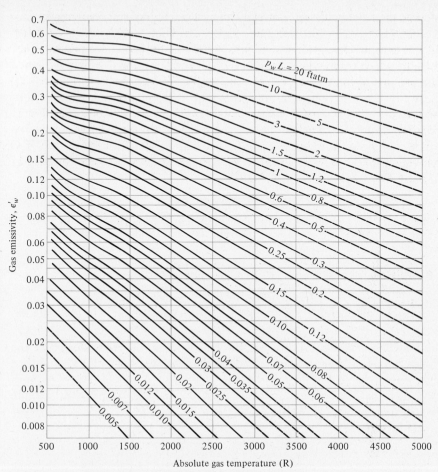

Figure 8.34 Emissivity data for H_2O at a total pressure of 1 atm and partial pressure of water vapor extrapolated to zero with p_wL as the parameter according to Jakob (14). Reproduced with permission of John Wiley & Sons, Inc.

Figure 8.37 permits estimation of the mutual absorption effect when mixtures of CO_2 and H_2O are being analyzed; the figure gives $\Delta\varepsilon$ which is to be subtracted from the sum of the emissivities determined from Figures 8.33 and 8.34. Thus

$$\varepsilon_{mix} = \varepsilon_{CO_2} + \varepsilon_{H_2O} - \Delta\varepsilon \qquad (8.94)$$

Equation 8.94 can also be used to estimate the absorption of a mixture α_{mix} by estimating a $\Delta\alpha$ from Figure 8.37.

So far we have just discussed emissivity determinations for gases. The absorptivity of the gas depends not only on the gas temperature, but also on the source temperature of the radiation being absorbed, T_s, in a complicated way.

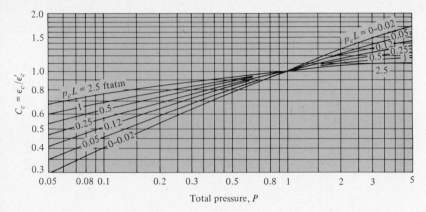

Figure 8.35 Correction factor for the influence of total pressure on the emissivity of carbon dioxide gas as given by Jakob (14). Reproduced with the permission of John Wiley & Sons, Inc.

For carbon dioxide gas Hottel (15) recommends

$$\alpha_{gCO_2}(T_g, T_s, p_c L) = \left(\frac{T_g}{T_s}\right)^{0.65} \varepsilon_{gc}\left(T_s, p_c L \frac{T_s}{T_g}\right) \tag{8.95}$$

The functional notation means that the emissivity of CO_2 is evaluated at the temperature of the source T_s and an adjusted optical depth $p_c L(T_s/T_g)$ and then multiplied by the ratio T_g/T_s raised to the 0.65 power. A note of caution requires recognition that the temperature ratios must always be evaluated with absolute units, degrees Rankine or Kelvin.

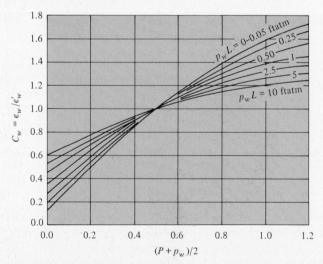

Figure 8.36 Correction factor for the influence of p_w and P on the emissivity of water vapor as given by Jakob (14). Reproduced with permission of John Wiley & Sons, Inc.

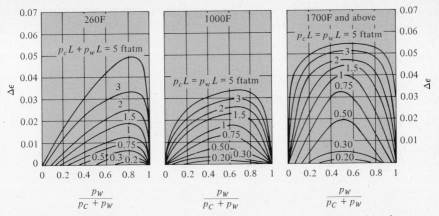

Figure 8.37 Correction factor for mutual absorption in mixtures of carbon dioxide and water vapor gases as given in Jakob (14). Reproduced with permission of John Wiley & Sons, Inc.

For water vapor the procedure is similar to that for CO_2 except that the ratio of absolute temperatures T_g/T_s is raised to a different power as follows:

$$\alpha_{gH_2O}(T_g, T_s, p_w L) = \left(\frac{T_g}{T_s}\right)^{0.45} \varepsilon_{gw}\left(T_s, p_w L \frac{T_s}{T_g}\right) \tag{8.96}$$

The evaluation of absorptivity for other gases is recommended by Hottel and Sarofim (11) to follow Eq. 8.95 using the 0.65 power on the absolute temperature ratio.

8.9.3 Heat Flux Calculations

When the gas emissivity and gas absorptivity can be evaluated by the methods described in the previous section, the heat flux per unit area can then be estimated from Eq. 8.93 for the mean beam length employed. The most effective way to consolidate the above information is to work through an example.

Example 8.13

We would like to estimate the radiant heat flux to the walls of the combustion chamber of a liquid oxygen–liquid hydrogen rocket engine operating at a chamber pressure of 500 psia and stiochiometric mixture ratio. The temperature of the combustion gases is 4540F and the wall temperature is 700F. The combustion chamber can be approximated as a cylinder 1 ft in diameter and 1 ft long with a surface emissivity of 0.8.

SOLUTION

At stiochiometric conditions we assume that the combustion products are all H_2O at a pressure of $500/14.7 = 34$ atm. From Table 8.3 (item 3b) one obtains the mean beam length $L = 0.6D$ for radiation to the whole surface which includes the combustion chamber walls; therefore, $L = 0.6$ ft and since $P_T = p_w$ $= 34$ atm the optical depth $p_w L = 34(0.6) = 20.4$ ft atm. The pertinent temperatures are

$$T_g = 4540 + 460 = 5000R$$

$$T_s = 700 + 460 = 1160R$$

From Figure 8.34 we obtain ε'_w at $p_w L$ and at $p_w L(T_s/T_g) = 20.4(1160/5000)$ $= 4.7$; therefore,

$$\varepsilon'_{gw}(T_g, p_w L) = \varepsilon'_{gw}(5000, 20.4) = 0.24$$

$$\varepsilon'_{gw}\left(T_s, p_w L \frac{T_s}{T_g}\right) = \varepsilon'_{gw}(1160, 4.7) = 0.45$$

Since we are at a fairly high pressure, we must estimate the pressure broadening effect from Figure 8.36. We note that $(P_T + p_w)/2 = (34 + 34)/2 = 34$ atm requires considerable extrapolation, but at $p_w L = 20.4$ ft atm the curve is fairly flat so that we estimate $C_w = 1.5$ so that

$$\varepsilon_{gw}(5000, 20.4) = 0.24(1.5) = 0.36$$

$$\varepsilon_{gw}(1160, 4.7) = 0.45(1.5) = 0.68$$

After we determine the absorptance of the gas for radiation coming from the surface at temperature $T_s = 1160R$ we can use Eq. 8.93 to estimate the unit area flux. Accordingly, from Eq. 8.96

$$\alpha_{gw}(T_g, T_s, p_w L) = \left(\frac{T_g}{T_s}\right)^{0.45} \varepsilon_{gw}\left(T_s, p_w L \frac{T_s}{T_g}\right)$$

$$= \left(\frac{5000}{1160}\right)^{0.45} (0.68) > 1$$

Since α can never be greater than 1 we must recognize that the limits of this simple representation have been exceeded. Eckert and Drake (16) show that water vapor has nearly continuous strong absorption bands from about 2.5 to 20 μ. At 1160R the wavelength of maximum emission is about 5 μ (see Wien's law, Eq. 8.12) and therefore, we estimate $\alpha_{gw} = 0.95$ since the gas in our example is optically very thick. With an effective emissivity $\varepsilon_e = (\varepsilon_s + 1)/2 = (0.8 + 1)/2$ $= 0.9$ we estimate the unit area heat flux as

$$\frac{q_{net}}{A} = \varepsilon_e \sigma \left(\varepsilon_{gw} T_g^4 - \alpha_{gw} T_s^4\right)$$

$$= 0.9(0.1714)\left[0.36(50)^4 - 0.95(11.6)^4\right]$$

$$= 344,000 \text{ Btu/hrft}^2$$

This radiant heat flux is roughly the order of magnitude of the convective flux at the chamber wall. In the nozzle section of the rocket motor the LOX-H$_2$

propellant system can produce heat fluxes as high as 10 to 12 million Btu/hrft^2, depending on the mass flux per unit area. This completes the example.

Example 8.14 illustrates the application of the method to mixtures and also completes this section.

Example 8.14

Methane, CH_4, is burned in a cylindrical combustor can that is 15 cm in diameter (base) and 30 cm long. The analysis of the combustion products shows 8 percent carbon dioxide and 17 percent water vapor by volume, the balance being excess oxygen and nitrogen at a gas temperature of 3040F and a total pressure of 1 atm. The injector face is at a temperature of 940F with a surface emissivity $\varepsilon_s = 0.8$. We want to estimate the radiant heat flux from the CO_2 and H_2O vapor in the combustion gases to the center of the base of the combustor can at the injector face.

SOLUTION

We recognize that at a pressure of 1 atm the combustion gases can be considered as ideal gases so that partial pressure ratio, mole fraction, and volume percent are all equal. From mean beam length information we can evaluate optical thickness pL and also mixture emissivity and absorptivity from which follows the unit area heat flux. Table 8.3 (item 5a) gives $L = 0.6D$ for the cylinder with height $= 2$ diameters so that the mean beam length $L = 0.3$ ft. The ideal gas assumption gives $p_c = 0.08$ atm and $p_w = 0.17$ atm for the partial pressures so that the optical thicknesses are

$$p_c L = 0.08(0.3) = 0.024 \text{ ft atm}$$

$$p_w L = 0.17(0.3) = 0.051 \text{ ft atm}$$

with temperatures

$$T_g = 3040 + 460 = 3500 \text{R}$$

$$T_s = 940 + 460 = 1400 \text{R}$$

We now evaluate emissivities and absorptivities according to Eqs. 8.94, 8.95, and 8.96.

For the Carbon Dioxide

From Figure 8.33: $\varepsilon'_{gc}(3500, 0.024) = 0.02$.
From Figure 8.35: $C_c = 1$ at $P_T = 1$ atm
 so that $\varepsilon_{gc} = C_c \varepsilon'_{gc} = 0.02$,
 $p_c L(T_s/T_g) = 0.024(1400/3500) = 0.01$.
From Figure 8.33: $\varepsilon'_{gc}(1400, 0.01) = 0.031$.
From Figure 8.35: $C_c = 1$ at $P_T = 1$ atm,
 therefore $\varepsilon_{gc}(1400, 0.01) = C_c \varepsilon'_{gc} = 0.031$.
From Eq. 8.95: $\alpha_{gc} = 0.031(3500/1400)^{0.65}$
$= 0.056$.

For the Water Vapor
From Figure 8.34: $\varepsilon'_{gw}(3500, 0.051) = 0.011$.
Pressure correction at $(P_T + p_w)/2 = 1.17/2 = 0.59$.
From Figure 8.36: $C_w = 1.15$
so that $\varepsilon_{gw} = C_w \varepsilon'_{gw} = 1.15(0.011) = 0.013$,
$p_w L(T_s/T_g) = 0.051(1400/3500) = 0.02$.
From Figure 8.34: $\varepsilon'_{gw}(1400, 0.02) = 0.023$.
From Figure 8.36 at $(P_T + p_w)/2 = 0.59$ read $C_w = 1.1$,
therefore $\varepsilon_{gw}(1400, 0.02) = 0.023(1.1) = 0.025$.
From Eq. 8.96: $\alpha_{gw} = (3500/1400)^{0.45}(0.025) = 0.038$.

We must now estimate the mutual broadening correction, if any, from Figure 8.37 for the sum of optical thicknesses $p_c L + p_w L = 0.024 + 0.051 = 0.075$ and $p_w/(p_c + p_w) = (0.17/0.25) = 0.68$; therefore,

for emission at $T_g = 3500\text{R}$: $\qquad \Delta\varepsilon = 0$

for absorption at $T_s = 1400\text{R}$: $\qquad \Delta\alpha = 0$

The results indicate that there are not enough molecules in the mixture to absorb radiant energy from the other species and that essentially all the interaction is with the surface. The mixture properties are calculated from Eq. 8.94 as follows:

$$\varepsilon_{gmix} = \varepsilon_{gc} + \varepsilon_{gw} + \Delta\varepsilon = 0.02 + 0.013 + 0 = 0.033$$

$$\alpha_{gmix} = \alpha_{gc} + \alpha_{gw} + \Delta\alpha = 0.056 + 0.038 + 0 = 0.094$$

so that we can calculate the unit area heat flux to the center of the base from Eq. 8.93 as

$$\frac{q_{net}}{A} = \varepsilon_e \sigma \left(\varepsilon_{gmix} T_g^4 - \alpha_{gmix} T_s^4 \right)$$

$$= \left(\frac{0.85 + 1}{2} \right)(0.1714)\left[0.033(35)^4 - 0.094(14)^4 \right]$$

$$= 7280 \text{ Btu/hrft}^2$$

This amount of flux represents somewhat less than 10 percent of the heat flux due to convection.

8.9.4 Network Approximation

When the geometry of the surfaces in contact with a radiating gas is simple enough that an interchange factor F_{12} can be evaluated, an approximate estimate of the influence of the gas on the radiant flux between the surfaces can be made using the equivalent dc network analysis method described above in Section 8.7. The application was described by Wohlenberg (17) for the case where the gas is inactive. By inactive we mean that there is no chemical reaction, no fusion or fission, and no electrical discharge that supplies energy to the gas to independently maintain the gas temperature. An inactive radiating gas will reach a temperature level that is just high enough to reradiate all the radiant energy that

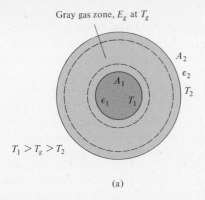

Gray gas zone, E_g at T_g

A_2
ϵ_2
T_2

A_1
ϵ_1 T_1

$T_1 > T_g > T_2$

(a)

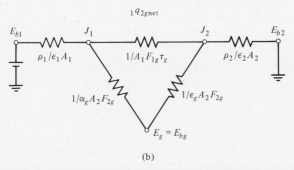

$_1q_{2gnet}$

E_{b1} J_1 J_2 E_{b2}

$\rho_1/\epsilon_1 A_1$ $1/A_1 F_{1g}\tau_g$ $\rho_2/\epsilon_2 A_2$

$1/\alpha_g A_2 F_{2g}$ $1/\epsilon_g A_2 F_{2g}$

$E_g = E_{bg}$

(b)

Figure 8.38 Network approximation of a gray gas between two long cylindrical surfaces. (a) Geometric orientation of a gray gas between two long concentric cylinders. (b) Analogous dc network for gray gas exchange for the geometry shown in (a).

the gas absorbs from the bounding surfaces. In that sense the gas acts similar to a shield, but not as effective, since the gas transmits whereas a shield is opaque.

The analysis requires the assumption that the gas is all at uniform temperature T_g. The geometry and the network we develop below is shown in Figure 8.38. The net radiant energy from surface A_1 and A_2 is transmitted through the gas and is given by

$$_1q_{2gnet} = A_1 F_{12}(J_1 - J_2)\tau_g$$

$$= \frac{J_1 - J_2}{1/A_1 F_{12}\tau_g} \tag{8.97}$$

We note immediately that the space resistance between J_1 and J_2 has been increased by the factor of $1/\tau_g$, thus reducing the net flux between J_1 and J_2 from what it would be without the gas present. At equilibrium the net energy received by the gas will be reradiated to the lower temperature surface. It is important to note that we are speaking of net energy which must include all reflections from surface A_1; that concept is embodied in the radiosity J_1. We could observe at this point that the gas also acts like a refractory surface in the sense just described in addition to the shield function we noted above.

The radiosity from the high-temperature surface A_1 absorbed by the gas is $A_1 F_{1g} J_1 \alpha_g$ and the energy returned to surface A_1 from the gas is $A_g F_{g1} \varepsilon_g E_{bg}$ so that the net exchange is

$$_1 q_{gnet} = A_1 F_{1g} J_1 \alpha_g - A_g F_{g1} \varepsilon_g E_{bg} \qquad (8.98)$$

The concept of a gas surface area has meaning only when we consider the gas as though it were a cylindrical area between (the shield concept) A_1 and A_2. Doing so also permits us to use the reciprocity relationship $A_1 F_{1g} = A_g F_{g1}$ so that

$$_1 q_{gnet} = A_1 F_{1g} \left(J_1 \alpha_g - \varepsilon_g E_{bg} \right) \qquad (8.99)$$

We can only write Eq. 8.99 in the form of a potential divided by a resistance if we assume the gas to be gray and such that $\alpha_g = \varepsilon_g$. We have seen that $\alpha_g(T_g, T_1, T_2)$ cannot be the same as $\varepsilon_g(T_g)$ at a given optical thickness, but since T_g lies between T_1 and T_2, the assumption is not quite so severe. When the gas is active and the gas temperature T_g is considerably different from the surface temperature, the $\alpha_g = \varepsilon_g$ analysis should not be used. Accordingly,

$$\begin{aligned} _1 q_{gnet} &= A_1 F_{1g} \varepsilon_g \left(J_1 - E_{bg} \right) \\ &= \frac{J_1 - E_{bg}}{1/A_1 F_{1g} \varepsilon_g} \end{aligned} \qquad (8.100)$$

A similar arrangement gives for the net energy exchange with the low-temperature surface.

$$_g q_{2net} = \frac{E_{bg} - J_2}{1/A_2 F_{2g} \varepsilon_g} \qquad (8.101)$$

The surface resistances for A_1 and A_2 are as we have noted before ($\rho/\varepsilon A$). Therefore, the analogous dc network can be represented as shown in Figure 8.38(b). The solution for the net flux between A_1 and A_2 is the potential difference $E_{b1} - E_{b2}$ divided by the equivalent resistance of the network. Determination of the equivalent resistance is a trial-and-error procedure because ε_g depends on the gas temperature which is unknown, and must be assumed and refined by solving for J_1, J_2, and E_{bg}, and then checking to see that the energy absorbed by the gas is equal to the energy emitted.

Wohlenberg (17) has worked out such a trial-and-error procedure for large flat plates at temperatures 2460 and 1460R with water vapor at an optical thickness of 1 ft atm (partial pressure of 0.1 atm) between the plates, and determined the equilibrium gas temperature to be 2085R and that the gas caused a 1.05 percent reduction in the radiant flux exchange that would exist without the gas.

We have mentioned above that the assumption of uniform gas temperature is not very good. Holman (18) has suggested dividing the gas up into zones of constant temperature and solving the resultant network for the pertinent radiosities and gas temperatures. The complexity is increased because the number of unknown gas temperatures has increased.

8.10 CLOSURE

In the analysis of gray surfaces there are two very important assumptions that are rarely met exactly. They are the requirement of uniform irradiation on all surfaces and the assumption that Kirchhoff's law ($\alpha = \varepsilon$) holds.

An exact analysis of nonuniform irradiation situations is very complicated because the resulting expression for the radiosity is an integral equation where the unknown radiosity is also under the integral sign in addition to being on the left side of the equation. Cases where the surface temperature is not uniform also belong in this category. Sparrow (13) and Wiebelt (3) give solutions for some simple geometries and show errors as high as 30 percent for parallel plates, depending on the surface conditions and orientation; in general, as the interchange factor and the surface emittance decrease the error becomes larger.

When the emittance of a surface is a function of temperature and Kirchhoff's law cannot be employed, the accuracy of the calculations can be improved by evaluating the absorptance of the surface at the temperature of the irradiating source instead of at the surface temperature. Wiebelt (3) calls such an analysis semigray and describes the procedure in some detail.

We have also seen that the gray gas assumption is not very good. A more precise approach is to actually calculate the energy content in each emitting and absorbing band of significance for the gas. The band approximation technique increases the amount of computational work by orders of magnitude, but if the calculations can be automated on a computer, the hard work need only be done in the programming. References 3, 5, and 11 discuss the band correlations and the method of application to radiant transfer in gases.

PROBLEMS

1. Radiation wavelengths are very small quantities in the visible portion of the spectrum. How many waves of krypton 86 are there in 30 mm; in 1.50 in?
2. Is an angstrom unit larger or smaller than a micron? How many angstroms are there in a meter? How many microns are there in a meter; in an inch?
3. How long does it take for radio and TV communication to reach the moon? What is the minimum communication time between the earth and Jupiter?
4. Solar irradiation at the edge of the earth's atmosphere is 444 Btu/hrft^2 on the average over a year's time. Considering the sun as a blackbody radiator estimate the temperature of the surface of the sun.
5. A flat plate that has a transmissivity of 0.25 and a reflectivity of 0.3 is irradiated on top at the rate of 320 Btu/hrft^2, on the bottom at the rate of

400 Btu/hrft^2, and both surfaces have emissive power of 200 Btu/hrft^2. Determine the top and bottom radiosities.

6. Show whether the plate temperature of Problem 5 is increasing, decreasing, or in equilibrium.

7. Determine the equilibrium temperature of a plate that has a reflectivity of 0.5, a transmissivity of 0.2 and is irradiated on top at a rate of 600 Btu/hrft^2 and on the bottom at 300 Btu/hrft^2.

8. A flat plate with a reflectivity of 0.4 and a transmissivity of 0.3 is at a temperature of 200F. The plate is irradiated on the bottom at a rate of 500 Btu/hrft^2. What level of irradiation is required on the top of the plate to maintain the plate temperature at 200F?

9. Determine the radiosities of the top and bottom surfaces of the plate in Problem 8 under equilibrium conditions.

10. A flat plate, 4 in. on a side, is at such a temperature that a detector with a 3-in.-diameter aperature aimed at the center of the plate (and located 36 in. from the center of the plate) indicates a 25 Btu/hr radiant energy flux entering the aperature. The angle between the normal to the 4-in. plate and the direction from the plate center to the aperature center is 45 degrees. Determine the radiation intensity as viewed by the detector.

11. Determine the maximum monochromatic emissive power of a black surface at a temperature of 3000F and compare with the value given in Figure 8.6.

12. Calculate a value for the Stefan-Boltzmann constant using Eq. 8.17 and compare with the value given in Eq. 8.4. At a value of $\lambda T = 9000 \mu R$ calculate the ratio $E_{b\lambda}/\sigma T^5$ from Eq. 8.18 using the value of σ computed above and compare with the result given in Table 8.1.

13. What fraction and amount of blackbody radiation is contained in the band width from 1 to 5 μ at temperatures of 500R, 1000R, 2000R, 5000R, and 10,000R?

14. A transparent plastic sheet is used as a cover on a solar collector. The sheet has a transmissivity of 0.75 in the band from 0.4 to 1 μ and 0.9 from 2 to 8 μ and is opaque at all other wavelengths. What amount of solar irradiation at a level of 320 Btu/hrft passes through the sheet? Assume the blackbody temperature of the sun to be 10,500R.

15. Determine the amount of radiant solar flux in Problem 14 that is transmitted through two sheets of plastic that are separated by a few centimeters distance. Do not consider any reflections to take place; consider only the transmission phenomena.

16. Write an expression for the fraction of energy emitted by a gray surface in the wavelength band from λ_1 to λ_2 compared to the total energy emitted by the surface over the entire spectrum. If the gray surface has an emissivity of 0.46 and is at a temperature of 940F how much energy in W/m^2 is contained in the wavelength band from 1 to 3 μ?

17. A pane of window glass is solar irradiated at the rate of 300 Btu/hrft^2; assume solar temperature to be 10,500R. The glass transmits 90 percent of irradiation in the 0.4–3 μ band width and is essentially opaque outside the

band width. The surface reflectivity is 5 percent over the entire spectrum. The air surrounding the plate is at an ambient temperature of 70F and a surface convection coefficient of $h = 1$ Btu/hrft^{2}F may be assumed to apply. Determine the equilibrium temperature of the glass under these conditions.

18. Consider a black surface that has a radiation intensity variation such that $I_b = I_0 \cos \phi$ where I_0 is the value of the intensity (a constant) when viewing normal ($\phi = 0$) to the surface. Following the development given in Section 8.4 beginning with Eq. 8.20, determine the relationship between I_0 and the emissive power E_b of the surface.

19. What solid angle (in steradians) does a circular disk of radius r subtend from a point r units along the normal to the center of the disk.

20. Apply Eq. 8.30 to configuration IV in Table 8.2 and work out the formula for the interchange factor for the infinitesimal area dA_1 to the washer of inner and outer radius R_0 and R_2 for the orientation shown. Follow the method shown in Example 8.2 and note that in the final solution the trick of adding and subtracting unity ($+1 - 1$) must be employed.

21. A large area A_2 is 10 by 20 m in size and has a small area dA_1, whose plane is parallel to the large area, placed 5 m directly below the geometric center of A_2. Determine the percent of black diffuse radiation leaving dA_1 which strikes A_2.

22. Consider an arbitrarily shaped plane area A_2 and a small area dA_1 whose plane is parallel to A_2 and separated by a distance D. The interchange factor $F_{dA_1 \to A_2}$ is obtained by integrating $dF_{dA_1 \to dA_2}$ over the area A_2; thus

$$F_{dA_1 \to A_2} = \int_{A_2} dF_{dA_1 \to dA_2}$$

which is the origin of Eq. 8.30. Split the plane up into N equal ΔA_2's; evaluate the terms of the integrand in Eq. 8.30; approximate the integral with the summation from $i = 1$ to N; and formulate an approximate expression for $F_{dA_1 \to A_2}$. Check out your result by using it to evaluate an $F_{dA_1 \to A_2}$ for the configuration of Figure 8.13. Lay out rectangular area A_2 on a blackboard or use a concrete block wall using the blocks as the ΔA_2's. You may find it convenient to work with a classmate when using a tape measure to determine the r_i's needed for your summation.

23. Calculate the interchange factor between two directly opposed disks whose planes are parallel using Eq. 8.39. The diameter of the top disk is 80 cm, the diameter of the bottom disk is 40 cm, and the disks are 20 cm apart. Determine the factor from the small disk to the larger disk and check your result with Figure 8.12.

24. Two rectangles, one 10 by 20 m, and the other 10 by 6 m, are mutually perpendicular and have the 10-m side as a common edge. Determine the interchange factors F_{12} and F_{21} as carefully and as precisely as possible and show that reciprocity holds.

25. A room is 10 by 14 ft with an 8-ft-high ceiling. The ceiling contains heating elements and is textured so that it acts essentially as a black diffuse surface. What percentage of the radiant energy leaving the ceiling strikes all four walls?

26. Two parallel black square surfaces 4 ft on a side are directly opposed and separated by a distance of 2 ft. The top plate is maintained at 1200R and the bottom plate is insulated on the under side (side facing away from the top plate). Assuming the plates to be in a vacuum and with surroundings at zero degrees Rankine, how much energy must be supplied or removed from the bottom plate to keep its temperature at 940F?

27. What is the net exchange of radiant energy between the two square plates in Problem 26 under the temperature conditions given? How much of the energy supplied to the plate at 940F goes to the surroundings?

28. Consider directly opposed equal square areas whose planes are parallel. What variables is the interchange factor a function of? Without making any numerical calculations sketch a curve of F_{12} as a function of the pertinent variable or variables on a linear scale and discuss your reasoning.

29. The black square areas in Problem 26 are connected by a refractory surface A_R to form an enclosure. Discuss the meaning of the quantities q_{net1}, q_{net12R}, and q_{net12} in this case compared to the situation in Problem 26. Calculate the amount of energy required by the bottom plate to maintain its temperature at 940F with the top plate at 1200R as before.

30. Determine the temperature of the refractory surface A_R in Problem 29.

31. An annealing furnace is in the shape of a half-cylinder (split along the axis). Half of the flat part is the fire bed A_1 and is 10×30 ft and sheet strips of hot rolled steel are moved through the other side on a conveyor A_2, also 10×30 ft in the same plane as A_1. The areas A_1 and A_2 are separated by a small 1-ft-high dam to keep the solid fuel from spilling onto the conveyor. The fire bed is at an average temperature of 2100F and the sheet strips are at an average temperature of 1000F. Estimate the maximum possible transfer of radiant heat from the fire bed to the sheets.

32. A right circular cylinder with height equal to diameter is 6 ft tall. Determine the reentrant interchange factor for the cylindrical surface and the interchange factor for the cylindrical surface to the end of the cylinder.

33. Determine the interchange factors F_{12} and F_{11} where A_1 is the inside area of a hemisphere and A_2 is the base circle of the hemisphere.

34. A small sphere 3 m in diameter is inside a large sphere 6 m in diameter. What percent of black diffuse radiation leaving the inner surface of the large sphere (a) strikes the outer surface of the small sphere, and (b) returns to the inner surface of the large sphere?

35. Areas A_i and A_j are equal, mutually perpendicular rectangles whose length is twice the height, the long side of each rectangle being on a common line. The rectangles are slipped along their length a distance of their height from a directly opposed orientation. Determine F_{ij} using interchange factor algebra.

36. Consider the areas $A_1 + A_2$ and $A_3 + A_4$ as directly opposed parallel rectangles as shown in the figure. Prove that the special reciprocity relationship $A_1 F_{13} = A_2 F_{24}$ holds by showing that the defining multiple integrals like Eqs. 8.59 and 8.59a are equal and also by arguing from symmetry and ordinary reciprocity.

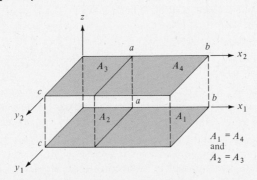

37 to 45. Determine the interchange factor F_{AB} for the areas A and B oriented as shown in the figures using interchange factor algebra. Where possible, check your result by reading a value from the appropriate chart.

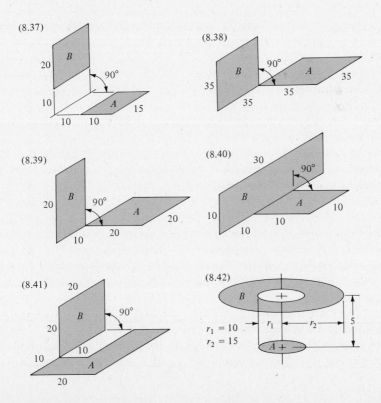

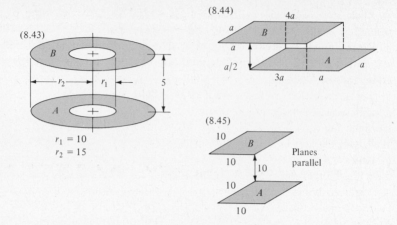

46. An unshielded thermometer is inserted into an air stream flowing in a 6-in. round duct. The actual air temperature is 510F at 30 psia and the wall temperature is 220F. The thermometer is positioned normal to the axis of the duct and is $\frac{1}{4}$ in. in diameter. What velocity of air is required to hold the temperature measurement error to 10 degrees?

47. In Problem 43 washer A has $\varepsilon = 0.5$ and is at 2000R, and washer B has $\varepsilon = 0.8$ and is at 1000R. How much energy must be supplied to the surface A to keep it at 2000R (with the bottom insulated) when the surroundings are at zero degrees Rankine? What percent of this energy goes to the surroundings?

48. Consider the analysis presented in Example 8.9 and perform an energy balance on surface A_1 using the radiosities and resistances calculated in the example.

49. Two equal, directly opposed circular disks each have an emissivity of 0.8, are 3 ft in diameter, and are separated by a cylindrical surface 3 ft high which acts as a no-flux refractory surface. The top disk is at 1200R. How much energy must be removed from the bottom disk to maintain it at 600R considering only radiation heat transfer?

50. A small area $A_1 = 1$ ft^2 is located 10 ft below the center of a large 20×20 ft square plate A_2; the planes of both areas are parallel. The small area is black and at 1000R, and the large square has an emissivity of 0.5 and is at 500R. Sketch the two-surface circuit and determine the net flux between the two surfaces. Compare this figure with the net flux calculated from basic concepts (note since A_1 is black there is only one reflection at A_2 to account for).

51. Two black circular disks A_1 and A_2 are both 10 ft in diameter, directly opposed, and 5 ft apart with planes parallel. A third disk A_3 5 ft in diameter, also black, is placed midway between A_1 and A_2 with its center on the same axis passing through the centers of A_1 and A_2, and also parallel to A_1 and

A_2. Disk A_1 is at 1600R and disk A_2 is at 1000R. Determine the percent reduction in net flux between A_1 and A_2 due to disk A_3.

52. A Dewar vessel is in the shape of a sphere and is fabricated from thin stainless-steel shells. The outer shell has 12-in. radius and the inner shell has 10-in. radius. Calculate the heat gain of the inner sphere when filled with LN_2 at atmospheric pressure and a spherical shield of stainless steel is placed equally spaced between the inner and outer shell (11-in. radius). The outside ambient temperature is 110F with an $h_0 = 2.5$ Btu/hrft^{2}F on the outer shell. Neglect the Dewar opening and assume all metallic surfaces have an emissivity of 0.3.

53. Two large black planes are at 1600R and at 1000R temperature. Two shields are inserted between the planes; one has $\varepsilon = 0.8$ on one side and $\varepsilon = 0.65$ on the other side, and the other shield has $\varepsilon = 0.5$ and 0.9 on the surfaces. Determine the percent reduction in radiant flux due to the presence of the shields.

54. Write the energy balance equations that would have to be solved for the gas temperature measurement problem described in Example 8.12 for the case where two concentric cylindrical shields are placed around the thermocouple. How many unknown temperatures must be determined? Comment on the technique required to solve for the gas temperature.

55. A test cell containing CO_2 at 540F and 1 atm pressure is 4 in. in diameter and 9.6 in. long. One end of the cell is exposed to a parallel beam of radiation coming from a blackbody source (a hohlraum) at 2000R. Determine the radiation flux absorbed and the strength of the beam leaving the other end of the cell. Assume the cell windows on each end to be completely transparent.

56. One method of utilizing solar energy is to form hydrogen and oxygen by an electrolysis process, then burn the H_2 and O_2 during the nighttime hours to form steam to run through a turbine to produce electricity in the usual manner. The H_2O combustion temperature is too high to use in a turbine so the combustion chambers walls could be film cooled with liquid H_2O in sufficient volume to bring the gas temperature down to a safe operating level. For an H_2–O_2 combustion temperature of 5000R, estimate the radiant heat flux to the center of the burner face at 600F. The combustion chamber is a 6-in.-diameter by 14 in. long cylinder with the stainless-steel injector burner at one end. Pressure of the combustion products is 150 psia.

57. An annular space between two concentric cylinders contains a CO_2–nitrogen gas mixture; the CO_2 has a partial pressure of $\frac{1}{2}$ atm and total pressure is 1.5 atm. The inner cylinder is 2 ft in diameter and the outer one is 2.5 ft in diameter; both are long enough to neglect end effects. The inner surface is at 1800F and the outer surface is at 800F. Determine the percent reduction in radiant flux due to the gas presence between the cylinders for one iteration at a gas temperature of 1400F. The inner cylinder has $\varepsilon = 0.7$ and the outer $\varepsilon = 0.4$.

References

1. Smith, R. A., Jones, F. E., and Chasmar, R. P., *The Detection and Measurement of Infra-Red Radiation*, Oxford University Press, 1957, Chapter I.
2. Gubareff, G. G., Janssen, J. E., and Torborg, R. H., *Thermal Radiation Properties Survey*, Honeywell Research Center, Minneapolis-Honeywell Regulator Company, Minneapolis, Minn., 1960.
3. Wiebelt, J. A., *Engineering Radiation Heat Transfer*, Holt, Rinehart and Winston, New York, 1966, pp. 59, 105.
4. Planck, M., *The Theory of Heat Radiation*, Dover, New York, 1959, English translation.
5. Love, T. J., *Radiative Heat Transfer*, Chas. E. Merrill, Columbus, Ohio, No. 9676, 1968, p. 23.
6. Dunkle, R. V., "Thermal Radiation Tables and Applications," *Transactions ASME*, Vol. 76, 1954, p. 549.
7. Hamilton, D. C. and Morgan, W. R., "Radiant-Interchange Configuration Factors," NACA Technical Note, TN 2836, 1952.
8. Mujahid, A. M. and Lee, P., Mechanical Engineering Department, University of Arkansas, Fayetteville, Ark., 1979.
9. Wills, L. D. and Wang, Y., Mechanical Engineering Department, University of Arkansas, Fayetteville, Ark., 1979.
10. Jinq, D. and Huang, R., Mechanical Engineering Department, University of Arkansas, Fayetteville, Ark., 1979.
11. Hottel, H. C. and Sarofim, A. F., *Radiative Transfer*, McGraw-Hill, St. Louis, Mo., 1967, pp. 23, 226, 277.
12. CINDAS Publications, 2595 Yeager Road, Purdue University, West Lafayette, Ind., 47906.
13. Sparrow, E. M., "Applications of Variational Methods to Radiation Heat Transfer Calculations," *Transactions ASME*, Vol. 82, Series C, 1960, p. 375.
14. Jakob, M., *Heat Transfer*, Vol. II, Wiley, New York, 1957, pp. 121, 130.
15. McAdams, W. H., *Heat Transmission*, McGraw-Hill, 3d Edition, 1954, p. 90.
16. Eckert, E. R. G., and Drake, R. M., *Heat and Mass Transfer*, McGraw-Hill, 2d Edition, 1959, p. 383.
17. Wohlenberg, W. J., "Heat Transfer by Radiation," Engineering Experiment Station Bulletin No. 75, August 1940, Purdue University, West Lafayette, Ind., p. 35.
18. Holman, J. P., *Heat Transfer*, McGraw-Hill, 3d Edition, 1972, p. 272.

Chapter 9
Heat Transfer with Phase Change

9.0 INTRODUCTION

During a change in phase of a substance an extra energy transfer is required to bring about the attendant change in molecular structure without changing temperature. The energy quantities involved are large for some liquids; for example, for water the latent heat of vaporization at 1 atm pressure is about 970 Btu/lb_m, for ammonia at 1 atm about 602 Btu/lb_m, and for Freon-12 at 1 atm about 71 Btu/lb_m. The phase change takes place at constant temperature at equilibrium, but during a dynamic process small changes in temperature exist that cause large amounts of energy to transfer to or from a surface. Accordingly, very high heat transfer coefficients are achieved and the process is an effective means of energy transfer.

Probably the most familiar documentation of phase change processes is by a temperature-specific volume diagram such as the one shown in Figure 9.1. The figure shows a constant-pressure line extending from the compressed solid region into the superheat region for a single component pure substance. When the substance contracts upon freezing, curve A is followed; when the substance expands upon freezing (like water does), then curve B is followed. All substances increase in specific volume in changing from a liquid to a vapor (boiling) and at 1 atm as we have seen above it takes about 970 Btu/lb_m or 2256 kJ/kg at 212F

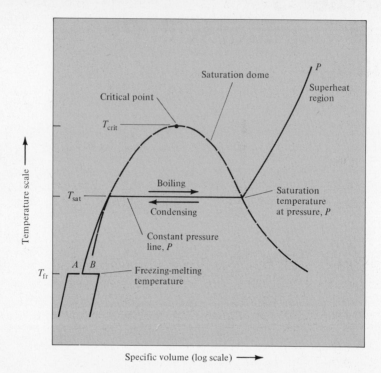

Figure 9.1   Temperature-specific volume diagram illustrating phase changes for a substance that (*A*) contracts on freezing and (*B*) expands on freezing.

or 100C to go from the saturated liquid to the saturated vapor state for water, As the temperature increases, the energy required to vaporize decreases and the saturation pressure increases. The locus of saturated liquid and saturated vapor states, shown in Figure 9.1, is called the saturation dome and the point where the specific volumes of the liquid and vapor are equal is called the critical point. At temperatures above the critical temperature (or pressures above the critical pressure) no change in phase from liquid to vapor occurs; the density is a continuous function with temperature or pressure. Although no boiling or condensing can take place, density fluctuations can cause motion that influences the heat transfer process.

There are also phase changes that take place in the compressed liquid region. At pressure up to 35,000 atm water has six known phases; there is, however, little information on heat transfer characteristics in liquid-liquid phase changes.

Sublimation is the change of phase in going directly from solid to vapor. The energy required to make the change of phase is larger than that required to go from liquid to vapor roughly by the amount required to melt the solid at the triple point. For water about 144 Btu/lb$_m$ are required to melt the solid at 32F, 1075 Btu/lb$_m$ are required to vaporize the liquid, and about 1220 Btu/lb$_m$ to sublimate the solid. Thermodynamic properties for water are given by Keenan and coauthors (1). At the other end of the temperature scale carbon sublimates at 3680C or 6650F, while most organic compounds sublimate somewhere between −50 and 200C.

We will consider means for correlating and calculating heat transfer for (a) boiling, (b) condensing, and (c) freezing and melting. It may be helpful to the reader to review the properties chapter in a beginning thermodynamics text in order to review these concepts.

9.1 BOILING

When the temperature of the fluid reaches the saturation temperature at the given system pressure, boiling is possible and depends on a number of factors. In the idealized, quasistatic, thermal equilibrium case the amount of vapor is formed according to the amount of energy supplied once the liquid reaches saturation. In the real dynamic situation vapor formation depends on the presence of what is called nucleation sites and the nature of the temperature distribution in the fluid near the surface. The nucleation sites are small crevices or pits that contain some adsorbed gas or vapor. Rough surfaces promote more bubble formation (hence better boiling heat transfer than smooth or polished surfaces. It has also been found that about 10 degrees or so of superheat of the liquid is required to initiate and maintain bubble formation. A considerable body of research has been devoted to the correlation of bubble dynamics for given surface and thermodynamic conditions, but no generally useful predictions have resulted due to the difficulty in assessing bubble population statistics. Figure 9.2 shows the response of heat flux to temperature driving potential, which is the surface temperature minus the saturation temperature at the system pressure, for the pool boiling of water on a horizontal platinum wire; other fluids show a similar response. When the excess temperature (surface temperature minus saturation temperature) is less than about 8 to 10 degrees, the heat transfer is by free convection; the liquid is superheated and rises due to density driven buoyant forces, for ΔT_x from about 10 to about 40 or 50 degrees nucleation takes place, and bubbles form and rise to the surface when the bulk of the liquid is at saturation temperature. When the surface characteristics (strength, melt point) are such that high temperature and stress can be tolerated, the boiling curve reaches a maximum that is called the DNB point (departure from nucleate boiling) and regular bubble formation ceases. Large numbers of bubbles join together and cover the surface with patches of vapor, thus decreasing the heat transfer because of the momentary insulating effect during the time the vapor patch is intact.

Such a situation is unstable and difficult to obtain experimentally. When the surface-fluid combination can tolerate the temperatures involved, the curve jumps (in the direction of the arrow in Figure 9.2) to point F when the heat flux is the controlling factor as is the case for joulian heating. At point F the film of vapor covering the surface is stable, and the liquid is vaporized above the surface and no liquid contacts the surface. In the stable film boiling region radiation is a

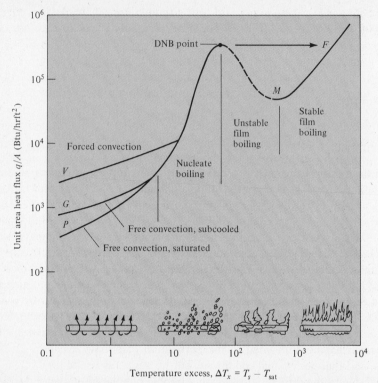

Figure 9.2   Variation of pool boiling heat flux with temperature excess for water on a platinum wire after Nukiyama (2) at 1-atm pressure and illustrating the physical appearance of the vapor-liquid interaction in the different boiling regions.

major factor in transporting the energy from the surface. If the heat flux is decreased, the curve drops below the DNB point then begins to approach the DNB point through the unstable boiling region.

For those situations where the heat flux is the independent variable the jump from the DNB point to point F brings the surface temperature to such a high level that the surface either melts or stress ruptures; therefore, the DNB point would correspond to the burnout point or burnout flux. Clearly operation of heat transfer devices near the DNB point is undesirable under such circumstances.

When the bulk liquid is at a temperature less than the saturation temperature (subcooled pool boiling), the free convection curve G (in Figure 9.2) is slightly higher than the case for saturated pool boiling P due to the larger ΔT's existing. Forced convection raises the nonboiling portion of the curve still further as indicated by curve V.

Since the removal of bubbles from the surface is necessary to prevent patches of vapor insulating the surface, the orientation of the surface with the gravity vector must be considered. Githinji and Sabersky (3) have shown that boiling with surfaces facing up as compared to facing down results in a burnout flux higher by a factor of about seven. As we shall see in the section to follow, the strength of the gravitational attraction toward the surface is also an important factor.

9.1.1 Pool Boiling

When the bulk of the liquid is not in aggregate motion, the phenomena is called pool boiling. The agitation caused by the violent bubble action causes local random velocities to exist near the surface, and in the bulk of the fluid, which contributes to the heat transfer process.

Another inherent factor in the boiling process is the ability of the fluid to wet the surface. That aspect depends on the fluid-surface combination and the cleanliness of the surface. Additives in the liquid also affect both surface tension and the wetting ability of the fluid. These remarks should give the reader an uneasy feeling about predicting heat transfer during boiling conditions for an industrial situation where complete control of fluid purity and surface conditions is not possible.

Although the correlations for boiling heat transfer do not involve a heat transfer coefficient like we used in convection, it is constructive to define such a coefficient, h_B, based on the excess temperature for use in parametric studies with a given surface-fluid combination. Accordingly, we define

$$h_B = \frac{q/A}{\Delta T_x} \tag{9.1}$$

where ΔT_x is the difference between the surface temperature and the saturation temperature at the system pressure. The reader will note that ΔT_x is small at the onset of boiling, especially when the bulk liquid is subcooled, so that the coefficients are very large.

Rohsenow (4) has correlated the data of several surface-fluid combinations for pool boiling in the nucleate region with the equation

$$\frac{c_{pl}\Delta T_x}{\mathbf{h}_{fg} N_{\text{Pr}l}^n} = C_{sf} \left\{ \frac{q/A}{\mu_l \mathbf{h}_{fg}} \left[\frac{g_c \sigma}{g(\rho_l - \rho_v)} \right]^{1/2} \right\}^{1/3} \tag{9.2}$$

$n = 1$ for water

$n = 1.7$ for other liquids

where the subscript l refers to saturated liquid and subscript v refers to saturated vapor. The density, viscosity, and specific heat symbols ρ, μ, and c_{pl} are as before. The latent heat of vaporization* at the system pressure is $\mathbf{h}_{fg}$ and the

---

*To avoid confusion in nomenclature between h for the heat transfer coefficient and h for enthalpy we will use the boldface $\mathbf{h}$ for the enthalpy change during the change of phase, $\mathbf{h}_{fg} = \mathbf{h}_g - \mathbf{h}_f$ and $\mathbf{h}_{sf} = |\mathbf{h}_f - \mathbf{h}_{\text{solid}}|$. The reader will recall that enthalpy for water at 32F is very close to zero, so that the enthalpy of ice at 32F is a negative number. In the correlations we use the absolute value or the magnitude of that negative number.

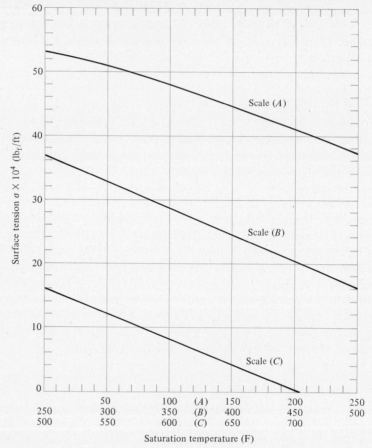

Figure 9.3 Surface tension for the water-air interface as a function of temperature.

Table 9.1 VALUES OF CORRELATION CONSTANT C_{sf} FROM REFERENCE 5 FOR USE IN EQ. 9.2

Isopropyl alcohol-copper	0.0025
Ethyl alcohol-chromium	0.0027
50 percent K_2CO_3-copper	0.0027
35 percent K_2CO_3-copper	0.0054
n-Butyl alcohol-copper	0.0030
Water-brass	0.006
Water-nickel	0.006
Benzene-Chromium	0.010
Water-platinum	0.013
Water-copper	0.013
Carbon tetrachloride-copper	0.013
N-pentane-chromium	0.015

liquid-vapor surface tension is σ. The surface tension for water is given in Figure 9.3 as a function of saturation temperature; values for some other liquids can be found in Reference 6 or other compendiums of physical properties. The constant of proportionality g_c (see Eq. 1.17) depends on the set of units employed; g is the acceleration due to gravity. The left and right side of 9.2 are dimensionless, a useful bit of information when checking out units. The correlation constant C_{sf} is determined by the fluid-surface combination. Rohsenow and Choi (5) give several values that have been determined from experiments by different investigators using their correlations; representative values are given in Table 9.1.

It is interesting to note that the constants fall roughly in groups of three values 0.003, 0.006, and 0.013, but there appears to be no correlation with type of surface or type of fluid.

Figure 9.4 illustrates the correlation given in Eq. 9.2 for the water-platinum surface combination over a wide range in pressure. The steepness of the correlation line (slope = 3) unfortunately adversely affects the precision of the correlation. For example, at a value of the abscissa of about 0.012 the spread in data for the ordinate extends from 0.3 to 3.5, more than a factor of 10.

It is important to note that Eq. 9.2 predicts neither the DNB point nor the burnout point. The peak flux or DNB point has been predicted by Zuber (7) on

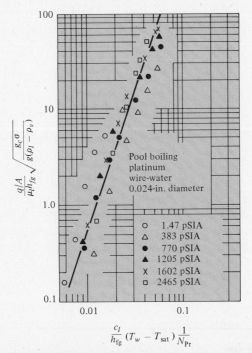

Pool boiling platinum wire-water 0.024-in. diameter

O	1.47 pSIA
$\triangle$	383 pSIA
●	770 pSIA
▲	1205 pSIA
X	1602 pSIA
□	2465 pSIA

Ordinate: $\dfrac{q/A}{\mu_l h_{fg}} \sqrt{\dfrac{g_c \sigma}{g(\rho_l - \rho_v)}}$

Abscissa: $\dfrac{c_l}{h_{fg}}(T_w - T_{sat})\dfrac{1}{N_{Pr}}$

Figure 9.4   Pool boiling data correlated according to the method of Rohsenow (4), reproduced from Reference 5 with the permission of Prentice-Hall, Inc., and the author.

the basis of a hydrodynamic stability analysis of jets of liquid approaching a surface and an equal mass flow of vapor flowing away from the surface between the jets. Zuber proposed the following equation for the peak flux, $(q/A)_{DNB}$

$$\frac{(q/A)_{DNB}}{\rho_v h_{fg}} = 0.18 \left[\frac{\sigma g g_c (\rho_l - \rho_v)}{\rho_v^2} \right]^{1/4} \left[\frac{\rho_l}{(\rho_l - \rho_v)} \right]^{1/2} \tag{9.3}$$

The time units of $(q/A)_{DNB}$ will depend on the time units employed for g and g_c. The constant 0.18 was recommended by Roshenow and Choi (5) rather than $\pi/24$ as originally given by Zuber because the equation consistently underpredicted the data.

The parameters in Zuber's equation to predict maximum boiling flux are pressure sensitive. Ciechelli and Bonilla (10) showed experimentally that pressure at the maximum DNB heat flux for clean and dirty surfaces was about one-third of the critical pressure for the fluid under consideration. There are numerous empirical pressure correlations in the literature dealing with burnout correlations for water as applied to boiling water reactors; Lottes (11) and Rohsenow and Hartnett (12) give an introduction to that body of information.

9.1.2 Pool Film Boiling

The hydrodynamic phenomena in film boiling are very difficult to describe. Zuber (7) has used a stability criteria for the vapor-liquid jet interaction to predict the minimum heat flux point (point M in Figure 9.2) for the film boiling region as follows:

$$\left(\frac{q}{A} \right)_{min} = 0.09 h_{fg} \rho_{vf} \left[\frac{g_c \sigma g (\rho_l - \rho_v)}{(\rho_l + \rho_v)^2} \right]^{1/4} \tag{9.4}$$

The coefficient 0.09 was suggested by Berenson (9) instead of $\pi/24$ as obtained by Zuber in order to better correlate the data. The quantity ρ_{vf} is the vapor density at the average film temperature $(T_s + T_v)/2$, and the other properties are evaluated at the saturation temperature for the system pressure.

It is interesting to note that the acceleration due to gravity is an important factor in both peak flux and minimum film boiling flux. With all other factors constant the fluxes are proportional to $g^{1/4}$, consequently any aspect that increases g, such as acceleration, would increase the peak flux. Such aspects as a rotational component to the flow direction (Reference 8) would also increase the peak flux as well as electrostatic or ultrasonic effects.

9.1.3 Forced Convection Boiling

When the bulk of the liquid is in motion, the convection action of the liquid removes energy as well as enhancing the motion of the bubbles away from the surface. Rohsenow (5) has suggested an additive method for predicting heat transfer in forced convection flow with nucleate boiling as

$$\frac{q}{A} = \left(\frac{q}{A}\right)_{\text{conv.}} + \left(\frac{q}{A}\right)_{\text{boil}} \tag{9.5}$$

where the convective portion of the heat flux is calculated from Eq. 6.17a using the constant 0.019 instead of 0.023, and evaluating the boiling flux with the use of Eq. 9.2. The method is restricted to the surface-fluid combinations given in Table 9.1.

The peak flux in subcooled forced convection nucleate boiling has been calculated by Zuber (7) as comprising a sum of four main effects.

$$\left(\frac{q}{A}\right)_{\text{peak}} = \frac{q_1}{A} + \frac{q_2}{A} + \frac{q_3}{A} + \frac{q_4}{A} \tag{9.6}$$

where

$$\frac{q_1}{A} = \text{latent heat effect}$$

$$\frac{q_2}{A} = \text{sensible heat effect}$$

$$\frac{q_3}{A} = \text{conduction effect}$$

$$\frac{q_4}{A} = \text{convection effect}$$

as given by the following expressions:

$$\frac{q_1}{A} = \frac{\pi}{24} \mathbf{h}_{fg} \rho_v \frac{\lambda_0}{\tau} \tag{9.7}$$

$$\frac{q_2}{A} = \frac{\pi}{24} \rho_v \frac{\lambda_0}{\tau} c_{pl}(T_{\text{sat}} - T_l) \tag{9.8}$$

$$\frac{q_3}{A} = \sqrt{\frac{2\pi}{\alpha_l \tau}} \, k_l(T_{\text{sat}} - T_l) \tag{9.9}$$

$$\frac{q_4}{A} = \sqrt{\frac{2\pi V}{\alpha_l \lambda_0}} \, k_l(T_{\text{sat}} - T_l) \tag{9.10}$$

The quantities λ_0 and τ are the Helmholtz stability wavelength in feet and the corresponding critical period in seconds. Accordingly, λ_0/τ is the velocity of the critical Helmholtz interface wave. Thus,

$$\lambda_0 = 2\pi \left[\frac{g_c \sigma}{g(\rho_l - \rho_v)} \right]^{1/2} \text{ft} \tag{9.11}$$

when g and g_c are in second time units, and

$$\frac{\lambda_0}{\tau} = \left[\frac{g(\rho_l - \rho_v)g_c \sigma}{\rho_v^2} \right]^{1/4} \left[\frac{\rho_l}{\rho_l + \rho_v} \right]^{1/2} \text{ft/s} \tag{9.12}$$

Zuber shows reasonable agreement with experiments for the above approach. In Eqs. 9.7 through 9.12 the subscript l means evaluation at the liquid temperature and the subscript v indicates evaluation at the saturation temperature; the surface tension is evaluated at the liquid temperature.

Recent experiments by Yilmaz and Westwater (26) have shown very good agreement between Zuber's theory and experiments. Reference 26 also indicates that data in the nucleate boiling, transition boiling, and film boiling regions can be correlated as unit area heat flux $q/A = a \, \Delta T^b$, but no generally applicable information with regard to the constants a and b was obtained.

Before turning our attention to condensation phenomena we will consider an example with regard to pool boiling calculations.

Example 9.1

We would like to determine the heat flux for water boiling on a clean copper surface at 1 atm pressure with a surface temperature of 227F. The water above the surface is not in bulk motion and is at saturation conditions.

SOLUTION

The phenomenon is pool boiling for which Eq. 9.2 applies. At 1 atm pressure $T_{sat} = 212F$ for water, which is the temperature for evaluating the physical properties.

$$\Delta T_x = (T_s - T_{sat}) = 227 - 212 = 15F$$

$$\sigma(212) = 40.3 \times 10^{-4} \text{ lb}_f/\text{ft (from Figure 9.3)}$$

$$N_{Pr}(212) = 1.74 \text{ (for water } n = 1)$$

$$c_{pl}(212) = 1.0 \text{ Btu/lb}_m\text{F}$$

$$\mu_l(212) = 0.68 \text{ lb}_m/\text{fthr}$$

$$\mathbf{h}_{fg}(212) = 970.3 \text{ Btu/lb}_m$$

$$\rho_l(212) = 59.82 \text{ lb}_m/\text{ft}^3$$

$$\rho_v(212) = 0.0373 \text{ lb}_m/\text{ft}^3$$

$$g_c = 32.174 \text{ lb}_m\text{ft}/\text{lb}_f\text{s}^2$$

$$g = 32.174 \text{ ft/s}^2$$

$$C_{sf} = 0.013 \text{ (from Table 9.1)}$$

Equation 9.2 can be written as

$$Y = C_{sf}\left[Q(G)^{1/2}\right]^{1/3}$$

for convenience in evaluation where

$$Y = \frac{c_{pl} \Delta T_x}{\mathbf{h}_{fg} N_{\text{Pr}}^1}$$

$$= \frac{(1)15}{970.3(1.74)^1}$$

$$= 0.00888 \text{ dimensionless}$$

$$G^{1/2} = \left[\frac{g_c \sigma}{g(\rho_l - \rho_v)} \right]^{1/2} \text{ft}$$

$$= \left[\frac{32.174(40.3)10^{-4}}{32.174(59.82 - 0.0373)} \right]^{1/2}$$

$$= 0.821 \times 10^{-2} \text{ ft}$$

Substituting in the abbreviated form of Eq. 9.2 above gives

$$Q = \frac{(Y/C_{sf})^3}{G^{1/2}} \text{ft}^{-1}$$

$$= \frac{(0.00888/0.013)^3}{0.821(10^{-2})} \text{ft}^{-1}$$

$$= 38.82 \text{ ft}^{-1}$$

$$\frac{q}{A} = 38.82(\mu_l \mathbf{h}_{fg}) \text{ Btu/hrft}^2$$

$$= 38.82(0.68)970.3$$

$$= 25,600 \text{ Btu/hrft}^2$$

It is interesting to note that the heat flux calculated in this example is about what takes place in a tea kettle on a kitchen stove. The heat flux is comparable to that obtained on a high-velocity, long-range cruise missle due to skin friction heating. Another perspective is that the boiling flux in the example is about 85 times as large as solar flux at noon on a hot cloudless July day. This completes the example.

We close our brief look at boiling heat transfer with the observation that for other than pool boiling the phenomena is really a very involved and difficult problem in two-phase flow. Wallis (14) gives an extensive treatment of the topic and describes the physical aspects involved. We now direct our attention to the process in the opposite direction on the thermodynamic $T - v$ diagram, that of condensation.

9.2 CONDENSATION

In this section we consider the transfer of heat to a surface which stems from the deposition of the latent heat of phase change, $\mathbf{h}_{fg}$, on a surface when a vapor condenses to a liquid. The condensed liquid will either wet the surface and drain away as a film according to the dictates of the geometry and fluid dynamics, or form droplets and run off the surface as droplets or rivulets depending on the condition of the surface. It is difficult to maintain droplet condensation conditions due to the leaching action of most liquids which eventually removes droplet promoters (some types of organics such as mercaptans and oils promote nonwetting of the surface) that have been introduced into the system. Droplet condensation heat transfer is about an order of magnitude larger than film condensation, hence desirable to achieve but unfortunately not reliable. Some noble metals such as gold and silver can maintain droplet formation without additives, but the cost is prohibitive even for plating. Design of condensers is based very conservatively on film type correlations to allow for performance degradation due to scale formation.

We will consider only the results of a simple analysis of laminar film condensation where no shear interaction takes place with the vapor. When such interaction takes place and there is significant vapor velocity, the problem becomes one of two-phase annular flow of the type described by Charvonia (15) where the nature of the interfacial surface becomes important to the heat transfer process.

9.2.1 Laminar Film Condensation

The pioneering analysis by Nusselt (16) remains the basis for film condensation correlations currently in use. The analysis assumes the vapor contains no noncondensable gases and moves only in a direction normal to the vertical surface on which it condenses, and then drains from the surface in a liquid film due to gravity. By performing a force balance on a fluid element at an arbitrary distance from the top of the vertical plate, shown in Figure 9.5, Nusselt formulated the following expression for the thickness of the condensate film at a distance x from the top of the plate.

$$\delta_x = \left[\frac{4k_l \mu_l (T_{\text{sat}} - T_s)x}{g\rho_l(\rho_l - \rho_v)\mathbf{h}'_{fg}} \right]^{1/4} \tag{9.13}$$

where $\mathbf{h}'_{fg}$ is the average enthalpy of the vapor in condensing to liquid and then subcooling to the wall temperature. Thus

$$\mathbf{h}'_{fg} = \mathbf{h}_{fg} + \tfrac{3}{8}c_{pl}(T_{\text{sat}} - T_s) \tag{9.14}$$

The properties in Eqs. 9.13 and 9.14 are evaluated at the average temperature of the condensate film $(T_{\text{sat}} + T_s)/2$. Nusselt assumed a linear temperature profile in the thin condensate layer so, therefore, the temperature gradient in the liquid at the surface is simply $dT/dy = (T_{\text{sat}} - T_s)/\delta_x$. Upon equating the Fourier

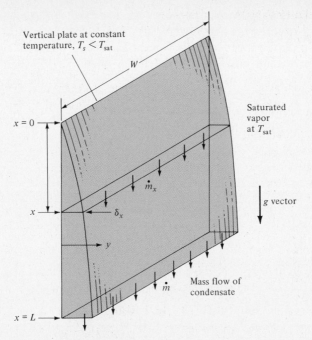

Figure 9.5 Laminar condensate film formed on a vertical plate at constant temperature that is exposed to saturated vapor.

conduction flux to the convection heat flux we can arrive at a heat transfer coefficient for condensation. Thus

$$\frac{q}{A} = \frac{k(T_{sat} - T_s)}{\delta_x} = h_x(T_{sat} - T_s)$$

or

$$h_x = \frac{k}{\delta_x} \tag{9.15}$$

Substituting for δ_x from Eq. 9.13 above gives an expression for the local condensation coefficient which can be integrated over the length of the plate to obtain the average condensation coefficient for the entire plate of height L as

$$\bar{h}_L = 0.94 \left[\frac{g\rho_l(\rho_l - \rho_v)k_l^3 h'_{fg}}{L\mu_l(T_{sat} - T_s)} \right]^{1/4} \tag{9.16}$$

When the plate is inclined an angle θ from the vertical, the effective gravity component $(g\sin\theta)$ should be used in Eq. 9.16 instead of the g term.

The amount of energy that must be removed from the plate can then be calculated from the definition of the convection equation as

$$q = \bar{h}_L A(T_{sat} - T_s) \tag{9.17}$$

We recall that the above result neglects shear between the vapor and the liquid surface. Since the liquid film is very thin, the result can also be applied to condensation on the inner or outer surface of vertical tubes. When applied to condensation on the inside of vertical tubes, there can be significant vapor velocities with respect to the liquid surface for which the simple analysis considered here will not hold. Rohsenow and Choi (5) discuss a means for accounting for interfacial shear stress.

Example 9.2 illustrates the basic concept.

Example 9.2

We would like to determine the thickness of the condensate layer draining from a vertical surface at 188F exposed to saturated water vapor at 1 atm pressure during steady-state condensation at a location 12 in. from the top of the plate.

SOLUTION

The average temperature of the film of condensate is $(188 + 212)/2 = 200F$, the temperature for evaluating the condensate liquid properties:

$$k_l = 0.392 \text{ Btu/hrftF} \qquad \mu_l = 0.74 \text{ lb}_m/\text{fthr}$$

$$\rho_l = 60.13 \text{ lb}_m/\text{ft}^3 \qquad \rho_v = 0.0373 \text{ lb}_m/\text{ft}^3$$

$$h_{fg} = 970.3 \text{ Btu/lb}_m \qquad c_{pl} = 1 \text{ B/lb}_m\text{F}$$

$$\mathbf{h}'_{fg} = \mathbf{h}_{fg} + \tfrac{3}{8}c_{pl}(T_{sat} - T_s)$$

$$= 970.3 + \tfrac{3}{8}1(212 - 188)$$

$$= 979.3 \text{ Btu/lb}_m$$

At the location $x = 1$ ft we calculate the thickness of the film from Eq. 9.13 to be

$$\delta_x = \left[\frac{4(0.392)0.74(212 - 188)(1)}{32.174(3600)^2 60.13(60.09)979.3} \right]^{1/4}$$

$$= 0.000371 \text{ ft} \equiv 0.0044 \text{ in.}$$

The thickness of 4.4 thousandths of an inch is very small, approximately a little more than the thickness of the paper this is printed on. This calculation assumes the film is laminar the entire height of the plate; a topic we explore in the next section.

9.2.2 Laminar-Turbulent Film Transition

When the momentum forces in the falling film of condensate on a vertical surface become larger than the viscous forces, the film becomes turbulent and ripples or waves start to cascade over the surface. Clearly such ripples and waves change the heat transfer characteristics through the condensate layer.

The ratio of the momentum to the viscous forces is determined by the Reynolds number of the condensate film and as we will see below, transition to turbulence takes place at a Reynolds number of about 1800. The Reynolds number is defined as $N_{Re} = D_h \rho u_b / \mu$ where D_h is the hydraulic diameter. Referring to Figure 9.5, at the location x, the cross-sectional area of the condensate flow film is the plate width W times the film thickness δ and the wetted perimeter is only the plate width W; therefore

$$D_h = \frac{4A_c}{W_p} = \frac{4W\delta}{W} = 4\delta \tag{9.18}$$

From the continuity equation we can express the ρu_b product in the Reynolds number as $\rho u_b = \dot{m}/A_c = \dot{m}/W\delta$ so that we have $N_{Re} = 4\,\delta\dot{m}/W\,\delta\mu = 4\dot{m}/W\mu$. Nusselt in his analysis formulated the mass flow per unit width in terms of the laminar film thickness as

$$\frac{\dot{m}}{W} = \frac{g\rho_l(\rho_l - \rho_v)\delta^3}{3\mu_l} \tag{9.18a}$$

so that the Reynolds number for laminar flow becomes

$$N_{Re} = \frac{4g\rho_l(\rho_l - \rho_v)\delta^3}{3\mu_l^2} \tag{9.19}$$

Experiments (Reference 17) have shown that a value for the Reynolds number of about 1800 is the limit for laminar flow in the condensate film. The properties of the liquid (for use in Eq. 9.19) should be evaluated at the average temperature of the liquid.

Example 9.3

In Example 9.2 we simply assumed the film was laminar and calculated δ_x from Eq. 9.13; we desire to verify that assumption.

SOLUTION
The properties of the liquid condensate film at 200F are given in Example 9.2. Since viscosity units are in $lb_m/fthr$ we must use g also in hour units. Accordingly, from Eq. 9.19 we have

$$N_{Re} = 4(32.174)(3600)^2 60.13(60.09)(0.000371)^3/3(0.74)^2$$
$$= 187$$

which is significantly less than 1800 so that we are sure the condensate film is laminar.

9.2.3 Turbulent Film Condensation

The experimental work of five investigations of condensation on the outside of vertical tubes was correlated by McAdams (17) for Reynolds numbers greater than 1800 by the equation

$$\bar{h}_t \left(\frac{\mu_f^2}{k_f^3 \rho_f^2 g} \right)^{1/4} = 0.0077 \left(\frac{4\dot{m}}{W\mu_f} \right)^{0.4} \tag{9.20}$$

McAdams recommends evaluating the properties in the turbulent correlation equation at the temperature $T_f = T_{sat} - 0.75(T_{sat} - T_s)$ which is slightly below the average condensate temperature. Use of Eq. 9.20 requires an iterative procedure. The value of $\dot{m}/W$ is first estimated from Eq. 9.18, then later refined by obtaining $\dot{m} = q/\text{h}_{fg}$ until agreement is reached. The values of $\bar{h}$ in the correlation Eq. 9.20 were obtained from experiments so that they include the laminar portion along the upper part of the vertical surface where $N_{Re} < 1800$. It is not necessary to make a two-part calculation for the laminar and turbulent portion of the condensate film when using Eq. 9.20.

9.2.4 Condensation on Horizontal Tubes

The condensate film on horizontal tubes travels only a distance of $\pi D/2$ before it falls off the surface at the bottom of the tube. For a single tube the analysis that leads to Eq. 9.16 must be modified by replacing the gravity vector by the local effective force, $g \sin \theta$, and integrating over θ from zero to 180 degrees so that the average coefficient for a single horizontal tube becomes

$$\bar{h}_D = 0.728 \left[\frac{g\rho_l(\rho_l - \rho_v)k_l^3 \text{h}'_{fg}}{D\mu_l(T_{sat} - T_s)} \right]^{1/4} \tag{9.21}$$

where h'_{fg} is evaluated by Eq. 9.14. For a series of n horizontal tubes arranged directly on top of each other in a vertical row Nusselt (16) showed the length criteria should be nD but his analysis predicts values of the condensing coefficient that are too low. Chen (18) theorized that since the liquid film was subcooled, a significant amount of vapor was condensed on the film. Accordingly, for n horizontal tubes stacked vertically Chen proposed

$$\bar{h}_{nD} = 0.728(1 + 0.2\overline{CT}) \left[\frac{g\rho_l(\rho_l - \rho_v)k_l^3 \text{h}'_{fg}}{nD\mu_l(T_{sat} - T_s)} \right]^{1/4} \tag{9.22}$$

where

$$\overline{CT} = (n - 1)c_{pl}(T_{sat} - T_s)/\text{h}_{fg} \tag{9.23}$$

Agreement with experimental results is good when the parameter $\overline{CT} < 2$.

There is a great deal of experimental and empirical information on condensation in the literature. Superheating the vapor gives a slight increase in the heat transfer coefficient for condensation. Some information is available for mixtures of condensable vapors and for mixtures of condensable vapors and noncondensable gases. Dropwise condensation has not been described analytically because the phenomenon is too complex; some data are available for steam.

We have discussed only the fundamental aspects involved in condensation to get an orientation acquaintance with the magnitude of the coefficients and heat fluxes associated with the phenomena.

Example 9.4

A horizontal tube 1 in. in diameter and 6 ft long is maintained at an average surface temperature of 94F. Saturated steam at 2 psia condenses on the outer surface. Determine the amount of steam condensed in 1 hr.

SOLUTION
From the heat transfer coefficient $\bar{h}_D$ we can determine the heat flux to the tube. That heat flux comes from the latent heat of condensation during the condensation process of $\dot{m}$ lb/hr. Accordingly, the properties needed for Eq. 9.21 must be evaluated at $T_l = (T_{sat} + T_s)/2 = (126 + 94)/2 = 110F$. Therefore,

$$\rho_l = 61.84 \text{ lb}_m/\text{ft}^3 \qquad k_l = 0.368 \text{ Btu/hrftF}$$

$$\rho_v = 0.0057 \text{ lb}_m/\text{ft}^3 \qquad \mu_l = 1.49 \text{ lb}_m/\text{hrft}$$

$$T_{sat} = 126F \qquad c_{pl} = 0.997 \text{ Btu/lb}_m\text{F}$$

From the *Steam Tables* (1) at 2 psia $h_{fg} = 1022.1$ Btu/lb$_m$ so that from Eq. 9.14 we determine

$$h'_{fg} = 1022.1 + \tfrac{3}{8}(0.997)(126 - 94)$$

$$= 1034.1 \text{ Btu/lb}_m$$

A careful units check will show that g must be in ft/hr^2 in Eq. 9.21; thus

$$\bar{h}_D = 0.728\left[\frac{32.174(3600)^2 61.84(61.83)(0.368)^3 1034.1}{\tfrac{1}{12}1.49(126 - 94)} \right]^{1/4}$$

$$= 2130 \text{ Btu/hrft}^2\text{F}$$

The surface area of the tube is $A = \pi D L = \pi \tfrac{1}{12} 6 = \pi/2$ ft^2 so that the heat flux is given by

$$q = \bar{h}_D A(T_{sat} - T_s)$$

$$= 2130\frac{\pi}{2}(126 - 94)$$

$$= 107{,}000 \text{ Btu/hr}$$

The amount of steam condensed to supply this heat flux to the surface is then

$$\dot{m} = \frac{Q}{h_{fg}} = \frac{107{,}000}{1022.1}$$

$$= 105 \text{ lb}_m/\text{hr}$$

or about 12.6 gal in an hour. If we had a series of tubes arranged in a vertical row, we would use Eq. 9.22 to determine the average heat transfer coefficient and then proceed as just above.

9.3 FREEZING AND MELTING

We have seen in Figure 9.1 that when the temperature decreases far enough at the system pressure P we reach the level where another change in phase takes place from the liquid to the solid which is commonly called freezing; the reverse change from solid to liquid is termed melting. The melting and freezing process is difficult to analyze exactly, because of a number of factors. The interface between the two phases moves with an uncertain motion as the phase change energy is absorbed or liberated. The thermal properties of the two phases are different and are also different functions of temperature. Convection effects in the liquid phase require inclusion of the equation of motion for the liquid together with the conduction equation for the solid phase; the matching conditions at the interface boundary make the problem nonlinear. A partial simplification of the complicated analysis has been proposed by Sfeir and Clumpner (19) with application to continuous casting of cylindrical ingots.

We will look at a very simple approach to the melting-freezing problem which does not consider any forced convection in the liquid phase. London and Seban (20) have analyzed the freezing process on the basis of a variable resistance analog to account for the motion of the interface and the varying thickness of the solid phase. Kreith (21) has suggested that the same equations obtained by London and Seban also apply to the melting case, if sensible heat storage in the liquid is negligible with respect to the latent heat of the phase change. We will develop the melting problem first then the corresponding freezing problem for the slab geometry, and conclude by looking at some results given by London and Seban for cylinders.

9.3.1 Melting; Solid Saturated

We consider a horizontal slab of solid material that is at the freezing temperature T_{fr} as shown in Figure 9.6. The ambient temperature T_∞ is greater than the freezing temperature T_{fr} and the heat transfer coefficient h_∞ that exists at the interface between the liquid is constant with time. The heat transfer q that enters the liquid surface is conducted through the liquid layer to supply the energy requirements of the latent heat of fusion, h_{sf}. We neglect the energy required to heat the liquid to temperatures above T_{fr}. The assumption of conduction only through the liquid is reasonable when the gravity vector is perpendicular to the solid-liquid interface because most fluids have a density that decreases as the temperature increases. Water shows a maximum density at 4C or 39F so that there is a thin layer of liquid near the solid interface where some convection can occur. The time intervals predicted for melting a given thickness of solid by the analysis we develop will be somewhat longer than actual due to the slightly enhanced energy transfer near the surface as compared to pure

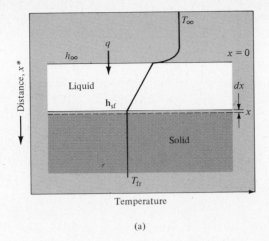

(a)

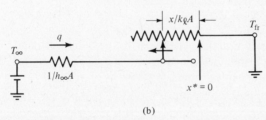

(b)

Figure 9.6   (a) Physical orientation of liquid and solid phases during the melting process on top of a solid slab, and (b) electrical analog of the melting process taking place on a horizontal solid slab.

conduction. The recession of the liquid-solid interface makes the thickness of the liquid layer x a function of time.

The heat conducted from the temperature potential T_∞ to T_{fr} is the temperature potential divided by the equivalent resistance, which is comprised of the convective resistance at the liquid-ambient surface, $1/h_\infty A$, and the changing resistance of the liquid layer, $x/k_l A$. Thus

$$q = \frac{T_\infty - T_{fr}}{[(1/h_\infty A) + (x/k_l A)]} \tag{9.24}$$

The mass of the solid that is melted in time $d\theta$ is $\rho_s A\, dx$ so that the energy flux requirement of the melted mass then is

$$q = \rho_s \mathbf{h}_{sf} A \frac{dx}{d\theta} \tag{9.25}$$

where $\mathbf{h}_{sf}$ is the latent heat of fusion for the solid. For example, for solid water (ice) 143.3 Btu/lb$_m$ is required at 32F to melt 1 lb$_m$ at constant pressure. Equating Eqs. 9.24 and 9.25 and separating variables gives us a relationship between the thickness of the melted layer x and the time θ. Thus

$$\left[\frac{h_\infty(T_\infty - T_{\text{fr}})}{\rho_s \mathbf{h}_{sf}}\right] d\theta = \left[1 + \frac{xh_\infty}{k_l}\right] dx \tag{9.25a}$$

We can recognize that xh_∞/k_l has the same form as a Biot number, but since x continually changes we simply designate that group as x^*, the dependent variable. It is possible to show that the bracketed term on the left side of Eq. 9.25a contains a Fourier number, but no particular advantage is gained by recasting the term in that form. Accordingly, we define the dimensionless quantities x^* and θ^* as

$$x^* = \frac{xh_\infty}{k_l} \tag{9.26}$$

$$\theta^* = \left[\frac{h_\infty^2(T_\infty - T_{\text{fr}})}{k_l \rho_s \mathbf{h}_{sf}}\right] \theta \tag{9.26a}$$

Inserting these quantities into Eq. 9.25a yields

$$d\theta^* = (1 + x^*)\, dx^* \tag{9.27}$$

as the governing differential equation with the boundary conditions that at time zero, $\theta^* = 0$, $x^* = 0$, and at time θ^* the nondimensional thickness is x^*. Expressed in mathematical terms, we have

$$\int_0^{\theta^*} d\theta^* = \int_0^{x^*} (1 + x^*)\, dx^* \tag{9.28}$$

which is straightforward to integrate. Thus

$$\theta^* = x^* + 0.5x^{*2} \tag{9.29}$$

or

$$x^* = (2\theta^* + 1)^{1/2} - 1 \tag{9.30}$$

which gives the depth of the melted layer at any time θ.

Example 9.5

A thick slab of ice at 32F is exposed to a mild breeze of air at 60F such that $h_\infty = 10$ Btu/hrft^{2}F at the interface between the liquid and the ambient air. We would like to know the thickness of the water layer at the end of 1.5 hr.

SOLUTION

This is a melting problem with the solid at saturation temperature; therefore, we can use Eq. 9.30 to calculate the thickness of solid that has melted. We estimate the thermal conductivity of the water at an average temperature of 40F to be $k_l = 0.332$ Btu/hrftF and the properties for the solid (ice) at 32F are

$$\rho_s = 57.24\ \text{lb}_\text{m}/\text{ft}^3$$

$$\mathbf{h}_{sf} = 143.3\ \text{Btu/lb}_\text{m}$$

The dimensionless time is given by Eq. 9.26a to be

$$\theta^* = \left[\frac{10^2(60 - 32)}{0.332(57.24)143.3} \right] 1.5 = 1.5423$$

Accordingly, Eq. 9.30 gives the dimensionless thickness as

$$x^* = [2(1.5423) + 1]^{1/2} - 1 = 1.021$$

so that the real thickness can be calculated from the definition of x^* as

$$x = \frac{k_l x^*}{h_\infty} = \frac{0.332(1.021)}{10}$$
$$= 0.0339 \text{ ft}$$
$$= 0.41 \text{ in.}$$

Doubling the heat transfer coefficient at the liquid-ambient interface would give a liquid layer of about 0.90 in. in $1\frac{1}{2}$ hr. We should really check the liquid surface temperature and verify our estimate of average liquid conductivity. The observation that the actual liquid thickness would be somewhat greater than 0.4 in. due to the stirring action of the breeze completes the example.

9.3.2 Freezing; Liquid Temperature Greater than Saturated

When a liquid at a temperature greater than the freezing temperature is exposed to ambient conditions below the freezing point, as shown in Figure 9.7, the solid phase will form at the surface and grow thicker with time. The sketch in the figure shows that the liquid at T_l must be cooled by extraction of the heat flux q_2, which then combines with the heat flux extracted from the solidification process to become the heat flux q_1 that is discharged to the ambient temperature potential T_∞.

The heat flux due to the solidification of an infinitesimally small layer dx in thickness in time $d\theta$ is

$$q_{fr} = \rho_s \mathrm{h}_{sf} A\, dx/d\theta \tag{9.31}$$

and the heat flux to cool the liquid is

$$q_2 = h_l A(T_l - T_{fr}) \tag{9.32}$$

The sum of these fluxes, q_1, must pass through the resistance offered by the solidified layer and the convection resistance at the interface. Therefore

$$q_1 = \frac{T_{fr} - T_\infty}{(x/k_s A) + (1/h_\infty A)} \tag{9.33}$$

The heat balance $q_1 = q_{fr} + q_2$ can then be expressed as

$$\frac{T_{fr} - T_\infty}{(x/k_s A) + (1/h_\infty A)} = \frac{T_l - T_{fr}}{(1/h_l A)} + \frac{\rho_s \mathrm{h}_{sf} A\, dx}{d\theta} \tag{9.34}$$

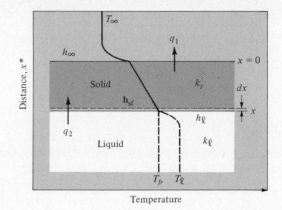

Temperature

(a)

(b)

Figure 9.7 (a) Physical orientation of solid and liquid during the freezing process of a slab and (b) electrical analog for the freezing of a plane horizontal slab exposed to a liquid at or above the freezing temperature.

We divide both sides of the equation by $h_\infty(T_{fr} - T_\infty)A$ and define

$$T^* = \frac{T_l - T_{fr}}{T_{fr} - T_\infty} \tag{9.35}$$

$$x^* = \frac{xh_\infty}{k_s} \tag{9.36}$$

$$\theta^* = \left[\frac{h_\infty^2(T_{fr} - T_\infty)}{k_s \rho_s h_{sf}}\right]\theta \tag{9.36a}$$

$$H^* = \frac{h_l}{h_\infty} \tag{9.36b}$$

The reader should carefully note the difference between Eqs. 9.36 and 9.26, and between 9.36a and 9.26a. In nondimensional form the energy balance as given by Eq. 9.34 becomes

$$1/(x^* + 1) = H^*T^* + dx^*/d\theta^* \tag{9.37}$$

Noting that H^*T^* is a constant for a given set of physical conditions let us separate the variables to obtain

$$d\theta^* = \frac{(x^* + 1)\,dx^*}{1 - H^*T^*(x^* + 1)} \tag{9.38}$$

which is of the form $\int z \, dz / (1 - az)$ which can be found in standard integral tables [see for instance Burington (22) integral no. 47]; therefore, integration from 0 to θ^* and 0 to x^* gives

$$\theta^* = \left(\frac{1}{H^*T^*} \right)^2 \ln \left[\frac{1 - H^*T^*}{1 - H^*T^*(1 + x^*)} \right] - \frac{x^*}{H^*T^*} \tag{9.39}$$

Equation 9.39 cannot be inverted to get x^* as a function of θ^* in closed form; the equation has been plotted in Figure 9.8 from which thicknesses for a given time can be determined.

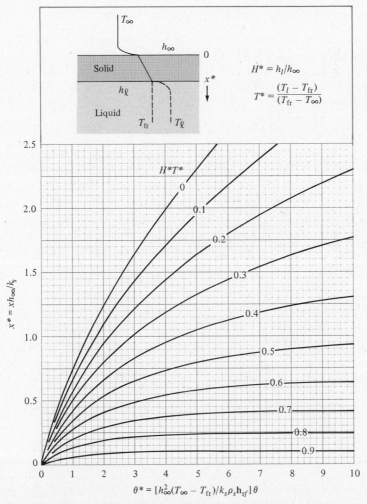

Figure 9.8   Variation of the freezing thickness x^* with dimensionless time θ^* during slab formation of the solid phase according to Eq. 9.39.

9.3.3 Freezing; Liquid Temperature Saturated

When the liquid temperature $T_l = T_{fr}$ there is no additional q_2 (see Figure 9.7) to be conducted through the solid, thus making the dimensionless parameter $T^* = 0$. According to Eq. 9.35, the differential equation for $d\theta^*$ in Eq. 9.38 reduces to

$$d\theta^* = (x^* + 1)\, dx^* \tag{9.40}$$

which is exactly the same as Eq. 9.27 for the melting case with the exception that x^* and θ^* are defined by Eqs. 9.36 and 9.36a instead of 9.26 and 9.26a. The solution to Eq. 9.40 is, however, identical to that for Eqs. 9.29 and 9.30 which is plotted in Figure 9.8 as the curve for $H^*T^* = 0$.

9.3.4 Freezing Internal and External to Cylinders

The basic analog shown in Figure 9.7 has been analyzed by London and Seban (20) also for cylindrical geometry by substituting for the slab resistance x/kA the resistance of the cylindrical solid $\ln(r/r_0)/2\pi k_s$ per unit length. For the case where the solid freezes internally ($r < r_0$), London and Seban obtained θ^* as a function of $r^* = r/r_0$ as

$$\theta^* = (0.5r^{*2})\ln r^* + \left(\frac{1}{2R^*} + 0.25 \right)(1 - r^{*2}) \tag{9.41}$$

and for the case where the solid freezes externally ($r > r_0$) on a tube

$$\theta^* = (0.5r^{*2})\ln r^* + \left(\frac{1}{2R^*} - 0.25 \right)(r^{*2} - 1) \tag{9.42}$$

The liquid temperature in each case is saturated, at the freezing temperature T_{fr}, and the parameters R^* and θ^* are defined as

$$R^* = \frac{h_\infty r_0}{k_s} \tag{9.43}$$

$$\theta^* = \frac{(T_{fr} - T_\infty)k_s\theta}{\rho_s h_{sf} r_0^2} \tag{9.44}$$

The heat transfer coefficient h_∞ is the coefficient that exists between the sink fluid at T_∞ and either the inside or the outside surface of the tube upon which the solid forms. Equations 9.41 and 9.42 are plotted in Figures 9.9 and 9.10, respectively, since it is not possible to invert the equations to obtain solid thickness as a function of time. We illustrate the above ideas with Example 9.6.

Example 9.6

An ethylene glycol-water mixture at $-22F$ flows inside a thin walled copper tube, 2-in. OD, which is immersed in water at 32F. The glycol mixture flows at such a rate that the heat transfer coefficient between the glycol mixture and the tube wall is 75 Btu/hrft^{2}F. We would like to estimate the time required to freeze a 1-in.-thick layer of ice on the outside of the tube.

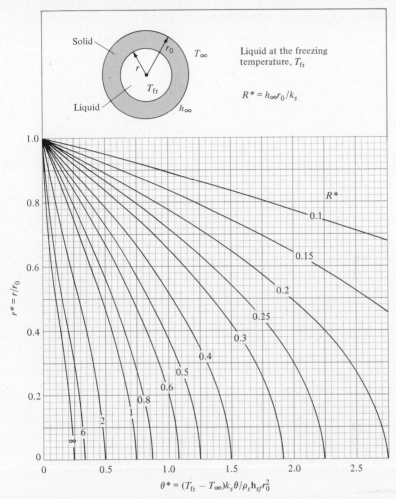

Figure 9.9   Variation of r^* with dimensionless time θ^* for freezing *inside* a tube of negligible thermal resistance having convection on the outside and ambient sink temperature T_∞ according to Eq. 9.41.

SOLUTION

The figure shows the physical situation at the end of the time under consideration.

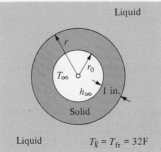

Figure for Example 9.6.

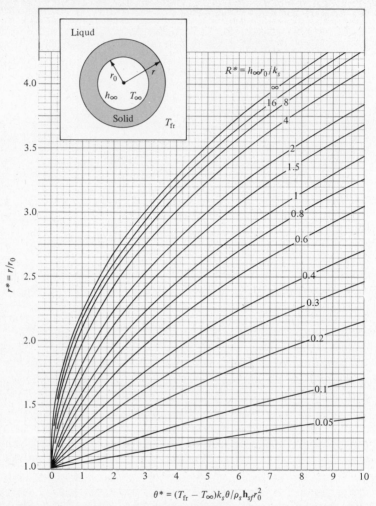

Figure 9.10 Variation of $r*$ with nondimensional time $\theta*$ for freezing on the *outside* of a tube of negligible thermal resistance having convection on the inside and sink temperature T_∞ according to Eq. 9.42.

The physical situation describes freezing on the outside of a cylinder with the liquid at the saturation temperature; therefore, Eq. 9.42 or Figure 9.10 applies. We will calculate $\theta*$ from the equation and check our result with Figure 9.10. Properties are as follows for the solid phase:

$$k_s = 1.28 \text{ Btu/hrftF} \qquad h_{sf} = 143.3 \text{ Btu/lb}_m$$

$$\rho_s = 57.24 \text{ lb}_m/\text{ft}^3 \qquad r_0 = \tfrac{1}{12} \text{ ft}$$

The parameter $R*$ is given by Eq. 9.43 to be

$$R* = \frac{h_\infty r_0}{k_s} = \frac{75(1)}{12(1.28)} = 4.88 \text{ dimensionless}$$

From Eq. 9.42 we calculate

$$\theta^* = \frac{4}{2}\ln(2) + \left(\frac{0.5}{4.88} - 0.25\right)(4 - 1) = 0.943 \text{ dimensionless}$$

We check from Figure 9.10 by reading θ^* slightly less than 0.95 at $r^* = 2$. The time is calculated from the definition of θ^* as given by Eq. 9.44; thus

$$\theta = \frac{\theta^* \rho_s \mathbf{h}_{sf} r_0^2}{k_s(T_{fr} - T_\infty)}$$

$$= \frac{0.943(57.24)143.3(1)}{144(1.28)[32 - (-22)]}$$

$$= 0.778 \text{ hr or about 47 min}$$

Increasing the flow velocity of the glycol mixture would increase h_∞ and thus decrease the time required to freeze the same thickness of ice on the outside of the tube.

9.4 FINAL REMARKS

The two-phase phenomena that we considered in the sections on boiling and condensing were essentially nonflow situations. Two-phase flow phenomena are very complex and difficult to analyze theoretically because of the chaotic nature of the vapor formation. The problem chiefly hinges about the difficulty in predicting void fractions. Experimental and theoretical work has been described by Tong (23), Wallis (14), and Lottes (11).

In the area of solidification two-phase problems we looked only at a very simple approach that neglected the thermal capacity of the phases in comparison to the latent heat of fusion. That approach is reasonable for substances other than metals. Solidification of molten metals is a transient phenomenon that requires consideration of the temperature and energy happenings inside the solid; that is, the thermal capacity of the solid cannot be neglected or treated as simply a linear temperature profile as we did. Muehlbauer (25) discusses some aspects of transient analysis in metal alloy solidification. Sadd and Didlake (24) have shown that the finite speed of heat propagation has an insignificant effect in the melting and freezing phenomena of metals; such may not be the case with very low conductivity materials.

PROBLEMS

1. Obtain a 6- or 8-in. diameter stainless-steel saucepan and fill to a depth of about 3 in. with cold water. Turn the burner on a stove (preferably electric) on high and put the water on the hot burner. Observe and describe the happenings in the water and on the heated surface until bulk boiling takes place.

2. Determine the correlation constant C_{sf} for Eq. 9.2 for the boiling data shown in Figure 9.4. Show how you arrived at your result.

3. Calculate the unit area heat flux for water in pool boiling on a clean nickel surface at a system pressure of 100 psia and a surface temperature of 340F.

4. Determine the DNB or peak flux point and the minimum flux point in the film boiling region for the water-nickel surface boiling conditions described in Problem 3.

5. Estimate the unit area peak flux with Zuber's method for pure water flowing over a smooth clean surface at 40 ft/s at a system pressure of 100 psia and a bulk liquid temperature of 227F in a standard gravity field normal to the surface. Compare your result with the data of Gunther (13) shown in his Figure 5, and also by his recommended equation.

6. Predict the maximum possible heat flux in nucleate boiling (maximum DNB point) for water on a horizontal surface under clean surface conditions.

7. Predict the heat flux per unit area to water flowing at a bulk velocity of 10 ft/s in a smooth clean nickel tube 1-in. ID for a bulk temperature of 200F and a surface temperature of 340F with a pressure of 100 psia inside the tube at the location where the bulk velocity is measured.

8. Under what conditions can the film Reynolds number for laminar condensate on a vertical plate be approximated by the expression $4g\delta_x^3/3\nu_l^2$?

9. Determine the amount of condensate dropping off the bottom of a 4-ft square vertical plate that is exposed to saturated steam on one side at 1 atm pressure. The plate surface is maintained at a temperature of 188F.

10. A large steam turbine requires 235,000 lb_m/hr of superheated steam to drive a generator. The steam enters the condenser at 1.6 psia and a quality of 92 percent. How many vertical stacks or rows of 25-in. OD tubes 10 ft long are required to handle the turbine exhaust? The average surface temperature of a stack of 25 tubes is 62F.

11. Compare the equation for the average condensing coefficient for a plate of height L with that for a horizontal tube of diameter D under identical conditions of temperature and properties (Eqs. 9.16 and 9.21) and determine the length L (in terms of D) required to have identical coefficients. Is your result the same as $\pi D/2$? Discuss.

12. Saturated steam at 6 psia condenses on the outer surface of a 2-in.-OD vertical tube that is 20 ft in height. The average surface temperature of the tube is held at 143F by a coolant flowing at a relatively high velocity inside the tube. Estimate the amount of steam condensed and the heat flux to the coolant inside the tube.

13. Freon-12 at 151 psia condenses on the outside of a horizontal tube $\frac{3}{4}$ in. in diameter and 6 ft long; the average surface temperature of the tube is 90F. How much F-12 is condensed in a 12-hr period of operation?

14. Show that the expression for $\bar{h}_L$ in Eq. 9.16 can be obtained from Eqs. 9.13 and 9.15 using the definition of the average as $\bar{h}_L = (1/L)\int_0^L h_x\,dx$.

15. For a given set of physical conditions such as h_∞, freezing temperature T_{fr}, and ambient temperature, would you expect the actual time to freeze for a given thickness of the solid phase to be more or less than that predicted by the analysis discussed in Section 9.3?

16. Water at 32F is on the inside of a thin walled $\frac{3}{4}$-in. metal tube that is positioned horizontally in still air at $-12F$. Estimate how long it would take to freeze the water solid inside the tube.

17. The mass flow of the ethylene glycol-water mixture in Example 9.7 is increased considerably such that the heat transfer coefficient on the inner surface of the tube is doubled. By what percentage is the time required to freeze a 1-in. thick cylinder of ice on the outside of the tube changed?

18. How long would it take to freeze ice on a pond to a thickness of 6 in. when the ambient air temperature is suddenly dropped to $-8F$ and the pond water is at 40F? Assume an h for the ice and water interface of 2.5 Btu/hrft^{2}F and an h between the air and surface of the ice of 10 Btu/hrft^{2}F.

19. Estimate the time required to freeze a stack of canned orange juice (assume properties same as water) in a 15 mph wind at 10F. The cans are 4 in. in diameter and are initially at 42F. Neglect the thermal resistance of container walls and also end effects.

20. Determine the thickness of ice that can be formed in a 1-hr period on a pond having a water temperature of 36F which is exposed to an ambient air temperature of 0F such that the heat transfer coefficient between the air and the solid phase is 30 Btu/hrft^{2}F. The coefficient between the solid and liquid phase is 3 Btu/hrft^{2}F.

21. Ice is sometimes manufactured in cans with a slight taper; the cans have an average diameter of 10 in. and are 3 ft long. If the cans are filled with water at 32F and an external heat transfer coefficient of 50 Btu/hrft^{2}F can be expected on the outer surface, estimate the ambient temperature necessary to freeze the cans solid overnight (a 12-hr period). Neglect end effects and also the thermal resistance of the container.

22. How thick a layer of ice can be frozen on the outside of a 2-in.-diameter thin walled tube in 6 hr? The tube is immersed in 32F water and has brine at $-10F$ flowing on the outside with a heat transfer coefficient between the brine and tube wall of 30 Btu/hrft^{2}F.

References

1. Keenan, J. H., Keye, F. G., Hill, P. G., and Moore, J. G., *Steam Tables; Thermodynamic Properties of Water*, Wiley, New York, 1969.

2. Nukiyama, S., "Maximum and Minimum Heat Transfer from a Metal to Boiling Water at Atmospheric Pressure," *Journal Japan Society Mechanical Engineering*, Vol. 37, 1934, p. 367.

3. Githinji, P. M., and Sabersky, R. H., "Some Effects of Orientation of Heating Surfaces in Nucleate Boiling," *Transactions ASME*, JHT, No. 85, 1963, p. 379.

4. Rohsenow, W. M., "A Method of Correlating Heat Transfer Data for Surface Boiling Liquids," *Transactions ASME*, Vol. 74, 1952, p. 969.

5. Rohsenow, W. M. and Choi, H. Y., *Heat, Mass, and Momentum Transfer*, Prentice-Hall, Englewood Cliffs, NJ, 1961, pp. 224, 225, 229.

6. *C.R.C. Handbook of Chemistry and Physics*, Chemical Rubber Company, 18901 Cranwood Pkwy., Cleveland, Ohio 44128.

7. Zuber, N., "On the Stability of Boiling Heat Transfer," *Transactions ASME*, JHT, Vol. 80, 1958, p. 711.

8. Gambill, W. R. and Greene, N. D., "A Preliminary Study of Boiling Burnout Heat Fluxes for Water in Vortex Flow," Preprint 29, ASME-AIChE Second Annual Heat Transfer Conf., Chicago, 1958.

9. Berenson, P., "Transition Boiling Heat Transfer from a Horizontal Surface," AIChE Paper No. 18, ASME-AIChE Heat Transfer Conference, Buffalo, NY, 1960.

10. Ciechelli, M. T. and Bonilla, C. F., "Heat Transfer to Liquids Boiling Under Pressure," *AIChE Transactions* Vol. 41, 1945.

11. Lottes, P. A., *Nuclear Reactor Heat Transfer*, Argonne National Laboratory Report ANL-6469, Dec. 1961.

12. Rohsenow, W. M. and Harnett, J. P., Eds., *Handbook of Heat Transfer*, McGraw-Hill, New York, 1973.

13. Gunther, F. C., "Photographic Study of Boiling Heat Transfer to Water with Forced Convection," *Transactions ASME*, Vol. 73, 1951, p. 115.

14. Wallis, G. B., *One-Dimensional Two-Phase Flow*, McGraw-Hill, New York, 1970.

15. Charvonia, D. A., "A Study of the Mean Thickness of Liquid Film and Characteristics of the Interfacial Surface in Annular Two-Phase Flow in a Vertical Tube, Including a Review of the Literature," Ph.D. Thesis, Purdue University, 1959.

16. Nusselt, W., "Die Oberflächenkondensation des Wasserdampfes," *Zeitschrift V.D.I.*, No. 60, 1916, p. 541.

17. McAdams, W. H., *Heat Transmission*, 3d Edition, McGraw-Hill, New York, 1954, pp. 334, 335.

18. Chen, M. M., "An Analytical Study of Laminar Film Condensation: Part 2—Single and Multiple Horizontal Tubes," *Transactions ASME*, JHT, Vol. 83, 1961, p. 55.

19. Sfeir, A. A. and Clumpner, J. A., "Continuous Casting of Cylindrical Ingots," *Transactions ASME*, JHT, Vol. 99, 1977, p. 29.

20. London, A. L. and Seban, R. A., "Rate of Ice Formation," *Transactions ASME*, Vol. 65, 1943, p. 771.

21. Kreith, F., *Principles of Heat Transfer*, 2d Edition, Harper & Row, New York, 1965, p. 472.

22. Burington, R. S., *Handbook of Math Tables and Formulas*, 4th Edition, McGraw-Hill, New York, 1965, p. 72.

23. Tong, L. S., *Boiling Heat Transfer and Two-Phase Flow*, Wiley, New York, 1965.

24. Sadd, M. H. and Didlake, J. E., "Non-Fourier Melting of a Semi-Infinite Solid," *Transactions ASME*, JHT, Vol. 99, 1977, p. 25.

25. Muehlbauer, J. C., Coauthors, "Transient Heat Transfer Analysis of Alloy Solidification," *Transactions ASME*, JHT, No. 95, 1973, p. 324.

26. Yilmaz, S. and Westwater, J. W., "Effect of Velocity on Heat Transfer to Boiling Freon-113," *Transactions ASME*, JHT, Vol. 102, 1980, p. 26.

Appendix A
Abbreviations, Multiples, General Information

Abbreviations for the commonly used British engineering (BE) units and the Système International (SI) units were given in Chapter 1, Table 1.3. Multiples of the basic units can be expressed by prefacing the unit with the abbreviations given in Table A.1. For example, in describing small distances such as the wavelength of light, the unit of measurement in common usage is the micron (μ) which is one millionth of a meter. To be strictly correct according to Table 1.3 and Table A.1 we should write μm and call it a micrometer, but the terminology micron has been so firmly established that it will probably remain so.

Another example that follows the recommended pattern is the description of the energy term electron-volts, eV. One million electron-volts is MeV, and one billion electron-volts is written as GeV. Along this same vein we are familiar with millivolts (mV), microvolts (μV), and nanovolts (nV) where these are understood to be:

$$1 \text{ mV} = 10^{-3} \text{ V}$$
$$1 \text{ } \mu\text{V} = 10^{-6} \text{ V}$$
$$1 \text{ nV} = 10^{-9} \text{ V}$$

Perhaps not so familiar are the SI pressure designations kilopascals (kPa) or megapascals (mPa) because they are not that commonly used. When scales that

Table A.1 UNIT MULTIPLES

Amount	Multiples and Submultiples	Prefixes	Symbols	Pronunciations	Means
1 000 000 000 000 000 000	10^{18}	exa	E	ĕx´a	One quintillion times
1 000 000 000 000 000	10^{15}	peta	P	pĕt´a	One quadrillion times
1 000 000 000 000	10^{12}	tera	T	tĕr´à	One trillion times
1 000 000 000	10^{9}	giga	G	jĭ´ga	One billion times
1 000 000	10^{6}	mega	M*	mĕg´à	One million times
1 000	10^{3}	kilo	k*	kĭl´o	One thousand times
100	10^{2}	hecto	h	hĕk´t o	One hundred times
10	10	deka	da	dĕk´à	Ten times
0.1	10^{-1}	deci	d	dĕs´ĭ	One tenth of
0.01	10^{-2}	centi	c*	sĕn´tĭ	One hundredth of
0.001	10^{-3}	milli	m*	mĭl´ĭ	One thousandth of
0.000 001	10^{-6}	micro	μ*	mī´kr o	One millionth of
0.000 000 001	10^{-9}	nano	n	năn´ o	One billionth of
0.000 000 000 001	10^{-12}	pico	p	pē´c o	One trillionth of
0.000 000 000 000 001	10^{-15}	femto	f	fĕm´t o	One quadrillionth of
0.000 000 000 000 000 001	10^{-18}	atto	a	ăt´t o	One quintillionth of

source: The above table was reproduced from ASME Booklet SI-7 with the permission of the publishers, ASME.
*Most commonly used.

are sold in department stores are calibrated in newtons, tire gauges come marked in kilopascals (kPa), and everyone knows that atmospheric pressure at sea level is about 101 kPa, the changeover to SI units will be complete.

The use of multiples or multipliers on table headings often causes confusion. Units such as viscosity in lb_m/fts for gases are very small numbers. Nitrogen (see Appendix C) has a viscosity of 0.00001222 lb_m/fts at 100F. We could also write the number as 1.222×10^{-5}, which would be somewhat shorter to write. A compact listing, such as given in the property tables in Appendix C, includes the multiplier in the table heading. For example, for nitrogen viscosity the tabulation is as follows:

T (F)	$\mu \times 10^5$ (lb_m/fts)
0	1.055
100	1.222
200	1.380
400	1.660
⋮	⋮

At the temperature of 100F the quantity $\mu \times 10^5 = 1.222$ and the actual value of the viscosity is $\mu = 1.222 \times 10^{-5}$ lb_m/fts as we have noted above. The convention just described is in almost universal use in the scientific and engineering literature.

One other aspect that receives little if any attention in the literature (that a beginning engineer is liable to encounter) is the technique of plotting on log-log paper. Graph paper with the gridwork laid out in a logarithmic scale is particularly useful when correlating or visualizing an exponential relationship. It is also useful when the occasion requires considerable compression of a large linear scale. For example, the relationship

$$y = x^a \tag{A.1}$$

can be written as follows by taking the natural logarithm of both sides.

$$\ln(y) = a \cdot \ln(x) \tag{A.2}$$

If we plot $\ln(y)$ against $\ln(x)$ on linear paper, the result is a straight line. We can also plot y against x on log-log paper and also obtain a straight line whose slope is a. We should note that a must be dimensionless (or empirically defined) and that cycle on the log-log paper must be a square. Accordingly, the markings on the logarithmic cycles can only be altered by a power of 10; arbitrary values cannot be assigned or else the logarithmic proportions are destroyed. Logarithmic paper is very convenient and useful in plotting convection results and some radiation data. Visualizing transient and steady-state conduction calculations can best be done by using semilog paper which is linear on one axis and logarithmic on the other. See for example Figures 3.7, 3.8, 3.9, and so on; 6.7, 6.12, and 6.15; and 8.10, 8.13, 8.17.

Appendix B
Fundamental Differential and
Integral Information

The information contained in this appendix can be found in all tables of mathematical formulas. Since the topic of conduction is one of applied mathematics at the differential and integral level, it is convenient to have some of the basic mathematical tools readily at hand to refresh one's memory.

Table B.1 gives the basic differentiation facts and Table B.2 gives the basic integration facts. Tables of more extensive integrals are given in Burrington, Dwight, or Pierce.

Often when evaluating a limit the result becomes indeterminate. In that case a single or multiple evaluation of L'Hôspital's rule is useful:

$$\lim_{x \to a} \frac{f(x)}{g(x)} = \lim_{x \to a} \frac{f'(x)}{g'(x)} = \lim_{x \to a} \frac{f''(x)}{g''(x)}$$

When integrals do not respond to substitution treatment, integration by parts sometimes results in a more simple formulation that can be attacked with tables:

$$\int_a^b (u)\, dv = [\, uv \,]_a^b - \int_a^b (v)\, du$$

Judicious choice of $u(x)$ and $v(x)$ often governs the success of the method.

Table B.1 BASIC DIFFERENTIATION FACTS

In this table ln represents the natural logarithm to the base $e = 2.718281828459045\ldots$ and $\log_a$ represents logarithm to the base a. The variables u and v are functions of x.

1. $\dfrac{d(ax)}{dx} = a \qquad \dfrac{d(a)}{dx} = 0$

2. $\dfrac{d}{dx}(uv) = u\dfrac{dv}{dx} + v\dfrac{du}{dx}$

3. $\dfrac{d}{dx}\left(\dfrac{u}{v}\right) = \left(v\dfrac{du}{dx} - u\dfrac{dv}{dx}\right)/v^2$

4. $\dfrac{d}{dx}e^u = e^u\dfrac{du}{dx}$

5. $\dfrac{d}{dx}\ln(u) = \dfrac{1}{u}\dfrac{du}{dx}$

6. $\dfrac{d}{dx}a^v = a^v\ln(a)\dfrac{dv}{dx}$

7. $\dfrac{d}{dx}(u^n) = nu^{n-1}\dfrac{du}{dx}$

8. $u'(x) = \dfrac{du}{dx} = \lim\limits_{\Delta x \to 0}\dfrac{u(x + \Delta x) - u(x)}{\Delta x}$

9. $\dfrac{d}{dx}\sinh(u) = \cosh(u)\dfrac{du}{dx}$

10. $\dfrac{d}{dx}\cosh(u) = \sinh(u)\dfrac{du}{dx}$

11. $\dfrac{d}{dx}\tanh(u) = \mathrm{sech}^2(u)\dfrac{du}{dx}$

12. $\dfrac{d}{dx}\sin(u) = \cos(u)\dfrac{du}{dx}$

13. $\dfrac{d}{dx}\cos(u) = -\sin(u)\dfrac{du}{dx}$

14. $\dfrac{d}{dx}\tan(u) = \sec^2(u)\dfrac{du}{dx}$

15. $\dfrac{d}{dx}\sin^{-1}(u) = \dfrac{1}{\sqrt{1 - u^2}}\dfrac{du}{dx}$

$-\dfrac{\pi}{2} \leqslant \sin^{-1}(u) \leqslant \dfrac{\pi}{2}$

Table B.2 BASIC INTEGRATION FACTS

In this table ln represents the natural logarithm to the base $e = 2.718281828459045\ldots$ and the variables u and v are functions of x.

1. $\int(A + B) = \int A + \int B$

2. $\int au\,dx = a\int u\,dx$

3. $\int v^n\,dv = \dfrac{v^{n+1}}{n + 1} + C \qquad n \neq -1$

4. $\int\dfrac{dv}{v} = \ln(v) + C$

5. $\int e^v\,dv = e^v + C$

6. $\int a^u\,dv = \dfrac{a^u}{\ln a} + C$

7. $\int\cos(v)\,dv = \sin(v) + C$

8. $\int\sin(v)\,dv = -\cos(v) + C$

9. $\int\cos^2(u)\,du = \dfrac{u}{2} + \dfrac{1}{2}\sin(u)\cos(u) + C$

10. $\int\sin^2(u)\,du = \dfrac{u}{2} - \dfrac{1}{2}\sin(u)\cos(u) + C$

Appendix C
Physical Property Tables

As heat transfer books have been written since Jakob's pioneering effort, physical property tables have gradually boiled down to a series of tables that are common to most heat transfer books on the market at the time of this writing (1982); this book is no exception. Only the tables on electrical resistivity for use in the Wiedemann-Franz law for thermal conductivity estimation will not be found in other heat transfer books.

Most engineers consider the numbers in "the back of the book" to be the last word and authoritative. For most applications such a viewpoint is satisfactory. For critical applications it would be wise to consult the literature, or easier, to contact the Center for *IN*formation and *D*ata *A*nalysis (CINDAS) at Purdue University for the latest information and recommended correlations. Cost of the information is reasonable and is supplied by CINDAS upon request; it might be apropos to quote CINDAS's motto "Correct information does not cost, it pays!" The following tables were collected from a number of sources which are specified at the end of each table.

Some years ago in 1960 at the Eleventh General Conference on Weights and Measures a common international set of units called the Système International d'Unités (designated SI units in English) was agreed upon by the 36 participating countries. No official action, by law, has been taken in the United

States but our enormous trade deficit has been a stimulus to manufacture to SI specifications. There is, however, a mitigating factor that will probably delay an official change over for quite some time and that is the productivity slump we are in at the beginning of the 1980 decade. A mandatory switch to SI units would result in a momentary decrease in productivity (which we can ill afford right now) until everyone has been able to acclimatize their thinking to the new numbers. That period of acclimation is anywhere from six months to several years depending on the frequency of contact. Engineers will accordingly need to be familiar with conversions so that they will be able to work in SI units if necessary. Since most industries still work with BE units the tables presented here are in those familiar units.

TABLE	TITLE
C.1	Properties of Nonmetallic Materials
C.2	Properties of Metals and Alloys
C.3	Electrical Resistivity of the Elements
C.4	Electrical Resistivity and Properties of Tungsten
C.5	Steel Pipe Dimensions
C.6	Thermodynamic Properties of Saturated Nitrogen
C.7	Thermal Properties of Dry Air at Atmospheric Pressure
C.8	Thermal Properties of Liquid Water at Saturation Pressure
C.9	Thermal Properties of Dry Gases at Atmospheric Pressure
C.10	Thermal Properties of Liquids at Saturation Pressure
C.11	Properties of WEMCO "C" Transformer Oil
C.12	Total Normal Emissivities of Various Surfaces

Table C.1 PROPERTIES OF NONMETALLIC MATERIALS

SUBSTANCE	c_p (Btu/lb$_m$F)		ρ (lb$_m$/ft^3)		t (F)	k (Btu/hrftF)		α (ft^2/hr)
Structural								
Asphalt					68	0.43	a	
Bakelite	0.38	b	79.5	b	68	0.134	b	0.0044
Bricks								
Common	0.20	d	100	d	68	0.40	a	0.02
Face			128	d	68	0.76	a	
Carborundum					1110	10.7	a	
brick					2550	6.4	a	
					392	1.34	a	0.036
Chrome brick	0.20	d	188	d	1022	1.43	a	0.038
					1652	1.15	a	0.031
Diatomaceous					400	0.14	a	
earth(fired)					1600	0.18	a	
					932	0.60	a	0.020
Fire clay brick	0.23	d	128	d	1472	0.62	a	0.021
(burnt 2426F)					2012	0.63	a	0.021
					932	0.74	a	0.022
Fire clay brick	0.23	d	145	d	1472	0.79	a	0.024
(burnt 2642F)					2012	0.81	a	0.024
					392	0.58	a	0.015
Fire clay brick	0.23	d	165	f	1112	0.85	a	0.022
(Missouri)					2552	1.02	a	0.027
					400	2.2	a	
Magnesite	0.27	d			1200	1.6	a	
					2200	1.1	a	
Cement, Portland			94			0.17	a	
Cement, mortar					75	0.67	a	
Concrete	0.21	b	119–144	b	68	0.47–0.81	b	0.019–0.027
Concrete, cinder					75	0.44	a	
Glass, plate	0.2	b	169	b	68	0.44	b	0.013
Glass, borosilicate			139	b	86	0.63	b	
Plaster, Gypsum	0.2	d	90	d	70	0.28	a	0.016
Plaster, metal lath					70	0.27	a	
Plaster, wood lath					70	0.16	a	
Stone								
Granite	0.195	d	165	d		1.0–2.3	a	0.0031–0.071
Limestone	0.217	d	155	d	210–570	0.73–0.77	a	0.022–0.023
Marble	0.193	b	156–169	b	68	1.6	b	0.054
Sandstone	0.17	b	135–144	b	68	0.94–1.2	b	0.041–0.049
Wood, cross grain:								
Balsa			8.8	a	86	0.032	a	
Cypress			29	d	86	0.056	a	
Fir	0.65	d	26.0	b	75	0.063	a	0.0037
Oak	0.57	d	38–30	b	86	0.096	a	0.0049
Yellow pine	0.67	d	40	d	75	0.085	a	0.0032
White pine			27	d	86	0.065	a	
Wood, radial								
Oak	0.57	b	38–30	b	68	0.10–0.12	b	0.0043 / 0.0047
Fir	0.65	b	26.0–26.3	b	68	0.08	b	0.0048
Insulating								
Asbestos			29.3	b	−328	0.043	b	
					32	0.090	b	

continued

Table C.1   continued

SUBSTANCE	c_p (Btu/lb$_m$F)		ρ (lb$_m$/ft^3)		t (F)	k (Btu/hrftF)		α (ft^2/hr)
Asbestos			36.0	b	32	0.087	b	
					212	0.111	b	
					392	0.120	b	
					752	0.129	b	
Asbestos			43.5	b	-328	0.09	b	
					32	0.135	b	
Asbestos cement						1.2	a	
Asbestos cement board					68	0.43	a	
Asbestos sheet					124	0.096	a	
Asbestos felt					100	0.033	a	
(40 laminations					300	0.040	a	
per inch)					500	0.048	a	
Asbestos felt					100	0.045	a	
(20 laminations					300	0.055	a	
per inch)					500	0.065	a	
Asbestos, corrugated					100	0.05	a	
(4 plies per inch)					200	0.058	a	
					300	0.069	a	
Balsam wool			2.2	a	90	0.023	a	
Cardboard, corrugated						0.037	a	
Celotex					90	0.028	a	
Corkboard			10	b	86	0.025	b	
Cork, expanded scrap	0.45	b	2.8–7.4	b	68	0.021	b	0.006–0.017
Cork, ground			9.4	b	86	0.025	b	
Diatomaceous earth (powdered)			10	e	200	0.029	e	
					400	0.038	e	
					600	0.048	e	
Diatomaceous earth (powdered)			14	e	200	0.033	e	
					400	0.039	e	
					600	0.046	e	
Diatomaceous earth (powdered)			18	e	200	0.040	e	
					400	0.045	e	
					600	0.049	e	
Felt, hair			8.2	c	20	0.0237	c	
					100	0.0269	c	
					200	0.0310	c	
Felt, hair			11.4	c	20	0.0212	c	
					100	0.0254	c	
					200	0.0299	c	
Felt, hair			12.8	c	20	0.0233	c	
					100	0.0262	c	
					200	0.0295	c	
Fiber insulating board			14.8	b	70	0.028	b	
Glass wool			1.5	c	20	0.0217	c	
					100	0.0313	c	
					200	0.0435	c	
Glass wool			4.0	c	20	0.0179	c	
					100	0.0239	c	
					200	0.0317	c	
Glass wool			6.0	c	20	0.0163	c	
					100	0.0218	c	
					200	0.0288	c	

continued

Table C.1 continued

SUBSTANCE	c_p (Btu/lb_mF)		ρ (lb_m/ft^3)		t (F)	k (Btu/hrftF)		α (ft^2/hr)
Kapok					86	0.020	a	
Magnesia, 85 percent			16.9	c	100	0.039	a	
					200	0.041	a	
					300	0.043	a	
					400	0.046	a	
Rock wool			4.0	c	20	0.0150	c	
					100	0.0224	c	
					200	0.0317	c	
Rock wool			8.0	c	20	0.0171	c	
					100	0.0228	c	
					200	0.0299	c	
Rock wool			12.0	c	20	0.0183	c	
					100	0.0226	c	
					200	0.0281	c	
Miscellaneous								
Aerogel, silica			8.5	b	248	0.013	b	
Clay	0.21	b	91.0	b	68	0.739	b	0.039
Coal, anthracite	0.30	b	75–94	b	68	0.15	b	0.005–0.006
Coal, powdered	0.31	b	46	b	86	0.067	b	0.005
Cotton	0.31	b	5	b	68	0.034	b	0.075
Earth, coarse	0.44	b	128	b	68	0.30	b	0.0054
Earth, wet					70	0.25		0.032
Ice	0.46	b	57	b	32	1.28	b	0.048
Rubber, hard			74.8	b	32	0.087	b	
Sawdust					75	0.034	a	
Silk	0.33	b	3.6	b	68	0.021	b	0.017

SOURCE: The above table is reproduced from Chapman, A. J., *Heat Transfer*, 2d. ed., Macmillan, New York, 1967, with the permission of the Macmillan Publishing Co.

Adapted from (a) A. I. Brown and S. M. Marco, *Introduction to Heat Transfer*, 3d ed., McGraw-Hill, New York, 1958; (b) E. R. G. Eckert, *Introduction to the Transfer of Heat and Mass*, McGraw-Hill, New York, 1950; (c) R. H. Heilman, *Ind. Eng. Chem.*, Vol. 28, 1936, p. 782; (d) L. S. Marks, *Mechanical Engineers' Handbook*, 6th ed., McGraw-Hill, New York, 1958; (e) R. Calvert, *Diatomaceous Earth*, Chemical Catalog Company, Inc., 1930; (f) H. F. Norton, *J. Am. Ceram. Soc.*, Vol. 10, 1957, p. 30.

Table C.2 PROPERTIES OF METALS AND ALLOYS

Metal	Properties at 68F				k (Btu/hrftF)									
	ρ (lb$_m$/ft^3)	c_p (Btu/lb$_m$F)	k (Btu/hrftF)	α (ft^2/hr)	-148F -100C	32F 0C	212F 100C	392F 200C	572F 300C	752F 400C	1112F 600C	1472F 800C	1832F 1000C	2192F 1200C
Aluminum														
Pure	169	0.214	132	3.665	134	132	132	132	132					
Al-Cu (Duralumin) 94-96 Al, 3-5 Cu, trace Mg	174	0.211	95	2.580	73	92	105	112						
Al-MG (Hydronalium) 91-95 Al, 5-9 Mg	163	0.216	65	1.860	54	63	73	82						
AL-Si (Silumin) 87 Al, 13 Si	166	0.208	95	2.773	86	94	101	107						
Al-Si (Silumin, copper bearing 86.5 Al; 12.5 Si; 1 Cu	166	0.207	79	2.311	69	79	83	88	93					
Al-Si (Alusil) 78-80 Al; 20-22 Si	164	0.204	93	2.762	83	91	97	101	103					
Al-Mg-Si 97 Al; 1 Mg; 1 Si; 1 Mn	169	0.213	102	2.859	—	101	109	118						
Lead	710	0.031	20	0.924	21.3	20.3	19.3	18.2	17.2					
Iron														
Pure	493	0.108	42	0.785	50	42	39	36	32	28	23	21	20	21
Wrought iron (C < 0.50%)	490	0.11	34	0.634	—	34	33	30	28	26	21	19	19	19
Cast iron (C ≈ 4%)	454	0.10	30	0.666										
Steel (C max ≈ 1.5%)														
Carbon steel C ≈ 0.5%	489	0.111	31	0.570	—	32	30	28	26	24	29	17	17	18
1.0%	487	0.113	25	0.452	—	25	25	24	23	21	19	17	16	17
1.5%	484	0.116	21	0.376	—	21	21	21	20	19	18	16	16	17
Nickel steel Ni ≈ 0%	493	0.108	42	0.785										
10%	496	0.11	15	0.279										
20%	499	0.11	11	0.204										
30%	504	0.11	7	0.118										
40%	510	0.11	6	0.108										
50%	516	0.11	8	0.140										
60%	523	0.11	11	0.182										
70%	531	0.11	15	0.258										
80%	538	0.11	20	0.344										

continued

Table C.2 continued

Metal	Properties at 68F				k (Btu/hrftF)									
	ρ (lb$_m$/ft^3)	c_p (Btu/lb$_m$F)	k (Btu/hrftF)	α (ft^2/hr)	−148F −100C	32F 0C	212F 100C	392F 200C	572F 300C	752F 400C	1112F 600C	1472F 800C	1832F 1000C	2192F 1200C
90%	547	0.11	27	0.452										
100%	556	0.106	52	0.892										
Invar Ni ≅ 36%	508	0.11	6.2	0.108										
Crome steel Cr = 0%	493	0.108	42	0.785	50	42	39	36	32	28	23	21	20	21
1%	491	0.11	35	0.645	—	36	32	30	27	24	21	19	19	
2%	491	0.11	30	0.559	—	31	28	26	24	22	19	18	18	
5%	489	0.11	23	0.430	—	23	22	21	21	19	17	17	17	17
10%	486	0.11	18	0.344	—	18	18	18	17	17	16	16	17	
20%	480	0.11	13	0.258	—	13	13	13	13	14	14	15	17	
30%	476	0.11	11	0.204										
Cr-Ni (chrome-nickel)														
15 Cr; 10 Ni	491	0.11	11	0.204										
18 Cr; 8 Ni (V2A)	488	0.11	9.4	0.172	—	9.4	10	10	11	11	13	15	18	
20 Cr; 15 Ni	489	0.11	8.7	0.161										
25 Cr; 20 Ni	491	0.11	7.4	0.140										
Ni-Cr (nickel-chrome)														
80 Ni; 15 Cr	532	0.11	10	0.172										
60 Ni; 15 Cr	516	0.11	7.4	0.129										
40 Ni; 15 Cr	504	0.11	6.7	0.118										
20 Ni; 15 Cr	491	0.11	8.1	0.151	—	8.1	8.7	8.7	9.4	10	11			
Cr-Ni-Al; 6 Cr; 1.5 Al; 0.5 Si (Sicromal 8)	482	0.117	13	0.237								13		
24 Cr; 2.5 Al 0.5 Si (Sicromal 12)	479	0.118	11	0.194										
Manganese steel Ma = 0%	493	0.118	42	0.784										
1%	491	0.11	29	0.538										
2%	490	0.11	22	0.376	—	22	21	21	21	20	19			
5%	487	0.11	13	0.247										
10%	493	0.11	10	0.194										
Tungsten steel W = 0%	494	0.108	42	0.785										
1%	497	0.107	38	0.720										
2%	497	0.106	36	0.677	—	36	34	31	28	26	21			

Metal														
5%	504	0.104	31	0.591										
10%	519	0.100	28	0.527										
20%	551	0.093	25	0.484										
Silicon steel Si = 0%	493	0.108	42	0.785										
1%	485	0.11	24	0.451										
2%	479	0.11	18	0.344										
5%	463	0.11	11	0.215										
Copper														
Pure	559	0.0915	223	4.353	235	223	219	216	—	210	204			
Aluminum bronze 95 Cu; 5 Al	541	0.098	48	0.903										
Bronze 75 Cu; 25 Sn	541	0.082	15	0.333										
Red brass 85 Cu; 9 Sn; 6 Zn	544	0.092	35	0.699	—	34	41							
Brass 70 Cu; 30 Zn	532	0.092	64	1.322	51	—	74	83	85	85				
German silver 62 Cu; 15 Ni; 22 Zn	538	0.094	14.4	0.290	11.1	—	18	23	26	28				
Constantan 60 Cu; 40 Ni	557	0.098	13.1	0.237	12	—	12.8	15						
Magnesium														
Pure	109	0.242	99	3.762	103	99	97	94	91					
Mg-Al (electrolytic) 6-8% Al; 1-2% Zn	113	0.24	38	1.397	—	30	36	43	48					
Mg-Mn 2% Mn	111	0.24	66	2.473	54	64	72	75	74					
Molybdenum	638	0.060	79	2.074	80	79	78	77	74	72	67	62	58	54*
Nickel														
Pure (99.9%)	556	0.1065	52	0.882	60	54	48	42	37	34	32	36	39	40
Impure (99.2%)	556	0.106	40	0.677	—	40	37	34	32	30				
Ni-Cr; 90 Ni; 10 Cr	541	0.106	10	0.172	—	9.9	10.9	12.1	13.2	14.2	13.0			
80 Ni; 20 Cr	519	0.106	7.3	0.129	—	7.1	8.0	9.0	9.9	10.9				
Silver														
Purest	657	0.0559	242	6.601	242	241	240	238	209					
Pure (99.9%)	657	0.0559	235	6.418	242	237	240	216		208				
Tungsten	1208	0.0321	94	2.430	—	96	87	82	77	73	65	44		
Zinc, pure	446	0.0918	64.8	1.591	66	65	63	61	58	54				
Tin, pure	456	0.0541	37	1.505	43	38.1	34	33						

SOURCE: Reproduced from Eckert, E. R. G. and Drake, R. M., *Heat and Mass Transfer*, 2d ed., McGraw-Hill, New York, 1959 with the permission of the publisher.
*Molybdenum data from Wills, Dr. L. D., Ph.D. thesis, University of Arkansas, 1982.

Table C.3 ELECTRICAL RESISTIVITY OF THE ELEMENTS

ELEMENT	TEMPERATURE (C)	$\mu\Omega$cm	TEMPERATURE COEFFICIENT PER C
Aluminum, 99.996 percent	20	2.6548	0.00429 [20§§]
Antimony	0	39.0	
Arsenic	20	33.3	
Beryllium*	20	4.0	0.025 [20§§]
Bismuth	0	106.8	
Boron	0	1.8×10^{12}	
Cadmium	0	6.83	0.0042 [0§§]
Calcium	0	3.91	0.00416 [0§§]
Carbon[†]	0	1375.0	
Cerium	25	75.0	0.00087 [0-25]
Cesium	20	20	
Chromium	0	12.9	0.003 [0§§]
Cobalt	20	6.24	0.00604 [0-100]
Copper	20	1.6730	0.0068 [20§§]
Dysprosium[‡]	25	57.0	0.00119 [0-25]
Erbium	25	107.0	0.00201 [0-25]
Europium	25	90.0	
Gadolinium	25	140.5	0.00176 [0-25]
Gallium[§]	20	17.4	
Germanium[¶]	22	46×10^6	
Gold	20	2.35	0.004 [0-100]
Hafnium	25	35.1	0.0038 [25§§]
Holmium	25	87.0	0.00171 [0-25]
Indium	20	8.37	
Iodine	20	1.3×10^{15}	
Iridium	20	5.3	0.003925 [0-100]
Iron, 99.99 percent	20	9.71	0.00651 [20§§]
Lanthanum	25	5.70	0.00218 [0-25]
Lead	20	20.648	0.00336 [20-40]
Lithium	0	8.55	
Lutetium	25	79.0	0.00240 [0-25]
Magnesium**	20	4.45	0.0165 [20§§]
Manganese α	23–100	185.0	
Mercury	50	98.4	
Molybdenum	0	5.2	
Neodymium	25	64.0	0.00164 [0-25]
Nickel	20	6.84	0.0069 [0-100]
Niobium (columbium)[††]	0	12.5	
Osmium	20	9.5	0.0042 [0-100]
Palladium	20	10.8	0.00377 [0-100]
Phosphorus, white	11	1×10^{17}	
Platinum, 99.85 percent	20	10.6	0.003927 [0-100]
Plutonium	107	141.4	
Potassium	0	6.15	
Praseodymium	25	68	0.00171 [0-25]
Rhenium	20	19.3	0.00395 [0-100]
Rhodium	20	4.51	0.0042 [0-100]
Rubidium	20	12.5	
Ruthenium	0	7.6	
Samarium	25	88.0	0.00184 [0-25]

continued

Table C.3   continued

ELEMENT	TEMPERATURE (C)	$\mu\Omega$cm	TEMPERATURE COEFFICIENT PER C
Scandium[‡‡]	22	61.0	0.00282^{0-25}
Selenium	0	12.0	
Silicon	0	10.0	
Silver	20	1.59	0.0041^{0-100}
Sodium	0	4.2	
Sulfur, yellow	20	2×10^{23}	
Strontium	20	23.0	
Tantalum	25	12.45	0.00383^{0-100}
Tellerium	25	4.36×10^{5}	
Thallium	0	18.0	
Thorium	0	13.0	0.0038^{0-100}
Thulium	25	79.0	0.00195^{0-25}
Tin	0	11.0	0.0047^{0-100}
Titanium	20	42.0	
Tungsten	27	5.65	
Uranium		30.0	
Vanadium	20	24.8–26.0	
Ytterbium	25	29.0	0.0013^{0-25}
Yttrium	25	57.0	0.0027^{0-25}
Zinc	20	5.916	0.00419^{0-100}
Zironium	20	40.0	$0.0044^{20§§}$

SOURCE: The above table is reproduced from the *Handbook of Chemistry and Physics*, 48th ed., The Chemical Rubber Publishing Co., with the permission of the publisher. Copyright The Chemical Rubber Co., CRC Press, Inc.
*Annealed, commercial pure.
[†]Graphite.
[‡]Polycrystalline.
[§]Hard wire.
[¶]Intrinsic Ge.
**Polycrystalline.
[††]High purity.
[‡‡]Zone refined bar.
[§§]Data not available to indicate range over which coefficient is valid.

Table C.4 ELECTRICAL RESISTIVITY AND PROPERTIES OF TUNGSTEN

TEMPER-ATURE (K)	RESIS-TIVITY ($\mu\Omega$cm)	ELECTRON EMISSION (A/cm^2)	EVAPORATION (g/cm^{2}s)	VAPOR PRESSURE (dyn/cm^2)	THERMAL EXPANSION PERCENT l_0 at 293°	ATOMIC HEAT cal/ g atom/C)
300	5.65				0.003	6.0
400	8.06				0.044	6.0
500	10.56				0.086	6.1
600	13.23				0.130	6.1
700	16.09				0.175	6.2
800	19.00				0.222	6.2
900	21.94				0.270	6.3
1000	24.93	1.07×10^{-15}	5.32×10^{-34}	1.98×10^{-29}	0.320	6.4
1100	27.94	1.52×10^{-18}	2.17×10^{-30}	1.22×10^{-25}	0.371	6.4
1200	30.98	9.73×10^{-12}	3.21×10^{-27}	1.87×10^{-22}	0.424	6.5
1300	34.08	3.21×10^{-10}	1.35×10^{-24}	8.18×10^{-20}	0.479	6.7
1400	37.19	6.62×10^{-9}	2.51×10^{-22}	1.62×10^{-17}	0.535	6.8
1500	40.36	9.14×10^{-8}	2.37×10^{-20}	1.54×10^{-15}	0.593	7.0
1600	43.55	9.27×10^{-7}	1.25×10^{-18}	8.43×10^{-14}	0.652	7.1
1700	46.78	7.08×10^{-6}	4.17×10^{-17}	2.82×10^{-12}	0.713	7.2
1800	50.05	4.47×10^{-5}	8.81×10^{-16}	6.31×10^{-11}	0.775	7.4
1900	53.35	2.28×10^{-4}	1.41×10^{-14}	1.01×10^{-9}	0.839	7.6
2000	56.67	1.00×10^{-3}	1.76×10^{-13}	1.33×10^{-8}	0.904	7.7
2100	60.06	3.93×10^{-3}	1.66×10^{-12}	1.28×10^{-7}	0.971	7.8
2200	63.48	1.33×10^{-2}	1.25×10^{-11}	9.88×10^{-7}	1.039	8.0
2300	66.91	4.07×10^{-2}	8.00×10^{-11}	6.47×10^{-6}	1.109	8.2
2400	70.39	1.16×10^{-1}	4.26×10^{-10}	3.52×10^{-5}	1.180	8.3
2500	73.91	2.98×10^{-1}	2.03×10^{-9}	1.71×10^{-4}	1.253	8.4
2600	77.49	7.16×10^{-1}	8.41×10^{-9}	7.24×10^{-4}	1.328	8.6
2700	81.04	1.63	3.19×10^{-8}	2.86×10^{-3}	1.404	8.7
2800	84.70	3.54	1.10×10^{-7}	9.84×10^{-3}	1.479	8.9
2900	88.33	7.31	3.30×10^{-7}	3.00×10^{-2}	1.561	9.0
3000	92.04	1.42×10	9.95×10^{-7}	9.20×10^{-2}	1.642	9.2
3100	95.76	2.64×10	2.60×10^{-6}	2.50×10^{-1}	1.724	9.4
3200	99.54	4.78×10	6.38×10^{-6}	6.13×10^{-1}	1.808	9.5
3300	103.3	8.44×10	1.56×10^{-5}	1.51	1.893	9.6
3400	107.2	1.42×10^2	3.47×10^{-5}	3.41	1.980	9.8
3500	111.1	2.33×10^2	7.54×10^{-5}	7.52	2.068	9.9
3600	115.0	3.73×10^2	1.51×10^{-4}	1.53×10	2.158	10.1
3655	117.1	4.79×10^2	2.28×10^{-4}	2.33×10	2.209	10.2

SOURCE: Data of Jones and Langmuir, General Electric Review, reproduced from the *Handbook of Chemistry and Physics*, 48th ed., 1968, The Chemical Rubber Publishing Co, with the permission of the publisher. Copyright The Chemical Rubber Co., CRC Press, Inc.

Table C.5   STEEL PIPE DIMENSIONS

NOMINAL PIPE SIZE (in.)	OUTSIDE DIA. (in.)	SCHEDULE NO.	WALL THICKNESS (in.)	INSIDE DIA. (in.)	METAL SECTIONAL AREA (in.^{2})	INSIDE CROSS-SECTIONAL AREA (ft^2)
$\frac{1}{8}$	0.405	40	0.068	0.269	0.072	0.00040
		80	0.095	0.215	0.093	0.00025
$\frac{1}{4}$	0.540	40	0.088	0.364	0.125	0.00072
		80	0.119	0.302	0.157	0.00050
$\frac{3}{8}$	0.675	40	0.091	0.493	0.167	0.00133
		80	0.126	0.423	0.217	0.00098
$\frac{1}{2}$	0.840	40	0.109	0.622	0.250	0.00211
		80	0.147	0.546	0.320	0.00163
$\frac{3}{4}$	1.050	40	0.113	0.824	0.333	0.00371
		80	0.154	0.742	0.433	0.00300
1	1.315	40	0.133	1.049	0.494	0.00600
		80	0.179	0.957	0.639	0.00499
$1\frac{1}{2}$	1.900	40	0.145	1.610	0.799	0.01414
		80	0.200	1.500	1.068	0.01224
		160	0.281	1.338	1.429	0.00976
2	2.375	40	0.154	2.067	1.075	0.02330
		80	0.218	1.939	1.477	0.02050
3	3.500	40	0.216	3.068	2.228	0.05130
		80	0.300	2.900	3.016	0.04587
4	4.500	40	0.237	4.026	3.173	0.08840
		80	0.337	3.826	4.407	0.07986
5	5.563	40	0.258	5.047	4.304	0.1390
		80	0.375	4.813	6.112	0.1263
		120	0.500	4.563	7.953	0.1136
		160	0.625	4.313	9.696	0.1015
6	6.625	40	0.280	6.065	5.584	0.2006
		80	0.432	5.761	8.405	0.1810
10	10.75	40	0.365	10.020	11.90	0.5475
		60	0.500	9.750	16.10	0.5185

Table C.6 THERMODYNAMIC PROPERTIES OF SATURATED NITROGEN*

TEMP. (R)	ABS. PRESS. ($lb_f/in.^2$) P	SPECIFIC VOLUME (ft^3/lb_m)			ENTHALPY (Btu/lb_m)			ENTROPY ($Btu/lb_m R$)		
		SAT. LIQUID v_f	EVAP. v_{fg}	SAT. VAPOR v_g	SAT. LIQUID h_f	EVAP. h_{fg}	SAT. VAPOR h_g	SAT. LIQUID s_f	EVAP. s_{fg}	SAT. VAPOR s_g
113.670	1.813	0.01845	23.793	23.812	0.000	92.891	92.891	0.00000	0.81720	0.81720
120.000	3.337	0.01875	13.570	13.589	3.113	91.224	94.337	0.02661	0.76020	0.78681
130.000	7.654	0.01929	6.3208	6.3401	8.062	88.432	96.494	0.06610	0.68025	0.74634
139.255	14.696	0.01984	3.4592	3.4791	12.639	85.668	98.306	0.09992	0.61518	0.71510
140.000	15.425	0.01989	3.3072	3.3271	13.006	85.436	98.443	0.10253	0.61026	0.71279
150.000	28.120	0.02056	1.8865	1.9071	17.945	82.179	100.124	0.13628	0.54786	0.68414
160.000	47.383	0.02132	1.1469	1.1682	22.928	78.548	101.476	0.16795	0.49093	0.65888
170.000	74.991	0.02219	0.7299	0.7521	28.045	74.383	102.427	0.19829	0.43754	0.63584
180.000	112.808	0.02323	0.4789	0.5021	33.411	69.478	102.889	0.22805	0.38599	0.61404
190.000	162.761	0.02449	0.3190	0.3435	39.153	63.582	102.735	0.25789	0.33464	0.59254
200.000	226.853	0.02613	0.2119	0.2380	45.283	56.474	101.757	0.28780	0.28237	0.57017
210.000	307.276	0.02845	0.1354	0.1639	52.061	47.475	99.536	0.31894	0.22607	0.54501
220.000	406.739	0.03249	0.0750	0.1075	60.336	34.536	94.872	0.35494	0.15698	0.51192
226.000	477.104	0.03806	0.0374	0.0755	68.123	20.423	88.546	0.38789	0.09037	0.47826

SOURCE: The above table was reproduced from Van Wylen, G. J. and Sonntag, R. E., *Fundamentals of Classical Thermodynamics*, 2d ed., Wiley, New York, 1973, with the permission of the publisher.

*Abstracted from National Bureau of Standards Technical Note 129A, The Thermodynamic Properties of Nitrogen from 114 to 540 R between 1 and 3000 psia. Supplement A (British Units) by Thomas R. Strobridge.

Table C.7 THERMAL PROPERTIES OF DRY AIR AT ATMOSPHERIC PRESSURE*

t (F)	ρ (lb$_\text{m}$/ft^3)	μ (lb$_\text{m}$/fthr)	k (Btu/hrftF)	c_p (Btu/lbF)	ν (ft^2/hr)	α (ft^2/hr)	N_{PR}	$g\beta/\nu^2$ (1/Rft^3)	$g\beta/\nu\alpha \times 10^{-6}$
−100	0.11028	0.03214	0.01045	0.2405	0.2914	0.3940	0.739	13.64	10.09
−80	0.10447	0.03365	0.01099	0.2404	0.3221	0.4377	0.736	10.58	7.783
−60	0.09924	0.03513	0.01153	0.2404	0.3540	0.4832	0.733	8.319	6.094
−40	0.09451	0.03658	0.01207	0.2403	0.3870	0.5315	0.728	6.629	4.827
−20	0.09021	0.03800	0.01260	0.2403	0.4212	0.5812	0.725	5.342	3.871
0	0.08629	0.03939	0.01312	0.2403	0.4565	0.6326	0.722	4.350	3.139
20	0.08269	0.04075	0.01364	0.2403	0.4928	0.6865	0.718	3.577	2.568
40	0.07938	0.04208	0.01416	0.2404	0.5301	0.7421	0.714	2.968	2.120
60	0.07633	0.04339	0.01466	0.2404	0.5685	0.7989	0.712	2.481	1.766
80	0.07350	0.04467	0.01516	0.2405	0.6078	0.8575	0.709	2.090	1.482
100	0.07087	0.04594	0.01566	0.2406	0.6482	0.9185	0.706	1.772	1.251
120	0.06843	0.04718	0.01615	0.2407	0.6895	0.9806	0.703	1.512	1.063
140	0.06614	0.04839	0.01664	0.2409	0.7316	1.0446	0.700	1.298	0.9094
160	0.06401	0.04959	0.01712	0.2411	0.7747	1.1095	0.698	1.121	0.7824
180	0.06201	0.05077	0.01759	0.2413	0.8187	1.1758	0.696	0.9720	0.6768
200	0.06013	0.05193	0.01806	0.2415	0.8636	1.2438	0.694	0.8471	0.5882
220	0.05836	0.05308	0.01853	0.2418	0.9095	1.3133	0.693	0.7413	0.5134
240	0.05669	0.05420	0.01899	0.2421	0.9561	1.3841	0.691	0.6516	0.4501
260	0.05512	0.05531	0.01945	0.2424	1.0034	1.4558	0.689	0.5772	0.3965
280	0.05363	0.05640	0.01990	0.2427	1.0517	1.5284	0.688	0.5094	0.3505
300	0.05221	0.05748	0.02034	0.2431	1.1009	1.6028	0.687	0.4527	0.3109
320	0.05087	0.05854	0.02079	0.2435	1.1508	1.6780	0.686	0.4037	0.2768
340	0.04960	0.05959	0.02122	0.2439	1.2014	1.7537	0.685	0.3623	0.2474
360	0.04839	0.06063	0.02166	0.2443	1.2529	1.8325	0.684	0.3239	0.2215
380	0.04724	0.06165	0.02208	0.2447	1.3050	1.9100	0.683	0.2915	0.1991

continued

Table C.7 continued

t (F)	ρ (lb$_m$/ft^3)	μ (lb$_m$/fthr)	k (Btu/hrftF)	c_p (Btu/lbF)	ν (ft^2/hr)	α (ft^2/hr)	N_{PR}	$g\beta/\nu^2$ (1/Rft^3)	$g\beta/\nu\alpha \times 10^{-6}$
400	0.04614	0.06266	0.02251	0.2452	1.3580	1.9902	0.682	0.2629	0.1794
420	0.04509	0.06366	0.02293	0.2457	1.4118	2.0695	0.682	0.2377	0.1622
440	0.04409	0.06464	0.02335	0.2462	1.4660	2.1521	0.681	0.2156	0.1468
460	0.04313	0.06561	0.02376	0.2467	1.5212	2.2331	0.681	0.1959	0.1334
480	0.04221	0.06657	0.02417	0.2472	1.5771	2.3174	0.680	0.1784	0.1214
500	0.04133	0.06752	0.02458	0.2478	1.6337	2.4004	0.680	0.1627	0.1108
520	0.04049	0.06846	0.02498	0.2483	1.6908	2.4856	0.680	0.1488	0.1012
540	0.03968	0.06939	0.02538	0.2489	1.7487	2.5688	0.680	0.1364	0.09282
560	0.03890	0.07031	0.02577	0.2495	1.8075	2.6540	0.681	0.1251	0.08522
580	0.03815	0.07122	0.02616	0.2501	1.8668	2.7421	0.681	0.1150	0.07832
600	0.03743	0.07212	0.02655	0.2507	1.9268	2.8305	0.681	0.1060	0.07213
620	0.03673	0.07301	0.02694	0.2513	1.9878	2.9187	0.681	0.09771	0.06655
640	0.03607	0.07389	0.02732	0.2519	2.0485	3.0055	0.682	0.09033	0.06157
660	0.03543	0.07477	0.02770	0.2525	2.1104	3.0950	0.682	0.08359	0.05700
680	0.03481	0.07563	0.02807	0.2531	2.1727	3.1862	0.682	0.07748	0.05284
700	0.03420	0.07649	0.02844	0.2538	2.2365	3.2765	0.683	0.07186	0.04905
720	0.03362	0.07734	0.02881	0.2544	2.3004	3.3696	0.683	0.06678	0.04559
740	0.03306	0.07818	0.02918	0.2550	2.3648	3.4615	0.683	0.06214	0.04245
760	0.03252	0.07901	0.02954	0.2557	2.4296	3.5505	0.684	0.05790	0.03962
780	0.03200	0.07984	0.02990	0.2563	2.4950	3.6463	0.684	0.05402	0.03696
800	0.03149	0.08066	0.03026	0.2570	2.5614	3.7404	0.685	0.05044	0.03454
820	0.03100	0.08147	0.03062	0.2576	2.6281	3.8323	0.686	0.04717	0.03234
840	0.03052	0.08227	0.03097	0.2582	2.6956	3.9302	0.686	0.04414	0.03028
860	0.03006	0.08307	0.03132	0.2589	2.7635	4.0257	0.686	0.04136	0.02840
880	0.02961	0.08386	0.03167	0.2595	2.8322	4.1237	0.687	0.03879	0.02664

464

900	0.02917	0.08464	0.03201	0.2601	2.9016	4.2174	0.688	0.03642	0.02505
920	0.02875	0.08542	0.03235	0.2608	2.9711	4.3133	0.689	0.03423	0.02358
940	0.02834	0.08620	0.03269	0.2614	3.0416	4.4116	0.690	0.03219	0.02220
960	0.02794	0.08696	0.03303	0.2620	3.1124	4.5123	0.690	0.03031	0.02091
980	0.02755	0.08772	0.03337	0.2626	3.1840	4.6155	0.690	0.02856	0.01970
1000	0.02717	0.08847	0.03370	0.2632	3.2562	4.7133	0.691	0.02694	0.01861
1050	0.02627	0.09034	0.03452	0.2648	3.4389	4.9598	0.693	0.02335	0.01619
1100	0.02543	0.09216	0.03533	0.2663	3.6241	5.2186	0.695	0.02035	0.01413
1150	0.02464	0.09396	0.03613	0.2677	3.8133	5.4742	0.697	0.01781	0.01241
1200	0.02390	0.09572	0.03691	0.2691	4.0050	5.7403	0.698	0.01566	0.01093
1250	0.02320	0.09746	0.03768	0.2706	4.2009	6.0000	0.700	0.01382	0.00967
1300	0.02254	0.09917	0.03844	0.2719	4.3997	6.2708	0.702	0.01224	0.00859
1350	0.02192	0.10085	0.03919	0.2732	4.6008	6.5426	0.703	0.01088	0.00765
1400	0.02133	0.10250	0.03993	0.2745	4.8054	6.8140	0.705	0.00971	0.00685
1450	0.02077	0.10414	0.04066	0.2758	5.0140	7.0960	0.707	0.00868	0.00614
1500	0.02024	0.10575	0.04137	0.2771	5.2248	7.3734	0.709	0.00779	0.00552

SOURCE: The above table is reproduced from Chapman, A. J., *Heat Transfer*, 2d ed., Macmillan, New York, 1967 with the permission of the publisher.

*Up to about 20 atm the properties μ, k, c_p, and N_{Pr} can be considered essentially independent of the pressure. The properties ρ, ν, and α are pressure dependent due to density.

$^\dagger\rho$: computed from ideal gas law. μ, k, c_p: computed from recommended equations in *Handbook of Supersonic Aerodynamics*, Vol. 5, Bureau of Ordinance, Department of the Navy, 1953.

Table C.8 THERMAL PROPERTIES OF LIQUID WATER AT SATURATION PRESSURE

t (F)	c_p^* (Btu/lb$_m$F)	$\rho^\dagger$ (lb$_m$/ft^3)	μ^* (lb$_m$/ft hr)	ν (ft^2/hr)	k^* (Btu/hr ft F)	α (ft^2/hr)	β (1/R) $\times 10^3$	N_{PR}	$g\beta/\nu^2$ (1/R ft^3)	$g\beta/\nu\alpha \times 10^{-6}$
32	1.009	62.42	4.33	0.0694	0.327	0.0052	0.03	13.37	2.597	34.66
40	1.005	62.42	3.75	0.0601	0.332	0.0053	0.045	11.36	5.195	58.91
50	1.002	62.38	3.17	0.0508	0.338	0.0054	0.070	9.41	11.31	106.4
60	1.000	62.34	2.71	0.0435	0.344	0.0055	0.10	7.88	22.04	174.3
70	0.998	62.27	2.37	0.0381	0.349	0.0056	0.13	6.78	37.34	254.1
80	0.998	62.17	2.08	0.0334	0.355	0.0057	0.15	5.85	56.07	328.5
90	0.997	62.11	1.85	0.0298	0.360	0.0058	0.18	5.13	84.52	434.2
100	0.997	61.99	1.65	0.0266	0.364	0.0059	0.20	4.52	117.9	531.4
110	0.997	61.84	1.49	0.0241	0.368	0.0060	0.22	4.04	157.9	634.4
120	0.997	61.73	1.36	0.0220	0.372	0.0060	0.24	3.65	206.8	758.1
130	0.998	61.54	1.24	0.0202	0.375	0.0061	0.27	3.30	275.9	931.7
140	0.998	61.39	1.14	0.0186	0.378	0.0062	0.29	3.01	349.5	1,049.0
150	0.999	61.20	1.04	0.0170	0.381	0.0063	0.31	2.72	447.3	1,207.0
160	1.000	61.01	0.97	0.0159	0.384	0.0063	0.33	2.53	544.3	1,374.0
170	1.001	60.79	0.90	0.0148	0.386	0.0064	0.35	2.33	666.3	1,541.0
180	1.002	60.57	0.84	0.0139	0.389	0.0064	0.37	2.16	798.5	1,734.0
190	1.003	60.35	0.79	0.0131	0.390	0.0065	0.39	2.03	947.6	1,910.0
200	1.004	60.13	0.74	0.0123	0.392	0.0065	0.41	1.90	1,130.0	2,138.0
210	1.005	59.88	0.69	0.0115	0.393	0.0065	0.43	1.76	1,356.0	2,399.0
220	1.007	59.63	0.65	0.0109	0.395	0.0066	0.45	1.66	1,579.0	2,608.0

230	1.009	59.38	0.62	0.0104	0.395	0.0066	0.47	1.58	1,812.0	2,855.0
240	1.011	59.10	0.59	0.0100	0.396	0.0066	0.48	1.51	2,001.0	3,032.0
250	1.013	58.82	0.56	0.0095	0.396	0.0066	0.50	1.43	2,310.0	3,325.0
260	1.015	58.51	0.53	0.0091	0.396	0.0067	0.51	1.36	2,568.0	3,488.0
270	1.017	58.24	0.50	0.0086	0.396	0.0067	0.53	1.28	2,988.0	3,835.0
280	1.020	57.94	0.48	0.0083	0.396	0.0067	0.55	1.24	3,329.0	4,124.0
290	1.023	57.64	0.46	0.0080	0.396	0.0067	0.56	1.19	3,648.0	4,356.0
300	1.026	57.31	0.45	0.0079	0.395	0.0067	0.58	1.17	3,875.0	4,569.0
350	1.044	55.59	0.38	0.0068	0.391	0.0067	0.62	1.01	5,591.0	5,674.0
400	1.067	53.65	0.33	0.0062	0.384	0.0068	0.72	0.91	7,810.0	7,121.0
450	1.095	51.55	0.29	0.0056	0.373	0.0066	0.93	0.85	12,366.0	10,492.0
500	1.130	49.02	0.26	0.0053	0.356	0.0064	1.18	0.83	17,516.0	14,506.0
550	1.200	45.92	0.23	0.0050	0.330	0.0060	1.63	0.84	27,187.0	22,656.0
600	1.362	42.37	0.21	0.0050	0.298	0.0052	—	0.96		

SOURCE: The above table is reproduced from Chapman, A. J., *Heat Transfer*, 2d ed., Macmillan, New York, 1967 with the permission of the publisher.

*From A. I. Brown and S. M. Marco, *Introduction to Heat Transfer*, 3d ed., New York, McGraw-Hill, 1958, as compiled from *International Critical Tables*, *ASHVE Guide*, 1940 and Schmidt and Selschopp, Forsch. a. d. Geb. d. Ingenieurves, Vol. 3, 1943, p. 227.
†From J. H. Keenan and F. G. Keyes, *Thermodynamic Properties of Steam*, New York, Wiley, 1936.

Table C.9 THERMAL PROPERTIES OF DRY GASES AT ATMOSPHERIC PRESSURE

T (F)	ρ (lb_m/ft^3)	c_p ($Btu/lb_m F$)	$\mu \times 10^5$ (lb_m/fts)	$\nu \times 10^3$ (ft^2/s)	k ($Btu/hrftF$)	Pr	α (ft^2/hr)	$\beta \times 10^3$ (1/R)	$\dfrac{g\beta\rho^2}{\mu^2}$ ($1/Fft^3$)
				Steam					
212	0.0372	0.451	0.870	0.234	0.0145	0.96	0.864	1.49	0.877×10^6
300	0.0328	0.456	1.000	0.303	0.0171	0.95	1.14	1.32	0.459
400	0.0288	0.462	1.130	0.395	0.0200	0.94	1.50	1.16	0.243
500	0.0258	0.470	1.265	0.490	0.0228	0.94	1.88	1.04	0.139
600	0.0233	0.477	1.420	0.610	0.0257	0.94	2.31	0.943	82×10^3
700	0.0213	0.485	1.555	0.725	0.0288	0.93	2.79	0.862	52.1
800	0.0196	0.494	1.700	0.855	0.0321	0.92	3.32	0.794	34.0
900	0.0181	0.50	1.810	0.987	0.0355	0.91	3.93	0.735	23.6
1000	0.0169	0.51	1.920	1.13	0.0388	0.91	4.50	0.685	17.1
1200	0.0149	0.53	2.14	1.44	0.0457	0.88	5.80	0.603	9.4
1400	0.0133	0.55	2.36	1.78	0.053	0.87	7.25	0.537	5.49
1600	0.0120	0.56	2.58	2.14	0.061	0.87	9.07	0.485	3.38
1800	0.0109	0.58	2.81	2.58	0.068	0.87	10.8	0.442	2.14
2000	0.0100	0.60	3.03	3.03	0.076	0.86	12.7	0.406	1.43
2500	0.0083	0.64	3.58	4.30	0.096	0.86	18.1	0.338	0.603
3000	0.0071	0.67	4.00	5.75	0.114	0.86	24.0	0.289	0.293
				Oxygen					
0	0.0955	0.2185	1.215	0.127	0.0131	0.73	0.627	2.18	4.33×10^6
100	0.0785	0.2200	1.420	0.181	0.0159	0.71	0.880	1.79	1.76
200	0.0666	0.2228	1.610	0.242	0.0179	0.722	1.20	1.52	0.84
400	0.0511	0.2305	1.955	0.382	0.0228	0.710	1.94	1.16	0.256
600	0.0415	0.2390	2.26	0.545	0.0277	0.704	2.79	0.943	0.103
800	0.0349	0.2465	2.53	0.725	0.0324	0.695	3.76	0.794	48.5×10^3
1000	0.0301	0.2528	2.78	0.924	0.0366	0.690	4.80	0.685	25.8
1500	0.0224	0.2635	3.32	1.480	0.0465	0.677	7.88	0.510	7.50

Nitrogen

0	0.0840	0.2478	1.055	0.125	0.0132	0.713	0.635	2.18	4.55 × 10^6
100	0.0690	0.2484	1.222	0.177	0.0154	0.71	0.898	1.79	1.84
200	0.0585	0.2490	1.380	0.236	0.0174	0.71	1.20	1.52	0.876
400	0.0449	0.2515	1.660	0.370	0.0212	0.71	1.88	1.16	0.272
600	0.0364	0.2564	1.915	0.526	0.0252	0.70	2.70	0.943	0.110
800	0.0306	0.2623	2.145	0.702	0.0291	0.70	3.62	0.794	52.0 × 10^3
1000	0.0264	0.2689	2.355	0.891	0.0330	0.69	4.65	0.685	27.7
1500	0.0197	0.2835	2.800	1.420	0.0423	0.676	7.58	0.510	8.12

Carbon Monoxide

0	0.0835	0.2482	1.065	0.128	0.0129	0.75	0.621	2.18	4.32 × 10^6
200	0.0582	0.2496	1.390	0.239	0.0169	0.74	1.16	1.52	0.860
400	0.0446	0.2532	1.670	0.374	0.0208	0.73	1.84	1.16	0.268
600	0.0362	0.2592	1.910	0.527	0.0246	0.725	2.62	0.943	0.109
800	0.0305	0.2662	2.134	0.700	0.0285	0.72	3.50	0.794	52.1 × 10^3
1000	0.0263	0.2730	2.336	0.887	0.0322	0.71	4.50	0.685	28.0
1500	0.0196	0.2878	2.783	1.420	0.0414	0.70	7.33	0.510	8.13

Helium*

0	0.012	1.24	1.140	0.950	0.078	0.67	5.25	2.18	77,800
200	0.00835	1.24	1.480	1.77	0.097	0.686	9.36	1.52	15,600
400	0.0064	1.24	1.780	2.78	0.118	0.707	14.2	1.16	4,840
600	0.0052	1.24	2.02	3.89	0.137	0.728	19.2	0.943	2,010
800	0.00436	1.24	2.285	5.24	0.155	0.745	25.3	0.794	932
1000	0.00377	1.24	2.520	6.69	0.173	0.760	31.7	0.685	494
1500	0.0028	1.24	3.160	11.30	0.221	0.791	51.4	0.510	129

continued

Table C.9 continued

T (F)	ρ (lb$_m$/ft^3)	c_p (Btu/lb$_m$F)	$\mu \times 10^5$ (lb$_m$/fts)	$\nu \times 10^3$ (ft^2/s)	k (Btu/hrftF)	Pr	α (ft^2/hr)	$\beta \times 10^3$ (1/R)	$\dfrac{g\beta\rho^2}{\mu^2}$ (1/Fft^3)
Hydrogen									
0	0.0060	3.39	0.540	0.89	0.094	0.70	4.62	2.18	86,600
100	0.0049	3.42	0.620	1.26	0.110	0.695	6.56	1.79	36,600
200	0.0042	3.44	0.692	1.65	0.122	0.69	8.45	1.52	18,000
500	0.0028	3.47	0.884	3.12	0.160	0.69	16.5	1.04	3,360
1000	0.0019	3.51	1.160	6.2	0.208	0.705	31.2	0.685	591
1500	0.0014	3.62	1.415	10.2	0.260	0.71	51.4	0.510	161
2000	0.0011	3.76	1.64	14.4	0.307	0.72	74.2	0.406	59
3000	0.0008	4.02	1.72	24.2	0.380	0.66	118.0	0.289	20
Carbon Dioxide									
0	0.132	0.184	0.88	0.067	0.0076	0.77	0.313	2.18	15.8×10^6
100	0.108	0.203	1.05	0.098	0.0100	0.77	0.455	1.79	6.10
200	0.092	0.216	1.22	0.133	0.0125	0.76	0.63	1.52	2.78
500	0.063	0.247	1.67	0.266	0.0198	0.75	1.27	1.04	0.476
1000	0.0414	0.280	2.30	0.558	0.0318	0.73	2.75	0.685	71.4×10^3
1500	0.0308	0.298	2.86	0.925	0.0420	0.73	4.58	0.510	19.0
2000	0.0247	0.309	3.30	1.34	0.050	0.735	6.55	0.406	7.34
3000	0.0175	0.322	3.92	2.25	0.061	0.745	10.8	0.289	1.85

SOURCE: The above table is reproduced from Kreith, F., *Principles of Heat Transfer*, 3d ed., Harper & Row, New York, 1973 with the permission of the publisher.
*Values for k, α, N_{Pr} have been reevaluated according to the recommendations given in RR-60-12, Reference 17 in Chapter 6.

Table C.10 THERMAL PROPERTIES OF LIQUIDS AT SATURATION PRESSURE

T (F)	ρ (lb$_m$/ft^3)	c_p (Btu/lb$_m$F)	$\mu \times 10^5$ (lb$_m$/fts)	$\nu \times 10^5$ (ft^2/s)	k (Btu/hrftF)	Pr	$\alpha \times 10^3$ (ft^2/hr)	$\beta \times 10^3$ (1/R)	$\dfrac{g\beta\rho^2}{\mu^2}$ (1/Fft^3)
					n-Butyl Alcohol				
60	50.5	0.55	226	4.48	0.097	46.6	3.49		
100	49.7	0.61	129	2.60	0.096	29.5	3.16		21.5×10^6
150	48.5	0.68	67.5	1.39	0.095	17.4	2.88		80
200	47.2	0.77	38.6	0.815	0.094	11.3	2.58	0.45	
300	—	—	19.0	—	—	—	—	0.48	
					Benzene				
60	55.1	0.40	46.0	0.835	0.093	7.2	4.22	0.60	0.3×10^9
80	54.6	0.42	39.6	0.725	0.092	6.5	4.01		
100	54.0	0.44	35.1	0.650	0.087	5.1	3.53		
150	53.5	0.46	26.0	0.480	—	4.5			
200	—	—	20.3	—	—	4.0			
					Light Oil				
60	57.0	0.43	5820	102	0.077	1170	3.14	0.38	1.17×10^4
80	56.8	0.44	2780	49	0.077	570	3.09	0.38	5.1
100	56.0	0.46	1530	27.4	0.076	340	2.95	0.39	16.7
150	54.3	0.48	530	9.8	0.075	122	2.88	0.40	1.34×10^6
200	54.0	0.51	250	4.6	0.074	62	2.69	0.42	6.4
250	53.0	0.52	139	2.6	0.074	35	2.67	0.44	21.0
300	51.8	0.54	83	1.6	0.073	22	2.62	0.45	56.5

continued

Table C.10 continued

T (F)	ρ (lb$_m$/ft³)	c_p (Btu/lb$_m$F)	μ×10⁵ (lb$_m$/fts)	ν×10⁵ (ft²/s)	k (Btu/hrftF)	Pr	α×10³ (ft²/hr)	β×10³ (1/R)	$\dfrac{g\beta\rho^2}{\mu^2}$ (1/Fft³)
					Commercial Aniline				
60	64.0	0.48	325.0	5.08	0.10	56.0	3.25		
100	63.0	0.49	170.0	2.70	0.10	30.0	3.24		
150	61.5	0.505	96.5	1.57	0.098	18.0	3.16		
200	60.0	0.515	61.1	1.02	0.096	11.8	3.11	0.49	21.6 × 10⁶
300	57.5	0.54	32.5	0.565	0.093	6.8	3.00	0.492	64.5
					Ammonia (Saturated Liquid)				
−20	42.4	1.07	17.6	0.417	0.317	2.15	6.94		
0	41.6	1.08	17.1	0.410	0.316	2.09	7.04		
10	40.8	1.09	16.6	0.407	0.314	2.07	7.08		
32	40.0	1.11	16.1	0.402	0.312	2.05	7.03		
50	39.1	1.13	15.5	0.396	0.307	2.04	6.95		
80	37.2	1.17	14.5	0.386	0.293	2.01	6.73	1.2	238 × 10⁶
120	35.2	1.22	13.0	0.355	0.275	1.99	6.40	1.3	266
					Freon-12, CCl₂F₂ (Saturated Liquid)				
−40	94.8	0.211	28.4	0.300	0.040	5.4	2.00	1.03	4.6 × 10⁹
−20	93.0	0.214	25.0	0.272	0.040	4.8	2.01	1.05	5.27
0	91.2	0.217	23.1	0.253	0.041	4.4	2.07	1.34	7.80
20	89.2	0.220	21.0	0.238	0.042	4.0	2.14	1.72	10.5
32	87.2	0.223	20.0	0.230	0.042	3.8	2.16		
60	83.0	0.231	18.0	0.213	0.042	3.5	2.19		
100	78.5	0.240	16.0	0.206	0.040	3.5	2.12	2.1	14.4
120	75.9	0.244	15.5	0.204	0.039	3.5	2.12	2.5	19.4

Glycerin

T (F)	ρ (lb$_m$/ft^3)	c_p (Btu/lb$_m$F)	$\mu \times 10^2$ (lb$_m$/fts)	$\nu \times 10^2$ (ft^2/s)	k (Btu/hrftF)	Pr	$\alpha \times 10^3$ (ft^2/hr)	$\beta \times 10^3$ (1/F)	$\dfrac{g\beta\rho^2}{\mu^2}$ (1/Fft^3)
50	79.3	0.554	256	3.23	0.165	31,000	3.76		
70	78.9	0.570	100	1.27	0.165	12,500	3.67	0.28	56
85	78.5	0.584	42.4	0.54	0.164	5,400	3.58	0.30	332
100	78.2	0.600	18.8	0.24	0.163	2,500	3.45		
120	77.7	0.617	12.4	0.16	—	$\approx$ 1,600			

Liquid Metals

Bismuth

T (F)	ρ (lb$_m$/ft^3)	c_p (Btu/lb$_m$F)	$\mu \times 10^3$ (lb$_m$/fts)	$\nu \times 10^6$ (ft^2/s)	k (Btu/hrftF)	Pr	α (ft^2/hr)	$\beta \times 10^3$ (1/R)	$\dfrac{g\beta\rho^2}{\mu^2}$ (1/Fft^3)
600	625	0.0345	1.09	1.74	9.5	0.014	0.44	0.065	0.687×10^9
800	616	0.0357	0.90	1.5	9.0	0.013	0.41	0.068	
1000	608	0.0369	0.74	1.2	9.0	0.011	0.40	0.070	
1200	600	0.0381	0.62	1.0	9.0	0.009	0.39		
1400	591	0.0393	0.53	0.9	9.0	0.008	0.39		

Mercury

T (F)	ρ (lb$_m$/ft^3)	c_p (Btu/lb$_m$F)	$\mu \times 10^3$ (lb$_m$/fts)	$\nu \times 10^6$ (ft^2/s)	k (Btu/hrftF)	Pr	α (ft^2/hr)	$\beta \times 10^3$ (1/R)	$\dfrac{g\beta\rho^2}{\mu^2}$ (1/Fft^3)
50	847	0.033	1.07	1.2	4.7	0.027	0.17	0.1	2.02×10^9
200	834	0.033	0.84	1.0	6.0	0.016	0.22	0.1	2.02
300	826	0.033	0.74	0.9	6.7	0.012	0.25		
400	817	0.032	0.67	0.8	7.2	0.011	0.27		
600	802	0.032	0.58	0.7	8.1	0.008	0.31		

continued

Table C.10

					Liquid Metals				
T	ρ	c_p	$\mu \times 10^3$	$\nu \times 10^6$	k	Pr	α	$\beta \times 10^3$	$\dfrac{g\beta\rho^2}{\mu^2}$
(F)	(lb$_m$/ft^3)	(Btu/lb$_m$F)	(lb$_m$/fts)	(ft^2/s)	(Btu/hrftF)		(ft^2/hr)	(1/R)	(1/Fft^3)
					Sodium				
200	58.0	0.33	0.47	8.1	49.8	0.011	2.6	0.150	73.5×10^6
400	56.3	0.32	0.29	5.1	46.4	0.007	2.6	0.20	243
700	53.7	0.31	0.19	3.5	41.8	0.005	2.5		
1000	51.2	0.30	0.14	2.7	37.8	0.004	2.4		
1300	48.6	0.30	0.12	2.5	34.5	0.004	2.4		

SOURCE: The above table is reproduced from Kreith, F., *Principles of Heat Transfer*, 3d ed., Harper & Row, New York, 1973 with the permission of the publisher.

Table C.11 PROPERTIES OF WEMCO* "C" TRANSFORMER OIL FOR THE TEMPERATURE RANGE 70–130F

PROPERTY	UNITS	EQUATIONS
Density	lb$_m$/ft^3	$\rho = 56.57 - 0.02186\,(T)$
Constant-pressure specific heat	Btu/lb$_m$F	$c_p = 0.439$
Thermal conductivity	Btu/hrftF	$k = 7.915(10^{-2}) - 2.644(10^{-5})T$
Coefficient of volume expansivity	1/R	$\beta = 3.856(10^{-4}) + 1.643(10^{-7})T$
Viscosity	lb$_m$/hrft	$\mu = 1650(T/10)^{-1.935}$
		T in degrees Fahrenheit

*WEMCO is a Westinghouse Corp. product trade name.

Table C.12   TOTAL NORMAL EMISSIVITIES FOR VARIOUS SURFACES

SURFACE[**]	t (F)[*]	EMISSIVITY	REFERENCE NUMBER
	A. Metals and Their Oxides		
Aluminum:			
Highly polished plate, 98.3 percent, pure	440–1070	0.039–0.057	26
Polished	212	0.095	1
Rough polish	212	0.18	1
Rough plate	100	0.055–0.07	25
Commercial sheet	212	0.09	1
Oxidized at 1110F	390–1110	0.11–0.19	23
Heavily oxidized	200–940	0.20–0.31	2
Aluminum oxide	530–930	0.63–0.42	21
Aluminum oxide	930–1520	0.42–0.26	21
Al-surfaced roofing	100	0.216	15
Aluminum alloys[†]			
Alloy 75 ST; A, B_1, C	75	0.11, 0.10, 0.08	36
Alloy 75 ST; A[‡]	450–900	0.22–0.16	36
Alloy 75 ST; B_1[‡]	450–800	0.20–0.18	36
Alloy 75 ST; C[‡]	450–930	0.22–0.15	36
Alloy 24 ST; A, B_1, C	75	0.09	36
Alloy 24 ST; A[‡]	450–910	0.17–0.15	36
Alloy 24 ST; B_1[‡]	450–940	0.20–0.16	36
Alloy 24 ST; C[‡]	450–860	0.16–0.13	36
Calorized surfaces, heated at 1110F			
Copper	390–1110	0.18–0.19	23
Steel	390–1110	0.52–0.57	23
Brass:			
Highly polished			
73.2 Cu, 26.7 Zn	476–674	0.028–0.031	26
62.4 Cu, 36.8 Zn, 0.4 Pb, 0.3 Al	494–710	0.033–0.037	26
82.9 Cu, 17.0 Zn	530	0.030	26
Hard-rolled, polished, but direction of polishing visible	70	0.038	25
Hard-rolled, polished, but somewhat attacked	73	0.043	25
Hard-rolled, polished, but traces of stearin from polish left on	75	0.053	25
Polished	212	0.06	1
Polished	100–600	0.10	15
Rolled plate, natural surface	72	0.06	25
Rolled plate, rubbed with coarse emery	72	0.20	25
Dull plate	120–660	0.22	32
Oxidized by heating at 1110F	390–1110	0.61–0.59	23

continued

Table C.12 continued

SURFACE**	t (F)*	EMISSIVITY	REFERENCE NUMBER
A. Metals and Their Oxides			
Chromium (see nickel alloys for Ni-Cr steels):			
Polished	100–2000	0.08–0.36	7–17
Polished	212	0.075	1
Copper:			
Carefully polished electrolytic copper	176	0.018	16
Polished	242	0.023	34
Polished	212	0.052	1
Commercial emeried, polished, but pits remaining	66	0.030	25
Commercial, scraped shiny, but not mirrorlike	72	0.072	25
Plate, heated long time, covered with thick oxide layer	77	0.78	25
Plate heated at 1110F	390–1110	0.57	23
Cuprous oxide	1470–2010	0.66–0.54	4
Molten copper	1970–2330	0.16–0.13	4
Dow metal:[†]			
A; B_1; C	75	0.15, 0.15, 0.12	36
A[‡]	450–750	0.24–0.20	36
B_1[‡]	450–800	0.16	36
C[‡]	450–760	0.21–0.18	36
Gold:			
Pure, highly polished	440–1160	0.018–0.035	26
Inconel:[†]			
Types X and B; surface A, B_2, C	75	0.19–0.21	36
Type X; surface A[‡]	450–1620	0.55–0.78	36
Type X; surface B_2[‡]	450–1575	0.60–0.75	36
Type X; surface C[‡]	450–1650	0.62–0.73	36
Type B; surface A[‡]	450–1620	0.35–0.55	36
Type B; surface B_2[‡]	450–1740	0.32–0.51	36
Type B; surface C[‡]	450–1830	0.35–0.40	36
Iron and steel (not including stainless):			
Metallic surfaces (or very thin oxide layer)			
Electrolytic iron, highly polished	350–440	0.052–0.064	26
Steel, polished	212	0.066	1
Iron, polished	800–1880	0.14–0.38	27
Iron, roughly polished	212	0.17	1
Iron, freshly emeried	68	0.24	25

continued

Table C.12   continued

SURFACE**	t (F)*	EMISSIVITY	REFERENCE NUMBER
A. Metals and Their Oxides			
Cast iron, polished	392	0.21	23
Cast iron, newly turned	72	0.44	25
Cast iron, turned and heated	1620–1810	0.60–0.70	22
Wrought iron, highly polished	100–480	0.28	32
Polished steel casting	1420–1900	0.52–0.56	22
Ground sheet steel	1720–2010	0.55–0.61	22
Smooth sheet iron	1650–1900	0.55–0.60	22
Mild steel;[†] A, B_2, C	75	0.12, 0.15, 0.10	36
Mild steel;[†] A[‡]	450–1950	0.20–0.32	36
Mild steel;[†] B_2[‡]	450–1920	0.34–0.35	36
Mild steel;[†] C[‡]	450–1950	0.27–0.31	36
Oxidized surfaces			
Iron plate, pickled, then rusted red	68	0.61	25
Iron plate, completely rusted	67	0.69	25
Iron, dark gray surface	212	0.31	1
Rolled sheet steel	70	0.66	25
Oxidized iron	212	0.74	28
Cast iron, oxidized at 1100F	390–1110	0.64–0.78	23
Steel, oxidized at 1100F	390–1110	0.79	23
Smooth, oxidized electrolytic iron	260–980	0.78–0.82	26
Iron oxide	930–2190	0.85–0.89	6
Rough ingot iron	1700–2040	0.87–0.95	22
Sheet steel			
Strong, rough oxide layer	75	0.80	25
Dense, shiny oxide layer	75	0.82	25
Cast plate, smooth	73	0.80	25
Cast plate, rough	73	0.82	25
Cast iron, rough, strongly oxidized	100–480	0.95	32
Wrought iron, dull oxidized	70–680	0.94	32
Steel plate, rough	100–700	0.94–0.97	15
Molten surfaces			
Cast iron	2370–2550	0.29	31
Mild steel	2910–3270	0.28	31
Steel, several different kinds with 0.25–1.2			

continued

Table C.12 continued

SURFACE**	t (F)*	EMISSIVITY	REFERENCE NUMBER
A. Metals and Their Oxides			
percent C (slightly oxidized surface)	2840–3110	0.27–0.39	3
Steel	2730–3000	0.42–0.53	14
Steel	2770–3000	0.43–0.40	18
Pure iron	2760–3220	0.42–0.45	8
Armco iron	2770–3070	0.40–0.41	18
Lead:			
Pure (99.96 percent), unoxidized	260–440	0.057–0.075	26
Gray oxidized	75	0.28	25
Oxidized at 300F	390	0.63	23
Magnesium:			
Magnesium oxide	530–1520	0.55–0.20	21
Magnesium oxide	1650–3100	0.20	10
Mercury	32–212	0.09–0.12	11
Molybdenum:			
Filament	1340–4700	0.096–0.202	37
Massive, polished	212	0.071	1
Monel metal:[†]			
Oxidized at 1110F	390–1110	0.41–0.46	23
K Monel 5700; A, B_2, C	75	0.23, 0.17, 0.14	36
K Monel 5700; A[‡]	450–1610	0.46–0.65	36
K Monel 5700; B_2[‡]	450–1750	0.54–0.77	36
K Monel 5700; C[‡]	450–1785	0.35–0.53	36
Nickel:			
Electroplated, polished	74	0.045	25
Technically pure (98.9% Ni, +Mn), polished	440–710	0.07–0.087	26
Polished	212	0.072	1
Electroplated, not polished	68	0.11	25
Wire	368–1844	0.096–0.186	29
Plate, oxidized by heating at 1110F	390–1110	0.37–0.48	23
Nickel oxide	1200–2290	0.59–0.86	5
Nickel alloys:			
Chromnickel	125–1894	0.64–0.76	29
Copper-nickel, polished	212	0.059	1
Nichrome wire, bright	120–1830	0.65–0.79	30
Nichrome wire, oxidized	120–930	0.95–0.98	30
Nickel-silver, polished	212	0.135	1
Nickelin (18–32 Ni; 55–68 Cu; 20 Zn), gray oxidized	70	0.262	25
Type ACI-HW (60 Ni; 12 Cr)			

continued

Table C.12 continued

SURFACE**	t (F)*	EMISSIVITY	REFERENCE NUMBER
A. Metals and Their Oxides			
Smooth, black, firm adhesive oxide coat from service	520–1045	0.89–0.82	24
Platinum:			
Pure, polished plate	440–1160	0.054–0.104	26
Strip	1700–2960	0.12–0.17	11
Filament	80–2240	0.036–0.192	9
Wire	440–2510	0.073–0.182	13
Silver:			
Polished, pure	440–1160	0.020–0.032	26
Polished	100–700	0.022–0.031	15
Polished	212	0.052	1
Stainless steels:[†]			
Polished	212	0.074	1
Type 301; A, B_2, C	75	0.21, 0.27, 0.16	36
Type 301; A[‡]	450–1740	0.57–0.55	36
Type 301; B_2[‡]	450–1725	0.54–0.63	36
Type 301; C[‡]	450–1650	0.51–0.70	36
Type 316; A, B_2, C	75	0.28, 0.28, 0.17	36
Type 316; A[‡]	450–1600	0.57–0.66	36
Type 316; B_2[‡]	450–1920	0.52–0.50	36
Type 316; C[‡]	450–1920	0.26–0.31	36
Type 347; A, B_2, C	75	0.39, 0.35, 0.17	36
Type 347; A[‡]	450–1650	0.52–0.65	36
Type 347; B_2[‡]	450–1610	0.51–0.65	36
Type 347; C[‡]	450–1650	0.49–0.64	36
Type 304 (8 Cr; 18 Ni)			
Light silvery, rough, brown, after heating	420–914	0.44–0.36	24
After 42 hr heating at 980F	420–980	0.62–0.73	24
Type 310 (25 Cr; 20 Ni)			
Brown, splotched, oxidized from furnace service	420–980	0.90–0.97	24
Allegheny metal No. 4, polished	212	0.13	1
Allegheny alloy No. 66, polished	212	0.11	1
Tantalum filament	2420–5430	0.19–0.31	37
Thorium oxide	530–930	0.58–0.36	21
Thorium oxide	930–1520	0.36–0.21	21
Tin:			
Bright tinned iron	76	0.043 and 0.064	25
Bright	122	0.06	30
Commercial tin-plated sheet iron	212	0.07, 0.08	1

continued

Table C.12 continued

SURFACE**	t (F)*	EMISSIVITY	REFERENCE NUMBER
A. Metals and Their Oxides			
Tungsten:			
Filament, aged	80–6000	0.032–0.35	12
Filament	6000	0.39	38
Polished coat	212	0.066	1
Zinc:			
Commercial 99.1%			
pure, polished	440–620	0.045–0.053	26
Oxidized by heating at			
750F	750	0.11	23
Galvanized sheet iron,			
fairly bright	82	0.23	25
Galvanized sheet iron,			
gray oxidized	75	0.28	25
Zinc, galvanized sheet	212	0.21	1
B. Refractories, Building Materials, Paints, and Miscellaneous			
Alumina (99.5–85 Al_2O_3;			
0–12 SiO_2; 0–1 Fe_2O_3).			
Effect of mean grain			
size, microns (μ)	1850–2850	—	20
10 μ	—	0.30–0.18	
50 μ	—	0.39–0.28	
100 μ	—	0.50–0.40	
Alumina-silica (showing			
effect of Fe)	1850–2850	—	20
80–58 Al_2O_3; 16–38 SiO_2;			
0.4 Fe_2O_3	—	0.61–0.43	
36–26 Al_2O_3; 50–60 SiO_2;			
1.7 Fe_2O_3	—	0.73–0.62	
61 Al_2O_3; 35 SiO_2; 2.9			
Fe_2O_3	—	0.78–0.68	
Asbestos:			
Board	74	0.96	25
Paper	100–700	0.93–0.94	15
Brick: [§]			
Red, rough, but no			
gross irregularities	70	0.93	25
Grog brick, glazed	2012	0.75	22
Building	1832	0.45	30
Fireclay	1832	0.75	30
Carbon:			
T-carbon (Gebrüder			
Siemens) 0.9% ash.			
This started with emis-			
sivity at 260F of 0.72,			
but no heating changed			
to values given	260–1160	0.81–0.79	26

continued

Table C.12 continued

SURFACE**	t (F)*	EMISSIVITY	REFERENCE NUMBER
B. Refractories, Building Materials, Paints, and Miscellaneous			
Filament	1900–2560	0.526	19
Rough plate	212–608	0.77	1
Rough plate	608–932	0.77–0.72	1
Graphitized	212–608	0.76–0.75	1
Graphitized	608–932	0.75–0.71	1
Candle soot	206–520	0.952	33
Lampblack-waterglass coating	209–440	0.96–0.95	16, 26
Thin layer of same on iron plate	69	0.927	25
Thick coat of same	68	0.967	25
Lampblack, 0.003 in. or thicker	100–700	0.945	15
Lampblack, rough deposit	212–932	0.84–0.78	1
Lampblack, other blacks.	122–1832	0.96	30
Graphite, pressed, filed surface	480–950	0.98	21
Carborundum (87 SiC; density 2.3)	1850–2550	0.92–0.82	20
Concrete tiles	1832	0.63	30
Enamel, white, fused on iron	66	0.90	25
Glass:			
Smooth	72	0.94	25
Pyrex, lead, and soda	500–1000	ca 0.95–0.85	21
Gypsum, 0.02 in. thick on smooth or blackened plate	70	0.903	25
Magnesite refractory brick	1832	0.38	30
Marble, light gray, polished	72	0.93	25
Oak, planed	70	0.90	25
Oil layers on polished nickel (lubricating oil)	68	—	25
Polished surface alone	—	0.045	
+0.001, 0.002, 0.005 in oil	—	0.27, 0.46, 0.72	
Thick oil layer	—	0.82	
Oil layers on aluminum foil (linseed oil)	—	—	28
Aluminum foil	212	0.087	
+1,2 coats oil	212	0.561, 0.574	
Paints, lacquers, varnishes:			
Snow-white enamel varnish on rough iron plate	73	0.906	25

continued

Table C.12 continued

SURFACE**	t (F)*	EMISSIVITY	REFERENCE NUMBER
B. Refractories, Building Materials, Paints, and Miscellaneous			
Black shiny lacquer, sprayed on iron	76	0.875	25
Black shiny shellac on tinned iron sheet	70	0.821	25
Black matte shellac	170–295	0.91	35
Black or white lacquer	100–200	0.80–0.95	15
Flat black lacquer	100–200	0.96–0.98	15
Oil paints, 16 different, all colors	212	0.92–0.96	28
Aluminum paints and lacquers:			
10 percent Al, 22 percent lacquer body, on rough or smooth surface	212	0.52	28
Other Al paints, varying age and Al content	212	0.27–0.67	28
Al lacquer, varnish binder, on rough plate	70	0.39	25
Al paint, after heating to 620F	300–600	0.35	26
Radiator paint; white, cream, bleach	212	0.79, 0.77, 0.84	1
Radiator paint, bronze	212	0.51	1
Lacquer coatings, 0.001–0.015 in. thick on aluminum alloys ‡	100–300	0.87 to 0.97	36
Clear silicon vehicle coatings, 0.001–0.015 in. thick: ‡			
On mild steel	500	0.66	36
On stainless steels, 316, 301, 347	500	0.68, 0.75, 0.75	36
On Dow metal	500	0.74	36
On Al alloys 24 ST, 75 ST	500	0.77, 0.82	36
Aluminum paint with silicone vehicle, two coats on Inconel ‡	500	0.29	36
Paper, thin, pasted on tinned or blackened plate	66	0.92, 0.94	25
Plaster, rough lime	50–190	0.91	32
Porcelain, glazed	72	0.92	25
Quartz:			
Rough, fused	70	0.93	25
Glass, 1.98 mm thick	540–1540	0.90–0.41	21
Glass, 6.88 mm thick	540–1540	0.93–0.47	21
Opaque	570–1540	0.92–0.68	21

continued

Table C.12   continued

SURFACE**	t (F)*	EMISSIVITY	REFERENCE NUMBER
B. Refractories, Building Materials, Paints, and Miscellaneous			
Roofing paper	69	0.91	25
Rubber:			
Hard, glossy plate	74	0.94	25
Soft, gray, rough			
(reclaimed)	76	0.86	25
Serpentine, polished	74	0.90	25
Silica (98 SiO_2; Fe-free),			
effect of grain size,			
microns (μ)	1850–2850	—	20
10 μ	—	0.42–0.33	
70–600 μ	—	0.62–0.46	
(See also Alumina-			
silica and quartz)			
Water	32–212	0.95–0.963	¶
Zirconium silicate	460–930	0.92–0.80	21
Zirconium silicate	930–1530	0.80–0.52	21

SOURCE: The above table is reproduced from Hottel, H. C. and Sarofim, A. F., *Radiative Transfer*, McGraw-Hill, New York, 1967 with the permission of the publisher.
*When temperatures and emissivities appear in pairs separated by dashes, they correspond; and linear interpolation is permissible.
[†] Identification of surface treatment: surface A, cleaned with toluene, then methanol; B_1, cleaned with soap and water, toluene, and methanol in succession; B_2 cleaned with abrasive soap and water, toluene, and methanol; C, polished on buffing wheel to mirror surface, cleaned with soap and water.
[‡] Results after repeated heating and cooling.
[§] See also under material type.
[¶] Calculated from spectral data.
**Table A-23, compiled by Hottel, from W. H. McAdams' *Heat Transmission*, 3d ed., McGraw-Hill, New York, 1954. For more extensive compilations refer to Gubareff, G. G., J. E. Janssen, and R. H. Torborg, *Thermal Radiation Properties Survey*, Honeywell Research Center, Minneapolis, Minn. (1960); and Goldsmith, A., T. E. Waterman, and J. H. Hirschhorn, *Thermophysical Properties of Solid Materials*, WADC, TR58–476, Vols. I-V, Wright-Patterson Air Force Base, Ohio (1960).

References
1. Barnes, B. T., W. E. Forsythe, and E. Q. Adams, *J. Opt. Soc. Am.*, 37(10), pp. 804–807 (1947).
2. Binkley, E. R., private communication, 1933.
3. Bacon, J. E., and J. W. James, "Proceedings of the General Discussion on Heat Transfer," pp. 117–121, Institution of Mechanical Engineers, London, and American Society of Mechanical Engineers, New York, 1952.
4. Burgess, G. K., *Natl. Bur. Standards, Bull. 6, Sci. Paper* 121, 111 (1909).
5. Burgess, G. K., and P. D. Foote, *Natl. Bur. Standards, Bull. 11 Sci. Paper* 224, 41–64 (1914).
6. Burgess, G. K., and P. D. Foote, *Natl. Bur. Standards, Bull. 12 Sci. Paper* 249, 83–89 (1915).
7. Coblentz, W. H., *Natl. Bur. Standards, Bull. 7*, 197 (1911).
8. Dastar, M. N., and N. A. Gokcen, *J. Metals*, 1(10), trans. 665–667 (1949).
9. Davisson, C., and J. R. Weeks, Jr., *J. Opt. Soc. Am.*, 8, 581–606 (1924).
10. Féry, C., *Ann. phys. chim.*, 27, 433 (1902).
11. Foote, P. D., *Natl. Bur. Standards, Bull. 11, Sci. Paper* 243, 607 (1914); *J. Wash. Acad. Sci.*, 5, 1 (1914).
12. Forsythe, W. E., and A. G. Worthing, *Astrophys. J.*, 61, 146–185 (1925).
13. Geiss, W., *Physica*, 5, 203 (1925).

14. Goller, G. N., *Trans. Am. Soc. Metals*, 32, 239 (1944).
15. Heilman, R. H., *Trans. ASME*, FSP51, 287–304 (1929).
16. Hoffmann, K., *Z. Physik*, 14, 310 (1923).
17. Hulbert, E. O., *Astrophys. J.*, 42, 205 (1915).
18. Knowles, D., and R. J. Sarjant, *J. Iron Steel Inst.* (*London*), 155, 577 (1947).
19. Lummer, O., *Elektrotech. Z.*, 34, 1428 (1913).
20. Michaud, M., Sc.D. Thesis, University of Paris, 1951.
21. Pirani, M., *J. Sci. Instr.*, 16, (12) (1939).
22. Polak, V., *Z. tech. Physik*, 8, 307 (1927).
23. Randolph, C. F., and J. J. Overholtzer, *Phys. Rev.*, 2, 144 (1913).
24. Rice, H. S., Chemical Engineering Thesis, Massachusetts Institute of Technology, 1931.
25. Schmidt, E., *Gesundh.-Ing.*, Beiheft 20, Reihe 1, 1–23 (1927).
26. Schmidt, E., and E. Furthmann, *Mitt. Kaiser-Wilhelm-Inst., Eisenforsch. Dusseldorf*, Abhandl., 109, 225 (1928).
27. Snell, F. D., *Ind. Eng. Chem.*, 29, 89–91 (1937).
28. Standard Oil Development Company, personal communication, 1928.
29. Suydam, V. A., *Phys. Rev.*, (2)5, 497–509 (1915).
30. Thring, M. W., *The Science of Flames & Furnaces*, Chapman & Hall, London, 1952.
31. Thwing, C. B., *Phys. Rev.*, 26, 190 (1908).
32. Wamsler, F., *Z. Ver. deut. Ing.*, 55, 599–605 (1911); *Mitt. Forsch.*, 98, 1–45 (1911).
33. Wenzl, M., and F. Morawe, *Stahl u. Eisen*, 47, 867–871 (1927).
34. Westphal, W., *Verhandl. deut. physik. Ges.*, 10, 987–1012 (1912).
35. Westphal, W., *Verhandl. deut. physik. Ges.*, 11, 897–902 (1913).
36. Wilkes, G. B., Final Report on Contract No. W33-038-20486, Air Material Command; Wright Field, Dayton, Ohio, DIC Report, Massachusetts Institute of Technology (1950).
37. Worthing, A. G., *Phys. Rev.*, 28, 190 (1926).
38. Zwikker, C., *Arch. néerland. sci.*, 9, (Pt. IIIA), 207 (1925).

Appendix D
Program and Data for
Figures 3.2(a), (b), and (c)

The equation relating the temperature ratio to time for a given Biot number is given in Appendix F as Eq. F.4 and the program outlined in Table D.1 was used to calculate the response tabulated in Table D.2 which is the basis for Figures 3.2(a), 3.2(b), and 3.2(c). The geometry of the infinite slab is illustrated in Figure 3.2 together with definitions of the temperature excesses employed in the calculated temperature ratios. The dictionary of the program explains the major symbols that are easily recognizable cognizants except perhaps for the eigenvalues. The symbol could have been better written as $D(H, N)$ in the dictionary since the parameters are the Biot number H and the number N of the nth root of the indicial equation

$$\delta_n \tan(\delta_n) = H \tag{D.1}$$

where

$$\delta_n = D(H, N) \tag{D.2}$$

Only three values of X/L were used, as shown in Table D.1, to determine the general shape of the curves for Biot numbers of 0.1, 1.0, and 10. The solutions presented in Table D.2 are accurate only for Fourier numbers greater than 0.2.

Table D.1

```
C       PROGRAM    7801    TRSB    H. WOLF AND C. WOLF
C
C       DICTIONARY:
C
C           TXTI=TEMPERATURE RATIO
C             H=BIOT NUMBER
C            XL=POSITION RATIO
C           DHN=EIGENVALUES, D(1,1), D(1,2), D(1,3), ETC.
C            FO=FOURIER NUMBER
C
C       THE PROGRAM IS DESIGNED TO CALCULATE TXTI FOR H'S OF 0.1, 1, 10
C       AT XL LOCATIONS OF 0, 0.5, AND 1.0 WITH FO AS THE MAIN ARGUMENT.
C
C       DATA
        DIMENSION D(3,6), H(3), XL(3), XTXTI(10,6)
C (DI  IS FOR H=0.1)
        D(1,1)=0.3111
        D(1,2)=3.1731
        D(1,3)=6.2991
        D(1,4)=9.4354
        D(1,5)=12.5743
        D(1,6)=15.7143
C (D2  IS FOR H=1)
        D(2,1)=0.8603
        D(2,2)=3.4256
        D(2,3)=6.4373
        D(2,4)=9.5293
        D(2,5)=12.6453
        D(2,6)=15.7713
C (D3  IS FOR H=10)
        D(3,1)=1.4289
        D(3,2)=4.3058
        D(3,3)=7.2281
        D(3,4)=10.2003
        D(3,5)=13.2142
        D(3,6)=16.2594
C
C POSITION RATIOS XL(K)
C
        XL(1)=0.0
        XL(2)=0.5
        XL(3)=1.0
C
C BIOT NUMBERS H(I)
C
        H(1)=0.1
        H(2)=1.0
        H(3)=10.0
C
C START MAIN CALCULATIONS
C
   10 FO=0.00
   20 PRINT 500
   25 PRINT 505, FO
   30 DO 190 I=1, 3
   40    DO 190 K=1, 3
   50    XTXTI(I,K)=0.00
```

continued

Table D.1 continued

```
60          DO 150 J = 1, 6
70          A = SIN(D(I, J))
80          B = COS(D(I, J)*XL(K))
90          XN = 2.*H(I)*A*B
100         XD = D(I, J)*(H(I) + (A*A))
112         XEXP = (D(I, J)*D(I, J)*FO)
114         IF(XEXP − 162) 120, 120, 122
120         EXT = EXP( − (XEXP))
121         GO TO 130
122         EXT = 0.0
130         TXTI = XN*EXT/XD
140         XTXTI(I, K) = XTXTI(I, K) + TXTI
150         CONTINUE
190      CONTINUE
195 PRINT 515, ((XTXTI(I, K), K = 1, 3), I = 1, 3)
200      FO = FO + 0.05
210      IF(FO − 2.0) 220, 220, 230
220      GO TO 25
230      STOP
C
C FORMAT STATEMENTS
C
500 FORMAT (1H1,/// 28X, 'TEMPERATURE RESPONSE OF THE'/25X,
    A'INFINITE SLAB INSULATED AT THE FACE'/25X,
    B'AT X/L = 0. AS A FUNCTION OF THE BIOT,'/25X,
    C'FOURIER NUMBER, FO, AND POSITION'///17X,'BIOT NO. = 0.1'
    D, 7X, 'BIOT NO. = 1.0', 7X, 'BIOT NO. = 10.'//22X, 'X/L',
    E 17X, 'X/L', 18X, 'X/L', / 16X, '---------------', 4X,
    F'-----------------', 4X,'-----------------' / 10X, 'FO',
    G 2X, '0.0   0.5   1.0   0.0   0.5   1.0   0.0   0.5   1.0'
    H//)
505 FORMAT (9X, F5.2)
515 FORMAT (' + ', 13X, 3F6.3, 6F7.4)
    END
```

Table D.2

TEMPERATURE RESPONSE OF THE
INFINITE SLAB INSULATED AT THE FACE
AT X / L = 0. AS A FUNCTION OF THE BIOT,
FOURIER NUMBER, FO, AND POSITION

	BIOT NO. = 0.1			BIOT NO. = 1.0			BIOT NO. = 10.		
	X / L			X / L			X / L		
FO	0.0	0.5	1.0	0.0	0.5	1.0	0.0	0.5	1.0
0.0	1.000	1.000	0.996	0.9968	1.0037	0.9635	0.9735	1.0334	0.6745
0.05	1.000	0.998	0.975	0.9998	0.9863	0.7904	0.9985	0.9324	0.2323
0.10	0.999	0.994	0.965	0.9931	0.9506	0.7236	0.9684	0.8101	0.1705
0.15	0.997	0.989	0.958	0.9756	0.9139	0.6784	0.9047	0.7152	0.1411
0.20	0.994	0.984	0.951	0.9507	0.8793	0.6434	0.8292	0.6390	0.1225
0.25	0.990	0.980	0.946	0.9220	0.8465	0.6144	0.7536	0.5743	0.1086
0.30	0.986	0.975	0.940	0.8918	0.8153	0.5889	0.6824	0.5175	0.0973
0.35	0.982	0.970	0.936	0.8613	0.7854	0.5658	0.6170	0.4669	0.0876
0.40	0.977	0.966	0.931	0.8310	0.7568	0.5442	0.5574	0.4214	0.0790
0.45	0.972	0.961	0.926	0.8014	0.7292	0.5239	0.5034	0.3804	0.0712
0.50	0.968	0.956	0.922	0.7726	0.7026	0.5046	0.4546	0.3435	0.0643
0.55	0.963	0.952	0.917	0.7447	0.6771	0.4861	0.4105	0.3101	0.0581
0.60	0.959	0.947	0.913	0.7177	0.6525	0.4683	0.3707	0.2800	0.0524
0.65	0.954	0.942	0.908	0.6917	0.6288	0.4513	0.3347	0.2529	0.0473
0.70	0.949	0.938	0.904	0.6666	0.6059	0.4348	0.3022	0.2283	0.0427
0.75	0.945	0.933	0.899	0.6424	0.5839	0.4190	0.2729	0.2062	0.0386
0.80	0.940	0.929	0.895	0.6191	0.5627	0.4038	0.2464	0.1862	0.0348
0.85	0.936	0.924	0.891	0.5966	0.5423	0.3891	0.2225	0.1681	0.0315
0.90	0.931	0.920	0.886	0.5749	0.5226	0.3750	0.2009	0.1518	0.0284
0.95	0.927	0.916	0.882	0.5540	0.5036	0.3614	0.1814	0.1370	0.0257
1.00	0.922	0.911	0.878	0.5339	0.4853	0.3482	0.1638	0.1237	0.0232
1.05	0.918	0.907	0.874	0.5145	0.4676	0.3356	0.1479	0.1117	0.0209
1.10	0.913	0.902	0.870	0.4958	0.4507	0.3234	0.1336	0.1009	0.0189
1.15	0.909	0.898	0.865	0.4778	0.4343	0.3116	0.1206	0.0911	0.0171
1.20	0.905	0.894	0.861	0.4605	0.4185	0.3003	0.1089	0.0823	0.0154
1.25	0.900	0.889	0.857	0.4437	0.4033	0.2894	0.0983	0.0743	0.0139
1.30	0.896	0.885	0.853	0.4276	0.3886	0.2789	0.0888	0.0671	0.0126
1.35	0.892	0.881	0.849	0.4121	0.3745	0.2688	0.0802	0.0606	0.0113
1.40	0.887	0.876	0.845	0.3971	0.3609	0.2590	0.0724	0.0547	0.0102
1.45	0.883	0.872	0.841	0.3827	0.3478	0.2496	0.0654	0.0494	0.0092
1.50	0.879	0.868	0.836	0.3688	0.3352	0.2405	0.0590	0.0446	0.0083
1.55	0.874	0.864	0.832	0.3554	0.3230	0.2318	0.0533	0.0403	0.0075
1.60	0.870	0.860	0.828	0.3425	0.3113	0.2234	0.0481	0.0363	0.0068
1.65	0.866	0.856	0.824	0.3300	0.3000	0.2152	0.0434	0.0328	0.0061
1.70	0.862	0.851	0.820	0.3180	0.2891	0.2074	0.0392	0.0296	0.0055
1.75	0.858	0.847	0.816	0.3065	0.2786	0.1999	0.0354	0.0268	0.0050
1.80	0.854	0.843	0.813	0.2953	0.2684	0.1926	0.0320	0.0242	0.0045
1.85	0.849	0.839	0.809	0.2846	0.2587	0.1856	0.0289	0.0218	0.0041
1.90	0.845	0.835	0.805	0.2743	0.2493	0.1789	0.0261	0.0197	0.0037
1.95	0.841	0.831	0.801	0.2643	0.2402	0.1724	0.0235	0.0178	0.0033
2.00	0.837	0.827	0.797	0.2547	0.2315	0.1661	0.0213	0.0161	0.0030

Appendix E
The Conduction
Differential Equations

The topic of heat conduction is essentially applied mathematics and the working equations presented in Chapters 2 and 3 represent solutions to the ordinary or partial differential equations for the geometry under consideration together with appropriate boundary conditions. The mathematical procedures for the solution of partial differential equations are involved and complicated and are generally beyond the immediate capabilities of the person at the junior or senior university level. Consequently, it is the writer's opinion that such topics belong in an advanced course in conduction and not in an introductory level course.

It is of value in an introductory text, such as this book, to show in an appendix the forms of the general conduction equation which serve as the basis of the material presented. Readers who have the interest and curiosity can see where their working equations and curves come from. Such a presentation can also serve as a convenient reference for future time. An excellent and detailed treatment of the topic is given by G. E. Myers in his book *Analytical Methods in Conduction Heat Transfer*, McGraw-Hill, New York, 1971.

The basic equation in one-dimensional, steady-state conduction is called Fourier's law:

$$q = -kA(x)\frac{dT}{dx} \tag{E.1}$$

$$q = -kA(r)\frac{dT}{dr} \tag{E.2}$$

for the Cartesian, cylindrical, and spherical coordinate systems, respectively.

When energy is stored or depleted at locations in the solid, then the response is time dependent and requires a storage term in the differential equation that includes the rate of change of temperature with time, dT/dt. The form of the equation depends on whether there is any internal generation present and whether the solid is isotropic (properties the same in all directions) or anisotropic. The general heat conduction equation for an isotropic solid does not have a particular name. The equation is given below for the three basic coordinate systems.

Cartesian (x, y, z):

$$\frac{\partial^2 T}{\partial x^2} + \frac{\partial^2 T}{\partial y^2} + \frac{\partial^2 T}{\partial z^2} + \frac{q'''}{k} = \frac{\partial T/\partial t}{\alpha} \tag{E.3}$$

Cylindrical (r, θ, z):

$$\frac{\partial^2 T}{\partial r^2} + \frac{\partial T/\partial r}{r} + \frac{\partial^2 T/\partial \theta^2}{r^2} + \frac{\partial^2 T}{\partial z^2} + \frac{q'''}{k} = \frac{\partial T/\partial t}{\alpha} \tag{E.4}$$

Spherical (r, θ, ϕ):

$$\frac{\partial^2(rT)/\partial r^2}{r} + \frac{\partial[\sin\phi(\partial T/\partial\phi)/\partial\phi]}{r^2\sin\phi} + \frac{\partial^2 T/\partial\theta^2}{r^2\sin^2\phi} + \frac{q'''}{k} = \frac{\partial T/\partial t}{\alpha} \tag{E.5}$$

where

T = temperature

t = time

$\alpha = \dfrac{k}{\rho c_p}$ thermal diffusivity

q''' = generation term, can be a function of spacial coordinates and time in the most general sense

θ = angle between the x axis and the projection of r on the xy plane

ϕ = angle between r and the z axis

When the generation term is zero, the heat conduction equation is called the Fourier equation and in Cartesian coordinates becomes:

Fourier equation (x, y, z):

$$\frac{\partial^2 T}{\partial x^2} + \frac{\partial^2 T}{\partial y^2} + \frac{\partial^2 T}{\partial z^2} = \frac{\partial T/\partial t}{\alpha} \tag{E.6}$$

When there are heat sources and steady-state conditions prevail such that $\partial T/\partial t = 0$ then the general conduction equation is called the Poisson equation and in Cartesian coordinates is given by:

Poisson equation (x, y, z):

$$\frac{\partial^2 T}{\partial x^2} + \frac{\partial^2 T}{\partial y^2} + \frac{\partial^2 T}{\partial z^2} + \frac{q'''}{k} = 0 \tag{E.7}$$

When there are no heat sources or no generation and steady state prevails the equation becomes the Laplace equation and for Cartesian coordinates:

Laplace equation (x, y, z):

$$\frac{\partial^2 T}{\partial x^2} + \frac{\partial^2 T}{\partial y^2} + \frac{\partial^2 T}{\partial z^2} = 0 \tag{E.8}$$

Many times the reader may encounter a shorthand notation in the literature that utilizes the vector differential operator ∇ (called "del" or "nabla") which is defined as

$$\nabla(\ ) = \frac{\mathbf{i}\partial(\ )}{\partial x} + \frac{\mathbf{j}\partial(\ )}{\partial y} + \frac{\mathbf{k}\partial(\ )}{\partial z} \tag{E.9}$$

where $\mathbf{i}, \mathbf{j}, \mathbf{k}$ are the unit vectors in the x, y, z directions. According to the rules of vector multiplication, the dot product of del, that is $\nabla \cdot \nabla$, is a scalar quantity so that we may write $\nabla \cdot \nabla = \nabla^2$ and operating on the temperature T one obtains

$$\nabla^2(T) = \frac{\partial^2 T}{\partial x^2} + \frac{\partial^2 T}{\partial y^2} + \frac{\partial^2 T}{\partial z^2} \tag{E.10}$$

The general heat conduction equation is the often written as

$$\nabla^2 T + \frac{q'''}{k} = \frac{\partial T/\partial t}{\alpha} \tag{E.11}$$

and the task of supplying the proper form for $\nabla^2 T$ in the particular coordinate system of interest is left for the readers to figure out if they do not have Eqs. E.3, E.4, and E.5 at hand. Accordingly the Fourier, Poisson, and Laplace equations can also be written in the shorter form; the Laplace equation becomes simply $\nabla^2 T = 0$, for example.

Appendix F
Transient Solutions for the Plate, Cylinder, and Sphere

Most heat transfer texts utilize the Heissler charts to present the transient response of the simple plate, cylinder, and sphere geometries. Unfortunately, the Heissler charts are not easy to read at small and moderate time intervals. Accordingly, the known solutions for the simple geometries were programmed and calculations were made which permitted plotting the temperature excess ratios as a function of the Fourier number on semilog paper rather than the log-log plots due to Heissler. Such calculations have also been made by Schneider (Reference 6, Chapter 3) but that excellent source is no longer in print. This appendix lists the solutions and the computer programs employed to calculate the temperature excess ratios are plotted in Figures 3.7 through 3.12 and for the semi-infinite solid in Figures 3.16 and 3.17.

THE INFINITE PLANE PLATE

The sketch in Figure 3.7 shows the geometry and the symbols; the partial differential equation describing the phenomena is Eq. E.6 in one dimension

$$\alpha \frac{\partial^2 T}{\partial x^2} = \frac{\partial T}{\partial t} \tag{F.1}$$

together with the boundary conditions

1. The convection boundary condition:

$$x = \pm L; \qquad \frac{\partial T}{\partial x} = -\frac{h}{k}(T_L - T_\infty) \qquad\qquad \text{(F.2)}$$

2. The initial condition:

$$t \leqslant 0; \qquad T = T_i, \qquad -L \leqslant x \leqslant +L \qquad\qquad \text{(F.3)}$$

3. The ambient condition, $T = T_\infty$, a constant value outside the boundary layer on the surface.

The solution to Eq. F.1 can be obtained by the method of separation of variables to be

$$\frac{T_x - T_\infty}{T_i - T_\infty} = \sum_{n=1}^{\infty} \frac{2H\sin(\delta_n)\cos(\delta_n x/L)}{\delta_n(H + \sin^2 \delta n)} e^{-\delta_n^2 N_{Fo}} \qquad\qquad \text{(F.4)}$$

where

T_x = temperature at location x in the plate

T_i = uniform initial temperature

H = the Biot number, $N_{Bi} = \dfrac{hL}{k}$

N_{Fo} = the Fourier number, $\dfrac{\alpha t}{L^2}$

δ_n = roots of the transcendental equation, $\delta_n \tan(\delta_n) = H$

Table F.1 presents a FORTRAN listing of the program constructed to calculate the center temperature excess ratio $(T_o - T_\infty)/(T_i - T_\infty)$ and the position correction ratio $(T_x - T_\infty)/(T_o - T_\infty)$ which are tabulated in Table F.2 for the value of $N_{Bi} = 1$ as an example of the 16 values of N_{Bi} calculated. The reader will note that the center excess ratio is the column of numbers under the quantity $x/L = 0$ and the position correction ratios are the columns of numbers under the remaining values of x/L. The reader who desires to compare Table F.2 with Table D.2 should note that Table D.2 contains all temperature excess ratios.

The introductory comments in Table F.1 state that the results are accurate down to $N_{Fo} = 0.2$ since six terms of the series were evaluated in all cases. The eigenvalues employed in the calculations for the plate, cylinder, and the sphere are those given in Appendix IV of the 1959 edition of Carslaw and Jaeger's book *Conduction of Heat in Solids*, The Clarendon Press, Oxford.

Table F.1 THE PLATE

```
C       PROGRAM 7802 TRSB H WOLF
C
C       DICTIONARY:
C
C               TXT=TEMP RATIO (TX − TI)/(TI − TINF)
C               RTX=POSN RATIO (TX − TI)/(TO − TI)
C                 H=BIOT NUMBER
C               D(I, J)=EIGENVALUES (6 FOR EACH H)
C                FO=FOURIER NUMBER
C
C       THIS PROGRAM CALCULATES THE CENTER TEMPERATURE EXCESS
C       RATIOS AND POSITION CORRECTION RATIOS AS DEFINED
C       ABOVE FOR THE 2L PLATE, THE CYLINDER, AND THE SPHERE
C       WITH CONVECTION ON THE OUTER SURFACE. THE PROGRAM
C       EVALUATES THE SERIES (6 TERMS MAX) FOR EACH CONFIGURATION
C       AND IS ACCURATE FOR FOURIER NUMBERS DOWN TO 0.2.
C
C       DATA AND DIMENSION STATEMENTS
C
        DIMENSION D(16, 6), H(16), TXT(9), RTX(9)
        DATA H/0.1, 0.2, 0.3, 0.4, 0.5, 0.6, 0.8.1.0,
      A1.5, 2.0, 3.0, 5.0, 10., 20., 100., 10000./
        DATA D/0.3111, 0.4328, 0.5218, 0.5932, 0.6533, 0.7051.
      A 0.7910, 0.8603, 0.9882, 1.0769, 1.1925, 1.3138, 1.4289,
      B 1.4961, 1.5552, 1.5708,
      C        3.1731, 3.2039, 3.2341, 3.2636, 3.2923, 3.3204,
      D 3.3744, 3.4256, 3.5422, 3.6436, 3.8088, 4.0336, 4.3058,
      E 4.4915, 4.6658, 4.7124,
      F        6.2991, 6.3148, 6.3305, 6.3461, 6.3616, 6.3770,
      G 6.4074, 6.4373, 6.5097, 6.5783, 6.7040, 6.9096, 7.2281,
      H 7.4954, 7.7764, 7.8540,
      I        9.4354, 9.4459, 9.4565, 9.4670, 9.4775, 9.4879,
      J 9.5087, 9.5293, 9.5081, 9.6296, 9.7240, 9.8928, 10.2003,
      K10.5117, 10.8871, 10.9956,
      L         12.5743, 12.5823, 12.5902, 12.5981, 12.6060, 12.6139,
      M12.6296, 12.6453, 12.6841, 12.7223, 12.7966, 12.9352, 13.2142,
      N13.5420, 13.9981, 14.1372.
      O        15.7143, 15.7207, 15.7270, 15.7334, 15.7397, 15.7460,
      P15.7587, 15.7713, 15.8026, 15.8336, 15.8945, 16.0107, 16.2594,
      Q16.5864, 17.0193, 17.2788/
C
C
C                    START MAIN CALCULATIONS
C
C    10 DO  190 I=1, 16
        20 PRINT 500, H(I)
        25 FO = 0.050000
```

continued

Table F.1   continued

```
        26 N = 14
        30      DO 180 M = 1, 3
        31       A = 10.**M/10.
        32       IF(M.GE.2)   N = 13
        35      DO 180 L = 1,N
        37      IF(((M.EQ.1).AND.(L.EQ.10)).OR.((M.GE.2).AND.
         A          (L.EQ.9)))      A = 2.*A
        40      FO = FO + 0.0500*A
C
        50      DO 170 K = 1, 9
        52       XL = (K − 1)
        54       IF(K − 1) 70, 70, 56
        56       XL = K/10.
        58       IF(K − 2) 70, 70, 60
        60       XL = (K + 1)/10.
C
        70              TXT(K) = 0.000000
        72              DO 150 J = 1, 6
        74              AA = SIN(D(I, J))
        80              BB = COS(D(I, J)*XL)
        90              XN = 2.*H(I)*AA*BB
       100              XD = D(I, J)*(H(I) + (AA*AA))
       112              XEXP = (D(I, J)*D(I, J)*FO)
       114      IF(XEXP − 162.) 120, 120, 155
       120              EXT = EXP( − (XEXP))
       130              TXTI = XN*EXT/XD
       140              TXT(K) = TXT(K) + TXTI
       150              CONTINUE
C
       155 IF(TXT(1).LT.1.E − 7) GO TO 190
       160      RTX(K) = TXT(K)/TXT(1)
       170 CONTINUE
       185 PRINT 510, FO, TXT(1), (RTX(K), K = 2, 9)
       180      CONTINUE
C
       190 CONTINUE
           STOP
       500 FORMAT (1H1,/// 32X, 'PLATE TEMPERATURE RESPONSE'/ 27X,
          A'(TO − TINF)/(TI − TINF) AT CENTER AND'/ 25X,
          B'(TXL − TINF)/(TO − TINF) AT X/L FOR THE'/ 24X,
          C'GIVEN BIOT NUMBER AS A FUNCTION OF FOURIER NO.'/// 37X,
          D'BIOT NO = ', F7.1//17X,'(CENTER)', 14X,'POSITION X/L', 19X,
          E'(SURFACE)'/17X, '--------------',
          F'------------------'/13X,'FO', 4X,'0.0',
          G4X,'0.2', 4X,'0.4', 4X,'0.5', 4X,'0.6', 4X,'0.7', 4X,'0.8',
          H4X,'0.9', 4X,'1.0'//)
       510   FORMAT(F17.2, F7.4, 8F7.4)
           END
```

Table F.2

PLATE TEMPERATURE RESPONSE
$(TO - TINF)/(TI - TINF)$ AT CENTER AND
$(TXL - TINF)/(TO - TINF)$ AT X /L FOR THE
GIVEN BIOT NUMBER AS A FUNCTION OF FOURIER NO.

BIOT NO $= 1.0$

	(CENTER)				POSITION X /L				(SURFACE)
FO	0.0	0.2	0.4	0.5	0.6	0.7	0.8	0.9	1.0
0.10	0.9931	0.9947	0.9751	0.9571	0.9316	0.8970	0.8521	0.7960	0.7285
0.15	0.9756	0.9908	0.9611	0.9368	0.9053	0.8657	0.8176	0.7608	0.6954
0.20	0.9507	0.9884	0.9527	0.9249	0.8901	0.8479	0.7982	0.7411	0.6768
0.25	0.9220	0.9871	0.9479	0.9181	0.8814	0.8377	0.7872	0.7299	0.6663
0.30	0.8918	0.9863	0.9451	0.9142	0.8764	0.8319	0.7809	0.7236	0.6603
0.35	0.8613	0.9858	0.9435	0.9119	0.8736	0.8286	0.7773	0.7199	0.6569
0.40	0.8310	0.9856	0.9426	0.9107	0.8719	0.8267	0.7752	0.7178	0.6549
0.45	0.8014	0.9854	0.9421	0.9099	0.8710	0.8256	0.7740	0.7166	0.6538
0.50	0.7726	0.9854	0.9418	0.9095	0.8705	0.8250	0.7733	0.7159	0.6531
0.60	0.7177	0.9853	0.9415	0.9091	0.8700	0.8244	0.7727	0.7152	0.6525
0.70	0.6666	0.9852	0.9414	0.9090	0.8698	0.8242	0.7725	0.7150	0.6523
0.80	0.6191	0.9852	0.9414	0.9089	0.8697	0.8241	0.7724	0.7150	0.6522
0.90	0.5749	0.9852	0.9414	0.9089	0.8697	0.8241	0.7724	0.7149	0.6522
1.00	0.5339	0.9852	0.9414	0.9089	0.8697	0.8241	0.7724	0.7149	0.6522
1.50	0.3688	0.9852	0.9414	0.9089	0.8697	0.8241	0.7724	0.7149	0.6522
2.00	0.2547	0.9852	0.9414	0.9089	0.8697	0.8241	0.7724	0.7149	0.6522
2.50	0.1759	0.9852	0.9414	0.9089	0.8697	0.8241	0.7724	0.7149	0.6522
3.00	0.1215	0.9852	0.9414	0.9089	0.8697	0.8241	0.7724	0.7149	0.6522
3.50	0.0839	0.9852	0.9414	0.9089	0.8697	0.8241	0.7724	0.7149	0.6522
4.00	0.0580	0.9852	0.9414	0.9089	0.8697	0.8241	0.7724	0.7149	0.6522
4.50	0.0400	0.9852	0.9414	0.9089	0.8697	0.8241	0.7724	0.7149	0.6522
5.00	0.0277	0.9852	0.9414	0.9089	0.8697	0.8241	0.7724	0.7149	0.6522
6.00	0.0132	0.9852	0.9414	0.9089	0.8697	0.8241	0.7724	0.7149	0.6522
7.00	0.0063	0.9852	0.9414	0.9089	0.8697	0.8241	0.7724	0.7149	0.6522
8.00	0.0030	0.9852	0.9414	0.9089	0.8697	0.8241	0.7724	0.7149	0.6522
9.00	0.0014	0.9852	0.9414	0.9089	0.8697	0.8241	0.7724	0.7149	0.6522
10.00	0.0007	0.9852	0.9414	0.9089	0.8697	0.8241	0.7724	0.7149	0.6522
15.00	0.0000	0.9852	0.9414	0.9089	0.8697	0.8241	0.7724	0.7149	0.6522
20.00	0.0000	0.9852	0.9414	0.9089	0.8697	0.8241	0.7724	0.7149	0.6522

THE INFINITE CYLINDER

The sketch in Figure 3.9 shows the geometry and the symbols; the partial differential equation governing the phenomena is Eq. E.4 in one dimension with $q''' = 0$ such that

$$\frac{\partial^2 T}{\partial r^2} + \frac{\partial T/\partial r}{r} = \frac{\partial T/\partial t}{\alpha} \tag{F.5}$$

together with the boundary conditions

1. The convection boundary condition:

$$r = r_0; \qquad \frac{\partial T}{\partial r} = -\frac{h}{k}\left(T_{r_0} - T_\infty\right) \tag{F.6}$$

2. The initial condition:

$$t \leqslant 0; \qquad T = T_i, \qquad 0 \leqslant r \leqslant r_0 \tag{F.7}$$

3. The ambient condition, $T = T_\infty$, a constant value outside the boundary layer on the surface.

The solution to Eq. F.5 can also be obtained by the separation of variables method to be

$$\frac{T_r - T_\infty}{T_i - T_\infty} = \sum_{m=1}^{\infty} \frac{2HJ_0\left(\lambda_m r/r_0\right)}{\left(\lambda_m^2 + H^2\right)J_0(\lambda_m)} e^{-\lambda_m^2 N_{Fo}} \tag{F.8}$$

where

T_r = temperature at the radial location r in the cylinder

T_i = uniform initial temperature

H = the Biot number, $N_{Bi} = \dfrac{hr_0}{k}$

N_{Fo} = the Fourier number, $\dfrac{\alpha t}{r_0^2}$

$J_0(\ )$ = Bessel function of zero order

λ_m = roots of the transcendental equation, $\lambda_m J_1(\lambda_m) = HJ_0(\lambda_m)$

Table F.3 presents the FORTRAN listing of the program constructed to calculate the center temperature excess ratio, $(T_o - T_\infty)/(T_i - T_\infty)$, and the position correction ratio, $(T_r - T_\infty)/(T_o - T_\infty)$, which are tabulated in Table F.4 for the value of $N_{Bi} = 1$ (of the 15 values calculated). As with the plate the center excess ratio is tabulated under R/RO = 0.0 and the position correction ratios are tabulated under the remaining values of R/RO.

Table F.3 THE CYLINDER

```
C       PROGRAM 7802 TRSB H WOLF
C
C DICTIONARY:
C
C                     TXT=TEMP RATIO (TX − TI)/(TI − TINF)
C                     RTX=POSN RATIO (TX − TI)/(TO − TI)
C                       H=BIOT NUMBER
C                     D(I, J)=EIGENVALUES (6 FOR EACH H)
C                       FO=FOURIER NUMBER
C
C       THIS PROGRAM CALCULATES THE CENTER TEMPERATURE EXCESS
C       RATIOS AND POSITION CORRECTION RATIOS AS DEFINED
C       ABOVE FOR THE 2L PLATE, THE CYLINDER, AND THE SPHERE
C       WITH CONVECTION ON THE OUTER SURFACE. THE PROGRAM
C       EVALUATES THE SERIES (6 TERMS MAX) FOR EACH CONFIGURATION
C       AND IS ACCURATE FOR FOURIER NUMBERS DOWN TO 0.2.
C
C       DATA AND DIMENSION STATEMENTS
C
        DIMENSION D(16, 6), H(16), TXT(9), RTX(9)
C
        DATA H          /0.1, 0.2, 0.3, 0.4, 0.5, 0.6, 0.8, 1.0,
      A   1.5, 2.0, 3.0, 5.0, 10.0, 20.0, 100.0/
        DATA D /.4417, .6170, .7465, .8516, .9408, 1.0184, 1.1490, 1.2558,
      1   1.4569, 1.5994, 1.7887, 1.9898, 2.1795, 2.2880, 2.3809, 2.2048,
      2   3.8577, 3.8835, 3.9091, 3.9344, 3.9594, 3.9841, 4.0325, 4.0795,
      2   4.1902, 4.2910, 4.4634, 4.7131, 5.0332, 5.2568, 5.4652, 5.5201,
      3   7.0298, 7.0440, 7.0582, 7.0723, 7.0864, 7.1004, 7.1282, 7.1558,
      3   7.2233, 7.2884, 7.4103, 7.6177, 7.9569, 8.2534, 8.5678, 8.6537,
      4   10.1833, 10.1931, 10.2029, 10.2127, 10.2225, 10.2322, 10.2516, 10.2710,
      4   10.3188, 10.3658, 10.4566, 10.6223, 10.9363, 11.2677, 11.6747, 11.7915,
      5   13.3312, 13.3387, 13.3462, 13.3537, 13.3611, 13.3686, 13.3835, 13.3984,
      5   13.4353, 13.4719, 13.5434, 13.6786, 13.9580, 14.2983, 14.7834, 14.9309,
      6   16.4767, 16.4828, 16.4888, 16.4949, 16.5010, 16.5070, 16.5191, 16.5312,
      6   16.5612, 16.5910, 16.6499, 16.7630, 17.0099, 17.3442, 17.8931, 18.0711/
C
C       START MAIN CALCULATIONS
C
     10 DO 190 I=1, 15
     20 PRINT 500, H(I)
     25 FO = 0.050000
     26 N =14
     30         DO 180 M =1,3
     31          A =10.**M/10.
     32          IF(M.GE.2)    N =13
     35          DO 180 L =1,N
     37          IF(((M.EQ.1).AND.(L.EQ.10)).OR.
      A          ((M.GE.2).AND.(L.EQ.9))) A = 2.*A
```

continued

Table F.3   continued

```
40         FO = FO + 0.0500*A
           DO 170 K = 1, 9
52         XL = (K − 1)
54         IF(K − 1) 70, 70, 56
55         XL = K /10.
58         IF(K − 2) 70, 70, 60
60         XL = (K + 1)/10.
70             TXT(K) = 0.000000
72              DO 150 J = 1, 6
           ERROR = .1**(6.0 − J/2)*SQRT(H(I) + 1.)
           EXARG = − D(I, J)**2*FO
           IF(EXARG.LT..−162) GO TO 155
           ELAM = EXP(EXARG)
219        ARG = D(I, J)*XL
           IF(ARG.LT.1.E − 7) GO TO 220
           CALL BESJ(ARG, 0, BJ1, ERROR, IER)
           IF(IER.NE.0) GO TO 225
221        ARG = D(I, J)
           IF(ARG,LT.1.E − 7) GO TO 222
           CALL BESJ(ARG,0, BJ2, ERROR, IER)
           IF(IER.NE.0) GO TO 225
           GO TO 150
220        BJ1 = 1.
           GO TO 221
222        BJ2 = 1.
150        TXT(K) = TXT(K) + 2.*H(I)*ELAM*BJ1/((D(I, J)**2 + H(I)**2)
  A        *BJ2)
155        IF(TXT(1).LT.1.E − 7) GO TO 190
160        RTX(K) = TXT(K)/TXT(1)
170            CONTINUE
185 PRINT 510, FO, TXT(1), (RTX(K), K = 2, 9)
180            CONTINUE
190 CONTINUE
      GO TO 200
225   WRITE(6, 520) IER, XL, D(I, J)
200 STOP
500 FORMAT(1H1, ///31X, 'CYLINDER TEMPERATURE RESPONSE'/27X,
    A'(TO − TINF)/(TI − TINF) AT CENTER AND'/25X,
    B'(TXL − TINF)/(TO − TINF) AT R /RO FOR THE'/24X,
    C'GIVEN BIOT NUMBER AS A FUNCTION OF FOURIER NO.'///37X,
    D'BIOT NO = ', F7.1//17X.'(CENTER)', 14X, 'POSITION R /RO', 19X,
    E'(SURFACE)'/17X,'--------------',
    F'------------------'/13X, 'FO', 4X, '0.0',
    G'4X, '0.2', 4X, '0.4', 4X, '0.5', 4X, '0.6', 4X, '0.7', 4X, '0.8', 4X, '0.9', 4X,
    H'1.0'//)
510 FORMAT(F17.2, F7.4, 8F7.4)
520 FORMAT('BESSEL FUNCTION ROUTINE DOES NOT CONVERGE', I3, 2F10.4)
    END
```

Table F.4

CYLINDER TEMPERATURE RESPONSE
(TO − TINF)/(TI − TINF) AT CENTER AND
(TX − TINF)/(TO − TINF) AT R/RO FOR THE
GIVEN BIOT NUMBER AS A FUNCTION OF FOURIER NO.

BIOT NO = 1.0

| | (CENTER) | | | | POSITION R/RO | | | | (SURFACE) |
FO	0.0	0.2	0.4	0.5	0.6	0.7	0.8	0.9	1.0
0.10	0.9768	0.9920	0.9652	0.9423	0.9118	0.8727	0.8244	0.7669	0.7008
0.15	0.9290	0.9880	0.9507	0.9218	0.8857	0.8422	0.7914	0.7336	0.6695
0.20	0.8702	0.9850	0.9439	0.9122	0.8736	0.8282	0.7763	0.7184	0.6553
0.25	0.8093	0.9851	0.9407	0.9078	0.8680	0.8217	0.7693	0.7114	0.6487
0.30	0.7501	0.9847	0.9392	0.9057	0.8653	0.8186	0.7660	0.7082	0.6457
0.35	0.6942	0.9845	0.9385	0.9047	0.8641	0.8172	0.7645	0.7066	0.6442
0.40	0.6420	0.9844	0.9382	0.9042	0.8635	0.8165	0.7638	0.7059	0.6435
0.45	0.5935	0.9843	0.9380	0.9040	0.8633	0.8162	0.7635	0.7056	0.6432
0.50	0.5486	0.9843	0.9380	0.9039	0.8631	0.8161	0.7633	0.7054	0.6431
0.60	0.4686	0.9843	0.9379	0.9039	0.8630	0.8160	0.7632	0.7053	0.6430
0.70	0.4002	0.9843	0.9379	0.9038	0.8630	0.8160	0.7632	0.7053	0.6429
0.80	0.3418	0.9843	0.9379	0.9038	0.8630	0.8159	0.7632	0.7053	0.6429
0.90	0.2920	0.9843	0.9379	0.9038	0.8630	0.8159	0.7632	0.7053	0.6429
1.00	0.2494	0.9843	0.9379	0.9038	0.8630	0.8159	0.7632	0.7053	0.6429
1.50	0.1133	0.9843	0.9379	0.9038	0.8630	0.8159	0.7632	0.7053	0.6429
2.00	0.0515	0.9843	0.9379	0.9038	0.8630	0.8159	0.7632	0.7053	0.6429
2.50	0.0234	0.9843	0.9379	0.9038	0.8630	0.8159	0.7632	0.7053	0.6429
3.00	0.0106	0.9843	0.9379	0.9038	0.8630	0.8159	0.7632	0.7053	0.6429
3.50	0.0048	0.9843	0.9379	0.9038	0.8630	0.8159	0.7632	0.7053	0.6429
4.00	0.0022	0.9843	0.9379	0.9038	0.8630	0.8159	0.7632	0.7053	0.6429
4.50	0.0010	0.9843	0.9379	0.9038	0.8630	0.8159	0.7632	0.7053	0.6429
5.00	0.0005	0.9843	0.9379	0.9038	0.8630	0.8159	0.7632	0.7053	0.6429
6.00	0.0001	0.9843	0.9379	0.9038	0.8630	0.8159	0.7632	0.7053	0.6429
7.00	0.0000	0.9843	0.9379	0.9038	0.8630	0.8159	0.7632	0.7053	0.6429
8.00	0.0000	0.9843	0.9379	0.9038	0.8630	0.8159	0.7632	0.7053	0.6429
9.00	0.0000	0.9843	0.9379	0.9038	0.8630	0.8159	0.7632	0.7053	0.6429
10.00	0.0000	0.9843	0.9379	0.9038	0.8630	0.8159	0.7632	0.7053	0.6429

THE SPHERE

The sketch in Figure 3.11 shows the geometry and some pertinent symbols; the partial differential equation governing the phenomena is Eq. E.5 in one dimension with $q''' = 0$ such that

$$\alpha \frac{\partial^2(rT)/\partial r^2}{r} = \frac{\partial T}{\partial t} \tag{F.9}$$

with the boundary and initial conditions the same as those given by Eqs. F.6 and F.7 together with the restraint of constant ambient temperatures, T_∞.

The solution to Eq. F.9 can be determined by substitution or by separation of variables to be

$$\frac{T_r - T_\infty}{T_i - T_\infty} = \sum_{j=1}^{\infty} \frac{2\sin(\lambda_j \zeta)}{\lambda_j \zeta} \frac{(\text{NUM})}{(\text{DEN})} e^{-\lambda_j^2 N_{Fo}} \tag{F.10}$$

where

$$(\text{NUM}) = (\sin \lambda_j - \lambda_j \cos \lambda_j)$$

$$(\text{DEN}) = (\lambda_j - \sin \lambda_j \cos \lambda_j)$$

H = the Biot number, $N_{\text{Bi}} = \dfrac{hr_0}{k}$

N_{Fo} = the Fourier number, $\dfrac{\alpha t}{r_0^2}$

ζ = the position ratio, $\dfrac{r}{r_0}$

λ_j = roots of the transcendental equation $\lambda_j = (1 - H)\tan(\lambda_j)$

The reader will note that calculating the center temperature excess ratio for $\zeta = 0$ requires determination of the limit of $\sin(\lambda_j\zeta)/\zeta$ as $\zeta \to 0$ which is λ_j according to L'Hôspital's rule. Table F.5 presents the FORTRAN listing of the program used to calculate the temperature ratios given in Table F.6. As before the column of numbers under the heading $R/RO = 0.0$ is the center temperature excess ratio, $(T_o - T_\infty)/(T_i - T_\infty)$, and the columns of numbers under the other values of R/RO are the position correction ratios, $(T_r - T_\infty)/(T_o - T_\infty)$, as before.

THE SEMI-INFINITE SOLID

The sketch in Figure 3.16 or 3.17 shows the geometry and the symbols; the partial differential equation governing the phenomena is the same as Eq. F.1 but with the boundary and initial conditions as follows:

1. The convection boundary condition:

$$x = 0; \qquad \frac{\partial T}{\partial x} = -\frac{h}{k}(T_{x=0} - T_\infty)$$

2. The initial condition:

$$t < 0; \qquad T = T_i, \qquad x > 0$$

3. The ambient condition, $T = T_\infty$, a constant value outside the boundary layer on the surface.

The solution to Eq. F.1 under these conditions is given by Eq. 3.32 and can be obtained by the Laplace transform method as described by Myers (Reference 3, Chapter 3) or Schneider (Reference 9, Chapter 3).

Table F.7 lists the FORTRAN program employed to compute the local temperature excess, $(T_x - T_\infty)/(T_i - T_\infty)$, with the local Biot number hx/k as the parameter; typical results are given in Table F.8 for $N_{\text{Bi}} = 1$.

Table F.9 lists the FORTRAN program employed to compute the local temperature excess with the quantity $h\sqrt{\alpha t}/k$ as the parameter; typical results are given in Table F.10 for $h\sqrt{\alpha t}/k = \text{HAK} = 1.0$.

The author would like to express his thanks and appreciation to Dr. L. D. Wills for his astute and able assistance in constructing some of the algorithms used in the computer programs.

Table F.5 THE SPHERE

```
C      PROGRAM 7802 TRSB H WOLF 8-8-78
C
C      DICTIONARY:
C
C              TXT = TEMP RATIO (TX − TI)/(TI − TINF)
C              RTX = POSN RATIO (TX − TI)/(TO − TI)
C                H = BIOT NUMBER
C              D(I, J) = EIGENVALUES (6 FOR EACH H)
C                FO = FOURIER NUMBER
C
C      THIS PROGRAM CALCULATES THE CENTER TEMPERATURE EXCESS
C      RATIOS AND POSITION CORRECTION RATIOS AS DEFINED
C      ABOVE FOR THE 2L PLATE, THE CYLINDER, AND THE SPHERE
C      WITH CONVECTION ON THE OUTER SURFACE. THE PROGRAM
C      EVALUATES THE SERIES (6 TERMS MAX) FOR EACH CONFIGURATION
C      AND IS ACCURATE FOR FOURIER NUMBERS DOWN TO 0.2.
C
C      DATA AND DIMENSION STATEMENTS
C
       DIMENSION D(16,6), H(16), TXT(9), RTX(9)
C
       DATA H      /0.1,0.2,0.3,0.4,0.5,0.6,0.8, 1.0,
      A 1.5, 2.0, 3.0, 5.0, 20.0, 100.0, 10000.0/
       DATA D/.5423,.7593,.9208, 1.0528, 1.1656, 1.2644, 1.4320, 1.5708,
      1 1.8366, 2.0288, 2.2889, 2.5704, 2.8363, 2.9857, 3.1102, 3.1416,
      2 4.5157, 4.5379, 4.5601, 4.5822, 4.6042, 4.6261, 4.6696, 4.7124,
      2 4.8158, 4.9132, 5.0870, 5.3540, 5.7172, 5.9783, 6.2204, 6.2832,
      3 7.7382, 7.7511, 7.7641, 7.7770, 7.7899, 7.8028, 7.8284, 7.8540,
      3 7.9171, 7.9787, 8.0962, 8.3029, 8.6587, 8.9831, 9.3308, 9.4248,
      4 10.9133, 10.9225, 10.9316, 10.9408, 10.9499, 10.9591, 10.9774, 10.9956,
      4 11.0409, 11.0856, 11.1727, 11.3349, 11.6532, 12.0029, 12.4414, 12.5664,
      5 14.0733, 14.0804, 14.0875, 14.0946, 14.1017, 14.1088, 14.1230, 14.1372,
      5 14.1724, 14.2075, 14.2764, 14.4080, 14.6870, 15.0384, 15.5521, 15.7080,
      6 17.2266, 17.2324, 17.2382, 17.2440, 17.2498, 17.2556, 17.2672, 17.2788,
      6 17.3076, 17.3364, 17.3932, 17.5034, 17.7481, 18.0887, 18.6632, 18.8496/
C
C          START MAIN CALCULATIONS
C
      10 DO 190 I = 1, 16
      20 PRINT 500, H(I)
      25 FO = 0.050000
      26  N = 14
```

continued

Table F.5 continued

```
30          DO 180 M = 1, 3
31           A = 10.**M /10.
32            IF(M.GE.2) N = 13
35          DO 180 L = 1,N
37          IF(((M.EQ.1).AND.(L.EQ.10)).OR.
 A          ((M.GE.2).AND.(L.EQ.9))) A = 2.*A
40          FO = FO + 0.0500*A
            DO 170 K = 1, 9
52          XL = (K − 1)
54          IF(K − 1) 70, 70, 56
56          XL = K /10.
58          IF(K − 2) 70, 70, 60
60          XL = (K + 1)/10.
70              TXT(K) = 0.000000
72              DO 150 J = 1, 6
                IF(XL.LT.1.E − 30) GO TO 226
                SLAM = SIN(D(I, J)*XL)/(D(I, J)*XL)
                GO TO 227
226             SLAM = 1.
227             EXARG = − D(I, J)**2*FO
                IF(EXARG.LT. −162) GO TO 155
                ELAM = EXP(EXARG)
150             TXT(K) = TXT(K) + 2.*((SIN(D(I, J)) − D(I, J)*COS(D(I, J)))/
 A              (D(I, J) − SIN(D(I, J))*COS(D(I, J))))*SLAM*ELAM
155          IF(TXT(1).LT.1.E − 7) GOTO 190
160          RTX(K) = TXT(K)/TXT(1)
170          CONTINUE
185 PRINT 510, FO, TXT(1), (RTX(K), K = 2, 9)
180          CONTINUE
190 CONTINUE
200     STOP
500     FORMAT(1H1,///32X,'SPHERE TEMPERATURE RESPONSE'/27X,
        A'(TO − TINF)/(TI − TINF) AT CENTER AND'/25X,
        B'(TXL − TINF)/(TO − TINF) AT R /RO FOR THE'/24X,
        C'GIVEN BIOT NUMBER AS A FUNCTION OF FOURIER NO.'///37X,
        D'BIOT NO = ', F7.1//17X, 'CENTER', 14X, 'POSITION R /RO', 19X
        E'(SURFACE)'/17X,'--------------',
        F'------------------'/13X,'FO', 4X,'0.0',
        G4X,'0.2', 4X,'0.4', 4X,'0.5', 4X,'0.6', 4X,'0.7', 4X,'0.8', 4X,'0.9', 4X,
        H'1.0'//)
510     FORMAT(F17.2,F7.4, 8F7.4)
        END
```

Table F.6

SPHERE TEMPERATURE RESPONSE
$(TO - TINF)/(TI - TINF)$ AT CENTER AND
$(TX - TINF)/(TO - TINF)$ AT R/RO FOR THE
GIVEN BIOT NUMBER AS A FUNCTION OF FOURIER NO.

BIOT NO $= 1.0$

	(CENTER)				POSITION R/RO			(SURFACE)	
FO	0.0	0.2	0.4	0.5	0.6	0.7	0.8	0.9	1.0
0.10	0.9493	0.9895	0.9559	0.9288	0.8941	0.8514	0.8007	0.7424	0.6775
0.15	0.8642	0.9858	0.9430	0.9108	0.8715	0.8254	0.7728	0.7146	0.6515
0.20	0.7723	0.9844	0.9383	0.9042	0.8632	0.8159	0.7627	0.7045	0.6421
0.25	0.6854	0.9839	0.9365	0.9018	0.8602	0.8124	0.7590	0.7008	0.6387
0.30	0.6068	0.9837	0.9359	0.9009	0.8591	0.8111	0.7576	0.6995	0.6374
0.35	0.5367	0.9837	0.9356	0.9005	0.8586	0.8106	0.7571	0.6989	0.6369
0.40	0.4745	0.9836	0.9355	0.9004	0.8585	0.8104	0.7569	0.6988	0.6367
0.45	0.4194	0.9836	0.9355	0.9003	0.8584	0.8104	0.7569	0.6987	0.6367
0.50	0.3708	0.9836	0.9355	0.9003	0.8584	0.8103	0.7568	0.6987	0.6366
0.60	0.2897	0.9836	0.9355	0.9003	0.8584	0.8103	0.7568	0.6986	0.6366
0.70	0.2264	0.9836	0.9355	0.9003	0.8584	0.8103	0.7568	0.6986	0.6366
0.80	0.1769	0.9836	0.9355	0.9003	0.8584	0.8103	0.7568	0.6986	0.6366
0.90	0.1382	0.9836	0.9355	0.9003	0.8584	0.8103	0.7568	0.6986	0.6366
1.00	0.1080	0.9836	0.9355	0.9003	0.8584	0.8103	0.7568	0.6986	0.6366
1.50	0.0314	0.9836	0.9355	0.9003	0.8584	0.8103	0.7568	0.6986	0.6366
2.00	0.0092	0.9836	0.9355	0.9003	0.8584	0.8103	0.7568	0.6986	0.6366
2.50	0.0027	0.9836	0.9355	0.9003	0.8584	0.8103	0.7568	0.6986	0.6366
3.00	0.0008	0.9836	0.9355	0.9003	0.8584	0.8103	0.7568	0.6986	0.6366
3.50	0.0002	0.9836	0.9355	0.9003	0.8584	0.8103	0.7568	0.6986	0.6366
4.00	0.0001	0.9836	0.9355	0.9003	0.8584	0.8103	0.7568	0.6986	0.6366
4.50	0.0000	0.9836	0.9355	0.9003	0.8584	0.8103	0.7568	0.6986	0.6366
5.00	0.0000	0.9836	0.9355	0.9003	0.8584	0.8103	0.7568	0.6986	0.6366
6.00	0.0000	0.9836	0.9355	0.9003	0.8584	0.8103	0.7568	0.6986	0.6366

Table F.7

```
C
C        DICTIONARY:
C                 X = X/(2*SQRT(ALPHA*TIME))
C          ALPHA = THERMAL DIFFUSIVITY
C             BI = LOCAL BIOT NUMBER, HX/K
C           TXTR = TEMP. RATIO, (T – TINF)/(TI – TINF)
C            ERF = ERROR FUNCTION
C              K = THERMAL CONDUCTIVITY
C              H = HEAT TRANSFER COEFFICIENT
C
         DOUBLE PRECISION X, BI, XBI2, XARG, XRF, CRF, TXTR, EXTRM,
        A Z, ZZ, ZZZ, SUM, SPI, ZLN1, ZLN2, ARGZ, EXARG, TERM2
    10 READ 500, BI
   130 IF(BI.GT.205.) GOTO 150
    20 PRINT 510, BI
    30 DO 120 I = 1, 31
    32    IF(I.LT.7) X = (I + 1)/200.
    34    IF(I.GT.5.AND.I.LT.13)     X = (I – 3)/100.
    36    IF(I.GT.12.AND.I.LT.19)    X = (I – 8)/50.
    38    IF(I.GT.18.AND.I.LT.23)    X = (I – 14)/20.
    40    IF(I.GT.22.AND.I.LT.29)    X = (I – 18)/10.
    42    IF(I.GT.28)                X = (I – 23)/5.
    45    XRF = DERF(X)
          XARG = BI*(1. + BI/(4.*X*X))
          XBI2 = X + BI/(2.*X)
          IF(XBI2.GT.12.) GOTO 200
          EXTRM = DEXP(XARG)
          CRF = DERFC(XBI2)
          TXTR = XRF + CRF*EXTRM
          PRINT 530, X, TXTR, XRF, CRF, EXTRM
          GOTO 120
   200    Z = XBI2
          ZZ = 1./Z – 0.5/(Z**3) + 0.750/(Z**5)
          ZZZ = 6.5625/(Z**9) – 1.875/(Z**7)
          SUM = ZZ + ZZZ
          SPI = (3.14159265389793)**.5
          ZLN1 = DLOG(1./SPI)
          ZLN2 = DLOG(SUM)
          ARGZ = ZLN1 + ZLN2 – Z*Z
          EXARG = ARGZ + XARG
          TERM2 = DEXP(EXARG)
          TXTR = XRF + TERM2
          PRINT 530, X, TXTR, XRF, ARGZ, XARG
   120 CONTINUE
   140 GOTO 10
   500 FORMAT (F12.6)
   510 FORMAT(1H1//////33X,'TEMPERATURE DISTRIBUTION'/ 31X,
        A'FOR THE SEMI-INFINITE SOLID'/ 30X,'WITH CONVECTION',
        B'ON THE SURFACE'// 37X, 'BIOT = 'F6.2// 16X, 'X',
        C 7X, 'TXTR', 9X, 'ERFX', 12X, 'CRF', 10X, 'EXTRM'/
        D 51X, '(ARGZ)', 9X, '(XARG)'//)
   530 FORMAT(13X, F5.3, 3X, F9.6, 3X, F11.8, 3X, E15.8, 3X, E15.8)
   150 STOP
       END
```

Table F.8

TEMPERATURE DISTRIBUTION
FOR THE SEMI-INFINITE SOLID
WITH CONVECTION ON THE SURFACE

BIOT = 1.00

X	TXTR	ERFX	CRF (ARGZ)		EXTRM (XARG)	
0.010	0.022562	0.01128341	− 0.25054859D	04	0.25010010D	04
0.015	0.033831	0.01692441	− 0.11161916D	04	0.11121116D	04
0.020	0.045087	0.02256457	− 0.62979327D	03	0.62600003D	03
0.025	0.056325	0.02820360	− 0.40457126D	03	0.40100005D	03
0.030	0.067541	0.03384122	− 0.28216808D	03	0.27877782D	03
0.035	0.078730	0.03947715	− 0.20831939D	03	0.20508167D	03
0.040	0.089889	0.04511111	− 0.16035605D	03	0.15725001D	03
0.050	0.112097	0.05637197	0.76260817D − 45		0.73071471D	44
0.060	0.134133	0.06762159	0.16953633D − 31		0.39231494D	31
0.070	0.155968	0.07885771	0.19721088D − 23		0.39100359D	23
0.080	0.177573	0.09007811	0.34918226D − 18		0.25057040D	18
0.090	0.198922	0.10128056	0.14163906D − 14		0.68937094D	14
0.100	0.219993	0.11246288	0.54937202D − 12		0.19573311D	12
0.120	0.261216	0.13475829	0.13421804D − 08		0.94218487D	08
0.140	0.301103	0.15694702	0.15313213D − 06		0.94138234D	06
0.160	0.339548	0.17901178	0.33893767D − 05		0.47364538D	05
0.180	0.376480	0.20093578	0.28777536D − 04		0.61000600D	04
0.200	0.411857	0.22270258	0.13433263D − 03		0.14081059D	04
0.250	0.493413	0.27632639	0.14627166D − 02		0.14841316D	03
0.300	0.565337	0.32862671	0.54143851D − 02		0.43718819D	02
0.350	0.628228	0.37938202	0.11893788D − 01		0.20922324D	02
0.400	0.682886	0.42839233	0.19624411D − 01		0.12968200D	02
0.500	0.770951	0.52049988	0.33894854D − 01		0.73890561D	01
0.600	0.836072	0.60385606	0.42658377D − 01		0.54436213D	01
0.700	0.883761	0.67780119	0.45489247D − 01		0.45276548D	01
0.800	0.918375	0.74210094	0.43878009D − 01		0.40173603D	01
0.900	0.943271	0.79690820	0.39545384D − 01		0.37011441D	01
1.000	0.961005	0.84270079	0.33894854D − 01		0.34903430D	01
1.200	0.982217	0.91031393	0.22236027D − 01		0.32336488D	01
1.400	0.992294	0.95228506	0.12956058D − 01		0.30880848D	01
1.600	0.996840	0.97634833	0.68369643D − 02		0.29971335D	01

Table F.9 THE SEMI-INFINITE SOLID

```
C       DICTIONARY:
C            X = X /(2*SQRT(ALPHA*TIME))
C         ALPHA = THERMAL DIFFUSIVITY
C            BI = LOCAL BIOT NUMBERY HX / —
C          TXTR = TEMP. RATIO, (T — TINF)/(TI — TINF)
C           ERF = ERROR FUNCTION
C             K = THERMAL CONDUCTIVITY
C             H = HEAT TRANSFER COEFFICIENT
C           HAK = (H / K)*SQRT(ALPHA*TIME)
C
        DOUBLE PRECISION X, BI, XBI2, XARG, XRF, CRF, TXTR, EXTRM,
       A Z, ZZ, ZZZ, SUM, SPI, ZLN1, ZLN2, ARGZ, EXARG, TERM2, HAK
     10 READ 500, HAK
    130 IF(HAK.GT.205.) GOTO 150
     20 PRINT 510, HAK
     30 DO120 I = 1, 31
     32    IF(I.LT.7) X = (I + 1)/200.
     34    IF(I.GT.6.AND.I.LT.13)     X = (I — 3)/100.
     36    IF(I.GT.12.AND.I.LT.19)    X = (I — 8)/50.
     38    IF(I.GT.18.AND.I.LT.23)    X = (I —14)/20.
     40    IF(I.GT.22.AND.I.LT.29)    X = (I —18)/10.
     42    IF(I.GT.28)                X = (I —23)/5.
     45    XRF = DERF(X)
     46    BI = 2.*X*HAK
           XARG = BI*(1. + BI/(4.*X*X))
           XBI2 = X + BI/(2.*X)
           IF(XBI2.GT.12) GOTO 200
           EXTRM = DEXP(XARG)
           CRF = DERFC(XBI2)
           TXTR = XRF + CRF*EXTRM
           PRINT 530, X, TXTR, XRF, CRF, EXTRM
           GOTO 120
    200    Z = XBI2
           ZZ = 1./Z — 0.5/(Z**3) + 0.750/(Z**5)
           ZZZ = 6.5625/(Z**9) —1.875/(Z**7)
           SUM = ZZ + ZZZ
           SPI = (3.141592653589793)**.5
           ZLN1 = DLOG(1./SPI)
           ZLN2 = DLOG(SUM)
           ARGZ = ZLN1 + ZLN2 — Z*Z
           EXARG = ARGZ + XARG
           TERM2 = DEXP(EXARG)
           TXTR = XRF + TERM2
           PRINT 530, X, TXTR, XRF, ARGZ, XARG
    120 CONTINUE
    140 GOTO 10
    500 FORMAT (F12.6)
    510 FORMAT (1H1///// 33X, 'TEMPERATURE DISTRIBUTION'/ 31X,
       A'FOR THE SEMI-INFINITE SOLID'/ 30X, 'WITH CONVECTION',
       B'ON THE SURFACE'// 37X, 'HAK = ',F6.2// 16X, 'X',
       C 7X, 'TXTR', 9X, 'ERFX', 12X, 'CRF', 10X, 'EXTRM'/
       D 51X, '(ARGZ)', 9X, '(XARG)'//)
    530 FORMAT(13X, F5.3, 3X, F9.6, 3X, F11.8, 3X, E15.8, 3X, E15.8)
    150 STOP
        END
```

Table F.10

TEMPERATURE DISTRIBUTION
FOR THE SEMI-INFINITE SOLID
WITH CONVECTION ON THE SURFACE

HAK = 1.00

X	TXTR	ERFX	CRF (ARGZ)		EXTRM (XARG)	
0.010	0.436108	0.01128341	0.15318950D	00	0.27731948D	01
0.015	0.440349	0.01692441	0.15116552D	00	0.28010658D	01
0.020	0.444576	0.02256457	0.14916198D	00	0.28292170D	01
0.025	0.448789	0.02820360	0.14717877D	00	0.28576511D	01
0.030	0.452988	0.03384122	0.14521579D	00	0.28863710D	01
0.035	0.457172	0.03947715	0.14327292D	00	0.29153795D	01
0.040	0.461342	0.04511111	0.14135005D	00	0.29446795D	01
0.050	0.469637	0.05637197	0.13756390D	00	0.30041660D	01
0.060	0.477872	0.06762159	0.13385641D	00	0.30648542D	01
0.070	0.486046	0.07885771	0.13022671D	00	0.31267683D	01
0.080	0.494159	0.09007811	0.12667385D	00	0.31899332D	01
0.090	0.502209	0.10128056	0.12319691D	00	0.32543740D	01
0.100	0.510196	0.11246288	0.11979494D	00	0.33201167D	01
0.120	0.525976	0.13475829	0.11321213D	00	0.34556131D	01
0.140	0.541491	0.15694702	0.10691768D	00	0.35966396D	01
0.160	0.556737	0.17901178	0.10090381D	00	0.37434211D	01
0.180	0.571708	0.20093578	0.95162588D − 01		0.38961929D	01
0.200	0.586397	0.22270258	0.89686025D − 01		0.40551999D	01
0.250	0.621864	0.27632639	0.77099872D − 01		0.44816891D	01
0.300	0.655488	0.32862671	0.65992065D − 01		0.49530320D	01
0.350	0.687225	0.37938202	0.56237810D − 01		0.54739470D	01
0.400	0.717051	0.42839233	0.47714884D − 01		0.60496472D	01
0.500	0.770951	0.52049988	0.33894854D − 01		0.73890561D	01
0.600	0.817312	0.60385606	0.23651620D − 01		0.90250129D	01
0.700	0.856482	0.67780119	0.16209542D − 01		0.11023176D	02
0.800	0.888984	0.74210094	0.10909500D − 01		0.13463737D	02
0.900	0.915467	0.79690820	0.72095715D − 02		0.16444646D	02
1.000	0.936656	0.84270079	0.46777350D − 02		0.20085537D	02
1.200	0.966132	0.91031393	0.18628480D − 02		0.29964089D	02
1.400	0.983062	0.95228506	0.68851525D − 03		0.44701150D	02
1.600	0.992089	0.97634833	0.23603517D − 03		0.66686255D	02

Appendix G
Properties of the
Standard Atmosphere*

Sea level temperature, $T_0 = 519.4R$ $(288.3K)$

Sea level pressure, $P_0 = 14.696$ psia $(1.01325 \times 10^5 \text{ N/m}^2)$

Sea level density, $\rho_0 = 0.07644 \text{ lb}_m/\text{ft}^3$ (1.2250 kg/m^3)

ALTITUDE		TEMPERATURE		DENSITY RATIO	PRESSURE RATIO	ACOUSTIC VELOCITY	
(ft $\times 10^{-3}$)	(km)	(R)	(K)	ρ/ρ_0	P/P_0	(m/s)	(ft/s)
−15	−4.57	572	318	1.5230	1.6812	357.4	1173
−10	−3.05	554	308	1.3273	1.4185	351.8	1154
−5	−1.52	537	298	1.1547	1.1944	346.1	1135
0	0.0	519	288	1.0000	1.0000	340.3	1116
5	1.52	500	278	0.8617	0.8321	334.4	1097
10	3.05	482	268	0.7386	0.6878	328.4	1077
15	4.57	464	258	0.6295	0.5646	322.3	1057
20	6.10	446	248	0.5332	0.4599	316.0	1037
25	7.62	430	239	0.4486	0.3716	309.7	1016
30	9.14	412	229	0.3747	0.2975	303.2	995
35	10.67	394	219	0.3106	0.2360	296.4	972
40	12.19	391	217	0.2471	0.1858	295.1	968
45	13.72	391	217	0.1945	0.1462	295.1	968
50	15.24	391	217	0.1531	0.1151	295.1	968
55	16.76	391	217	0.1205	0.0906	295.1	968
60	18.29	391	217	0.0949	0.0714	295.1	968
65	19.81	391	217	0.0748	0.0562	295.1	968
70	21.34	392	218	0.0586	0.0443	296.0	971
75	22.86	394	219	0.0459	0.0350	297.0	974
80	24.38	398	221	0.0361	0.0276	298.0	978

*U.S. Standard Atmosphere 1976, NASA-NOAA-S/T 76-1562, USGPO, Washington, D.C., 20402.

Appendix H
Answers to Selected Problems

CHAPTER 1

1. 329 W
4. 475 W
6. 15 to 132,000 Btu/hrft^2
8. 87.8F, 12,000 Btu/hrft^2
9. 1450 W/m^2, 440 Btu/hrft^2, increasing
11. ~ 9708R
12. 479 Btu/hr

CHAPTER 2

1. 124F max, 6 ft; $q/A = -38.1$ Btu/hrft^2 to left
3. Graphical analysis, -37.6 Btu/hrft^2 at right face
5. $2.60(10^{-8})(E/K)^2$
8. 56.4, 120, 7.8 Btu/hrft^2; 178, 378, 24.6 W/m^2
9. 211,000 Btu/hrft^2, 666,000 W/m^2
11. 1.41 Btu/hrft^2; 4.45 W/m^2
13. 39.7 lb$_m$/hr, 80 percent
18. 355 Btu/hr; 104 W

23. 850F
25. 1570 A
29. 10.9 Btu/hr, 3.2 W; 14.2 times
31. 163 W
33. 725 Btu/hrft^{2}F; 4114 W/m^{2}K
35. $S = 9.6$ ft, 2.93 m; 5.9 percent low
37. 121 Btu/hrft, 104 W/m
39. 0.024F, 0.013C
41. 61.5 Btu/hr, 18.02 W/hr, 1.95 hr

CHAPTER 3

1. Sketch, 90.9F, 109.1F
3. Sketch, 4.8F, 195F
8. 9.3 min
9. 10.52 Btu/hrF
11. 1.3 hr, 55,860 Btu
13. 259 s slow response
15. 15.4 min
17. 12.7 Btu/hrft^{2}F, 71.9 W/m^{2}K
19. 2.44 hr
21. 710F, 772F
23. About 16 in.
25. 2250F, 355F
27. 34 min
29. 16.5 min
31. One term is: $[1 + (h\,\Delta x/k) - M/2]T_5^k$
35. 400F
38. About 130 min

CHAPTER 4

1. 72.08 gpm; 9.63 cfm
3. 10.16(10^5) lb$_m$/hrft^2; 5.04 ft/s
12. 602 lb$_m$/s, 74.7 lb$_m$/s
13. 90C
15. Air 0.114 in., H$_2$O 0.382 in.
17. 78.4 lb$_f$, 42 HP
19. 0.59 psi; 0.03 HP
21. $N_{Re} = 18,695$, $f = 0.0225$, $\Delta P = 0.066$ psi
23. $y = 0.293\ r_0$
25. 42F
29. $h = 20$ Btu/hrft^{2}F

CHAPTER 5

1. Air 0.811 in., water 0.425 in.
4. $\delta_L^* = 0.562\ \delta_L$
7. 204
9. Approximately 4.5 hr
11. 4.9 V
13. 20.5, 18.2
15. 136 Btu/hr or 40 W
17. 0.015F, 17.6F
19. 4.1 hr

CHAPTER 6

1. 7.4
3. 854 ft, discussion
6. 262F, 292F
7. 21.8, 11.2
9. 5.26 lb_m/hr
14. 1000F
15. 5190; 259,000 Btu/hrft^2
18. 2450F
21. 16 percent
23. 5195 Btu/hrft^{2}F, 29,500 W/m^{2}K
25. None
29. 2.56 s
31. 336 ft
33. 102F
35. 52F
37. 0.0242 psi
39. 360F, discussion
41. 20F
43. Turbulent free convection

CHAPTER 7

1. 7.32 Btu/hrft^{2}F, 805 Btu/hrft^2
3. Sketch, LMTD = 232F, $\Delta T = 308$F
5. LMTD = 200F
8. 39,600 Btu/hr
11. 510,000 Btu/hr
13. 1.94 lb_m/hr
15. 277F
17. 960 lb_m/hr
19. 60 ft^2, 65F, 155F

CHAPTER 8

1. 49,523 waves
3. 1.28 s, 35 min
5. $J_1 = 396$ Btu/hrft^2
7. $T = 256F$
9. $J_{top} = 307.5$ Btu/hrft^2
11. $107 (10^3)$ Btu/hrft$^2\mu$
13. 5000R; 73.89 percent; 791; 546 Btu/hrft^2
15. 122.2 Btu/hrft^2
17. 96F
19. 1.84 Ω
21. 66.4 percent
25. 69 percent
27. 51,000 Btu/hr; 105,000 Btu/hr
29. Discussion, 86,900 Btu/hr
31. 9.87×10^6 Btu/hr
33. 0.5
35. 0.141
37. 0.115
39. 0.032
41. 0.172
43. 0.40
45. 0.045
47. 64.5 percent
49. 1040 Btu/hr
51. 25 percent reduction
55. 258 Btu/hr, 2135 Btu/hr.
57. 1.46 percent decrease

CHAPTER 9

2. 0.013
3. 564,000 Btu/hrft^2
5. 1077 Btu/ft^{2}s
7. 828,000 Btu/hrft^2
9. 395 lb$_m$/hr
13. 398 lb$_m$
16. About 2 hr
17. 16 percent reduction
19. 8.5 hr
21. 6F or less

Index